Selected Titles in This Series

3 **János Kollár, Editor,** Complex Algebraic Geometry, 1997

2 **Robert Hardt and Michael Wolf, Editors,** Nonlinear Partial Differential Equations in Differential Geometry, 1996

1 **Daniel S. Freed and Karen K. Uhlenbeck, Editors,** Geometry and Quantum Field Theory, 1995

Ω IAS/PARK CITY
MATHEMATICS SERIES

Volume 3

Complex Algebraic Geometry

János Kollár
Editor

American Mathematical Society
Institute for Advanced Study

IAS/Park City Mathematics Institute runs mathematics education programs that bring together high school mathematics teachers, researchers, graduate students, and undergraduates to participate in distinct but overlapping programs of research and education. This volume contains the lecture notes from the Graduate Summer School program on algebraic geometry held at the Regional Geometry Institute in Park City, Utah, in 1993.

Supported by the National Science Foundation.

2000 *Mathematics Subject Classification.* Primary 14-01, 14-02, 14J10, 14C20, 14D05, 14J32; Secondary 14F17, 81T30, 14-06, 14F45.

Library of Congress Cataloging-in-Publication Data

Complex algebraic geometry / János Kollár, editor.
p. cm.—(IAS/Park City mathematics series, ISSN 1079-5634 ; v. 3)
Lecture notes from a summer program held in Park City, Utah, in 1993.
Includes bibliographical references.
ISBN 0-8218-0432-4 (alk. paper)
1. Geometry, Algebraic—Congresses. I. Kollár, János. II. Series.
QA564.C655 1997
516.3′5—dc20 96-41826
CIP

∞ The paper used in this book is acid-free and falls within the guidelines
established to ensure permanence and durability.

Visit the AMS home page at URL: http://www.ams.org/

10 9 8 7 6 5 4 3 2 04 03 02 01 00

Contents

Preface

The Institute for Advanced Study/Park City Mathematics Institute (PCMI) was founded in 1991 as part of the "Regional Geometry Institute" initiative of the National Science Foundation. In mid 1993 it found an institutional home at the Institute for Advanced Study (IAS) in Princeton. The PCMI Summer Session will alternate between Park City, Utah, and Princeton, New Jersey.

The IAS/Park City Mathematics Institute encourages research and education in mathematics and fosters interaction between the two. The Summer Session offers programs for researchers and postdoctoral scholars, college and university faculty, graduate students, undergraduates, and high school teachers. One of our main goals is to make all of the participants aware of the total spectrum of activities that occur in mathematics education and research: we wish to involve professional mathematicians in education and to bring modern concepts in mathematics to the attention of educators. To that end the Summer Session features general sessions designed to encourage interaction among the various groups. In-year activities at sites around the country form an integral part of the Program for High School Teachers.

Each summer a different topic is chosen as the focus of the Research Program and Graduate Summer School. Activities in the Undergraduate Program deal with this topic as well. Lecture notes from the Graduate Summer School are being published each year in this series. The first three volumes are:

Volume 1: *Geometry and Quantum Field Theory* (1991)
Volume 2: *Nonlinear Partial Differential Equations in Differential Geometry* (1992)
Volume 3: *Complex Algebraic Geometry* (1993)

Future volumes from the 1994 Summer School, *Gauge Theory and the Topology of Four-Manifolds*, and the 1995 Summer School, *Nonlinear Wave Phenomena*, are in preparation. The 1996 Research Program and Graduate Summer School topic is *Probability* and the 1997 topic is *Symplectic Geometry and Topology*.

We plan to publish material from other parts of the IAS/Park City Mathematics Institute in the future. This will include material from the Undergraduate Program and publications documenting the interactive activities which are a primary focus of the PCMI. At the Summer Session late afternoons are devoted to seminars of common interest to all participants. Many deal with current issues in education; others treat mathematical topics at a level which encourages broad participation. The PCMI has also spawned interactions between universities and high schools at a local level. We hope to share these activities with a wider audience in future volumes.

Dan Freed, Series Editor
September 1996

Introduction

The third Summer Session of the Regional Geometry Institute in Park City, Utah, in 1993 was devoted to algebraic geometry. The main part of the program consisted of the following six series of lectures, listed in the order they were given:

Introduction to algebraic geometry —	Miles Reid
Linear series on varieties —	Robert Lazarsfeld
Intersection homology of algebraic varieties —	Eduard Looijenga
Mathematical Aspects of Mirror Symmetry —	David Morrison
Vanishing theorems and Shafarevich maps —	János Kollár
Semi-stable flips —	Shigefumi Mori

The lectures of Reid are intended as a quick introduction for graduate students who have had very little previous experience with algebraic geometry. The notes start with the study of cubic surfaces and scrolls. Both of these are beautiful subjects, where much of the theory can be developed with very little technical machinery. Historically, such concrete examples started the development of the theory of algebraic surfaces and of higher dimensional varieties. Their study still provides the student with a very clear motivation for the development of the basic machinery of algebraic geometry. Two of these technical tools – cohomology theory and intersection theory – are discussed in subsequent chapters. The rest of the notes is devoted to giving an overview of the state of the theory of surfaces. For the beginning reader, this provides one of the best introductions available to this still active area of research.

Lazarsfeld and Looijenga aim their lectures at students who have had previous encounters with algebraic geometry, roughly at the level of a one year introductory course. The intention is to concentrate on a single topic which provides relatively quick entry to a field of current research. Fortunately, there are a few questions where it is feasible to start at a rather elementary level and to reach a point where contemporary research problems can be presented and appreciated. Both of the series succeed in achieving this rather difficult goal.

The lectures of Lazarsfeld deal with linear series. One of the simplest questions that one can ask in algebraic geometry is: how can one decide that a linear series is very ample or at least base point free. This very classical question has been actively investigated recently. Several of the ideas and techniques that were introduced in higher dimensional geometry can be applied to this problem. In many cases the concepts can be elucidated by working on surfaces. This gets rid of most of the technical problems while preserving the essential features. The lectures of Lazarsfeld provided many people with their first glimpse of the ideas coming from higher dimensional geometry and it is one of the the best introductions to the subject.

Looijenga chooses as his topic the topology of algebraic varieties. One of the prettiest ways to learn about this subject is through Lefschetz theory, and this is the approach selected in the lectures. Anyone who reads the classical book of Lefschetz is struck by its great depth and beauty and also by how far its treatment is from the modern standards of rigor. Looijenga presents a very careful treatment along the lines pioneered by Lefschetz. The aim is to use the theory of Lefschetz pencils and vanishing cycles to reach the weak and hard Lefschetz theorems. The proof of the hard Lefschetz theorem presented here is somewhat new. It has the advantage that it connects very naturally with the ideas of intersection homology. The author's approach to this topic is a very welcome addition to the literature.

The lectures of Morrison and myself were designed for a more advanced audience. My notes expanded beyond the scope of this volume; they appeared recently as a book (Shafarevich maps and automorphic forms, Princeton University Press).

Morrison's lectures describe recent interactions between theoretical physics and algebraic geometry. His first topic is the quantum cohomology of algebraic varieties. The theory associates a "quantum cohomology ring" to a variety using the space of maps from curves to the variety. Beside its significance in physics, this construction is also very useful in enumerative questions. The second part of his notes outlines the development of mirror symmetry. The physical theory predicts a duality for threefolds with trivial canonical class, which translates very difficult questions about one threefold to much easier questions about the dual variety. At the moment even the definitions are unsettled, and both theories are full of exciting and deep open problems.

The lectures of Mori on semi-stable flips were the most advanced. Their aim was to present a self-contained treatment of semi-stable flips. At the end it turned out that this is still beyond our reach; the necessary prerequisites are too numerous. The complete treatment is being written up as a book and will soon be published by Iwanami Shoten.

Thus, the present volume does not provide a complete picture of the lectures delivered during the 1993 Summer Session. Fortunately, the other lectures are or will shortly be available for the interested reader.

Finally, I would like to express my sincere appreciation to the speakers whose great devotion to the Regional Geometry Institute made its success possible. All the speakers felt that they found a very interested and receptive audience, and the current forms of the lectures owe a considerable amount to their input.

János Kollár, Volume Editor
September 1996

Chapters on Algebraic Surfaces

Miles Reid

IAS/Park City Mathematics Series
Volume **3**, 1997

Chapters on Algebraic Surfaces

Miles Reid

Introduction

This is a first graduate course in algebraic geometry. It aims to give the student a lift up into the subject at the research level, with lots of interesting topics taken from the classification of surfaces, and a human-oriented discussion of some of the technical foundations, but with no pretence at an exhaustive treatment. I hope that graduate students can use some of these chapters as a reader through the subject, maybe in parallel with a conventional textbook. The early chapters introduce topics that are useful throughout projective and algebraic geometry, make little demands, and lead to fun calculations. The intermediate chapters introduce elements of the technical language gradually, whereas the later chapters get into the substance of the classification of surfaces.

I have given several sets of lectures on algebraic geometry at the graduate or advanced undergraduate level over the last few years. The 1993 Park City summer school, with the challenge of taking a numerous audience through from a basic level up to the sky, was perhaps the most fun. I have thus accumulated a considerable body of notes, without ever having time to organise the material properly in book form. The format of this series allows me to present lecture notes, which are at the same time (somewhat preliminary) extracts of chapters of a more ambitious book on surfaces. A number of the chapters here are based on the notes and examples sheets handed out in class at Park City, with only minor editing. I hope that the reader will excuse their many obvious shortcomings. One of these is the absence of any pictures, a consequence of the way I have put together the notes as computer files; I would not dream of lecturing any part of this material without drawing scribbles every minute or two, and you must make up your own figures if you want the argument to reach deep down to the parts of your intellect that words cannot reach. Some of the chapters have a serious structural fault, namely the tendency to break off into an incoherent list of further topics just as they seem to be getting somewhere. Chapter 4 on singularities should of course be broken up into an introductory discussion of Du Val singularities and the various games we want

[1]Math Institute, University of Warwick Coventry CV4 7AL, England
E-mail address: `e-mail: Miles@Maths.Warwick.Ac.UK`

to play with them, backed up by numerous exercises and worked examples, and a separate chapter giving a serious treatment of resolutions of surface singularities, with adjunction and cohomology; similar remarks apply to Chapter 3 on K3s. I would be very grateful if you would notify me of other possible improvements, and buy my subsequent masterpiece.

Thanks

Some of the material of this book has been lectured at graduate courses at University of Warwick, East China Normal University, Shanghai, University of Tokyo, University of Utah, as well as the Park City summer school. In addition to debts of gratitude stretching back 20 years that it would be tedious to enumerate, I am particularly grateful to Koji Ohno for his work with my Tokyo lecture notes. I also thank Gavin Brown for help preparing and proofreading the Park City notes, and Nick Shepherd-Barron for providing several arguments used in Chapter E.

CHAPTER 1
The Cubic Surface

There are several ways of treating the cubic surface $X \subset \mathbb{P}^3$ and its 27 lines, for example, in terms of elementary coordinate geometry in $\mathbb{P}^3$, or as a blowup of $\mathbb{P}^2$ in 6 points, as a blowup of $\mathbb{P}^1 \times \mathbb{P}^1$ in 5 points, etc. The difficulty with approaches in this style is that because they start by finding one or more lines of X, they do not give the full symmetry of the configuration of lines. The symmetry group of the configuration has order

$$51,840 = 2^7 \cdot 3^4 \cdot 5 = 27 \cdot 5! \cdot 2^4 = 27 \cdot 16 \cdot 10 \cdot 6 \cdot 2 = \cdots$$

The configuration of lines does, however, have an entirely symmetric description in terms of a certain lattice—the divisor class group or middle cohomology $\operatorname{Pic} X = H^2(X, \mathbb{Z})$. The lines $L \subset X$ are the solutions in this lattice of the equations $HL = 1$, $L^2 = -1$ where H is the class of the hyperplane section.

As well as discussing the cubic surface, this section introduces and gives examples of the following notions: -1-curve, blowup, conic bundle, $\mathbb{P}^1$-bundle, divisor class group $\operatorname{Pic} X$, intersection numbers, linear system and rational map, del Pezzo surface, normal surface singularity and its resolution.

1.1. Main tricks

Let $X \subset \mathbb{P}^3$ be a nonsingular cubic surface. I'm interested in the lines and the triangles of X. Here a *triangle* is a set of 3 distinct coplanar lines $L_1, L_2, L_3 \subset X$, so that $L_1 + L_2 + L_3 = X \cap H$ is a hyperplane section. The following 5 statements allow me to find and organise all the lines of X.

0. X contains at least one line L. (This is nontrivial, see [**UAG**], §7 or Shafarevich [**Sh**], Chapter I.)
1. Any two intersecting lines determine a triangle. (Obvious.)
2. If L_1, L_2, L_3 is a triangle and M a fourth line of X, then M meets exactly one of L_1, L_2, L_3. (Almost obvious.)
3. Main trick: there are exactly 10 lines meeting L, falling into 5 coplanar pairs L_i, L_i'; the pairs are disjoint, that is, $(L_i \cup L_i') \cap (L_j \cup L_j') = \emptyset$ for $i \neq j$. (See [**UAG**] or [**Sh**], Chapter IV or [**Beauville**], IV.15.)
4. In particular, there exist two disjoint lines L and M.

1.2. All the lines of X

Here I find and give names to all the lines of X. Fix disjoint lines L and M. By 1.1.3, L takes part in exactly 5 triangles L, L_i, L'_i for $i = 1, \dots, 5$, and M meets exactly one of L, L_i, L'_i for each i. By renumbering, let these be $L_1, \dots, L_5$. Then the 10 lines meeting M are $L''_1, \dots, L''_5$ as in the figure.

Claim. *There are* 10 *further lines* L_{klm} *which meet* L_k, L_l, L_m *and not* L_i, L_j, *where* $\{i, j, k, l, m\} = \{1, 2, 3, 4, 5\}$.

Proof. Notice that if $i \neq j$ then L'_i does not meet M nor L_j, so it must intersect the third line of that triangle, that is, L'_i meets L''_j for $i \neq j$.

Choose some L'_i, for clarity L'_1, and consider the 5 triangles involving L'_1. These are L, L_1, L'_1, and 4 others determined by the intersecting pairs L'_1, L''_j for $j = 2, \dots, 5$. Consider for example the triangle L'_1, L''_2, N. Then by 1.1.3, N is disjoint from the line pairs of the other 4 triangles, in particular N does not intersect L''_3, L''_4, L''_5. Therefore N intersects L_3, L_4 and L_5.

It is easy to see that the lines L, M, L_i, L'_i, L''_i and L_{klm} exhaust all the lines of X.

1.3. The lattice $A(X)$

I define a lattice $A(X)$, a kind of miniature version of the divisor class group of any nonsingular variety. $A(X)$ is defined as an Abelian group with generators and relations. The generators are the 27 lines of X; the relations are simply "triangle = constant". More formally, A is the free Abelian group on the 27 lines, modulo the set of relations $L + L' + L'' = M + M' + M''$ whenever L, L', L'' and M, M', M'' are triangles.

Proposition. $A = \mathbb{Z}^7$; *a basis is* $L_1, \dots, L_4, L'_5, L''_5, L_5$

Proof. The triangles containing L are $L + L_i + L'_i$ for $i = 1, \dots, 5$. Thus I get relation $L + L_i + L'_i = L + L_5 + L'_5$ for $i = 1, \dots, 5$, so that

$$L'_i = L_5 + L'_5 - L_i. \tag{1}$$

Arguing in the same way on the triangles containing M gives $L''_i = L_5 + L''_5 - L_i$.

I showed in 1.2 that $L'_i + L''_j + L_{klm}$ is a triangle when $\{i, j, k, l, m\}$ is a permutation of $\{1, 2, 3, 4, 5\}$, and hence

$$L_{klm} = L + L_i - L''_j. \tag{2}$$

Finally, $L_1 + L_{123} + L_{145}$ is another triangle, so that

$$L + L_1 + L'_1 = L_1 + L_{123} + L_{145} = L_1 + 2L + L_4 - L''_5 + L_2 - L''_3,$$

and therefore

$$L = L'_1 - L_4 + L''_5 - L_2 + L''_3 = 2(L_5 + L'_5 + L''_5) - L_1 - L_2 - L_3 - L_4 - L'_5. \tag{3}$$

Therefore, all the lines, hence everything in $A(X)$ can be written as integral combinations of the seven classes $L_1, \dots, L_4, L'_5, L''_5, L_5$. I prove that these elements are linearly independent in $A(X)$ in the next section by introducing a scalar product on $A(X)$.

1.4. The scalar product

I claim that there is a scalar product $A(X) \times A(X) \to \mathbb{Z}$ such that:

1. Two distinct lines L, L' have $LL' = 0$ or 1 according as they are disjoint or intersect.
2. $L^2 = -1$ for any line L.
3. $L(M + M' + M'') = 1$ for any line L and triangle M, M', M''.

Indeed, formally, I can set LL' and L^2 to be anything I wish, provided that my choice is compatible with the equivalence relation defining A. If M, M', M'' is a triangle and L a line distinct from M, M', M'' then we know that L meets exactly one of M, M', M'', so that (1) implies (3). If on the other hand $L = M$ then $LM' = LM'' = 1$ and $L^2 = -1$ again implies (3). This proves that the scalar product is well defined.

Proposition. *The scalar products of the 7 elements*

$$e_0 = L_5 + L_5' + L_5'', \quad e_1 = L_1, \ldots, e_4 = L_4, \quad e_5 = L_5' \quad \textit{and} \quad e_6 = L_5''$$

is given by $e_0^2 = 1$, $e_i^2 = -1$ for $i = 1, \ldots, 6$ and $e_ie_j = 0$.

In particular, $e_0, \ldots, e_6$ form a basis of $A(X)$.

Proof. $e_1 = L_1$, ... , $e_4 = L_4$, $e_5 = L_5'$ and $e_6 = L_5''$ are 6 disjoint lines, and the first 4 are also disjoint from L_5, so that most of the multiplication table follows at once by construction of the scalar product. The only assertions still to prove concerns e_0. Thus

$$e_0e_5 = e_0L_5' = (L_5 + L_5' + L_5'')L_5' = 1 - 1 + 0 = 0,$$

and similarly $e_0e_6 = e_0L_5'' = 0$. Finally

$$e_0^2 = e_0(L_5 + L_5' + L_5'') = e_0L_5 = (L_5 + L_5' + L_5'')L_5 = -1 + 1 + 1 = 1. \quad \text{Q.E.D.}$$

1.5. Symmetric treatment of the lines

Claim. *Write h for the class of a triangle. Then $h^2 = 3$ and $hx \equiv x^2 \bmod 2$ for every $x \in A(X)$.*

Proof. If $L+L'+L''$ is any triangle then $h = L+L'+L''$ and $h^2 = h(L+L'+L'') = 1+1+1 = 3$.

For the second part, note that for any lattice, $x^2 \bmod 2$ is a *linear* function $A \to \mathbb{F}_2$, because $(x + y)^2 = x^2 + 2xy + y^2 \equiv x^2 + y^2 \bmod 2$. Therefore it is enough to check that $hx \equiv x^2 \bmod 2$ holds for any set of generators of A. But $A(X)$ is generated by lines L, and we know $hL = 1$ and $L^2 = -1$.

To conclude: a cubic surface X has an associated lattice $A(X) \cong \mathbb{Z}^7$, with a scalar product that can be diagonalised to $\operatorname{diag}(1, -1, \ldots, -1)$, and an element $h \in A(X)$ such that $h^2 = 3$ and $hx \equiv x^2 \bmod 2$ for every $x \in A$. It is an easy result of lattice theory that the pair $A(X)$ and H is uniquely determined up to isomorphism by the stated properties.

Inside $A(X)$, the lines are the solutions of the equations $hL = 1$ and $L^2 = -1$. This is a completely symmetric description of the configuration of lines.

Exercise

In the basis $e_0, \dots, e_6$ of 1.3–4, show that

$$h = 3(L_5 + L_5' + L_5'') - L_1 - L_2 - L_3 - L_4 - L_5' - L_5'' = 3e_0 - e_1 - \cdots - e_6.$$

and that the 27 classes of lines are

6 classes e_i for $i = 1, \dots, 6$;
15 classes $e_0 - e_i - e_j$ for $i \neq j = 1, \dots, 6$;
6 classes $2e_0 - e_{i_1} - \cdots - e_{i_5}$ for 5 distinct $i_1, \dots, i_5 = 1, \dots, 6$.

What is $A(X)$? It was derived above in terms of lines, but maybe it has some more natural description in terms of X. One answer is that if $k = \mathbb{C}$, then

$$A(X) = H_2(X, \mathbb{Z}) = H^2(X, \mathbb{Z}).$$

It is not hard to see by topological arguments that this is a lattice of rank 7, with a scalar product defined by cup product of signature $1, -6$ and discriminant 1. Moreover, the hyperplane section $H = X \cap \mathbb{P}^2$ defines a cohomology class $h \in H^2(X, \mathbb{Z})$ with $h^2 = 3$ and $hx \equiv x^2 \bmod 2$ for every $x \in H^2(X, \mathbb{Z})$.

Taking a line L into the homology class $[L]$ defines a map $A(X) \to H_2(X, \mathbb{Z})$, well defined because all hyperplane sections of X are homologous. The proof that this is injective and surjective are rather similar to the arguments of the next section.

1.6. The divisor class group $\operatorname{Pic} X$

I now use the above results on lines on the cubic surface to motivate general ideas of divisors on varieties. A sum of lines $\sum n_i L_i$ is a particular case of a divisor, the relation "all triangles are equal" $L + L' + L'' = M + M' + M''$ a particular case of linear equivalence of divisors, the group $A(X)$ a particular case of the divisor class group $\operatorname{Pic} X$, and the scalar product on $A(X)$ given by $L_1 L_2 = 0, 1$ or -1 a particular case of intersection numbers.

I run briefly through the theory, which is in use throughout these notes; if you have not seen this material before, you should read for example [**Sh**], Chapter III for more details. Let X be a normal variety (for example, nonsingular). A *prime divisor* of X is an irreducible codimension 1 subvariety $\Gamma \subset X$. A *divisor* $D = \sum n_i \Gamma_i$ is a formal linear combination of prime divisors $\Gamma_i \subset X$ with coefficients $n_i \in \mathbb{Z}$. The group of divisors $\operatorname{Div} X$ is the free Abelian group generated by prime divisors. D is *effective*, written $D \geq 0$, if all $n_i \geq 0$.

Recall that if $\Gamma \subset X$ is a prime divisor then $\mathcal{O}_{X,\Gamma}$, the *local ring* at Γ, is the subring of $f \in k(X)$ regular at some point of Γ (therefore regular on a dense open set); $\mathcal{O}_{X,\Gamma}$ is a discrete valuation ring, that is, there is a valuation $v_\Gamma \colon k(X) \setminus 0 \to \mathbb{Z}$ such that $\mathcal{O}_{X,\Gamma} = \{f \mid v_\Gamma(f) \geq 0\} \cup \{0\}$. Here for $0 \neq f \in k(X)$, we say that

$v_\Gamma(f) > 0$ if and only if f has a zero along Γ of order $v_\Gamma(f)$;
$v_\Gamma(f) < 0$ if and only if f has a pole along Γ of order $-v_\Gamma(f)$;
$v_\Gamma(f) = 0$ if and only if f and f^{-1} are regular along Γ

If $0 \neq f \in k(X)$, the expression $\operatorname{div} f = \sum_\Gamma v_\Gamma(f)\Gamma$ defines the divisor of f; the sum runs over all prime divisors, but is finite, since both f and f^{-1} are regular on dense open sets of X. A *principal divisor* is a divisor of the form $\operatorname{div} f$. Two divisors D and D' are *linearly equivalent*, written $D \overset{\text{lin}}{\sim} D'$ or $D \sim D'$ if $D - D'$ is a principal divisor, that is,

$$D - D' = \operatorname{div} f \qquad \text{for some } 0 \neq f \in k(X).$$

If X is a nonsingular variety, its *divisor class group* $\operatorname{Pic} X$ is the group of divisors modulo linear equivalence, $\operatorname{Pic} X = \operatorname{Div} X / \sim$.

Beware that there are two or three other equivalence relations on divisors (and many others on algebraic cycles) in common use. To get the idea of linear equivalence, suppose that D, D' are effective with no common components and $D - D' = \operatorname{div} f$. This means that f has zeros on D and poles on D'. I can view f as a rational map $f\colon X \dashrightarrow \mathbb{P}^1$, and $D = f^{-1}(0)$, $D' = f^{-1}(\infty)$ (the locus of indeterminacy of f is a subset of X of codimension ≥ 2). Thus $D_t = f^{-1}(t)$ is a divisor that moves with $t \in \mathbb{P}^1$ from D at $t = 0$ to D' at $t = \infty$.

Example. Let $X \subset \mathbb{P}^3$ be a nonsingular cubic surface and $L+L'+L''$ and $M+M'+M''$ two triangles cut out on X by the planes $A = 0$ and $B = 0$ where A and B are linear forms on $\mathbb{P}^3$. Then $A/B \in k(X)$ and $\operatorname{div}(A/B) = L+L'+L''-M-M'-M''$, so that any two triangles are linearly equivalent. Since all the defining relations for A were of this form, this implies that there is a well-defined map $\alpha\colon A(X) \to \operatorname{Pic} X$.

I will prove in the next two sections that α is injective and surjective.

1.7. Intersection numbers

Here are some basic facts about divisors on a projective nonsingular surface Y, which are discussed in more detail in Chapter A below. One can define intersection numbers of divisors $D_1 D_2$ such that

1. $D_1 D_2$ is bilinear in each factor and symmetric;
2. $D_1 D_2$ only depends on D_1, D_2 up to linear equivalence: that is,

$$\mathcal{O}_Y(D_1) \cong \mathcal{O}_Y(D_1') \implies D_1 D_2 = D_1' D_2;$$

3. if $D_1, D_2 \geq 0$ and have no common components then $D_1 D_2 = \sum_P (D_1 D_2)_P$, where the sum runs over all $P \in D_1 \cap D_2$ and

$$(D_1 D_2)_P = \dim_k \mathcal{O}_{Y,P}/(I_{D_1} + I_{D_2})\mathcal{O}_{Y,P} = \dim_k \mathcal{O}_{Y,P}/(f_1, f_2),$$

   where D_1, D_2 are locally defined by f_1, f_2;
4. if C is an irreducible curve then $CD = \deg_C \mathcal{O}_C(D)$.

The properties (1–3) uniquely define the intersection number as a bilinear pairing $\operatorname{Div} Y \times \operatorname{Div} Y \to \mathbb{Z}$ (see, for example, Hartshorne [**H1**], Chapter V). The point of (2) is that $D_1 D_2$ is well defined on $\operatorname{Pic} Y$, not just on $\operatorname{Div} Y$. The bilinear pairing I gave on $A(S)$ satisfies (1–3) so is the intersection pairing on $\operatorname{Pic} S$ (under the map α defined at the end of 1.6).

Notice the useful fact that for irreducible curves C, C', the only way $CC' < 0$ is possible is for $C = C'$. The selfintersection C^2 can be interpreted as the degree of the normal bundle $N_{C/Y}$.

1.8. Conic bundles and the cubic surface

Let $L \subset S$ be a line and $\varphi = \varphi_L \colon S \dashrightarrow \mathbb{P}^1$ the projection away from L. I can realise this map as follows. Take $M = \mathbb{P}^1 \subset \mathbb{P}^3$ disjoint from L. If $P \in S \setminus L$ then P and L span a unique plane $\Pi = \mathbb{P}^2 \subset \mathbb{P}^3$. Say $\Pi \cap M = \pi \in \mathbb{P}^1$ and define $\varphi(P) = \pi$. φ is not *a priori* defined at points in L but in fact I have

Lemma. *$\varphi\colon S \setminus L \to \mathbb{P}^1$ extends to a morphism $\varphi\colon S \to \mathbb{P}^1$.*

I give two proofs.

First Proof. Let $L = (x = y = 0)$, so that S is defined by $Ax + By = 0$ where A and B are forms of degree 2 in $\{x, y, z, t\}$ which have no common zero on L (by nonsingularity). So the ratio $x : y = B : A$ is everywhere well defined. This ratio defines a morphism to $\mathbb{P}^1$ and is clearly just the map described above. Q.E.D.

Second Proof. This proof is by intersection numbers. Let $H = L + F$ be a hyperplane section of S through L. Moving H to a linearly equivalent section I can see that $HL = 1$ and $FL = FH = 2$ so $H^2 = 3$ and $F^2 = F(H - L) = 0$. Since the F are effective and distinct they must be disjoint so φ is well defined. Q.E.D.

By construction, the fibres of φ are the plane conics residual to L. By the Main Trick 1.1.3, five of these are line pairs, the remainder nonsingular conics. Also, $F^2 = 0$, as we saw in the second proof. In particular, if $F = L_1 + L_2$, then $L_1L_2 = 1$, so that $L_i^2 = -1$, which agrees with the choice made in 1.4.2.

Remark. L is not a section of the bundle. Since $LF = 2$, it cuts a generic fibre twice.

1.9. Other birational models for S

The conic bundle structure described in 1.8 can be constructed starting from any line L lying on S. I have two disjoint lines L and M lying on S, provided by 1.1.4, so I can construct

$$\varphi = \varphi_L \times \varphi_M \colon S \to \mathbb{P}^1 \times \mathbb{P}^1.$$

Since φ_L is the conic bundle obtained as the linear projection away from L, it is clear from 1.1.3 that it takes the 5 line pairs meeting L to 5 distinct points of $\mathbb{P}^1$. Thus both φ_L and φ_M are constant on the 5 lines $L_1, \ldots, L_5$ meeting both of them, so that $\varphi = \varphi_L \times \varphi_M$ contracts these 5 lines $L_1, \ldots, L_5$ to points $Q_1, \ldots, Q_5$ of $\mathbb{P}^1 \times \mathbb{P}^1$.

In fact S is $\mathbb{P}^1 \times \mathbb{P}^1$ blown up in five distinct points (see below for blowup).

If I identify $\mathbb{P}^1 \times \mathbb{P}^1$ with the quadric $Q \subset \mathbb{P}^3$, and compose the map $\varphi\colon S \to \mathbb{P}^1 \times \mathbb{P}^1$ with the linear projection $Q \dashrightarrow \mathbb{P}^2$ from Q_5 then I obtain a morphism $\psi\colon S \to \mathbb{P}^2$ that contracts $L_1, \ldots, L_4, L_5', L_5''$ to 6 points of $\mathbb{P}^2$. See Ex. 1.7 for this map in coordinates.

1.10. Blowup

We're going to see that S is isomorphic to $\mathbb{P}^2$ blown up in 6 points "in general position", with the basis of $A(S)$ corresponding to $e_0 =$ a line in $\mathbb{P}^2$ (with $e_0^2 = 1$), and the 27 classes of lines given as:

$$e_i \text{ the inverse image of the blowup of } P_i;$$
$$e_0 - e_i - e_j \text{ the line through } P_i, P_j;$$
$$2e_0 - e_1 - \cdots - e_5 \text{ the conic through } P_1, \dots, P_5.$$

For this I need to say what blowup means. (Compare also [**Sh**], Chapter II, §4.)

Let $P \in S$ be a point of a nonsingular surface. Then there exists a nonsingular surface and a morphism $\sigma\colon S_1 \to S$ such that

(i) σ is an isomorphism $S_1 \setminus \sigma^{-1}(P) \cong S \setminus P$;

(ii) $\sigma^{-1}(P) \cong \mathbb{P}^1$.

S_1 and σ are in fact uniquely determined by these properties.

Construction

Suppose first that $S = \mathbb{A}^2$ with coordinates x, y, and $P = (0,0)$. Define $S_1 \subset \mathbb{A}^2 \times \mathbb{P}^1$ to be the closed graph of the rational map $(x, y) \mapsto (x : y)$. If $(u : v)$ are homogeneous coordinates of $\mathbb{P}^1$ then S_1 is defined by $x/u = y/v$, that is $xv = yu$.

So if $(x, y) \neq (0,0)$, the ratio $(u : v)$ is well defined, which gives (i). Also $\sigma^{-1}(0,0)$ is clearly the whole of $\mathbb{P}^1$. Finally, to see that S_1 is nonsingular, note that since $(u : v)$ are homogeneous coordinates of $\mathbb{P}^1$, if $u \neq 0$ then I can assume $u = 1$, so that the open set $(u \neq 0) \subset S_1$ is the surface in $\mathbb{A}^3$ with coordinates (x, y, v) defined by $y = xv$.

More generally, if $P \in S$ is any nonsingular point of a surface, let x, y be local coordinates at P; by passing to a small enough open set, I can assume that S is affine, and $m_P = (x, y)$ (where m_P is the maximal ideal of functions vanishing at P in the coordinate ring of S). The same construction works. The curve C that comes out of blowing up a nonsingular point on a surface satisfies $C \cong \mathbb{P}^1$ and $C^2 = -1$. I call such a curve a -1-*curve.*

Castelnuovo's criterion. *Let Y be a nonsingular surface and C a curve in Y. Then there is a morphism $\varphi\colon Y \to X$ to a nonsingular surface X contracting C to a point and an isomorphism outside C if and only if C is a -1-curve.*

One proof of this is given in Contraction Theorem 4.15 below.

Remark. A blowup is also called a monoidal transformation, sigma process, etc.; blowups play a star role in birational geometry of surfaces and in resolution of singularities. For example, if $C \subset S = \mathbb{A}^2$ is a curve which has a singularity of multiplicity m at the origin, $\sigma\colon S_1 \to S$ the blowup, then $\sigma^{-1}(C) = L \cup C'$ where $C' \subset S_1$ is "less singular" than C.

Exercise. Show that $C : (x^2 = y^3) \subset \mathbb{A}^2$ is resolved by a single blowup, whereas $(x^2 = y^5)$ needs two.

1.11. The cubic surface as $\mathbb{P}^2$ blown up 6 times

Theorem. *Suppose $\Phi = \{P_1, \dots, P_6\} \subset \mathbb{P}^2$ is a set of 6 points with the following "general position" properties: (i) no 2 points coincide; (ii) no 3 collinear; (iii) not all 6 on a conic. Let S be the blowup of $\mathbb{P}^2$ at the 6 points Φ.*

Then the vector space of cubic forms on $\mathbb{P}^2$ vanishing at Φ is 4-dimensional, and if $F_1, \dots, F_4$ is a basis then the rational map $\mathbb{P}^2 \dashrightarrow \mathbb{P}^3$ defined by $P \mapsto$

$(F_1(P):\cdots:F_4(P))$ *induces an isomorphism of* S *with a nonsingular cubic surface* $S\cong S_3\subset\mathbb{P}^3$.

This construction is a 2-sided inverse to the map ψ *described at the end of 1.9 (see also Ex. 1.7).*

The fact that $S_3(P_1,\dots,P_6)$ is 4-dimensional is proved in [**UAG**], §1. The theorem can be proved by similar arguments, but I omit it for lack of time.

1.12. Final remarks on cubic surface

All the topics below have been extensively studied:

1. Weyl group of E_6 and Galois theory of cubic surfaces

If S is defined over an algebraically nonclosed field K, the 27 lines will not in general be defined over K, but will usually require a field extension L/K. Then $\mathrm{Gal}(L/K)$ acts by permuting the 27 lines, so defines a homomorphism φ: $\mathrm{Gal}(L/K)\to W(E_6)$ to the symmetry of the configuration of lines, which by what I said in 1.5 is the Weyl group $W(E_6)$ of the root system in the lattice $h^\perp\subset A(X)$ (compare Ex. 1.4). For example if $K=\mathbb{R}$ and $L=\mathbb{C}$ then complex conjugation interchanges some of the lines; the lines of the real locus $S(\mathbb{R})\subset S(\mathbb{C})$ are just the fixed lines.

The image of φ is a measure of how complicated the arithmetic properties of S are. The question of whether S is rational over the given field K (that is, does there exist birational map $S\dashrightarrow\mathbb{P}^2$ defined over K?) can be discussed in terms of $\operatorname{im}\varphi$.

2. Monodromy

If $\{S_t\}_{t\in T}$ is a family of cubic surfaces, say with S_t nonsingular for every $t\in T_0\subset T$ then the 27 lines of S_t define a 27-to-1 cover $L\to T_0$. Moving round a closed loop in T_0 permutes the lines among themselves, which defines a homomorphism $\psi\colon\pi_1(T_0,t_0)\to W(E_6)$.

3. Singular cubic surfaces

If S is a cubic surface with an ordinary double point $P\in S$ as its only singularity, it turns out that it still has 15 lines not through P, and 6 lines through P (which should be "counted with multiplicity 2"). S can still be obtained from a linear system of cubics in $\mathbb{P}^2$ through 6 points, but these must be in "special position", for example 3 collinear: then the line joining these 3 is contracted to the singularity of S (see Ex. 1.12).

4. Cubic surfaces and del Pezzo surfaces in general

Cubic surfaces are the best known of the series of del Pezzo surfaces. A *del Pezzo surface* is a nonsingular projective surface S with $-K_S$ ample. In addition to the cubic S_3, examples include the complete intersection of two quadrics $S_4=Q_1\cap Q_2\subset\mathbb{P}^4$, the double plane $S_2\to\mathbb{P}^2$ with quartic branch curve, and $S_6\subset\mathbb{P}^6$, the section of $\mathbb{P}^1\times\mathbb{P}^1\times\mathbb{P}^1\subset\mathbb{P}^7$ by a general hyperplane $\mathbb{P}^6$. Just as the cubic is obtained by blowing up $\mathbb{P}^2$ in 6 general points (then making the anticanonical embedding), the other del Pezzo surfaces are obtained by blowing up $\mathbb{P}^2$ in k general points for some $k=0,\dots,8$ (except $\mathbb{P}^1\times\mathbb{P}^1$, which is also a del Pezzo).

Del Pezzo surfaces have been studied in many contexts of geometry, singularity theory and number theory (see for example [**Manin**], Manin and Tsafsman [**M–Ts**]). They have recently acquired extra significance: in Mori theory (of minimal models, compare C.3 and Chapter D), a principal role is played by *Mori fibre spaces* $X \to Y$, that is, morphisms whose fibres S are varieties with ample $-K_S$. Thus del Pezzo surfaces and 1-parameter families of del Pezzo surfaces appear in a fundamental way in the classification of surfaces and 3-folds.

Exercises to Chapter 1

1. Determine the entire configuration of lines in terms of the notation L, M, L_i, L_i', L_i'', L_{klm}. Compare [**UAG**], §7.

2. A *double six* is two sextuples of disjoint lines $l_1, \dots, l_6$ and $m_1, \dots, m_6$ such that

$$l_i \cap m_j = \begin{cases} \text{one point} & \text{if } i \neq j, \\ \emptyset & \text{otherwise.} \end{cases}$$

Let $l_1 = L_1, \dots, l_4 = L_4, l_5 = L_5', l_6 = L_5''$. Find $m_1, \dots, m_6$ forming a double six.

3. If $l_1, \dots, l_6, m_1, \dots, m_6$ form a double six, and have classes $l_i = e_i$ and $m_i = 2e_0 -$ (the five the other e_j), show how to index the remaining 15 lines so that n_{ij} meets l_i, l_j, m_i, m_j. When does n_{ij} meet n_{kl}?

4. Prove that the six classes $f_1 = L_1 - L_2, f_2 = L_2 - L_3, f_3 = L_3 - L_4, f_4 = L_4 - L_5', f_5 = L_5' - L_5''$ and $f_6 = L_4 - L_5$ base the orthogonal complement of H in $A(X)$, where H is the hyperplane section (that is, the class of a triangle), and that $f_i^2 = -2$ and $f_i f_j = 0$ or 1 for $i \neq j$. (If we define a graph with nodes $\{f_i\}$ and an edge joining f_i and f_j if and only if $f_i f_j = 1$ then we obtain the graph E_6, a Dynkin diagram which appears in many areas of math.)

5. This is an exercise on $\mathbb{P}^1 \times \mathbb{P}^1$. Show that $\operatorname{Pic}(\mathbb{P}^1 \times \mathbb{P}^1) = \mathbb{Z} \oplus \mathbb{Z}$. Writing an element of $\operatorname{Pic}(\mathbb{P}^1 \times \mathbb{P}^1)$ as (a, b), work out an intersection formula for two curves lying on $\mathbb{P}^1 \times \mathbb{P}^1$.

6. Suppose that $L : (x = y = 0)$, $M : (z = t = 0)$ and $L_5 : (y = t = 0)$ lie on a nonsingular cubic $X \subset \mathbb{P}^3$, and let L_5', L_5'' be as in 1.2–3. Write $F = L_5 + L_5' + L_5''$. Let $\mathcal{L}(F) \subset k(X)$ be the vector space of rational functions whose divisor of poles is at most F. Prove that $\{\frac{x}{y}, \frac{z}{t}, 1\}$ is a basis of $\mathcal{L}(F)$. Write out in coordinates the rational map $\alpha : X \dashrightarrow \mathbb{P}^2$ to the projective plane that this defines.

7. As in the preceding question, suppose that $L : (x = y = 0)$, $M : (z = t = 0)$ and $L_5 : (y = t = 0)$ lie on a nonsingular cubic $X \subset \mathbb{P}^3$. Prove that

$$\alpha : X \dashrightarrow \mathbb{P}^2, \quad \text{defined by } (x, y, z, t) \mapsto (xt : yz : yt)$$

extends to a morphism. [Hint: $(x : y)$ is well defined at any point of L and $(y : t)$ is well defined at any point of L_5.]

8. Find three lines M_1, M_2, M_3 such that

$$L_1 + L_2 + M_3 = L_2 + L_3 + M_1 = L_1 + L_3 + M_2 = 2H - L - M - M_5$$

in $A(X)$. Prove that

$$(L_1 + L_2 + M_3) \cap (L_2 + L_3 + M_1) \cap (L_1 + L_3 + M_2) = \emptyset,$$

and use this to give an alternative proof of Ex. 1.7.

9. Show that α in Ex. 1.7 contracts the six lines $L_1, \dots, L_4, L_5', L_5''$ to points and is an isomorphism outside these.

10. Let L be a line in $\mathbb{P}^2$ and P_1, P_2, P_3 noncollinear points. Describe the map to projective space determined by the linear system $|2L - P_1 - P_2 - P_3|$ and its inverse in terms of blowing up points and contracting -1-curves. This map is called the *elementary quadratic transformation.*

11. Let S_m be a nonsingular surface of degree m in $\mathbb{P}^3$. Generalise the proof of 1.8 to show that if $L = \mathbb{P}^1$ lies on S_m then $L^2 = 2 - m$. Compare this to what you know on S_2 and S_3. The surface S_4 is an example of a K3 surface, and is studied in greater detail later.

12. Suppose that $\Phi = \{P_1, \dots, P_6\} \subset \mathbb{P}^2$ is a set 6 points which is in general position except that P_1, P_2, P_3 are collinear, on a line L. Prove that cubics through Φ define a birational map $\varphi\colon \mathbb{P}^2 \dashrightarrow S \subset \mathbb{P}^3$ to a singular cubic surface. It blows up the 6 points of Φ, then contracts the line L to a singular point $P \in S$, and is otherwise an isomorphism.

13. Let $S \subset \mathbb{P}^3$ be a cubic surface with an ordinary double point $P \in S$. Study the projection map $\pi\colon S \dashrightarrow \mathbb{P}^2$ from P, and show that it contracts 6 lines of S through P to a set of 6 points $\Phi \subset \mathbb{P}^2$ lying on a conic. Study the inverse map $\varphi\colon \mathbb{P}^2 \dashrightarrow S$ of π as in the preceding question.

CHAPTER 2
Rational Scrolls

This chapter describes scrolls, especially the rational normal scrolls. These varieties occur throughout projective and algebraic geometry, and the student will never regret the investment of time studying them. One reason for presenting them here is that they can be discussed with very little background, and can be used to illustrate many constructions of algebraic geometry with substantial examples. I use them here to give simple examples of rational surfaces, K3 surfaces, elliptic surfaces, and surfaces with pencils of curves of genus 2, 3, 4, etc.

In intrinsic terms, a scroll is a $\mathbb{P}^{n-1}$-bundle $F \to C$ over a curve C, that is, an algebraic fibre bundle, isomorphic to $U_i \times \mathbb{P}^{n-1}$ over small Zariski open sets $U_i \subset C$, and glued by transition functions given by morphisms $U_i \cap U_j \to \mathrm{PGL}(n)$. It can be written as the projectivisation $F = \mathbb{P}(E)$ of a vector bundle E, and the study of general scrolls is essentially equivalent to that of vector bundles over curves. In the case $C = \mathbb{P}^1$ everything is much simpler, because the base curve $\mathbb{P}^1$ is a very explicit object, and every vector bundle is a direct sum of line bundles. Thus any question about scrolls can be solved in very explicit terms. I give some examples in the text, and many more in the exercises (see for example Ex. 2.6–2.9).

As well as discussing scrolls, this section introduces and gives examples of the following notions: linear system, free linear system, very ample linear system and projective embedding, quadrics of rank 3 and 4, determinantal variety, base locus of linear system, divisor class group $\operatorname{Pic} X$, intersection numbers, Veronese surface and cones over it, vector bundles over curves, projectivised bundle $\mathbb{P}_C(E)$, Chern numbers, Harder–Narasimhan filtration, K3 surface, elliptic surface, Weierstrass normal form, surface with a pencil of curves of genus $g = 1, 2, 3, 4, \dots$.

2.1. Reminder: $\mathbb{P}^{l-1} \times \mathbb{P}^{m-1}$

I start by recalling the product of two projective spaces as treated in elementary textbooks (compare, for example, [**Sh**], Chapter I or [**UAG**], §5), which is a very useful analogy for scrolls. $\mathbb{P}^{l-1} \times \mathbb{P}^{m-1}$ can be defined as the quotient of $(\mathbb{C}^l \setminus 0) \times (\mathbb{C}^m \setminus 0)$ by the action of two copies of the multiplicative group $\mathbb{C}^* \times \mathbb{C}^*$ acting separately on the two factors:

$$(x_1, \dots, x_l; y_1, \dots, y_m) \mapsto (\lambda x_1, \dots, \lambda x_l; \mu y_1, \dots, \mu y_m) \qquad \text{for } (\lambda, \mu) \in \mathbb{C}^* \times \mathbb{C}^*.$$

Subvarieties of $\mathbb{P}^{l-1} \times \mathbb{P}^{m-1}$ are defined by *bihomogeneous* polynomials, that is, polynomials that are homogeneous separately in $x_1, \dots, x_l$ and $y_1, \dots, y_m$, and rational functions on $\mathbb{P}^{l-1} \times \mathbb{P}^{m-1}$ as quotients of two bihomogeneous polynomials of the same bidegree. Next $\mathbb{P}^{l-1} \times \mathbb{P}^{m-1}$ has the Segre embedding into usual projective space

$$\mathbb{P}^{l-1} \times \mathbb{P}^{m-1} \hookrightarrow \mathbb{P}^{lm-1}$$

defined by bilinear forms

$$(x_1, \dots, x_l; y_1, \dots, y_m) \mapsto \left(u_{ij} = x_i y_j\right)_{\substack{i=1\dots l \\ j=1\dots m}} .$$

The image is defined by equations $\operatorname{rank}(u_{ij}) \le 1$.

Remark. As an algebraic geometer, I should say that $\mathbb{P}^{l-1} \times \mathbb{P}^{m-1}$ is the quotient of the variety $(\mathbb{A}^l \setminus 0) \times (\mathbb{A}^m \setminus 0)$ by the action of the algebraic group $\mathbb{G}_m \times \mathbb{G}_m$. Here $\mathbb{A}^l$ is affine space, the variety corresponding to k^l for a field k, and $\mathbb{G}_m$ is the algebraic group corresponding to the multiplicative group k^*. If it makes life simpler, you can replace $\mathbb{A}^l$ by $\mathbb{C}^l$ and $\mathbb{G}_m$ by $\mathbb{C}^*$ throughout.

2.2. Definition of $\mathbb{F}(a_1, \dots, a_n)$

Let $a_1, \dots, a_n$ be integers. I define the scroll $\mathbb{F} = \mathbb{F}(a_1, \dots, a_n)$ as the quotient of $(\mathbb{A}^2 \setminus 0) \times (\mathbb{A}^n \setminus 0)$ by an action of $\mathbb{G}_m \times \mathbb{G}_m$, the product of two multiplicative groups. Write t_1, t_2 for coordinates on $\mathbb{A}^2$ and $x_1, \dots, x_n$ on $\mathbb{A}^n$, and λ and μ for elements of the two factors of $\mathbb{G}_m \times \mathbb{G}_m$, that is, $(\lambda, \mu) \in \mathbb{G}_m \times \mathbb{G}_m$. The action is given as follows:

$$\begin{aligned} (\lambda, 1)\colon (t_1, t_2; x_1, \dots, x_n) &\mapsto (\lambda t_1, \lambda t_2; \lambda^{-a_1} x_1, \dots, \lambda^{-a_n} x_n); \\ (1, \mu)\colon (t_1, t_2; x_1, \dots, x_n) &\mapsto (t_1, t_2; \mu x_1, \dots, \mu x_n). \end{aligned}$$

Note first that the ratio $t_1 : t_2$ is preserved by the action of $\mathbb{G}_m \times \mathbb{G}_m$, so that the projection to the first factor defines a morphism $\pi\colon \mathbb{F} \to \mathbb{P}^1$:

$$\begin{array}{ccc} (\mathbb{A}^2 \setminus 0) \times (\mathbb{A}^n \setminus 0) & \longrightarrow & \mathbb{F}(a_1, \dots, a_n) \\ {\scriptstyle p_1}\downarrow & & \downarrow{\scriptstyle \pi} \\ (\mathbb{A}^2 \setminus 0) & \longrightarrow & \mathbb{P}^1 \end{array}$$

Remark. Compared to 2.1, I have restricted to the case $l = 2$ (so that the first factor is $\mathbb{P}^1$), and generalised the group action to allow it to mix up the two factors, so that $\mathbb{F} \to \mathbb{P}^1$ can be a nontrivial $\mathbb{P}^{n-1}$ fibre bundle. The material of 2.2–7 follows exactly the remaining steps of 2.1.

2.3. $\mathbb{F}(a_1, \dots, a_n)$ as a fibre bundle

Above any given ratio $(t_1 : t_2) \in \mathbb{P}^1$, I can normalise to fix the values of t_1, t_2 with the given ratio, and this takes care of the action of the first factor $\mathbb{G}_m$; after this, the fibre of π over $(t_1 : t_2)$ consists of the set of ratios $(x_1 : \cdots : x_n)$, forming a copy of $\mathbb{P}^{n-1}$. Thus every fibre of the projection map $\pi\colon \mathbb{F} \to \mathbb{P}^1$ is isomorphic to $\mathbb{P}^{n-1}$. As I show in Theorem 2.5 below, a good 19th century way of understanding $\mathbb{F}$ is to embed it in a projective space so that the fibres of π are linearly embedded $(n-1)$-planes.

Remark. $\pi\colon \mathbb{F} \to \mathbb{P}^1$ is an example of a *fibre bundle* with fibre $\mathbb{P}^{n-1}$ and structure group the diagonal subgroup of $\mathrm{PGL}(n)$. More explicitly, on the affine piece $U_0 = (t_2 \neq 0) \subset \mathbb{P}^1$, I set $t_2 = 1$, so that $\pi^{-1}(U_0) = \mathbb{A}^1 \times \mathbb{P}^{n-1}$, with $t_1 = t_1/t_2$ the affine coordinate in the first factor and $(x_1 : \cdots : x_n)$ homogeneous coordinates in the second. Similarly, over $U_\infty = (t_1 \neq 0)$ I get $\pi^{-1}(U_\infty) = U_\infty \times \mathbb{P}^{n-1}$. The affine coordinates $t_1 = t_1/t_2$ and $t_2 = t_2/t_1$ on U_0 and U_∞ are related on the overlap $U_0 \cap U_\infty$ in the usual way by $t_1 = 1/t_2$, and the two open sets $\pi^{-1}(U_0) = U_0 \times \mathbb{P}^{n-1}$ and $\pi^{-1}(U_\infty) = U_\infty \times \mathbb{P}^{n-1} \subset \mathbb{F}$ are glued together as follows:

$$t_1 : 1; x_1^{(0)} : \cdots : x_n^{(0)} \xrightarrow{t_1^{-1}} 1, 1/t_1; t_1^{a_1} x_1^{(0)} : \cdots : t_1^{a_n} x_n^{(0)} = 1, t_2; x_1^{(\infty)} : \cdots : x_n^{(\infty)}.$$

(The arrow is the action of $\lambda = t_1^{-1} \in \mathbb{G}_m$). In brief, $\mathbb{F}$ is the union of two copies of $\mathbb{A}^1 \times \mathbb{P}^{n-1}$ glued together by $t_1 \mapsto t_1^{-1}$ in the first factor and $\mathrm{diag}(t_1^{a_1}, \ldots, t_1^{a_n})$ in the second.

2.4. Bihomogenous polynomials

Rational functions on $\mathbb{F}$ are defined as ratios of bihomogenous polynomials, that is, eigenfunctions of the action of $\mathbb{G}_m \times \mathbb{G}_m$. I write down some vector spaces of bihomogeneous functions. I've already given one, the space $\langle t_1, t_2 \rangle$ of homogeneous polynomials of degree 1 in t_1, t_2 and degree 0 in $x_1, \ldots, x_n$; the ratio $t_1 : t_2$ defines the projection $\pi\colon \mathbb{F} \to \mathbb{P}^1$.

Next, consider the functions that are linear in $x_1, \ldots, x_n$. It's clear that this means that the second factor of $\mathbb{G}_m \times \mathbb{G}_m$ acts by $(1, \mu)\colon h \mapsto \mu h$. Consider polynomials that are also invariant under the action of the first factor, that is, $(\lambda, 1)\colon h \mapsto h$. Obviously, to cancel the group action $x_i \mapsto \lambda^{-a_i} x_i$, the linear term x_i must be accompanied by a monomial $t_1^b t_2^c$ with $b + c = a_i$, and hence the vector space of μ-invariant polynomials is based by

$$S^{a_1}(t_1, t_2)x_1, \ldots, S^{a_n}(t_1, t_2)x_n,$$

where $S^a(t_1, t_2) = \{t_1^a, t_1^{a-1}t_2, \ldots, t_2^a\}$ is the set of monomials of degree a in t_1, t_2. Of course, $S^a = \emptyset$ if $a < 0$ and $S^0 = \{1\}$. The notation S^a stands for the ath symmetric tensor power: if $\langle \Sigma \rangle$ denotes the vector space spanned by a set Σ and Sym^a the ath symmetric tensor power of a vector space then $\langle S^{a_1}(t_1, t_2) \rangle = \mathrm{Sym}^a \langle t_1, t_2 \rangle$.

In the same way, the space of polynomials that are linear in $x_1, \ldots, x_n$ and in the λ^e eigenspace of the first factor is based by

$$S^{a_1+e}(t_1, t_2)x_1, \ldots, S^{a_n+e}(t_1, t_2)x_n;$$

its dimension is $\sum_{i=1,\ldots,n}^{+}(a_i + e + 1)$, where $\sum^+$ means you only take the sum of the terms that are ≥ 0.

There is a similar description of the bihomogeneous polynomials of any bidegree, that is, of degree d in $x_1, \ldots, x_n$ and *extra degree* e in the t_i; this is the vector space based by the monomials $t_1^{e_1} t_2^{e_2} x_1^{d_1} \cdots x_n^{d_n}$ with $\sum d_i = d$, and $e_1 + e_2 = \sum_{i=1}^n d_i a_i + e$. I discuss this in more detail later.

2.5. Theorem (Linear embeddings $\mathbb{F} \hookrightarrow \mathbb{P}^N$). *Suppose that $a_1, \ldots, a_n > 0$; then the ratios between the bihomogeneous polynomials*

$$S^{a_1}(t_1, t_2)x_1, \ldots, S^{a_n}(t_1, t_2)x_n \qquad (2.5.1)$$

define an embedding $\varphi\colon \mathbb{F}(a_1,\dots,a_n) \hookrightarrow \mathbb{P}^N$ (where $N = \sum_{i=1}^n (a_i+1) - 1$) in such a way that every fibre $\mathbb{P}^{n-1}$ of π goes into a linearly embedded $(n-1)$-plane.

The image is the subvariety of $\mathbb{P}^N$ defined by the determinantal equations

$$\operatorname{rank}\begin{pmatrix} u_1 & u_2 & \dots & u_{a_1} & u_{a_1+2} & \dots & u_{a_1+a_2+1} & \dots & u_N \\ u_2 & u_3 & \dots & u_{a_1+1} & u_{a_1+3} & \dots & u_{a_1+a_2+2} & \dots & u_N+1 \end{pmatrix} \le 1.$$

The matrix here has n blocks of size $2\times a_i$; in each block, the $(1,j)$th entry for $j\ge 2$ repeats the $(2,j-1)$st entry. The meaning of the determinantal equation is that if the monomials (2.5.1) are listed as $t_1^{a_1}x_1, t_1^{a_1-1}t_2x_1,\dots,t_1^{a_2}x_2,\dots$ then the ratio $t_1:t_2$ equals the ratio between the first and second rows, that is

$$\frac{t_1}{t_2} = \frac{t_1^{a_1}x_1}{t_1^{a_1-1}t_2x_1} = \frac{t_1^{a_1-1}t_2x_1}{t_1^{a_1-2}t_2^2x_1} = \dots = \frac{t_1^{a_2}x_2}{t_1^{a_2-1}t_2x_2} = \dots = \frac{t_1t_2^{a_n-1}x_n}{t_2^{a_n}x_n}.$$

Proof. The proof is very similar to that for the Segre embedding $\mathbb{P}^{l-1}\times\mathbb{P}^{m-1}\hookrightarrow \mathbb{P}^{lm-1}$ (see [**Sh**], Chapter I).

$\mathbb{F}$ is covered by a number of open sets U_{ij} : $(t_i\ne 0, x_j\ne 0)$ for $i=1,2$ and $j=1,\dots n$, each isomorphic to $\mathbb{A}^n$. The piece U_{11} is typical. The n affine coordinates on it are t_2/t_1 and $t_1^{a_i-a_1}x_i/x_1$ for $i=2,\dots,n$.

The set of monomials include $t_1^{a_1}x_1, t_1^{a_1-1}t_2x_1, t_1^{a_2}x_2,\dots,t_n^{a_n}x_n$. The first of these is nonzero everywhere on U_{11}, so that the ratio is well defined there. The n affine coordinates of U_{11} are precisely given by the ratios between $t_1^{a_1}x_1$ and the n succeeding monomials, so that these embed U_{11}.

In the given determinantal equations, clearly if $u_1=1$ then all the remaining u_i are determined by u_2 and $u_{a_1+2}, u_{a_1+a_2+3},\dots$ corresponding to the n affine coordinates of U_{11}. Q.E.D.

Remarks. (a) "Linear generation" of scrolls. The image variety $\mathbb{F}(a_1,\dots,a_n)\subset\mathbb{P}^N$ has the following description in projective geometry. Consider a fixed copy of $\mathbb{P}^1$ with homogeneous coordinates t_1,t_2, and n embeddings $v_{a_i}\colon\mathbb{P}^1\hookrightarrow\mathbb{P}^{a_i}$ defined by

$$(t_1:t_2)\mapsto (t_1^{a_i}:t_1^{a_i-1}t_2:\dots:t_2^{a_i});$$

this is the a_ith *Veronese embedding*, and the image $\Gamma_i = v_{a_i}(\mathbb{P}^1)$ is called the *rational normal curve* of degree a_i. Embed all the projective spaces $\mathbb{P}^{a_i}\hookrightarrow\mathbb{P}^N$ as linearly independent subspaces of a common $\mathbb{P}^N$ with $N=\sum_{i=1}^n(a_i+1)-1$. Now the curves Γ_i are all identified with $\mathbb{P}^1$, so that it makes sense to take the linear span of corresponding points. This is $\mathbb{F}(a_1,\dots,a_n)\subset\mathbb{P}^N$; prove this using the determinantal equations as an exercise. (See Ex. 2.14.)

(b) If the assumption of the theorem is weakened to $a_i\ge 0$ then the ratio of the monomials (2.5.1) still defines a morphism $\varphi\colon\mathbb{F}\to\mathbb{P}^N$ that embeds each fibre of π as an $(n-1)$-plane, but if some $a_i=0$ then x_i appears only in a single monomial $S^{a_i}(t_1,t_2)x_i=\{x_i\}$, and $\varphi(\mathbb{F})$ is a cone. The determinantal equations still make perfectly good sense, but the coordinate corresponding to x_i does not appear in the equations. (See Ex. 2.15.)

(c) "Linear equations of scrolls". There is a classical description of the determinantal equations of Theorem 2.5 as $c=\operatorname{codim}(\mathbb{F}\subset\mathbb{P}^N)$ quadrics through a $(N-2)$-plane. (See Ex. 2.16.)

2.6. Particular cases

$\mathbb{F}(1,0)$ is a surface scroll, $\varphi\colon \mathbb{F}(1,0) \to \mathbb{P}^2$ is the blowup of a point. More generally $\varphi\colon \mathbb{F}(1,0,\dots,0) \to \mathbb{P}^n$ is the pencil of hyperplanes through a given codimension 2 linear subspace.

$\mathbb{F}(2,1) \subset \mathbb{P}^4$ is the cubic scroll.

$\mathbb{F}(1,1) \cong \mathbb{P}^1 \times \mathbb{P}^1 \cong Q \subset \mathbb{P}^3$ is the nonsingular quadric surface with a choice of projection.

$\mathbb{F}(2,0) \to Q' \subset \mathbb{P}^3$ is the standard resolution of the ordinary quadric cone (blowup).

More generally $\mathbb{F}(1,1,\underbrace{0,\dots,0}_{n-2}) \to Q_4 \subset \mathbb{P}^{n+1}$ is a resolution of a quadric of rank 4 associated with a chosen family of generators, and $\mathbb{F}(2,\underbrace{0,\dots,0}_{n-1}) \to Q_3 \subset \mathbb{P}^{n+1}$ is the standard resolution of a quadric of rank 3.

$\varphi\colon \mathbb{F}(a,0) \to \overline{\mathbb{F}}_a \subset \mathbb{P}^a$ is the blowup of the cone over a rational normal curve. The surface scroll $\mathbb{F}(a,0) \cong \mathbb{F}(a+b,b)$ for any $b \in \mathbb{Z}$ is usually called $\mathbb{F}_a$. As I discuss below, the exceptional curve of the resolution $B = \varphi^{-1}(0) \subset \mathbb{F}_a$ is a section of $\pi\colon \mathbb{F}_a \to \mathbb{P}^1$ with $B^2 = -a$.

$\mathbb{F}(0,\dots,0) \cong \mathbb{P}^1 \times \mathbb{P}^{n-1}$ with $\varphi\colon \mathbb{F}(0,\dots,0) \to \mathbb{P}^{n-1}$ the second projection. This is the only case with all $a_i \geq 0$ for which φ is not birational.

2.7. Lemma. *The divisor class group of the scroll $\mathbb{F}$ is the free Abelian group*

$$\operatorname{Pic}\mathbb{F} = \mathbb{Z}L \oplus \mathbb{Z}M,$$

with generators the following two divisor classes: L is the class of a fibre of π, and M the class of any monomial $t_1^b t_2^c x_i$ with $b+c=a_i$. (If all the $a_i > 0$, then M is the divisor class of the hyperplane section under the embedding $\mathbb{F} \subset \mathbb{P}^N$ of Theorem 2.5.)

Proof. First note that any two fibres of $\pi\colon \mathbb{F} \to \mathbb{P}^1$ are linearly equivalent: because a fraction $\alpha(t_1,t_2)/\beta(t_1,t_2)$, where α,β are linear forms, is a rational function on $\mathbb{F}$ with divisor the difference of two fibres. Thus the divisor class L of a fibre is well defined.

To see M more clearly, let $F_i \subset \mathbb{F}$ be the locus defined by $x_i = 0$; this is clearly the subscroll $F_i = \mathbb{F}(a_1,\dots,\widehat{a_i},\dots,a_n)$. Then the divisors $a_iL + F_i$ are all linearly equivalent, and define the divisor class M. Indeed, the fraction $t_1^{a_i}x_i/t_1^{a_j}x_j$ is a rational function on $\mathbb{F}$ with divisor $(a_iL+F_i) - (a_jL+F_j)$.

L and M are linearly independent in $\operatorname{Pic}\mathbb{F}$, since if $aL+bM \overset{\text{lin}}{\sim} 0$ then restricting to any fibre $\mathbb{P}^{n-1}$ of π gives $b=0$, and then clearly $a=0$. Finally, I have to prove that every divisor of $\mathbb{F}$ is linearly equivalent to $aL+bM$ for some $a,b \in \mathbb{Z}$. Indeed, any irreducible codimension 1 subvariety $C \subset \mathbb{F}$ is defined by a single bihomogeneous polynomial equation in the sense of 2.4; to see this, take the inverse image in $\mathbb{A}^2 \times \mathbb{A}^n$, and argue as in the case of usual projective space. If C is defined by f with given bidegree, it is obvious how to fix up a monomial $t_1^{a_1d+e}x_1^d$ with the same bidegree, so that $f/t_1^{a_1d+e}x_1^d$ is a rational function, and $C \overset{\text{lin}}{\sim} eL+dM$. Q.E.D.

Remark. In this notation, the canonical class of $\mathbb{F}(a_1,\dots,a_n)$ is given by

$$K_{\mathbb{F}} \overset{\text{lin}}{\sim} -2L - \sum F_i \overset{\text{lin}}{\sim} (-2+\sum a_i)L - nM.$$

See A.10 and Ex. A.13 for details.

2.8. Negative subscrolls $B_b \subset \mathbb{F}$ and the base locus of linear systems

Linear systems on general varieties are discussed below. Here I treat from an elementary point of view the linear system $|eL + dM|$ on $\mathbb{F}$, the family of divisors of $\mathbb{F}$ parametrised by the vector space of bihomogeneous polynomials of degree d in the x_i and extra degree e in the t_i. I assume $d \geq 1$.

Definition. The *subscroll* corresponding to a subset $\{a_{i_1}, \ldots, a_{i_m}\} \subset \{a_1, \ldots, a_n\}$ is the subvariety $\mathbb{F}(a_{i_1}, \ldots, a_{i_m}) \subset \mathbb{F}(a_1, \ldots, a_n)$ defined by the equations $x_j = 0$ for $j \notin \{a_{i_1}, \ldots, a_{i_m}\}$. It is clearly a scroll in its own right with bihomogeneous coordinates $t_1 : t_2; x_{i_1} : \cdots : x_{i_m}$.

For any b, define the *negative subscroll* $B_b \subset \mathbb{F}(a_1, \ldots, a_n)$ to be the subscroll corresponding to the subset $\{a_i \mid a_i \leq b\}$. Suppose now for convenience that $a_1 \leq a_2 \leq \cdots \leq a_n$. Then

$$B_b = \mathbb{F}(a_1, \ldots, a_m) \subset \mathbb{F}(a_1, \ldots, a_n),$$

where m is determined by $a_1 \leq \cdots \leq a_m \leq b < a_{m+1} \leq \cdots \leq a_n$.

As shown by (1) of the following theorem, the point of the definition is that the B_b have a tendency to be base locuses of linear systems.

Theorem. *(1) The base locus of $|-(b+1)L + M|$ is exactly B_b.*

(2) Suppose that $b = a_m$. Then B_b is contained with multiplicity $< \mu$ in the base locus of $|eL + dM|$ if and only if

$$e + bd + (a_n - b)(\mu - 1) \geq 0. \tag{2.8.1}$$

Proof. (1) An element of the linear system $|eL+M|$ is a hypersurface in $\mathbb{F}$ defined by a form f which is a sum of monomials $S^{a_i+e}(t_1, t_2)x_i$ for $i = 1, \ldots, n$. Clearly, if $a_i \leq b < -e$ then x_i doesn't appear in any such monomial; therefore f vanishes identically on the locus $x_{m+1} = \cdots = x_n = 0$. I told you so!

(2) The proof of (2) will make more sense after thinking about the worked examples 2.10–11 and drawing the corresponding Newton polygons.

An element of $|eL+dM|$ is defined by a bihomogeneous polynomial of bidegree d, e. Monomials having degree $\geq \mu$ in $x_{m+1}, \ldots, x_n$ vanish with multiplicity μ along B_b. Thus the assertion is that there exists a monomial of bidegree d, e of degree $< \mu$ in $x_{m+1}, \ldots, x_n$ if and only if (2.8.1) holds. To make a monomial of extra degree e, the term $x_m^{d-\mu+1}x_n^{\mu-1}$ must be accompanied by a monomial of degree

$$e + a_m(d - \mu + 1) + a_n(\mu - 1),$$

which equals the left-hand side of (2.8.1), since $b = a_m$. This is obviously the highest accompanying degree of any of the allowed monomials. Q.E.D.

2.9. Special case: the surface scroll $\mathbb{F}_a = \mathbb{F}(0, a)$

Here $a \geq 0$, and $\mathbb{F}_a = \mathbb{F}(0, a)$ is the surface scroll. I adopt the notation $B = B_0 : (x_2 = 0)$ for the negative section (the point $(1, 0)$ on every fibre $\mathbb{P}^1$), and $A = L$ for the fibre. Then $\operatorname{Pic} \mathbb{F}_a = \mathbb{Z}A \oplus \mathbb{Z}B$, and the intersection numbers are

$$A^2 = 0, AB = 1, B^2 = -a.$$

Proof. From Theorem 2.5, the morphism $\varphi\colon \mathbb{F}_a \to \overline{\mathbb{F}}_a \subset \mathbb{P}^a$ defined by $|M| = |aA+B|$ is the natural resolution of the cone over the rational normal curve of degree a, with B contracting to the vertex. The curve $(x_1 = 0) : M \overset{\text{lin}}{\sim} aA + B$ maps to a hyperplane section, and is disjoint from B. Hence $B(aA+B) = 0$. Since $A^2 = 0$ and $AB = 1$ are obvious, this completes the proof. Q.E.D.

In this case $|eL+dM| = |(e+ad)A+dB|$, so that the conclusions of Theorem 2.8, (1) and (2) have very simple interpretations:

$$B \text{ is fixed in } |(e+ad)A+dB| \iff e<0 \iff B\big((e+ad)A+dB\big)<0$$

and

$$\mu B \text{ is fixed in } |(e+ad)A+dB| \iff e+a(\mu-1)<0$$
$$\iff B\big((e+ad)A+(d-\mu+1)B\big)<0.$$

2.10. Worked example: the Maroni invariant of a trigonal curve

A curve C (of genus $g \ge 3$, assumed to be nonhyperelliptic) is *trigonal* if it has a 3-to-1 map $C \to \mathbb{P}^1$, or equivalently, if it has a g^1_3, a free linear system $|D|$ with $\dim |D| = 1$ and $\deg D = 3$. Consider the canonical model $C \subset \mathbb{P}^{g-1}$. Then geometric RR says that 3 points $P_1+P_2+P_3$ on C move in a g^1_3 if and only if they are collinear in $\mathbb{P}^{g-1}$. (Compare 3.2 and [**4 authors**], Chapter III, §3.) It follows at once that the canonical model of a trigonal curve is contained in a rational normal surface scroll $C \subset \mathbb{F}(a_1,a_2) \subset \mathbb{P}^{g-1}$ where $g = a_1+a_2+2$ (or cone $\overline{\mathbb{F}(0,a_2)}$), and the pencil $|A|$ on $\mathbb{F}$ cuts out the g^1_3 on C.

Order the a_i as $a_1 \le a_2$, set $a = a_2 - a_1$, and, as before, write A for the fibre of $\mathbb{F}_a \to \mathbb{P}^1$ and $B \subset \mathbb{F}_a$ for the negative section. Then $\mathbb{F}(a_1,a_2) \subset \mathbb{P}^{g-1}$ is $\mathbb{F}_a$ embedded by a_2A+B. The canonical curve $C \subset \mathbb{F}_a$ is linearly equivalent to $\alpha A + 3B$ for some α, and computing the degree gives $\alpha = a + a_2 + 2$. By Theorem 2.8, the surface scroll $\mathbb{F}_a$ contains a nonsingular curve $C \subset \mathbb{F}_a$ linearly equivalent to $\alpha A + 3B$ if and only if $\alpha - 3a = B(\alpha A + 3B) \ge 0$. Therefore $\alpha \ge 3a$, that is, $a + a_2 + 2 \ge 3a$, which works out finally as

$$3a \le g+2, \quad \text{or} \quad 3a_2 \le 2g-2, \quad \text{or} \quad 3a_1 \ge g-4.$$

Thus the quantity a is a further invariant of trigonal curves, the *Maroni invariant.* The final inequality says in particular that $\mathbb{F}(a_1,a_2)$ can only be a cone if $g = 4$.

2.11. Worked example: elliptic surfaces $X \subset \mathbb{F} = \mathbb{F}(a_1,a_2,a_3)$ and Weierstrass fibration

I give a typical application of Theorem 2.8. Let $\mathbb{F} = \mathbb{F}(a_1,a_2,a_3) \to \mathbb{P}^1$ be a 3-fold scroll with $a_1 \le a_2 \le a_3$ and $X \subset \mathbb{F}$ a surface meeting the general fibre of $\mathbb{F} \to \mathbb{P}^1$ in a nonsingular cubic curve. Then $X \in |(k+2-\sum a_i)L + 3M|$ for some $k \in \mathbb{Z}$; note that the class of X is unchanged if I change $(a_1,a_2,a_3) \mapsto (a_1-\nu, a_2-\nu, a_3-\nu)$ for some $\nu \in \mathbb{Z}$ and $M \mapsto M - \nu L$. To tidy up the calculation, I will assume later that $a_1 = 0$. (The class of X is arranged so that the canonical class of X is $K_X = kL_{|X}$, by the adjunction formula, compare A.11 below.)

I write out the equation of X as a relative cubic

$$\sum_{i+j+k=3} c_{ijk}(t_1,t_2)x_1^i x_2^j x_3^k,$$

and keep track of the degrees $\deg c_{ijk} = (k+2-\sum a_i) + ia_1 + ja_2 + ka_3$ of the accompanying homogeneous terms in the Newton polygon:

$$\begin{array}{ccccccc} & & & k+2+2a_1-a_2-a_3 & & & \\ & & k+2+a_1-a_3 & & k+2+a_1-a_2 & & \\ & k+2+a_2-a_3 & & k+2 & & k+2-a_2+a_3 & \\ k+2-a_1+2a_2-a_3 & & k+2-a_1+a_2 & & k+2-a_1+a_3 & & k+2-a_1-a_2+2a_3 \end{array} \qquad (*)$$

In order for X to be nonsingular at the general fibre of $\mathbb{F} \to \mathbb{P}^1$, its base locus is restricted by two conditions: $B_{a_2} \not\subset X$ and $2B_{a_1} \not\subset X$. These conditions just say that the general fibre of $X \to \mathbb{P}^1$ does not break up as the line $x_3 = 0$ plus a conic (so at least one of the degrees on the left-hand side of the Newton polygon is ≥ 0, that is, $k+2-a_1+2a_2-a_3 \geq 0$), and does not have $(1,0,0)$ as a double point (so that at least one of the degrees in the top corner is ≥ 0, that is, $k+2+a_1-a_2 \geq 0$).

The criterion of Theorem 2.8 is of the form

$$(k+2-\sum a_i) + 3b + (a_n - a_m)(\mu - 1) \geq 0,$$

which works out as follows:

Condition $B_{a_2} \not\subset X$:

$b = a_2$, $\mu = 1$, so that $(k+2-\sum a_i) + 3a_2 + 0(a_3 - a_2) \geq 0$; setting $a_1 = 0$ and repeating the usual assumptions on the a_i gives

$$a_2 + k + 2 \geq a_3 - a_2 \geq 0. \qquad (1)$$

Condition $2B_{a_1} \not\subset X$:

$b = a_1$, $\mu = 2$, so that $k+2-\sum a_i + 3a_1 + (a_3 - a_1) \geq 0$; setting $a_1 = 0$ gives

$$k + 2 \geq a_2 \geq 0. \qquad (2)$$

For fixed value of k, (1) and (2) have solutions

$$\begin{aligned} a_2 &= 0, \dots, k+2; \\ a_3 &= a_2, \dots, 2a_2 + k + 2. \end{aligned}$$

It is fun to consider the extreme cases of these inequalities. Referring to the Newton polygon, one sees that:

1. If $k+2 < a_2 + a_3$ then the curve $B_{a_1} \subset \operatorname{Bs}|(k+2-\sum a_i)L + 3M|$.
2. If $k+2 < a_3$ then every $X \in |(k+2-\sum a_i)L + 3M|$ contains B_{a_1} and is tangent along it to B_{a_2}. In this case, the general X has singularities on B_{a_1} at the $k+2-a_2$ zeros of $c_{201}(t_1,t_2)$.
3. If $k+2 < a_3 - a_2$ then every $X \in |(k+2-\sum a_i)L + 3M|$ contains B_{a_1} and has a flex along B_{a_2}.

The extreme case of the inequalities (1–2) are $a_2 = k+2$, $a_3 = 3(k+2)$. In this case the critical coefficients of x_2^3 and $x_1^2x_3$ are homogeneous forms in t_1, t_2 of degree zero, that is constants, so that X has equation

$$1 \cdot x_2^3 + 1 \cdot x_1^2 x_3 + \text{other terms}.$$

In other words, $X \subset \mathbb{F}$ is a nonsingular surface, with every fibre the Weierstrass normal form of an elliptic curve.

2.12. Final remarks on scrolls

1. Minimal degree

Scrolls occur throughout projective algebraic geometry as projective varieties of minimal degree: del Pezzo's theorem (from the early 1880s) says that an irreducible d-dimensional variety V spanning $\mathbb{P}^n$ has degree $\geq n-d+1$, and if equality holds then V is a linearly embedded scroll (as in Theorem 2.5), a cone over a scroll (as in Remark 2.5, (b)), or one of the sporadic cases: $\mathbb{P}^n$ itself, a quadric hypersurface $Q \subset \mathbb{P}^n$, the Veronese surface $W \subset \mathbb{P}^5$ or a cone over W. See [**Bertini**] or Eisenbud and Harris [**E–H**] for proofs of different vintages, or do it for yourself (Ex. 2.19).

Hypersurfaces in scrolls play a similar role in the study of curves whose degree is small compared to the genus, or surfaces of general type with K^2 small compared to p_g. Compare Ex. 2.24 or [**Harris**] or [**4 authors**], Chapter 3 (including the exercises).

2. Surfaces with a pencil of curves

Many surfaces come with a natural pencil of curves of small genus; for example, Castelnuovo and Horikawa showed that surfaces with $K^2 = 3p_g - 7$ for which the canonical map φ_K is birational (and $p_g \geq 7$) are naturally relative quartics in a scroll $\varphi_K(X) \subset \mathbb{F} \subset \mathbb{P}^{p_g - 1}$. These surfaces can therefore be studied as hypersurfaces in an explicit rational 3-fold. Compare Ex. 2.24–25.

3. Scrolls over curves of genus ≥ 1

An n-dimensional scroll can more generally be defined as a $\mathbb{P}^{n-1}$-bundle $F \to C$ over any curve C, that is, a fibre bundle with fibre $\mathbb{P}^{n-1}$ and structure group $\mathrm{PGL}(n)$. It can be proved that every scroll F is the projectivisation $F = \mathbb{P}(\mathcal{E})$ of a rank n vector bundle $\mathcal{E}$ over C (compare Tsen's theorem in C.4 below). The assumption that the base curve is $\mathbb{P}^1$ is a major simplifying feature, which makes it possible to give a completely elementary self-contained treatment: in this case there is no effort in saying what the base curve $\mathbb{P}^1$ and the vector bundle $\mathcal{E}$ is: every vector bundle over $\mathbb{P}^1$ is a direct sum of $\mathcal{O}_{\mathbb{P}^1}(a_i)$. (This is a famous theorem, traditionally attributed to Grothendieck, Atiyah, Birkhoff, Hilbert, Dedekind, Gauss, Euler, Archimedes, ...).

However, for the knowledgeable reader, essentially each part of the discussion here carries over to the more general case. This was one of the prime motivations of the theory of algebraic vector bundles over curves in the 1950s. The positive subsheaves $\bigoplus_{a_i \geq c} \mathcal{O}_{\mathbb{P}^1}(a_i)$ that correspond to the negative sections of the scrolls generalise to the Harder–Narasimhan filtration of $\mathcal{E}$, etc. See for example [**H1**], Chapter V, §2.

Exercises to Chapter 2

1. Prove that $\mathbb{F}(a) \cong \mathbb{P}^1$ for any $a \in \mathbb{Z}$.

2. Prove that $\mathbb{F}(0,0) \cong \mathbb{P}^1 \times \mathbb{P}^1$. Generalise to $\mathbb{F}(0,\dots,0)$ (with n zeros).

3. Show how to cover $\mathbb{F}(a_1,\dots,a_n)$ by $2n$ standard affine pieces isomorphic to $\mathbb{A}^n$, and write down the transition functions glueing any two pieces.

4. Prove that $\mathbb{F}(3,1) \cong \mathbb{F}(2,0)$ by comparing coordinates patches, and that $\mathbb{F}(2,0) \cong \mathbb{F}(1,-1)$.

5. Prove that in general

$$\mathbb{F}(a_1,\dots,a_n) \cong \mathbb{F}(a_1+b,\dots,a_n+b)$$

for any $b \in \mathbb{Z}$. [Hint: Every element of the group $\mathbb{C}^* \times \mathbb{C}^*$ can be written as a product of $(\lambda,1)$ and $(1,\mu)$ (for suitable λ,μ), or alternatively as a product of (λ,λ^b) and $(1,\mu')$ (for suitable λ,μ'). In other words, the two actions of $\mathbb{C}^* \times \mathbb{C}^*$ only differ by an automorphism.] Deduce that the assumption $a_1 = 0$ is harmless if you're only interested in $\mathbb{F}$ up to isomorphism.

How is $\mathbb{F}(1,1) \cong \mathbb{F}(-1,-1)$ reconciled with Theorem 2.5?

6. Use the description of $\operatorname{Pic}\mathbb{F}$ and Theorem 2.8, (1) to prove that

$$\begin{aligned}\mathbb{F}(a_1,\dots,a_n) &\cong \mathbb{F}(b_1,\dots,b_n)\\ &\iff \{a_1,\dots,a_n\} = \{b_1+c,\dots,b_n+c\} \text{ for some } c \in \mathbb{Z}.\end{aligned}$$

7. Which of the following are rational functions on the named scrolls?

1. x_1 on $\mathbb{F}(0)$.
2. t_1x_2/x_1, t_1x_1/x_2 and t_2x_1/x_2 on $\mathbb{F}(1,0)$.
3. $(x_1^2+x_2)/t_1x_3$ on $\mathbb{F}(1,2,3)$.
4. $(x_1^3+t_1t_2x_1x_2^2)/(t_2x_1^2x_2+t_1^9x_3^3)$ on $\mathbb{F}(1,2,4)$.

Decide which of the following are bihomogeneous polynomials on $\mathbb{F}(0,3,5)$:

$$x_1^2+x_2,\quad t_1^3x_2+x_1^2,\quad t_1^3x_2+x_1,\quad x_3,\quad t_1t_2x_3+x_2.$$

8. Just as for projective space, a nonzero bihomogeneous polynomial g of bidegree $(d,e) \neq (0,0)$ is not a well-defined function on a scroll $\mathbb{F}$. Prove that for $P \in \mathbb{F}$ the condition $g(P) = 0$ is well defined, so that g defines a hypersurface in $\mathbb{F}$. [Hint: If you don't see this, prove it by lifting $P \in \mathbb{F}$ to different representatives $\widetilde{P} \in (\mathbb{C}^2-0)\times(\mathbb{C}^n-0)$, and evaluating g at these points.]

9. Convince yourself that any two curves of the same degree in $\mathbb{P}^2$ are linearly equivalent. Now prove the same for any two hypersurfaces $X_{d,e} \subset \mathbb{F}$ of the same bidegree on a scroll $\mathbb{F}$. [Hint: Because the ratio of their equations is a rational function.]

Let $\mathbb{F}_a = \mathbb{F}(0,a)$ be the surface scroll, and D_1, D_2 the sections defined by $(x_1 = 0)$ and $(x_2 = 0)$. Find all divisors linearly equivalent to D_1 and containing D_2.

10. Let $Q_3 \subset \mathbb{P}^3$ be a quadric of rank 3 and $\mathbb{F}_2 \to Q_3$ its natural resolution (see 2.6). Study curves in Q_3 in terms of $\mathbb{F}_2$.

11. (a) Using bihomogeneous polynomials of bidegree $(2,-4)$ on $\mathbb{F}(1,2,3)$, write down a nonsingular surface $X_{2,-4} \subset \mathbb{F}(1,2,3)$ of bidegree $(2,-4)$. How many singular fibres does the conic bundle $X_{2,-4} \to \mathbb{P}^1$ have?

(b) The same question for $X_{2,-3} \subset \mathbb{F}(1,2,3)$.

(c) Describe (in terms of its fibres) the locus $(t_1^2x_2^3 + t_1t_2x_1^2x_3 = 0) \subset \mathbb{F}(1,2,4)$.

12. Let $X = X_{2,e} \subset \mathbb{F}(a_1,a_2,a_3)$ be a surface of bidegree $(2,e)$. The fibres of $X \to \mathbb{P}^1$ are plane conics. Prove that, if X is nonsingular, then every fibre is either a nonsingular conic or line pair. [Hint: You have to show that a double line leads to a singularity of X.]

Deduce a formula for the number of line pairs. [Hint: Singular conics are detected by a determinant, and you have to find its degree in t_1, t_2. Compare [**UAG**], proof of Proposition 7.3.]

13. Consider $\mathbb{F}(a_1,\dots,a_n)$ with some of the $a_i < 0$. When is the rational map of Theorem 2.5 defined? When is it the constant map? What is the dimension of its image? When is it in fact a morphism? Compare this with Theorem 2.8.

14. Prove the statement on linear generation of scrolls given in Remark 2.5, (a). [Hint: Write down the equation of all the rational normal curves of degree a_i, then the condition that corresponding points are joined up. Compare with the equations in Theorem 2.5.]

15. Generalise Theorem 2.5 to the case that some of the $a_i = 0$; compare Remark 2.5, (b).

16. Let $\Pi = \mathbb{P}^{N-2} : (x_0 = x_1 = 0) \subset \mathbb{P}^N$ be a codimension 2 linear subspace, and let $Q_1,\dots,Q_{N-n}$ be linearly independent quadrics containing Π. Prove that the intersection $\bigcap Q_i$ consists of Π together with an n-dimensional variety $\overline{\mathbb{F}}$ that is the image of a scroll under a linear embedding. [Hint: The ratio $(x_0 : x_1)$ defines a rational map $\mathbb{P}^N \to \mathbb{P}^1$, whose fibres are the pencil of hyperplanes through Π. Fibre-by-fibre, each Q_i defines a hyperplane. If the statement is true for some c then Q_{c+1} is a divisor in the scroll for c.]

17. Let $\mathbb{F}_a = \mathbb{F}(a,0)$ be the surface scroll as in 2.9, and consider the linear system $|D| = |\alpha A + 4B|$ for suitable $\alpha \in \mathbb{Z}$. Prove that $|D|$ is very ample for $\alpha > 4a$, and find out what happens when $\alpha = 4a$. Use the Newton polygon argument to prove that for $3a \le \alpha < 4a$, the general element of $|\alpha A + 4B|$ is of the form $B + C$ where B is the negative section and C is a nonsingular curve having $\alpha - 3a$ transverse points of intersection with B.

If α is even, study the elliptic surface obtained as double cover of $\mathbb{F}_a$ branched in a general element of $|\alpha A + 4B|$.

18. Suppose that $a_1 < \dots < a_n$. Show that an automorphism of $\mathbb{F}(a_1,\dots,a_n)$ compatible with the projection $\mathbb{F} \to \mathbb{P}^1$ is of the form

$$\begin{pmatrix} x_1 \\ \vdots \\ x_n \end{pmatrix} \mapsto M \begin{pmatrix} x_1 \\ \vdots \\ x_n \end{pmatrix},$$

where $M = \{m_{ij}\}$ is an uppertriangular matrix with entries $m_{ij}(t_1,t_2)$ homogeneous polynomials of degree $a_j - a_i$.

What happens if $a_1 \le \dots \le a_n$, with some equalities allowed?

19. Prove del Pezzo's theorem: an irreducible surface spanning $\mathbb{P}^n$ and of degree $n-1$ is either a scroll $\mathbb{F}(a_1,a_2)$ with $a_1 + a_2 = n-1$, or a cone $\overline{\mathbb{F}}(n-1,0)$, or $\mathbb{P}^2$ if $n = 2$ or the Veronese surface if $n = 5$.

20. Suppose that $a_1 \le \dots \le a_n$ and $b_1 \le \dots \le b_m$. Prove that there exists a surjective homomorphism

$$\mathcal{O}(a_1) \oplus \dots \oplus \mathcal{O}(a_n) \twoheadrightarrow \mathcal{O}(b_1) \oplus \dots \oplus \mathcal{O}(b_m)$$

if and only if $m \le n$ and for every i,

$$a_i \le b_i, \quad \text{and if } (a_1, \dots, a_i) \ne (b_1, \dots, b_i) \text{ then also } b_{i+1} \le a_i.$$

If $0 < a_1$, deduce necessary and sufficient conditions for $\mathbb{F}(b_1, \dots, b_{n-1})$ to be a hyperplane section of $\mathbb{F}(a_1, \dots, a_n)$.

21. Problem: find necessary and sufficient conditions for the existence of a short exact sequence

$$0 \to \mathcal{O}(c_1) \oplus \dots \oplus \mathcal{O}(c_{n-m}) \to \mathcal{O}(a_1) \oplus \dots \oplus \mathcal{O}(a_n) \to \mathcal{O}(b_1) \oplus \dots \oplus \mathcal{O}(b_m) \to 0.$$

22. If $a_1 \le a_2$ and $a'_1 \le a'_2$, prove that $\mathcal{O}(a_1) \oplus \mathcal{O}(a_2)$ has a small deformation isomorphic to $\mathcal{O}(a'_1) \oplus \mathcal{O}(a'_2)$ if and only if $a_1 + a_2 = a'_1 + a'_2$ and $a_1 \le a'_1 \le a'_2 \le a_1$. [Hint: You can find small deformations of $\mathbb{F}(a_1, a_2)$ by taking it as a "special" hyperplane section of a 3-fold $\mathbb{F}(b_1, b_2, b_3)$, then varying the hyperplane.]

23. Problem: find necessary and sufficient conditions for $\mathcal{O}(a_1) \oplus \dots \oplus \mathcal{O}(a_n)$ to have a small deformation isomorphic to $\mathcal{O}(b_1) \oplus \dots \oplus \mathcal{O}(b_m)$.

24. By analogy with the relative cubics of 2.10, consider the scroll $\mathbb{F}(a_1, a_2, a_3)$ with $0 \le a_1 \le a_2 \le a_3$, and relative quartic surfaces $X \in |(2 - \sum a_i)A + 4M|$. (These are the surfaces of general type on the Castelnuovo–Horikawa line $K^2 = 3p_g - 7$.)

25. By analogy with the elliptic surface of Ex. 2.17 obtained as double cover of the surface scroll $\mathbb{F}_a = \mathbb{F}(a, 0)$ branched in a curve $|D| = |2\alpha A + 4B|$, study the linear system $|D| = |2\alpha A + 6B|$ on $\mathbb{F}_a$ and the double cover branched in $|D| = |2\alpha A + 6B|$. (These are the surfaces of general type on the Max Noether–Horikawa line $K^2 = 2p_g - 4$.)

CHAPTER A
Intersection Numbers

This chapter discusses intersection numbers, and gives practice at calculating them. At its most primitive, the idea is to count the number of points of intersection of two distinct irreducible curves C and C' on a surface:

$$CC' = \#\{C \cap C'\}.$$

For example, Bezout's theorem (due to Sir Isaac Newton, as everyone knows) states that two plane curves of degree n and m intersect in nm points counted with multiplicities. I give many simple examples of calculating intersection numbers. I prove that curves on a surface that can be contracted to a point by a nonconstant morphism have negative selfintersection. I also discuss the canonical class of an algebraic surface and the formula for the genus of a curve.

Formal treatments of the material of this section can be found in [**Sh**], Chapter IV, [**H1**], Chapter V or in much more rigour, generality and detail in Fulton [**F2–F3**]. Fulton's introductory book [**F1**] contains an extremely concrete discussion of all the issues discussed here in the particular case of plane curves.

A.1. The formal statement

I start with the formal statement from 1.7, the ingredients of which are discussed later. X is a nonsingular projective surface throughout this section.

Theorem (see [**H1**], Chapter V or [**Sh**], Chapter IV). *One can define an intersection pairing* $\operatorname{Div} X \times \operatorname{Div} X \to \mathbb{Z}$, *written* $(D_1, D_2) \mapsto D_1D_2$, *with the following four properties:*

1. D_1D_2 *is bilinear in each factor and symmetric.*
2. D_1D_2 *depends on* D_1, D_2 *only up to linear equivalence: that is,*

$$D_1 \overset{\text{lin}}{\sim} D_1' \implies D_1D_2 = D_1'D_2.$$

3. *If* $D_1, D_2 \geq 0$ *are effective divisors and have no common components then* $D_1D_2 = \sum_P (D_1D_2)_P$, *where the sum runs over all points of intersection* $P \in D_1 \cap D_2$, *and the local intersection number* $(D_1D_2)_P$ *is defined as follows:*

$$(D_1D_2)_P = \dim_k \mathcal{O}_{X,P}/(f_1, f_2) = \dim_k \mathcal{O}_{X,P}/(I_{D_1} + I_{D_2});$$

here $\mathcal{O}_{X,P}$ *is the local ring of* $P \in X$, *and* D_1, D_2 *are defined near* P *by local equations* f_1, f_2, *or (equivalently) by ideals* $I_{D_1}, I_{D_2} \subset \mathcal{O}_{X,P}$.

4. *If* C *is a nonsingular curve then* $C^2 = \deg_C N_{X|C}$ *is the degree of the normal line bundle* $N_{X|C}$ *to* X *along* C. *More generally, if* C *is irreducible, then* $C^2 = \deg_C \mathcal{O}_C(C)$, *where* $\mathcal{O}_C(C)$ *is the invertible sheaf on* C *(normal sheaf) discussed in the next section.*

The properties (1–3), or (1, 3–4), uniquely define the intersection number as a bilinear pairing $\operatorname{Div} X \times \operatorname{Div} X \to \mathbb{Z}$. The point of (2) is that D_1D_2 is well defined on the divisor class group $\operatorname{Pic} X = \operatorname{Div} X / \overset{\text{lin}}{\sim}$, not just on $\operatorname{Div} X$. I say a few words in A.8–9 about how the pairing can be constructed and the theorem proved.

A.2. Discussion

Notice that if $D_1, D_2 \geq 0$ are effective divisors with no common components, then each local contribution $(D_1D_2)_P$ is the dimension of a vector space, therefore ≥ 0. In fact $(D_1D_2)_P > 0$ if $P \in D_1 \cap D_2$, since then $f_1, f_2 \in m_P$. Therefore $D_1D_2 \geq 0$, and $D_1D_2 = 0$ implies that D_1 and D_2 are disjoint.

For an irreducible curve, C^2 is called its *selfintersection*. If C moves in a linear system then $C \overset{\text{lin}}{\sim} C'$, so that $C^2 = CC' \geq 0$. Or you can argue geometrically that if C moves (more generally in an algebraic family) then the normal bundle $N_{X|C}$ has a section, so again $C^2 = \deg_C N_{X|C} \geq 0$.

However, as I said in 1.4, a line on a cubic surface $L \subset X$ has $L^2 = -1$, and we will see presently that $C^2 < 0$ happens in lots of interesting cases (see Theorem A.7). A curve having negative selfintersection $C^2 < 0$ may be the cause of psychological discomfort to students, so it's worth underlining the following point, which follows immediately from the above discussion.

Proposition. *If* $D_1, D_2 \geq 0$ *are effective divisors and* $D_1D_2 < 0$ *then* D_1 *and* D_2 *have at least one common component* C *that is a curve with* $C^2 < 0$.

A.3. First example, the surface scroll $\mathbb{F}_a = \mathbb{F}(0, a)$

Here $\mathbb{F}_a = \mathbb{F}(0, a)$ is the surface scroll, with $a \geq 0$. Write $D_1 : (x_1 = 0)$ and $D_2 : (x_2 = 0)$ for the sections defined by x_1, x_2; then D_1 and D_2 are disjoint, since $(x_1, x_2) \neq (0, 0)$ at every point of $\mathbb{F}_a$. Therefore $D_1D_2 = 0$.

Also, x_1 and $S^a(t_1, t_2)x_2$ are both bihomogeneous of the same bidegree $1, 0$, it follows that $D_1 \overset{\text{lin}}{\sim} D_2 + aL$ where L is the divisor class of a fibre (compare Lemma 2.7). Obviously $D_1L = D_2L = 1$ (a single point transversally), so that

$$D_1^2 = D_1(D_2 + aL) = a \qquad \text{and} \qquad D_2^2 = D_2(D_1 - aL) = -a.$$

I write $A = L$ and $B = D_2$ in this example, which gives the traditional basis for $\operatorname{Pic} \mathbb{F}$, with $A^2 = 0$, $AB = 1$ and $B^2 = -a$. Notice that I can now write down the intersection number of any two curves on $\mathbb{F}_a$ by using this basis of $\operatorname{Pic} \mathbb{F}$.

By Remark 2.5, (b), x_1 and $S^a(t_1, t_2)x_2$ define a morphism $\varphi\colon \mathbb{F}_a \to \overline{\mathbb{F}}_a \subset \mathbb{P}^a$ such that φ contracts B to a point Q, and is an isomorphism $\mathbb{F}_a \setminus B \cong \overline{\mathbb{F}}_a \setminus Q$. The equations defining $\overline{\mathbb{F}}_a$ are the relations holding between the $a + 1$ monomials

$S^a(t_1, t_2)x_2$, that is, the determinantal equations

$$\text{rank}\begin{pmatrix} u_0 & u_1 & \dots & u_{a-1} \\ u_1 & u_2 & \dots & u_a \end{pmatrix} \le 1.$$

(I've renumbered the u_i slightly. $u_{-1} = x_1$ is a linear form on $\mathbb{P}^a$ that doesn't appear in the equations.) Thus $\overline{\mathbb{F}}_a$ is the cone over the rational normal curve of degree a. The singularity $Q \in \overline{\mathbb{F}}_a$ (given by the same equations in affine variables $u_0, \dots, u_a$) is the cyclic quotient singularity $\mathbb{A}^2/(\mathbb{Z}/a)$, where the cyclic group $\mathbb{Z}_a$ acts on $\mathbb{A}^2$ by $(x, y) \mapsto (\varepsilon x, \varepsilon y)$ for a primitive ath root ε.

Thus $\mathbb{F}_a$ is a resolution of the singularity of $\overline{\mathbb{F}}_a$, and the curve B with $B^2 = -a$ is the exceptional locus.

A.4. Intersection numbers on the n-fold rational scroll $\mathbb{F}(a_1, \dots, a_n)$

As in Chapter 2, I write M for the divisor class of any polynomial in the vector space spanned by

$$S^{a_1}(t_1, t_2)x_1, \dots, S^{a_n}(t_1, t_2)x_n.$$

and L for the class of a fibre. $L^2 = 0$, since any fibre is linearly equivalent to a disjoint fibre.

Claim. *$M^{n-1}L = 1$ and $M^n = \sum a_i$.*

Indeed, let $D_i : (x_i = 0)$, so that $D_i \overset{\text{lin}}{\sim} -a_iL + M$. On each fibre $\mathbb{P}^{n-1}$, the D_i for $i = 1, \dots, n$ are the coordinate hyperplanes $x_i = 0$, and, as before, $\bigcap D_i = \emptyset$; therefore $D_1D_2 \cdots D_n = 0$. Also, any $n-1$ of the D_i (say $D_2, \dots, D_n$) intersect transversally in a curve section of $\mathbb{F} \to \mathbb{P}^1$ (the point $(1, 0, \dots, 0)$ in each fibre). Therefore

$$1 = LD_2 \cdots D_n = L(-a_2L + M) \cdots (-a_nL + M) = LM^{n-1}$$

(because $L^2 = 0$), and

$$\begin{aligned} 0 = D_1D_2 \cdots D_n &= (-a_1L + M)(-a_2L + M) \cdots (-a_nL + M) \\ &= -\sum a_iLM^{n-1} + M^n. \end{aligned}$$

Remark. The argument here is essentially Grothendieck's construction of Chern classes: if E is a vector bundle of rank n over any variety X, $\mathbb{P}(E)$ the corresponding $\mathbb{P}^{n-1}$ bundle, and M the tautological hyperplane divisor class, then M satisfies the relation $M^n - c_1(E)M^{n-1} + \cdots = 0$. In our case, this just says $M^n = \sum a_i = c_1(\mathcal{O}(a_1) \oplus \cdots \oplus \mathcal{O}(a_n))$.

A.5. Fibred surfaces

Let $X = X_{2,e} \subset \mathbb{F}(a_1, a_2, a_3)$ be a nonsingular surface which is a divisor of bidegree $(2, e)$ in a 3-fold scroll. It can be shown (see Ex. 2.12) that every fibre of $X \to \mathbb{P}^1$ is either a nonsingular conic or a line pair. Any two distinct fibres F and F' of $X \to \mathbb{P}^1$ are disjoint, so that $CF = CF' = 0$ for every component $C \subset F$.

If $F = L_1 + L_2$ is a line pair, I have

$$0 = L_1F = L_1^2 + L_1L_2 = L_1^2 + 1.$$

Therefore $L_1^2 = -1$, and similarly, $L_2^2 = -1$.

More generally, if $X = X_{d,e} \subset \mathbb{F}(a_1, a_2, a_3)$ is a nonsingular surface fibred in curves of degree d, and one of the fibres of $X \to \mathbb{P}^1$ breaks up as a line L plus a curve C of degree $d-1$, then

$$0 = LF = L(L+C) = L^2 + d - 1, \qquad \text{so that } L^2 = -(d-1).$$

A.6. Another example

I discussed in 2.11 the example of the nonsingular surface

$$X = X_{3,-3k-6} \subset \mathbb{F}(0, k+2, 3k+6)$$

turning up as the extreme case of the nonsingularity inequalities $a_2 \le k+2$, $a_3 \le 2a_2 + k + 2$. It has inflectional tangency along the curve $C : (x_2 = x_3 = 0)$ to the hyperplane $D_3 : (x_3 = 0)$. Therefore D_3 restricts on X to $3C$.

To calculate the selfintersection C^2 of C on X, note first that $D_3 + (3k+6)L \overset{\text{lin}}{\sim} D_1$ on the scroll $\mathbb{F}$, because the difference is the divisor of $f_{3k+6}(t_1, t_2)x_3/x_1$, with f homogeneous of degree $3k+6$. The rational function $f_{3k+6}(t_1, t_2)x_3/x_1$ does not have X as zero or pole, so it restricts to a rational function on X having as zeros $3C$ plus $3k+6$ fibres L_X of $X \to \mathbb{P}^1$, and as pole a divisor $D_{1,X}$ supported on the locus $(x_1 = 0) \cap X$. Therefore also on X I have $3C + (3k+6)L_X \overset{\text{lin}}{\sim} D_{1,X}$. However, D_1 is disjoint from C (since C meets each fibre in $(1,0,0)$ and D_1 is defined by $x_1 = 0$, so that $C(3C + (3k+6)L_{|X}) = CD_{1,X} = 0$, and therefore $C^2 = -k-2$. (Ex. A.8 provides an independent check of this.)

A.7. Contracted divisors are negative

In A.3, I discussed an example of a curve B with negative selfintersection $B^2 = -a$ which is contracted to a point by a birational morphism. Also, in A.5, I gave an example of a curve L_1 with negative selfintersection which is a component of a fibre of a morphism $X \to C$. This is typical. Part (1) of the following theorem applies directly to resolutions of surface singularities.

Theorem (Negativity of exceptional locus).

1. *Let $\{\Gamma_i\}_{i=1}^k$ be a bunch of curves on a nonsingular surface Y contracted to points by a projective birational morphism $f\colon Y \to X$. Then for every $(n_1, \dots, n_k) \in \mathbb{Z}^k \setminus 0$*

$$q(n_1, \dots, n_k) = \left(\sum n_i\Gamma_i\right)^2 = \sum_{i=1}^k \sum_{j=1}^k n_i n_j \Gamma_i\Gamma_j < 0.$$

In other words, the quadratic form q represented by the symmetric matrix $(\Gamma_i\Gamma_j)$ is negative definite.

2. *Let $\{\Gamma_i\}_{i=1}^k$ be a bunch of k curves such that $\Sigma = \bigcup \Gamma_i$ is connected, and suppose that there is a surjective morphism $f\colon X \to C$ to a nonsingular curve C which contracts Σ to a point $Q \in C$. Then every $(n_1, \dots, n_k) \in \mathbb{Z}^k \setminus 0$ satisfies*

$$q(n_1, \dots, n_k) = \left(\sum n_i\Gamma_i\right)^2 \le 0.$$

In other words, q is negative semidefinite. Moreover, $q(n_1, \dots, n_k) = 0$ holds if and only if $\sum n_i\Gamma_i$ is proportional to the fibre. More precisely, if

$t \in m_Q \subset \mathcal{O}_{C,Q}$ is a local parameter at Q, and $g = f^*(t) \in k(X)$ is the rational function on X obtained as the pullback of t, then $\sum n_i\Gamma_i$ and the connected component of $\operatorname{div} g$ at $f^{-1}Q$ are rational multiples of one another.

Proof. For simplicity, I concentrate on proving (1) under the additional assumption that the bunch of curves $\{\Gamma_i\}_{i=1}^k$ has connected union $\Sigma = \bigcup \Gamma_i$, and that $f\colon Y \to X$ contracts Σ to a point Q, and is an isomorphism $Y \setminus \Sigma \to X \setminus Q$.

The proof breaks up into two steps; the first is geometric and the second an argument in the algebra of quadratic forms. (There is an alternative proof using the Hodge algebraic index theorem, see Ex. A.16.)

Claim. *There is an effective divisor $D = \sum a_i\Gamma_i$ such that $D\Gamma_i \le 0$ for every i (that is, $-D$ is nef on the Γ_i), and $D\Gamma_i < 0$ for at least one Γ_i.*

Proof. Let $g \in m_{X,Q} \subset \mathcal{O}_{X,Q}$ be a function that is regular at $Q = f(\Gamma_i)$ and vanishes there, and write $C = \operatorname{div} g \subset X$ for its divisor in X; then the function g on Y (in other words, $f^*(g) \in \Gamma(Y, \mathcal{O}_Y)$) is regular and its divisor $\operatorname{div}_Y g = f^*C$ on Y is of the form $D + C'$, where $D = \sum a_i\Gamma_i$ is exceptional, all $a_i > 0$, and C' is the birational transform of C. Now $(\operatorname{div}_Y g)\Gamma_i = 0$, so the first part of the claim follows: C' has no exceptional components, so that $C'\Gamma_i \ge 0$, therefore $D\Gamma_i \le 0$.

There is a tricky bit to the last part of the claim which I explain below. To get the result with the minimum of pain, choose first any curve $C'_0 \subset Y$ which is not exceptional, but intersects at least one of the Γ_i (for example, take C'_0 to be a hyperplane section of Y). Now assume that $g \in \mathcal{O}_{X,Q}$ vanishes on $C_0 = f(C'_0)$, and write as before $C = \operatorname{div} g$ and $f^*C = C' + D$. Then $C'\Gamma_i \ge 0$ for every i, and > 0 for at least one i (because $C'_0 \subset C'$). Then the claim follows by $(f^*C)\Gamma_i = 0$. Q.E.D.

Remark. In fact, it is true in general that for any $g \in m_{X,Q} \subset \mathcal{O}_{X,Q}$, the birational transform C' of the curve $C = \operatorname{div} g$ intersects every connected component of $f^{-1}Q$; however, this depends on a hard theorem of Zariski: first replace X by its normalisation $X' = \operatorname{Spec}_X(f_*\mathcal{O}_Y)$; the resolution f factors through a morphism $f'\colon Y \to X'$ to a normal variety X'. Then Zariski's connectedness theorem ([**EGA** **III$_1$**], 4.3.1 or [**H1**], Corollary III.11.4) asserts that every fibre of f' is connected (so that the birational transform of a connected divisor C must intersect every component of the exceptional locus): the basic reason is that if $f'^{-1}P$ has several connected components then $(f'_*\mathcal{O}_Y)_P$ has a function that vanishes on one and not on the other.

Proof of Claim $\implies$ Theorem A.7. I get a contradiction from $E^2 \ge 0$, where $E = \sum b_i\Gamma_i$ with some $b_i \ne 0$. First, D contains every curve Γ_i (that is, every $a_i > 0$): this is obvious if $D\Gamma_i < 0$ (see A.2), and also if $D\Gamma_i = 0$ and D contains some curve Γ_j intersecting Γ_i; by assumption D contains a curve in every connected component of $\bigcup\Gamma_i$, which gives what I said. Next, I can assume that $E > 0$. For if $E = E_1 - E_2$ with $E_1, E_2 > 0$ and E_1, E_2 having no common components then $0 \le E^2 = E_1^2 + E_2^2 - 2E_1E_2 \le E_1^2 + E_2^2$, so that, say $E_1^2 \ge 0$.

If some curves Γ_i do not appear in E (that is, $b_i = 0$), I can just delete them from the bunch $\{\Gamma_i\}$: by doing this, I decrease D at every deleted curve, and therefore the assumption that $D\Gamma_i < 0$ for at least one Γ_i in every connected component of $\bigcup\Gamma_i$ is preserved. The contradiction is now by induction on the number of components Γ_i appearing in E with $b_i > 0$. Write $c = \min\{b_i/a_i\}$, and set $E' = E - cD \geq 0$; then by construction, at least one Γ_i has coefficient $b'_i = b_i - ca_i = 0$ in E'. Now $-D$ nef implies

$$E'^2 = (E - cD)E' \geq EE' = (E - cD)E \geq E^2 \geq 0.$$

By induction, I have a contradiction unless $E' = 0$, but then $E = cD$ and $E^2 = cED < 0$, because D is negative on at least one curve of E. Q.E.D.

Sketch proof of (2). First, a brief discussion of the background: a surface fibration $f : X \to C$ is a projective morphism of a surface onto a curve with connected general fibre. A particular fibre may be reducible or have multiple components: write $f^{-1}P = \bigcup \Gamma_i$ for the set theoretic fibre and $f^*P = F = \sum a_i\Gamma_i$ for the scheme theoretic fibre over P.

Let $t \in m_Q \subset \mathcal{O}_{C,Q}$ be a local parameter on C at P and $g = f^*(t) \in k(X)$ be as in the statement of (2), and set $Z = \operatorname{div} g$. Then

$$Z = \sum a_i\Gamma_i - \text{part disjoint from } \Sigma = F - F'$$

satisfies $a_i > 0$ and

$$Z\Gamma_i = 0 \qquad \text{for all } \Gamma_i. \tag{**}$$

This is the geometric part of the proof of (2), with F the analog of D in the proof of (1).

I omit the algebraic part (but see Ex. A.17). If $E^2 > 0$ then the same kind of argument gives a contradiction. The idea for treating the case of equality $E > 0$, $E^2 = 0$ is to rework the preceding proof with only weak inequalities ≤ 0. This is just a matter of being careful about the logic.

A.8. How to construct the pairing

There are two different strategies to the proof of Theorem A.1. One is the rather straightforward idea of defining D_1D_2 by the property (3). First, the local intersection multiplicity $(D_1D_2)_P$ is bilinear whenever it makes sense (that is, for effective divisors without common components at P). For this, you have to prove that if f_1, f_2 and $g \in \mathcal{O}_{X,P}$ are such that f_1f_2 and g have no common factors, then the finite dimensional k vector space $\mathcal{O}_{X,P}/(f_1f_2, g)$ can be made up as an extension of $\mathcal{O}_{X,P}/(f_1, g)$ and $\mathcal{O}_{X,P}/(f_2, g)$. More precisely, you can prove the algebraic fact that multiplication by f_2 takes $\mathcal{O}_{X,P}/(f_1, g)$ isomorphically to the kernel of the reduction map $\mathcal{O}_{X,P}/(f_1f_2, g) \to \mathcal{O}_{X,P}/(f_2, g)$. In other words, there is a short exact sequence

$$0 \to \mathcal{O}_{X,P}/(f_1, g) \to \mathcal{O}_{X,P}/(f_1f_2, g) \to \mathcal{O}_{X,P}/(f_2, g) \to 0.$$

The definition can then be extended to noneffective divisors by bilinearity. In order to extend the definition to divisors with common components, one must prove that every curve C moves under linear equivalence to a divisor (not necessarily effective) not involving C. This is easy. However, in order that this definition be well defined, you must prove (2). This boils down eventually to arguing on the number of zeros and poles of a rational function on a nonsingular curve. This is the construction given in [**Sh**], Chapter IV.

A.9. Bezout's theorem

The other strategy for proving Theorem A.1 is more abstract and cohomological in nature, but is extremely convenient to use if you know how: you just define the pairing by

$$D_1D_2 = \chi(\mathcal{O}_X(H-D_1-D_2))-\chi(\mathcal{O}_X(H-D_1))-\chi(\mathcal{O}_X(H-D_2))+\chi(\mathcal{O}_X(H)) \quad (!)$$

where H is any divisor, for example a large multiple of a hyperplane section (I discuss the Euler–Poincaré characteristic χ in B.9, viii.) This is right because $\chi(D)$ depends on D as an inhomogeneous quadratic polynomial (by RR on a surface, see B.9, ix), and the formula (!) is just the associated homogeneous bilinear form obtained from the leading term (see Ex. A.15). The properties of Theorem A.1 then follow from easy cohomological manipulations (see [**H1**], Chapter V or [**Beauville**], I.4).

It is illuminating to see how this definition works in the particular case of the projective plane.

Bezout's theorem. *In $\mathbb{P}^2$, two curves of degree n and m with no common components meet in nm points, counted with the multiplicity of intersection in Theorem A.1, (3).*

Sketch proof (see [**F1**]). Suppose that $C : (f = 0)$ and $D : (g = 0)$, where f and g are homogeneous polynomials of degree n and m. Then since the polynomial ring $k[x, y, z]$ and the local rings $\mathcal{O}_{X,P}$ are UFDs and f and g have no common factors, it is easy to see that the sequence of sheaves

$$0 \to \mathcal{O}(-n-m) \xrightarrow{-g,f} \mathcal{O}(-n) \oplus \mathcal{O}(-m) \xrightarrow{\binom{f}{g}} \mathcal{O} \to \bigoplus_P \mathcal{O}_P/(f,g) \to 0$$

is exact. Straightforward manipulations from this show that for any large d we have the exact sequence of finite dimensional vector spaces

$$0 \to V_{d-n-m} \xrightarrow{-g,f} V_{d-n} \oplus V_{d-m} \xrightarrow{\binom{f}{g}} V_d \to \bigoplus_P \mathcal{O}_P/(f,g) \to 0$$

where $V_d = k[x, y, z]_d$ is the space of homogeneous polynomials in $k[x, y, z]$ of degree d. Now we know the dimension of all the first 3 terms, since $V_d = \binom{d+2}{2}$. Therefore (applying the alternating sum formula to the preceding display, with $d \mapsto d-2$) I get the triumphant conclusion

$$\dim_k \bigoplus_P \mathcal{O}_P/(f,g) = \binom{d-n-m}{2} - \binom{d-n}{2} - \binom{d-m}{2} + \binom{d}{2} = nm. \quad \text{Q.E.D.}$$

A.10. Enter the Hero: the canonical divisor class K_X

Let X be a nonsingular n-dimensional algebraic variety or complex manifold. If $z_1, \dots, z_n$ are local algebraic or complex analytic coordinates then I want to think of $\mathrm{d}z_1 \wedge \cdots \wedge \mathrm{d}z_n$ as a *(complex) volume element.* Now, in the algebraic case, take any elements $f_1, \dots, f_n \in k(X)$ that form a (separable) transcendence basis, that is, such that $k(f_1, \dots, f_n) \subset k(X)$ is a finite algebraic extension. In the complex case, choose global meromorphic functions $f_1, \dots, f_n$ that are algebraically independent.

Also pick any $0 \neq g \in k(X)$. I write formally $s = g\mathrm{d}f_1 \wedge \dots \wedge \mathrm{d}f_n$, and call it a *rational (meromorphic) n-form.* I'm not going to worry too much about what it *is*. The point is just that locally, I can compare it to the local volume element $\mathrm{d}z_1 \wedge \dots \wedge \mathrm{d}z_n$ by means of the Jacobian determinant:

$$s = g\mathrm{d}f_1 \wedge \dots \wedge \mathrm{d}f_n = J \cdot g\mathrm{d}z_1 \wedge \dots \wedge \mathrm{d}z_n, \quad \text{where} \quad J = J\left(\frac{f_1, \dots, f_n}{z_1, \dots, z_n}\right) = \det\left|\frac{\partial f_i}{\partial z_j}\right|.$$

(The reader is invited to provide his or her own interpretation of $\partial f_i / \partial z_j$ in the algebraic case, as an exercise.) Note that by the chain rule, a different choice of local coordinates $z_1, \dots, z_n$ multiplies J by an invertible function, so that the zeros and poles of J are well defined. This makes it possible to determine the zeros and poles of s: namely, define the valuation of s at a prime divisor Γ by

$$v_\Gamma(s) := v_\Gamma(J \cdot g), \qquad \text{and set} \quad \operatorname{div} s = \sum_\Gamma v_\Gamma(s)\Gamma$$

Note that this is *not the same* as the valuation of a function: we don't know $s = g\mathrm{d}f_1 \wedge \dots \wedge \mathrm{d}f_n$ as a function, and so we can't compare it with the unit function 1. It is a rational n-form, so we can compare it with the volume element $\mathrm{d}z_1 \wedge \dots \wedge \mathrm{d}z_n$.

Definition. The *canonical class* of X is the divisor class $K_X = \operatorname{div} s$ where s is a rational n-form. It is a well-defined divisor class, because two different n-forms s and s' are related by $s = hs'$ with $0 \neq h \in k(X)$ a rational function, and then, plainly, $\operatorname{div} s = \operatorname{div} s' + \operatorname{div} h$.

A.11. The adjunction formula

The canonical class K_X is the great hero of classification of varieties. There are many important things to say about it, and I return to this in several of the subsequent chapters. Here I just restrict myself to the single point that it is easy to calculate if you know X well enough. Here and in the exercises, we're going to see lots of cases where it can be calculated by a single trick.

Adjunction formula. *Let $X \subset Y$ be a nonsingular hypersurface in a nonsingular variety. Then $K_X = (K_Y + X)_X$.*

Here the restriction of divisor classes means that I first take a divisor D on Y with $D \overset{\text{lin}}{\sim} K_Y + X$ and such that D does not contain X, and then intersect D with X to get the divisor class $D_X = (K_Y + X)_X$. The construction of the restriction will be treated more formally in 3.1, where I give the traditional proof of the adjunction formula. (See also Ex. 3.26 for a more high-brow proof in terms of Serre duality.)

A.12. The genus formula

It is known that for a nonsingular curve C of genus g, the canonical class K_C has degree $2g - 2$. It thus follows from the adjunction formula that for a nonsingular curve $C \subset X$ on a nonsingular surface,

$$2g - 2 = (K_X + C)C.$$

This formula can be checked against almost every selfintersection number calculated in this section (see Ex. A.1–8). For this, I first need a few direct calculations as a starting point.

A.13. Proposition. *The canonical class of projective space* $\mathbb{P}^n$ *is given by*

$$K_{\mathbb{P}^n} = -(n+1)H,$$

where H *is the class of a hyperplane* $\mathbb{P}^{n-1} \subset \mathbb{P}^n$. *The canonical class of the scroll* $\mathbb{F}(a_1, \dots, a_n)$ *is given by*

$$K_{\mathbb{F}} = -2L - \sum D_i \overset{\text{lin}}{\sim} (-2 + \sum a_i)L - nM \overset{\text{lin}}{\sim} (-2 + \sum (a_i - a_1))L - nD_1;$$

here, as in 2.7, L *is the class of a fibre,* $D_i : (x_i = 0)$*, and* M *the divisor class of* $a_iL + D_i$.

Ex. A.9–14 outline several alternative proofs of the proposition. The crudest statement and proof for $\mathbb{P}^n$ is that if $x_1, \dots, x_n$ are affine coordinates on the 0th coordinate piece $\mathbb{A}^n \subset \mathbb{P}^n$ then $s = \mathrm{d}x_1 \wedge \cdots \wedge \mathrm{d}x_n$ has a pole of order $n+1$ along the hypersurface at infinity x_0. This can be checked by a direct change of variables argument (see Ex. A.9).

A slightly more sophisticated approach coming from toric geometry is to write down the n-forms

$$s^{(i)} = (-1)^i \frac{\mathrm{d}x_0^{(i)}}{x_0^{(i)}} \wedge \cdots \wedge \begin{pmatrix} \text{omit } i\text{th} \\ \text{factor} \end{pmatrix} \wedge \cdots \wedge \frac{\mathrm{d}x_n^{(i)}}{x_n^{(i)}}$$

on the ith piece $\mathbb{A}^n_{(i)}$ (here $x_j^{(i)} = x_j/x_i$ are the usual affine coordinates). This visibly has simple poles along the n coordinate hyperplanes of $\mathbb{A}^n_{(i)}$. However, one can check easily that all the $s^{(i)}$ are equal, so define a single rational n-form s_{best} on $\mathbb{P}^n$ with simple poles along all $n+1$ coordinate hyperplanes (see Ex. A.10).

Implicit in this argument is the observation that the complement of the $n+1$ coordinate hyperplanes in $\mathbb{P}^n$ is a big torus $\mathbb{T} = \mathbb{G}_m^n$ (where $\mathbb{G}_m$ is the algebraic group corresponding to the multiplicative group of the field k^*), and that any n of the $n+1$ vector fields $x_i \frac{\partial}{\partial x_i}$ (they sum to zero on $\mathbb{P}^n$ by the Euler relation) is a basis for the tangent bundle of $\mathbb{T}$, with logarithmic zeros along the coordinate hyperplanes.

Similar remarks apply to the scroll $\mathbb{F}(a_1, \dots, a_n)$ (see Ex. A.11–A.13).

A.14. Problem

The $(n+1)$-form on $\mathbb{A}^{n+1} \setminus 0$

$$t = \frac{\mathrm{d}x_0}{x_0} \wedge \cdots \wedge \frac{\mathrm{d}x_n}{x_n}$$

has simple poles along each of the coordinate hyperplanes, and is invariant under the action of the multiplicative group $\mathbb{G}_m$. It should be possible to prove that s_{best} is in an intrinsic way a contraction of t against the Euler vector field

$$\sum_{i=0}^{n} x_i \frac{\partial}{\partial x_i}.$$

Exercises to Chapter A

1. Let $X = X_d \subset \mathbb{P}^3$ be a nonsingular surface of degree d, and $L \subset X$ a line. Calculate L^2 in X. Check that this is what you know for $d = 1, 2, 3$.

2. X and L as in Ex. A.1. Prove that planes through L cut out a pencil $|F|$ of curves with $F^2 = 0$. Deduce that the linear projection $\mathbb{P}^3 \to \mathbb{P}^1$ away from L extends to a morphism $X \to \mathbb{P}^1$.

3. Let $X = X_d \subset \mathbb{P}^3$ be a nonsingular surface of degree d, and suppose that X has a plane section Π that decomposes as a union of two curves $\Pi \cap X = A + B$ of degrees a and b. Prove that $(AB)_{\mathbb{P}^2} = (AB)_X$ (that is, AB calculated in the plane Π or in X is the same thing). Calculate A^2 and B^2 in X.

4. Let $X = X_{3,e} \subset \mathbb{F}(a_1, a_2, a_3)$ be a nonsingular surface in the scroll of the indicated bidegree. Every fibre F of $X \to \mathbb{P}^1$ is a plane cubic curve. Prove that if F is a union of a line L and conic Q then $L^2 = Q^2 = -2$. If F is a union of 3 lines L_1, L_2, L_3, calculate L_i^2.

5. Use Proposition A.13 and the adjunction formula A.11 to find the canonical class of a surface $X_d \subset \mathbb{P}^3$ of degree d. Do the same for the hypersurface in a 3-fold scroll $X_{d,e} \subset \mathbb{F}(a_1, \dots, a_3)$ of the indicated bidegree.

6. Let $L \subset X_d \subset \mathbb{P}^3$ be a line on a nonsingular surface, as in Ex. A.1. Check your answers to Ex. A.1 and Ex. A.5 against the genus formula A.12.

7. A nonsingular plane curve A of degree a has genus $\binom{a-1}{2}$. If $A \subset X_d$ is as in Ex. A.3, check your answers to Ex. A.3 and Ex. A.5 against the genus formula.

8. Consider the example $X = X_{3,-3k-6} \subset \mathbb{F}(0, k+2, 3k+6)$ treated in 2.12 and A.6. Show that $K_X = kL$ and use the genus formula to calculate the selfintersection of any section of $X \to \mathbb{P}^1$. Compare with the result of the calculation in A.6.

9. Consider affine coordinates $x_1, \dots, x_n$ on $\mathbb{A}^n \subset \mathbb{P}^n$ where $x_0 \neq 0$, and verify that, as stated in A.13, $s = dx_1 \wedge \cdots \wedge dx_n$ has a pole of order $n + 1$ along the hypersurface at infinity x_0. [Hint: Use coordinates $y_0, y_2, \dots, y_n$ with $y_0 = 1/x_1$ and $y_i = x_i/x_1$.]

10. Check that $s^{(0)} = s^{(1)}$ in the discussion after Proposition A.13.

11. Let $\mathbb{F}(a_1, \dots, a_n)$ be the scroll. On the 1, 1th coordinate patch $\mathbb{A}^n_{11}$ (see Ex. 2.3) $t_1 = 1, x_1 = 1$, write down the n-form

$$s = dt_2 \wedge dx_2 \wedge \cdots \wedge dx_n.$$

Check that, viewed as a rational n-form on $\mathbb{F}$, this has a pole of order n along the "horizontal" hyperplane at infinity $D : (x_1 = 0)$, and a zero (or pole) of order $2 + \sum(a_i - a_1)$ along the "vertical" fibre at infinity $L_\infty : (t_1 = 0)$. [Hint: For the first statement, work on $\mathbb{A}^n_{12}$ given by $t_1 = 1, x_2 = 1$; then t_1 plays no role in the calculation, which is the same as in Ex. A.9. For the second statement, work on $\mathbb{A}^n_{21}$ given by $t_2 = 1, x_1 = 1$; the change of variable $t_2 = 1/t_1$ gives the factor t_1^{-2} and the change $y_j = x_j t_1^{-a_j}$ does the rest.]

12. By analogy with the more sophisticated argument given after Proposition A.13, write down a rational n-form $s^{(ij)}$ on each of the $2n$ affine pieces of $\mathbb{F}(a_1, \dots, a_n)$ with simple poles along each coordinate hyperplane, and check that they coincide on overlaps. This defines a rational n-form on $\mathbb{F}$ with simple poles along every coordinate hyperplane $x_i = 0$ and $t_i = 0$, proving the statement in Proposition A.13.

13. $\mathbb{F}(a_1, \ldots, a_n)$ is the quotient of $\mathbb{C}^2 \setminus 0 \times \mathbb{C}^n \setminus 0$ by a certain action of $\mathbb{C}^* \times \mathbb{C}^*$. Write down the two Euler relations holding between the $n+2$ vector fields

$$t_i \frac{\partial}{\partial t_i}, \quad \text{and} \quad x_j \frac{\partial}{\partial x_j}.$$

14. Solve Problem A.14, and generalise to the scroll $\mathbb{F}(a_1, \ldots, a_n)$. Try to get your solution published. [Hint: Choose a journal that doesn't use me as a referee.]

15. Let V be a k-vector space, $\varphi\colon V \to k$ a quadratic polynomial map (possibly inhomogeneous), and $h \in V$ any element. Define ψ by

$$\psi(x, y) = \tfrac{1}{2}\{\varphi(h) - \varphi(h - x) - \varphi(h - y) + \varphi(h - x - y))\}.$$

Prove that $\psi(x, y)$ is independent of h, that it is a symmetric bilinear form, and that $\psi(x, x)$ is the leading term of φ.

16. Give an alternative proof of negative definiteness (Theorem A.7) using the Hodge algebraic index theorem (Corollary D.2.2.). [Hint: Assume X and Y are projective, and let $H \in \operatorname{Pic} Y$ be the pullback of an ample divisor on X. Then $H^2 > 0$ and H is orthogonal to all the exceptional curves.]

17. Prove negative semidefiniteness for a fibre $F = \sum a_i \Gamma_i$ of a surface fibration (Theorem A.7, (2)). [Hint: In more detail, assume that $\bigcup \Gamma_i$ is connected. Then $F\Gamma_i = 0$; if $\operatorname{hcf}\{a_i\} = m$ then $F = mF'$, where $F' = \sum a'_i \Gamma_i$, and the $a'_i = a_i/m$ have no common factor. Now adapt the argument of A.7 to prove that $E = \sum b_i \Gamma_i$ has $E^2 \le 0$, with equality if and only if E is an integer multiple of F'. Compare also [**Beauville**], Corollary VIII.4.]

18. Prove the same statement as in the preceding exercise using the Hodge algebraic index theorem (Corollary D.2.2.).

CHAPTER B
Sheaves and Coherent Cohomology

Sheaf theory is a language to treat geometric data (functions, vector fields, etc.) on a space X in terms of the same kind of data on open sets $U \subset X$. You probably know that in Molière's play *Le Bourgeois Gentilhomme*, M. Jourdain was amazed to learn from his grammar teacher that he had been speaking all of his life *in prose*! In the same way, I hope to persuade you that you have been using some of the ideas of sheaf theory without knowing it ever since your first calculus course.

It is of course out of the question to give a reasonable treatment of sheaves and cohomology in a single lecture. Instead, I give examples of some of the main classes of sheaves occurring in algebraic geometry, and discuss their role in the foundational crises of the subject through the ages. I also try to explain the definition of coherent sheaves, and to highlight the specific features of coherent cohomology which make it different from other cohomology theories used in topology, differential geometry and algebraic geometry.

If all the definitions in this section are intimidating to the younger reader, I assure you that with coherent cohomology, the whole of this type of algebraic geometry becomes a game with fixed rules and just a few standard gambits. This section concludes with a list of "Rules of the Game" for coherent cohomology, which can be taken as axioms, or read up in several references. Remember the Zariski quote[1] we heard a few nights ago: before Serre's [**FAC**], just a few *maestri* who had spent all their lives contemplating the intricacies of the black arts could say when some restriction map was surjective, and all you could do was to believe them; after [**FAC**], any idiot could write down exact sequences and deduce any number of such statements.

B.1. You already know lots of sheaves

When we talk of functions on a space X, we often actually mean functions defined on some open subset $U \subset X$. For example, in elementary calculus, a function $f(x)$ of a single real variable might be defined on some interval $(a, b) \subset \mathbb{R}$. If f is defined on some big interval (A, B), the property that f is continuous (or differentiable, or real analytic) is defined locally at every point $P \in (A, B)$, and for P, only depends

[1]Compare [**Parikh**], p. 138; Carol Parikh gave an interesting evening talk on Zariski's life and opinions at the Park City summer school.

on the behaviour of f on any smaller interval $(a,b) \ni P$. *Voilà, M. Jourdain,* a sheaf!

In the same way, in a first course on manifolds, we learn that a vector field on a manifold X is a thing that can be written $\sum u_i \frac{\partial}{\partial x_i}$ in terms of local coordinates $x_1, \dots, x_n$ on a coordinate patch $U \subset X$. It is natural in discussing manifolds to work locally over small open sets (for example, coordinate patches), and very awkward to insist that the only vector fields we use are global vector fields on the whole of X. For any open subset $V \subset X$, write

$$\Gamma(V, \mathcal{T}_X) = \{\text{vector fields on } V\}$$

for the set of all vector fields on V. The *tangent sheaf* $\mathcal{T}_X$ is this data: all possible vector fields defined on all possible open subsets $V \subset X$. Or in other words, the assignment $V \mapsto \Gamma(V, \mathcal{T}_X)$.

B.2. The structure sheaf $\mathcal{O}_X$ of a variety

Let X be an irreducible quasiprojective variety with its Zariski topology, and $k(X)$ its field of rational functions. For $P \in X$, we know what it means for a rational function $f \in k(X)$ to be *regular* at P (see for example [**UAG**], §5), and the set of rational functions that are regular at P is the *local ring* $\mathcal{O}_{X,P} \subset k(X)$. For any Zariski open $U \subset X$, define

$$\Gamma(U, \mathcal{O}_X) = \{f \in k(X) \mid f \text{ is regular at all } P \in U\} = \bigcap_{P \in U} \mathcal{O}_{X,P} \subset k(X).$$

The *structure sheaf* of $\mathcal{O}_X$ is the assignment $U \mapsto \Gamma(U, \mathcal{O}_X)$. In this case everything is very simple, because all the $\Gamma(U, \mathcal{O}_X)$ are subrings of the fixed function field $k(X)$.

For $\emptyset \neq V \subset U$ a smaller Zariski open set, regular functions on U obviously restrict to regular functions on V, defining a inclusion $\Gamma(U, \mathcal{O}_X) \subset \Gamma(V, \mathcal{O}_X)$. Also, since every function regular at P is regular on some open neighbourhood of P, it follows that

$$\mathcal{O}_{X,P} = \bigcup_{U \ni P} \Gamma(U, \mathcal{O}_X) \subset k(X).$$

In sheaf theory, the inclusion $\Gamma(U, \mathcal{O}_X) \subset \Gamma(V, \mathcal{O}_X)$ is called a *restriction map* $\mathrm{Res}_{U,V}$ (or sometimes $\rho_{U,V}$), and $\mathcal{O}_{X,P}$ the *stalk* of $\mathcal{O}_X$ at P.

B.3. The definition of a sheaf

In general, a *presheaf* $\mathcal{F}$ on a topological space X is a way of assigning to every open subset $U \subset X$ a set $\Gamma(U, \mathcal{F})$ (or ring, or group, or object of any category; the symbol $\Gamma(U, \mathcal{F})$ is pronounced "the sections of $\mathcal{F}$ over U"), and to every inclusion $V \subset U$ a restriction map $\mathrm{Res}_{U,V} \colon \Gamma(U, \mathcal{F}) \to \Gamma(V, \mathcal{F})$ (or ring homomorphism, or group homomorphism, or morphism of any category). Restriction is supposed to be transitive, in the obvious sense that $\mathrm{Res}_{V,W} \circ \mathrm{Res}_{U,V} = \mathrm{Res}_{U,W}$ whenever $W \subset V \subset U$.

A *sheaf* is a presheaf with "glueing conditions": a section $s \in \Gamma(U, \mathcal{F})$ is uniquely determined by its restrictions $\mathrm{Res}_{U,U_\alpha}(s)$ to open sets U_α that cover U; and given an open cover, and sections $s_\alpha \in \Gamma(U_\alpha, \mathcal{F})$ having equal restrictions to all overlaps $U_\alpha \cap U_\beta$, the s_α come by restricting a section $s \in \Gamma(U, \mathcal{F})$.

If $\mathcal{F}$ is a presheaf whose $\Gamma(U, \mathcal{F})$ are sets of maps defined on U (say, with values in some set Σ), then the first glueing condition holds trivially, and the second holds only if the restrictions placed on maps $U \to \Sigma$ are "local in nature". For example, continuous or differentiable functions on a C^∞ manifold form a sheaf, since continuity and differentiability are local properties of a function. In the same way, sections of a vector bundle $\pi\colon E \to X$, or more generally, the continuous sections of any map $\pi\colon F \to X$ form a sheaf; see Ex. B.4–5 for details.

B.4. The definition of a variety

The first key service that the language of sheaves performs is to give a satisfactory definition of variety: irreducible affine algebraic sets and their sheaf of regular functions were discussed in B.2. A *variety* $X, \mathcal{O}_X$ is a topological space X with a sheaf $\mathcal{O}_X$ of rings of functions $U \to k$ such that $X, \mathcal{O}_X$ is locally isomorphic to an irreducible affine algebraic set with its sheaf of regular functions. (This can be easily generalised to allow reducible varieties, or indeed general k-schemes.)

Thus sheaves solve an important foundational problem of algebraic geometry, the intrinsic definition of variety. You know from a first course in algebraic geometry (for example, [**UAG**] or [**Sh**]) that quasiprojective varieties are quite convenient to define and treat using the tricks of homogeneous coordinates. However, it's unsatisfactory to take this as the formal definition of variety, because you only define X together with extrinsic data of an embedding $X \subset \mathbb{P}^N$.

Compare how the topologist, in defining a manifold, gets away without using sheaves: a topological manifold M is a topological space locally homeomorphic to a ball in $\mathbb{R}^n$; continuity of functions is completely determined by the topology (of course), so that there is no need to specify the sheaf of continuous functions. For a differentiable manifold, you require in addition that the local charts $\varphi_i\colon U_i \to$ (ball in $\mathbb{R}^n$) satisfy $\varphi_i \circ \varphi_j^{-1}$ is differentiable (wherever defined). Because of this assumption, the condition that a function on an open subset $V \subset M$ is differentiable is well defined. An equivalent definition would be to specify the sheaf $\mathcal{E}_{M,\mathrm{diff}}$ of differentiable functions on M, and assume that the local charts are isomorphisms of ringed spaces.

In case anyone still hasn't got the idea of what all the fuss is about, I repeat: you can't just define an algebraic variety X to be a point set, or a set with a Zariski topology, because all plane curves would be homeomorphic. If you define it as embedded in a space (affine or projective), you get the notion of rational function $f \in k(X)$ and regular function, and hence isomorphism of varieties, but you also get extrinsic stuff which may have less to do with X than with the embedding. The sheaf $\mathcal{O}_X$ specifies the regular functions on every open set. A variety is a space X together with a notion of regular function on opens of X. By giving $\mathcal{O}_X$, you give every possible isomorphism of opens of X with subvarieties of affine or projective space.

B.5. Other sheaves on algebraic varieties

I continue the theme that you already know lots of sheaves. If X is a normal variety and $D = \sum n_\Gamma \Gamma$ an effective divisor on X, rational functions on X with poles at worst D form a sheaf $\mathcal{O}_X(D)$, the *divisorial sheaf* of D. Recall that zeros and poles of rational functions are interpreted in terms of discrete valuations v_Γ on the local rings $\mathcal{O}_{X,\Gamma}$ at every prime divisor Γ (compare 1.6). The condition that $0 \neq f \in k(X)$ has a pole of order at worst n along Γ reads $v_\Gamma(f) \geq -n$. To define the sheaf $\mathcal{O}_X(D)$, I tell you what its sections are on every open set $U \subset X$:

$$\Gamma(U, \mathcal{O}_X(D)) = \{f \in k(X) \mid v_\Gamma(f) \geq -n_\Gamma \text{ for every } \Gamma \text{ such that } \Gamma \cap U \neq \emptyset\}.$$

These are local conditions, so that $\mathcal{O}_X(D)$ is a sheaf. Since $\operatorname{div} f = \sum_\Gamma v_\Gamma(f)\Gamma$, the condition on the right can also be written $\operatorname{div} f + D \geq 0$ on U.

Historical discussion. In the case $U = X$ a projective variety (for example a nonsingular projective curve), the space

$$\mathcal{L}(D) = \Gamma(X, \mathcal{O}_X(D))$$

is traditionally called the *RR space* of D. The RR theorem on a curve gives its dimension:

$$\ell(D) = \dim \mathcal{L}(D) \geq 1 - g + \deg D \qquad \text{with equality if } \deg D > 2g - 2, \qquad \text{(RR)}$$

where $g = g(X)$ is the genus of X. This result was the subject of an earlier foundational crisis in algebraic geometry. It follows from the maximum principle that the only global holomorphic functions on a compact Riemann surface are the constants. To find some maps from X to $\mathbb{P}^1 = \mathbb{C} \cup \infty$, Riemann allowed functions which are everywhere meromorphic, with a finite number of poles, which he wrote as a divisor D. He then claimed a proof of (RR) based on a Dirichlet principle from electrostatics: imagine that your Riemann surface is made of beaten copper, and the poles are point electric charges; then physical intuition says that the potential equations for the electric field must have a solution. Riemann used (RR) to prove that a compact Riemann surface can be embedded in a projective space $\mathbb{P}^N_{\mathbb{C}}$, and is hence an algebraic curve (the *Riemann existence theorem*). Unfortunately, Weierstrass pointed out at once that Riemann's Dirichlet principle was false as stated; in fact, 30 or 40 years later Hilbert gave a revised statement and claimed proof of the Dirichlet principle that was also erroneous, although, by all accounts, Hilbert was so famous by then that nobody dared tell him. In any case, Clebsch, Max Noether and Brill proved (RR) for algebraic curves by purely algebraic means shortly after Riemann.

On an irreducible variety, $\mathcal{O}_X(D)$ is defined as a subsheaf of the constant sheaf $k(X)$. If $\operatorname{div} f = D - D'$ then multiplying by f in $k(X)$ clearly takes $\mathcal{O}_X(D)$ into $\mathcal{O}_X(D')$. Thus linearly equivalent divisors D give rise to isomorphic divisorial sheaves $\mathcal{O}_X(D)$. See Ex. B.13 for more details.

In A.10 I introduced the canonical divisor class K_X as the divisor $\operatorname{div} s$ of a rational n-form $s = g\,\mathrm{d}f_1 \wedge \cdots \wedge \mathrm{d}f_n$. And K_X is defined up to linear equivalence, because any other rational n-form s' is of the form $s' = hs$ for some $h \in k(X)$, so that $\operatorname{div}(s') = \operatorname{div} h + \operatorname{div} s$. Therefore, there is a well-defined divisorial sheaf $\mathcal{O}_X(K_X)$ on X. In fact it is easier and more intrinsic to introduce the sheaf $\omega_X = \Omega^n_X$ first, prove that it is divisorial (on a nonsingular X it is *locally free of rank* 1, or *invertible*), and then to define K_X as any divisor such that $\mathcal{O}_X(K_X) = \Omega^n_X$. See Ex. B.16. In fact, although the divisor K_X is only a divisor class, the sheaf $\omega_X = \Omega^n_X = \mathcal{O}_X(K_X)$ is canonical.

B.6. Subsheaves, stalks and quotient sheaves

There are two very different kinds of definitions and arguments in sheaf theory, those that take place at the level of the spaces of sections $\Gamma(U, \mathcal{F})$, and those that are local at every point $P \in X$ and involve stalks. I start with a few instances of the first. A *homomorphism* of sheaves $\varphi\colon \mathcal{F} \to \mathcal{G}$ is a collection of morphisms $\varphi_U\colon \Gamma(U, \mathcal{F}) \to \Gamma(U, \mathcal{G})$ for all opens U that commutes with restrictions $\mathrm{Res}_{U,V}$ of $\mathcal{F}$ and $\mathcal{G}$:

$$\begin{array}{ccc} \Gamma(U,\mathcal{F}) & \to & \Gamma(U,\mathcal{G}) \\ \downarrow & © & \downarrow \\ \Gamma(V,\mathcal{F}) & \to & \Gamma(V,\mathcal{G}) \end{array}$$

Similarly, a *subsheaf* $\mathcal{F} \subset \mathcal{G}$ is a collection of subobjects $\Gamma(U, \mathcal{F}) \subset \Gamma(U, \mathcal{G})$ that themselves form a sheaf under the same restriction maps. If $f\colon X \to Y$ is a continuous map of topological spaces, and $\mathcal{F}$ a sheaf on X, then it is immediate to check that $U \mapsto \Gamma(f^{-1}U, \mathcal{F})$ for open sets $U \subset Y$ gives a sheaf on Y, the sheaf-theoretic *pushforward* $f_*\mathcal{F}$. All of these definitions and constructions can be made for presheaves just as well as for sheaves.

The *stalk* of a sheaf $\mathcal{F}$ at a point $P \in X$ is the direct limit $\mathcal{F}_P = \varinjlim_{U \ni P} \Gamma(U, \mathcal{F})$. The limit looks intimidating, but this is just another case of M. Jourdain's prose. Namely, the direct limit is the set of all sections $s \in \Gamma(U, \mathcal{F})$ over all open sets containing U, modulo the equivalence relation $s = s'$ if they coincide on some smaller neighbourhood of P; in other words, $\mathcal{F}_P$ consists of *germs* of sections at P. For example, if $\mathcal{O}_{\mathrm{an}}$ is the sheaf of holomorphic functions on $\mathbb{C}$ then the stalk $\mathcal{O}_{\mathrm{an},0}$ consists of all power series with positive radius of convergence, and a germ is an analytic function on some neighbourhood of 0; different germs are defined on different neighbourhoods. The stalks of the structure sheaf $\mathcal{O}_X$ of a variety X are the local rings $\mathcal{O}_{X,P}$, and in this case the direct limit is simply a union, as mentioned in B.2. Another example: a common definition of a tangent vector to a manifold M at a point P is as a derivation of functions defined near P. The derivation acts on germs of smooth functions: it looks at the function only in an arbitrarily small neighbourhood of P.

Now we say that a homomorphism of sheaves $\varphi\colon \mathcal{F} \to \mathcal{G}$ is *surjective* if it induces surjective maps $\varphi_P\colon \mathcal{F}_P \to \mathcal{G}_P$ on each stalk. Surjectivity onto $\Gamma(U, \mathcal{G})$ for all U is the wrong requirement. Another way of saying the same thing is as follows: if $s \in \Gamma(U, \mathcal{G})$, I don't require that s itself comes from some $t \in \Gamma(U, \mathcal{F})$, but only that this holds in a small neighbourhood of any $P \in U$. I now give a baby example (see also Ex. B.10).

Let $P_1, P_2, P_3 \in \mathbb{P}^n$ be 3 distinct points, not in the hyperplane $(x_0 = 0)$. On $\mathbb{P}^n$, consider the sheaf $\mathcal{O}_{\mathbb{P}^n}(1)$ of linear forms, which is the sheaf defined by

$$\Gamma(U, \mathcal{O}_{\mathbb{P}^n}(1)) = \left\{ \frac{f}{g} \in k(x_0, \dots, x_n) \;\middle|\; \begin{array}{l} f, g \in k[x_0, \dots, x_n] \text{ homog. of degree} \\ d+1 \text{ resp. } d, \text{ and } g(P) \neq 0 \text{ at } P \in U \end{array} \right\}.$$

Now I can find a linear form not vanishing at P_i, so that the evaluation map $\mathcal{O}_{\mathbb{P}^n}(1) \to k_P$ defined by $s \mapsto (s/x_0)(P)$ is surjective. Here k_P is the "skyscraper sheaf" with sections over U either zero if $P \notin U$ or a copy of k if $P \in U$. The kernel is the sheaf of linear forms vanishing at P, that is, $m_P \cdot \mathcal{O}_{\mathbb{P}^n}(1)$. Now consider the evaluation map at all 3 points at once:

$$0 \to \mathcal{I} \cdot \mathcal{O}_{\mathbb{P}^n}(1) \to \mathcal{O}_{\mathbb{P}^n}(1) \to k_{P_1} \oplus k_{P_2} \oplus k_{P_3} \to 0, \qquad (*)$$

where for brevity I write $\mathcal{I} = m_{P_1} \cdot m_{P_2} \cdot m_{P_3}$ for the ideal sheaf of $\{P_1, P_2, P_3\}$, that is, the subsheaf of $\mathcal{O}_{\mathbb{P}^n}$ consisting of regular functions on an open U vanishing at $\{P_1, P_2, P_3\} \cap U$. Now in sheaf theory we say that the evaluation map in $(*)$ is surjective, because it is surjective locally at every point. Is it surjective on global sections? The global map evaluates linear forms in $x_0, \dots, x_n$ on 3 points, which is surjective if P_1, P_2, P_3 span a plane in $\mathbb{P}^n$, and not surjective if they are collinear. I thus get the exact sequence

$$\begin{aligned} 0 &\to \text{linear forms vanishing at } P_1, P_2, P_3 \to \langle x_0, \dots, x_n \rangle \to k_{P_1} \oplus k_{P_2} \oplus k_{P_3} \to \\ &\to \text{linear dependences among } P_1, P_2, P_3 \to 0 \end{aligned}$$

In other words, the homomorphism of sheaves is surjective, but it gives a homomorphism on global sections which is not necessarily surjective. In more general language, the last display is written

$$\begin{aligned} 0 &\to H^0(\mathbb{P}^n, \mathcal{I} \cdot \mathcal{O}_{\mathbb{P}^n}(1)) \to H^0(\mathbb{P}^n, \mathcal{O}_{\mathbb{P}^n}(1)) \to H^0(\mathbb{P}^n, k_{P_1} \oplus k_{P_2} \oplus k_{P_3}) \to \\ &\to H^1(\mathbb{P}^n, \mathcal{I} \cdot \mathcal{O}_{\mathbb{P}^n}(1)) \to 0, \end{aligned}$$

where $H^1(\mathbb{P}^n, \mathcal{I} \cdot \mathcal{O}_{\mathbb{P}^n}(1))$ is the first cohomology group of the sheaf $\mathcal{I} \cdot \mathcal{O}_{\mathbb{P}^n}(1)$.

I mention two other constructions of sheaf theory that use stalks: the *quotient sheaf* $\mathcal{G}/\mathcal{F}$ is defined to have stalk $\mathcal{G}_P/\mathcal{F}_P$ at every $P \in X$; and the *sheaf-theoretic pullback* $f^{-1}\mathcal{F}$ of a sheaf on Y by a morphism $f\colon X \to Y$ is defined to have stalk $\mathcal{F}_{f(P)}$ at $P \in X$. I omit the definition, which involves the notion of the associated sheaf of a presheaf. (This material is completed in Ex. B.27–29.)

B.7. Coherent sheaves

The sheaves of algebraic geometry $\mathcal{O}_X$, $\mathcal{O}_X(D)$, $\mathcal{O}_{\mathbb{P}^n}(r)$ we have met so far are all coherent sheaves; so is the ideal sheaf $\mathcal{I}_Y \subset \mathcal{O}_X$ of a subvariety $Y \subset X$. The adjective coherent means that they are sheaves of modules over $\mathcal{O}_X$, with a finiteness condition, and closely related to the structure sheaf $\mathcal{O}_X$.

The general progression is presheaf, sheaf, sheaf of $\mathcal{O}_X$-modules, (quasi-) coherent sheaf, locally free sheaf. I have not been through the general definitions particularly carefully; it should be clear what the definition of sheaf of $\mathcal{O}_X$-modules is. If you have trouble see, for example, [**FAC**] or [**H1**], Chapter II. The definition of (quasi-)coherent involves tension between the requirements of generality and explicitness: namely, the definition is that $\mathcal{F}$ should be a sheaf of $\mathcal{O}_X$-modules, and $\mathcal{F}$ should be locally isomorphic to the cokernel of a homomorphism between free sheaves. In other words, on local pieces U, there should exist a resolution

$$\mathcal{O}_U^{\oplus N} \to \mathcal{O}_U^{\oplus M} \to \mathcal{F}_{|U} \to 0 \tag{1}$$

(that is, an exact sequence of sheaves of $\mathcal{O}_X$-modules). The tension comes because to have an intrinsic definition you want the condition for all sufficiently small neighbourhoods U, but to have an explicit construction you want only that there exists a cover of X by opens U with the property.

The tension is solved in the best possible way: thanks to cohomology, we can have it both ways! In other words, if I have an open cover $X = \bigcup U_i$ of X by affine sets U_i such that (1) holds for each U_i, then the same condition (1) holds for every affine open set $U \subset X$. This is the content of Rule ii below. It is proved in [**FAC**], [**H1**], Chapter II and [**Sh**], Chapter VII.

Now *quasicoherent* is condition (1) for all U of some (any) open cover with arbitrary cardinals N and M. *Coherent* is the same with finite N and M. Notice that for a fixed affine U, (1) and the vanishing of H^1 (see Rule ii) gives a sequence

$$k[U]^N \to k[U]^M \to \Gamma(U,\mathcal{F}) \to 0, \qquad (2)$$

where $k[U] = \Gamma(U, \mathcal{O}_X)$ is the affine coordinate ring of U. In other words, $\Gamma(U,\mathcal{F}) = F$ is just an arbitrary $k[U]$-module; moreover, (1) implies that $\mathcal{F}$ is determined on U by F and localisation. This is the construction of the sheaf $\widetilde{F}$ on U from a $k[U]$-module F.

It is instructive to compare the condition in (1) with the topologist's notion of a map of vector bundles. (1) is a homomorphism of free sheaves $\mathcal{O}_U^{\oplus N} \to \mathcal{O}_U^{\oplus M}$, and so for a choice of bases is determined by a $N \times M$ matrix A with coefficients in $\Gamma(U, \mathcal{O}_X)$. In algebraic geometry, we must allow the rank of the matrix A to vary from point to point. It is upper semicontinuous in any case, since $\operatorname{rank} A \leq r$ is a closed condition.

B.8. Examples

Example 1. If $Y \subset X$ is a subvariety (subscheme) of an affine variety defined by $f_1 = \cdots = f_n = 0$ then the structure sheaf $\mathcal{O}_Y$ is determined by the exact sequence

$$\mathcal{O}_X^n \xrightarrow{F} \mathcal{O}_X \to \mathcal{O}_Y \to 0, \qquad \text{where } F = (f_1,\dots,f_n).$$

Obvious F has rank 1 outside Y and rank 0 along Y.

Example 2. Now let $X = \mathbb{A}^n$ and $Y = \mathbb{A}^{n-1} : (x_n = 0) \subset X$, and consider the surjective homomorphism $p\colon \mathcal{O}_X^r \to \mathcal{O}_Y$ determined by $(f_1,\dots,f_r) \mapsto f_{r|Y}$. Obviously the kernel of p is the subsheaf $\mathcal{O}_X^{r-1} \oplus \mathcal{I}_Y \subset \mathcal{O}_X^r$. Since $\mathcal{I}_Y = x_n \cdot \mathcal{O}_X$ is itself a free sheaf, $\ker p \cong \mathcal{O}_X^r$. Using this isomorphism I get a short exact sequence

$$0 \to \mathcal{O}_X^r \xrightarrow{A} \mathcal{O}_X^r \to \mathcal{O}_Y \to 0, \qquad \text{where } A = \operatorname{diag}(1,\dots,1,x_n). \qquad (3)$$

The construction $\ker p$ passes from a locally free sheaf $\mathcal{E}$ to $\mathcal{E}' = \ker p \subset \mathcal{E}$, where p is the composite of restricting to a divisor $Y \subset X$ and a projection of $\mathcal{E}_{|Y}$ to a quotient bundle. This construction is well known as a standard elementary transformation of vector bundles.

Notice that the homomorphism A of sheaves is injective even at points of Y where A drops rank. The point is that the map of sheaves only looks at sections over opens, or stalks, and does not look at the *fibre* of the vector bundle. The stalk looks like $\mathcal{O}_{X,P}^r$, a free module over the local ring, whereas the fibre looks like the quotient $\mathcal{E}_P/m_P\mathcal{E}_P$, which is a k-vector space.

Example 3. A similar example. Suppose that $X = \mathbb{A}^2$ and that $Y \subset X$ is the subscheme defined by $f = g = 0$; suppose for simplicity that Y only lives at one point, that is $V(f,g) = \{P\}$. The subscheme Y is the point P with structure sheaf the finite dimensional ring $\mathcal{O}_Y = \mathcal{O}_{X,P}/(f,g)$. Then $\mathcal{O}_{X,P}$ is a UFD, so it's easy to check that the following sequence

$$0 \to \mathcal{O}_{X,P} \xrightarrow{-g,f} \mathcal{O}_{X,P} \oplus \mathcal{O}_{X,P} \xrightarrow{\substack{f\\g}} \mathcal{O}_{X,P} \to \mathcal{O}_Y \to 0$$

is exact. It's called the *Koszul complex* of f, g; its construction only depends on the fact that f, g forms a regular sequence in $\mathcal{O}_{X,P}$.

Now the point of this example is that every section of a locally free sheaf of rank 2 with only zeros in codimension 2 looks like this.

As a rule, traditional topologists have only allowed maps of constant rank between vector bundles, which is equivalent to saying that the kernel, image and cokernel are locally direct summands. As we have seen in Examples 1–3, the more general notion of sheaf homomorphism between locally free sheaves is very useful in algebraic geometry.

B.9. Rules of coherent cohomology

This table of rules states the main useful results of coherent cohomology at a fairly simple level of generality. I will take them as axioms throughout. For the proofs, see [**FAC**]. Anyone complaining that the paper is in French will receive a blast of unpleasant sarcasm. Together with the axioms, it is also very useful to know some of their immediate consequences, such as the cohomology of line bundles on $\mathbb{P}^n$ worked out in Ex. 3.1-2.

Actually, the hard thing is not to get used to these rules, but to understand what a coherent sheaf is.

Data 1

For any variety X over k and any (quasi-) coherent sheaf $\mathcal{F}$ on X there is a k-vector space $H^i(X, \mathcal{F})$, that is functorial in $\mathcal{F}$. In other words a homomorphism of sheaves of $\mathcal{O}_X$-modules $a\colon \mathcal{F} \to \mathcal{G}$ gives rise to a linear map $a_*\colon H^i(X, \mathcal{F}) \to H^i(X, \mathcal{G})$, with obvious compatibilities. (To answer the topologist's immediate question: there is no functoriality for morphisms of varieties X as yet. Sheaf cohomology is a property of the category of sheaves over a fixed X.)

Data 2

If $0 \to \mathcal{F}' \to \mathcal{F} \to \mathcal{F}'' \to 0$ is a short exact sequence of quasicoherent sheaves on X then there is a coboundary map

$$d_i\colon H^i(X, \mathcal{F}'') \to H^{i+1}(X, \mathcal{F}'),$$

again functorial in exact sequences.

So far, $H^*(X, \text{blank})$ is a *cohomological δ-functor*, if you like that kind of thing. This data satisfies the following conditions:

i. Sections H^0

$$H^0(X, \mathcal{F}) = \Gamma(X, \mathcal{F}),$$

the space of sections of a sheaf, as in the definition of a sheaf.

ii. Affine varieties X

If X is affine then

$$H^i(X, \mathcal{F}) = 0 \qquad \text{for all } i > 0.$$

Moreover, $H^0(X, \mathcal{F})$ is a sufficiently big module over the affine coordinate ring $k[X] = H^0(X, \mathcal{O}_X)$, so that the following localisation works:

$$H^0(U, \mathcal{F}) = H^0(X, \mathcal{F}) \otimes_{k[X]} H^0(U, \mathcal{O}_X) \qquad \text{for every open } U \subset X;$$
$$\mathcal{F}_P = H^0(X, \mathcal{F}) \otimes_{k[X]} \mathcal{O}_{X,P} \qquad \text{for every point } P \in X.$$

Actually, you have to prove all this before the notion of coherent sheaf is reasonably intrinsic (compare B.7).

iii. **Dimension**

$$H^i(X, \mathcal{F}) = 0 \qquad \text{for all } i > \dim X$$

The topologist who at last finds some mild satisfaction should beware that I mean the dimension of X as an algebraic variety, e.g., an algebraic curve has dimension 1 (although over the complex numbers it's a Riemann surface).

iv. **Long exact sequence**

If $0 \to \mathcal{F}' \to \mathcal{F} \to \mathcal{F}'' \to 0$ is a short exact sequence of quasicoherent sheaves on X then the functoriality homomorphisms of Data 1 and the coboundary homomorphisms of Data 2 give a cohomology long exact sequence

$$\cdots \to H^i(X, \mathcal{F}') \to H^i(X, \mathcal{F}) \to H^i(X, \mathcal{F}'') \to$$
$$\to H^{i+1}(X, \mathcal{F}') \to \cdots$$

v. **Finite dimensionality**

If $\mathcal{F}$ is coherent and X is proper (for example, projective) then

$$H^i(X, \mathcal{F}) \qquad \text{is finite dimensional over } k \text{ for any } i.$$

One traditionally writes $h^i(X, \mathcal{F}) = \dim_k H^i(X, \mathcal{F})$.

vi. **Ample line bundle, Serre vanishing**

Suppose that $X \subset \mathbb{P}^n$ is a closed subvariety. Let $\mathcal{O}_X(1) = \mathcal{O}_X \otimes \mathcal{O}_{\mathbb{P}^n}(1)$ be the invertible sheaf obtained by restricting $\mathcal{O}_{\mathbb{P}^n}(1)$ to X; this is the sheaf having the homogeneous coordinates $x_0, \dots, x_n$ as sections (compare B.6 for a definition of $\mathcal{O}_{\mathbb{P}^n}(1)$). Write $\mathcal{O}_X(r)$ for the r times tensor product of $\mathcal{O}_X(1)$, and for any quasicoherent sheaf $\mathcal{F}$ on X, write $\mathcal{F}(r) = \mathcal{F} \otimes \mathcal{O}_X(r)$.

Then given a coherent sheaf $\mathcal{F}$, there exists N such that all the following hold for all $r \geq N$: the space of global sections $H^0(X, \mathcal{F}(r))$ is big enough so that

$$\operatorname{Im}\{H^0(X, \mathcal{F}(r)) \to H^0(U, \mathcal{F}(r))\} \qquad \text{generates } H^0(U, \mathcal{F}(r))$$

as a module over $H^0(U, \mathcal{O}_X)$ for every open $U \subset X$, and

$$\operatorname{Im}\{H^0(X, \mathcal{F}(r)) \to \mathcal{F}(r)_P\} \qquad \text{generates } \mathcal{F}(r)_P$$

as a module over $\mathcal{O}_{X,P}$ for every point $P \in X$. In other words, $\mathcal{F}(r)$ is *generated by its* H^0. Moreover,

$$H^i(X, \mathcal{F}(r)) = 0 \qquad \text{for all } i > 0.$$

This is called *Serre vanishing*.

Actually everything in (vi), apart from the language, is a trivial consequence of (ii) applied to the affine cone C_X over X. A coherent sheaf $\mathcal{F}$ on X corresponds to a finitely generated graded module $\bigoplus_{r\geq 0} H^0(X,\mathcal{F}(r))$ over the usual homogeneous coordinate ring of $X \subset \mathbb{P}^n$, the affine coordinate ring of C_X. Thus the language is just a formal way of saying the usual correspondence between homogeneous polynomials and functions on $X \subset \mathbb{P}^n$.

vii. Serre duality

Let X be a nonsingular projective n-fold and K_X its canonical divisor class, so that $\mathcal{O}_X(K_X) = \Omega^n_X = \bigwedge^n \Omega^1_X$. Then

$$H^n(X,\mathcal{O}_X(K_X)) \qquad \text{is a 1-dimensional vector space} \cong k.$$

Assuming nobody objects, I pick a generator and write $= k$.

For any invertible sheaf $\mathcal{L} = \mathcal{O}_X(D)$, write $\mathcal{L}^{-1} = \mathcal{H}om_{\mathcal{O}_X}(\mathcal{L},\mathcal{O}_X) = \mathcal{O}_X(-D)$. Then there is a canonical pairing

$$H^i(X,\mathcal{L}) \times H^{n-i}(X,\mathcal{L}^{-1} \otimes_{\mathcal{O}_X} \mathcal{O}_X(K_X)) \to k,$$

or

$$H^i(X,\mathcal{O}_X(D)) \times H^{n-i}(X,\mathcal{O}_X(K_X - D)) \to k,$$

which establishes a duality between the two groups.

Remark. You can ignore this remark *en première lecture.* Of course Serre duality can be generalised to singular X and arbitrary coherent sheaves $\mathcal{F}$. However, in the same way that Poincaré duality for singular cohomology requires a manifold, or at least a space satisfying a suitable local duality, the general form is a bit complicated. If X is Cohen–Macaulay and $\dim X = n$ then there exists a sheaf ω_X, the *Grothendieck dualising sheaf,* such that $H^n(X,\omega_X) = k$, and for any coherent sheaf $\mathcal{F}$ there is a canonical pairing

$$H^i(X,\mathcal{F}) \times \operatorname{Ext}^{n-i}_{\mathcal{O}_X}(\mathcal{F},\omega_X) \to H^n(X,\omega_X) = k$$

which establishes a duality between the two groups.

If X is not Cohen–Macaulay, for example, if it has components of different dimension, then you can't expect a duality that works in a single dimension (i against $n-i$), and ω_X is replaced by a complex.[2]

viii. Euler–Poincaré characteristic $\chi(X,\mathcal{F})$ and Hilbert polynomial

Whenever the dimensions are finite, I write $h^i(X,\mathcal{F}) = \dim_k H^i(X,\mathcal{F})$. Define the Euler–Poincaré characteristic of $\mathcal{F}$ by $\chi(X,\mathcal{F}) = \sum_{i=0}^{\infty}(-1)^i h^i(X,\mathcal{F})$. Although its definition involves all the cohomology groups $H^i(X,\mathcal{F})$, this alternating sum is in fact a much more elementary quantity. For example, from the cohomology long exact sequence (iv), it follows at once that

$$\chi(X,\mathcal{F}) = \chi(X,\mathcal{F}') + \chi(X,\mathcal{F}'').$$

For $X \subset \mathbb{P}^n$, the numerical function given by $r \mapsto \chi(X,\mathcal{F}(r))$ is a polynomial, called the *Hilbert polynomial* of $\mathcal{F}$.

---

[2] I have colloquial lecture notes on this topic which I may include in a later edition. See also, for example, [**R1**], App. to §2.

ix. Riemann–Roch

For a divisor D on an algebraic curve C

$$\chi(\mathcal{O}_C(D)) = h^0(\mathcal{O}_C(D)) - h^1(\mathcal{O}_C(D)) = \chi(\mathcal{O}_C) + \deg D,$$
$$\text{and} \quad \chi(\mathcal{O}_C) = 1 - g, \qquad \text{where } g = g(C) \text{ is the genus.}$$

For a divisor D on an algebraic surface X

$$\chi(\mathcal{O}_X(D)) = h^0(\mathcal{O}_X(D)) - h^1(\mathcal{O}_X(D)) + h^2(\mathcal{O}_X(D)) = \chi(\mathcal{O}_X) + \tfrac{1}{2}D(D - K_X),$$
$$\text{and} \quad \chi(\mathcal{O}_X) = \frac{1}{12}(c_1^2 + c_2) = \frac{1}{12}(K_X^2 + e(X)),$$

where K_X is the canonical class of X and $e(X) =$ the topological Euler number of X, the alternating sum of Betti numbers. These formulas are discussed further in Ex. B.22–24 and in 3.2 and 3.4.

For an arbitrary sheaf on a projective variety X,

$$\chi(X, \mathcal{F}) = \int \operatorname{ch} \mathcal{F} \cdot \operatorname{Td}_X,$$

where $\operatorname{ch} \mathcal{F}$ and Td_X are certain characteristic classes of the sheaf $\mathcal{F}$ and the tangent sheaf of X, and the integral sign (also pronounced "evaluate on the fundamental class of X") means in practice that you take the sum of homogeneous terms of degree $\dim X$, interpret them as a zero dimensional cycle on X, and then as an integer.

You can't be a grown-up algebraic geometer until you have memorised these formulas. Eventually you have to learn what they mean, and how to calculate with them as well.

Exercises to Chapter B

1. Prove that the constant sheaf $\mathbb{Z}$ cannot be made into a sheaf of $\mathcal{O}_X$-modules.

2. Recall that the stalk $\mathcal{F}_P$ of a sheaf $\mathcal{F}$ at a point P is defined as the direct limit of the sets of sections $\Gamma(U, \mathcal{F})$ taken over all $U \ni P$. If $\varphi \colon \mathcal{F} \to \mathcal{G}$ is a morphism of sheaves, show how to define $\varphi_P \colon \mathcal{F}_P \to \mathcal{G}_P$, and check that it is well defined.

3. If X is a variety with structure sheaf $\mathcal{O}_X$ and $P \in X$ a point, prove that the stalk $\mathcal{O}_{X,P}$ is a local ring.

4. Let $\pi \colon F \to X$ be a continuous map of topological spaces. A section of π over an open set $U \subset X$ is a map $s \colon U \to F$ such that $\pi \circ s = \mathrm{id}_U$. Prove that sections of F form a sheaf $\mathcal{F}$.

5. Let $\pi \colon F \to X$ be a vector bundle of rank r (in the continuous, differentiable, complex analytic or algebraic categories). Prove that $\mathcal{F}$ constructed in the preceding exercise is a locally free sheaf of rank r over the appropriate structure sheaf of X.

6 (harder). Prove that there is an equivalence of categories between vector bundles and locally free sheaves. You'll need to choose one of the continuous, differentiable, complex analytic or algebraic categories, and be careful to ensure that the two sides of your equivalence have the same morphisms; one (boring) possibility is to allow only isomorphisms as morphisms.

7. An *affine structure* on an n-dimensional manifold M is an atlas consisting of charts $\varphi_i \colon U_i \xrightarrow{\sim}$ ball in $\mathbb{R}^n$ such that the glueing maps $\varphi_i \circ \varphi_j^{-1}$ are affine linear transformations $x \mapsto Ax + B$. Show how to introduce a sheaf of affine linear functions on M, and to give an alternative definition of manifold with affine structure based on an affine linear structure sheaf.

8. If $\varphi \colon \mathcal{F} \to \mathcal{G}$ is a morphism of sheaves, define $\operatorname{Im} \varphi \subset \mathcal{G}$ and prove that $\varphi \colon \mathcal{F} \to \operatorname{Im} \varphi$ is surjective. [Hint: As in B.6, use stalks $\mathcal{F}_P$.]

9. Let X be a projective variety, D a divisor on X, and $\mathcal{O}_X(D)$ the divisorial sheaf of D (see B.5). Choose a basis $f_0, \dots, f_n$ of the RR space $\mathcal{L}(D)$, that is, rational functions $f_i \in k(X)$ with $\operatorname{div} f_i + D \geq 0$; write $\varphi_{|D|} \colon X \to \mathbb{P}^n$ for the map defined by the ratio $f_0 : \cdots : f_n$. On the other hand, there is a map $\varphi_{\mathcal{L}} \colon X \to \mathbb{P}^n$ defined by the global sections of $\mathcal{L}$. Think through the definitions, and show that these two maps are identical.

10. On the complex plane $\mathbb{C}$ (with the complex topology), let $\mathcal{O}_{\text{an}}$ be the sheaf of holomorphic functions, and $\mathcal{O}^*_{\text{an}}$ the sheaf of invertible holomorphic functions; check you have mastered the language by writing down displayed formulas with $\{\,|\,\}$ for the sections of $\mathcal{O}_{\text{an}}$ and $\mathcal{O}^*_{\text{an}}$ over an appropriate domain.

Show that the exponential map $f \mapsto \exp(f)$ defines a morphism of sheaves $\exp \colon \mathcal{O}_{\text{an}} \to \mathcal{O}^*_{\text{an}}$, and that it is surjective. Prove that the kernel is the constant sheaf $2\pi i\mathbb{Z}$. Consider the exact sequence

$$0 \to 2\pi i\mathbb{Z} \to \mathcal{O}_{\text{an}} \xrightarrow{\exp} \mathcal{O}^*_{\text{an}} \to 0.$$

We agreed earlier that exp is surjective as a morphism of sheaves. Show that if $U \subset \mathbb{C}$ is the annular region $0 < a < |z| < b$ then $\Gamma(U, \mathcal{O}_{\text{an}}) \to \Gamma(U, \mathcal{O}^*_{\text{an}})$ is not surjective.

Find a necessary and sufficient condition on an open set $U \subset \mathbb{C}$ such that $\Gamma(U, \mathcal{O}_{\text{an}}) \to \Gamma(U, \mathcal{O}^*_{\text{an}})$ is surjective.

(In this question, the sheaves are not coherent algebraic sheaves. Only $\mathcal{O}_{\rm an}$ is a coherent analytic sheaf. The two sheaves $2\pi i\mathbb{Z}$ and $\mathcal{O}^*_{\rm an}$ are sheaves of Abelian groups, but obviously cannot be made into $\mathcal{O}_{\rm an}$-modules.)

11. Show that

$$0 \to \mathcal{I}_Y \to \mathcal{O}_X \to \mathcal{O}_Y \to 0$$

is an exact sequence of sheaves on X whenever Y is a subvariety.

12. (a) Let C be a projective curve. Prove that

$$C \cong \mathbb{P}^1 \iff g(C) = 0.$$

[Hint: Use RR.]

(b) Let $C \subset Y$ be an irreducible curve in a nonsingular surface. Show that if $K_Y C = 0$ and $C^2 = -2$ then $C \cong \mathbb{P}^1$ so that C is a -2-curve.

13. Prove that any locally free sheaf $\mathcal{L}$ of $\mathcal{O}_X$-modules of rank 1 (invertible sheaf) on a nonsingular variety X is of the form $\mathcal{L} \cong \mathcal{O}_X(D)$. Prove that $\mathcal{O}_X(D) \cong \mathcal{O}_X(D')$ as sheaves of $\mathcal{O}_X$-modules if and only if $D \overset{\rm lin}{\sim} D'$ (linear equivalence was defined in 1.6).

14. Let X be a nonsingular n-fold. Define a *rational 1-form* to be an expression $\sum f_i {\rm d}g_i$ with $f_i, g_i \in k(X)$ modulo the Leibnitz rules ${\rm d}a = 0$ for $a \in k$ and ${\rm d}(fg) = f{\rm d}g + g{\rm d}f$. Write $\Omega^1_{k(X)/k}$ for the set of rational 1-forms. Prove that it is an n-dimensional vector space over $k(X)$ with basis ${\rm d}g_1, \dots, {\rm d}g_n$, where $g_1, \dots, g_n$ is any (separable) transcendence basis of $k(X)/k$.

15. The sheaf Ω^1_X of regular 1-forms is defined by imposing regularity conditions on rational 1-forms; in other words, if $s \in \Omega^1_{k(X)/k}$, then s is regular at a point $P \in X$ if and only if it can be written $\sum f_i {\rm d}g_i$ with $f_i, g_i \in \mathcal{O}_{X,P}$. Prove that if $z_1, \dots, z_n$ are local coordinates at a point $P \in X$ then ${\rm d}z_1, \dots, {\rm d}z_n$ are local generators of Ω^1_X in a neighbourhood of P.

If you're happy with the tangent sheaf $\mathcal{T}_X$ or tangent bundle T_X of X, show that Ω^1_X can be identified with the sheaf of linear forms on $\mathcal{T}_X$ or T_X. That is, $\Omega^1_X = \mathcal{H}om_{\mathcal{O}_X}(\mathcal{T}_X, \mathcal{O}_X)$ is the sheaf Hom, defined either by setting the stalk at P equal to $\Omega^1_{X,P} = {\rm Hom}_{\mathcal{O}_{X,P}}(\mathcal{T}_{X,P}, \mathcal{O}_{X,P})$, or by setting the sections $\Gamma(U, \Omega^1_X)$ over U equal to the set of morphisms $T_{X|U} \to k \times U$ that commute with the projection to U and are linear in each fibre.

16. Set $\Omega^n_X = \bigwedge^n \Omega^1_X$, the sheaf of regular n-forms. Prove that $\Omega^n_X = \mathcal{O}_X(K_X)$.

17. Show that $\mathcal{O}_{\mathbb{P}^n}(H) \cong \mathcal{O}_{\mathbb{P}^n}(1)$ where H is any hyperplane. Extend to $\mathcal{O}_{\mathbb{P}^n}(kH)$ for any $k \in \mathbb{Z}$.

18. Give a definition of $\mathcal{O}_{\mathbb{P}^n}(r)$ in terms of ratios f/g of homogeneous polynomials in $x_1, \dots, x_n$ of degree $d + r$ and d respectively (compare B.6). Do the same for $\mathcal{O}_{\mathbb{F}}(eL + dM)$ on the scrolls in terms of bihomogeneous polynomials. Notice that here the space is constructed in terms of a group action, its structure sheaf in terms of invariant rational functions, and the other eigenspaces (character spaces) of rational functions correspond to divisorial sheaves (locally free sheaves of rank 1).

19. In the notation of Chapter 2, let $\mathbb{F} = \mathbb{F}(a_1, \dots, a_n)$ be the scroll, and M the divisor class linearly equivalent to $D_i + a_i L$. Prove that the pushforward of $\mathcal{O}_X(M)$ (defined in B.6) is a sheaf of $\mathcal{O}_{\mathbb{P}^1}$-modules isomorphic to $\mathcal{O}_{\mathbb{P}^1}(a_1) \oplus \dots \oplus \mathcal{O}_{\mathbb{P}^1}(a_n)$.

20. Construct an example of an invertible sheaf $\mathcal{L}$ on a variety X generated by its H^0, but whose sections do not separate points. Construct an example so that the sections separate points but do not separate tangent vectors.

21. State and prove Bezout's theorem on the surface scroll $\mathbb{F}_a = \mathbb{F}(a, 0)$; in other words, if C and D are curves of bidegree d, e and d', e' with no common components, state a formula for the number of points of $C \cap D$ counted with multiplicities, and prove it by the argument sketched in A.9 of the notes. You'll need to figure out the dimension of the space of forms of bidegree d, e.

22. From now on C is a nonsingular projective curve. I assume known that $\deg \operatorname{div} f = 0$ for any rational function $f \in k(C)$, that is, a rational function has the same number of zeros and poles (counted with multiplicities).

(a) Prove that if $\deg D < 0$ then $H^0(C, \mathcal{O}_C(D)) = 0$.

(b) Prove that $h^0(\mathcal{O}_C(D - P)) = h^0(\mathcal{O}_C(D))$ or $h^0(\mathcal{O}_C(D)) - 1$.

(c) Prove that $h^0(\mathcal{O}_C(D)) \leq \deg D + 1$, and that equality holds, with $\deg D \geq 1$, if and only if $C \cong \mathbb{P}^1$.

23. Given that $\chi(\mathcal{O}_C) = 1 - g$, use induction and short exact sequences of the form

$$0 \to \mathcal{O}_C(D - P) \to \mathcal{O}_C(D) \to k_P \to 0$$

to prove RR. [Hint: If $0 \to V_0 \to \cdots \to V_n \to 0$ is an exact sequence of finite dimensional vector spaces then $\sum_{i=0}^{n} (-1)^i \dim V_i = 0$. This exercise is carried out in [**H1**], Chapter IV.]

24. Use RR and Serre duality to prove that $\deg K_C = 2g - 2$ and $h^0(K_C) = g$.

To prove that the number g appearing in RR is the same as the number g in the famous picture of the surface with g holes, you have to use that T_C is the dual of K_C, so has degree $2 - 2g$, and some form of Gauss–Bonnet: the number of zeros of a regular vector field, counted with their indexes, equals the Euler characteristic. Any argument involving coherent cohomology (polynomials) on one side and topology on the other is automatically deeper than anything purely in algebraic geometry or purely in topology.

25. For points $P_1, \cdots, P_k \in \mathbb{P}^2$, write $h^0(\mathbb{P}^2, \mathcal{I}_{P_1 + \cdots + P_k} \cdot \mathcal{O}(2))$ for the vector space of conics through $P_1, \cdots, P_k$, and $h^1(\mathbb{P}^2, \mathcal{I}_{P_1 + \cdots + P_k} \cdot \mathcal{O}(2))$ for the space of linear dependence relations between the conditions $P_1, \ldots, P_k$ impose on conics (compare the example in B.6). State and prove the results of [**UAG**], §1 on the dimension of the space conics through points $P_1, \cdots, P_k$ in terms of coherent cohomology groups

$$h^1(\mathbb{P}^2, \mathcal{I}_{P_1 + \cdots + P_k} \cdot \mathcal{O}(2)).$$

26. Let $P_1, \ldots, P_9 \in C \subset \mathbb{P}^2$ be 9 distinct points contained in a nonsingular cubic curve. Suppose that $h^1(\mathbb{P}^2, \mathcal{I}_{P_1 + \cdots + P_9} \cdot \mathcal{O}(3)) \neq 0$. Prove that the surface $S = \operatorname{Bl}_{P_1, \ldots, P_9} \mathbb{P}^2$ obtained by blowing up $P_1, \ldots, P_9$ has an elliptic fibration $S \to \mathbb{P}^1$.

27. The sheafication $\operatorname{sh}(\mathcal{F})$ of a presheaf. If $\mathcal{F}$ is a presheaf, there is an *associated sheaf* or *sheafication* $\operatorname{sh}(\mathcal{F})$ which satisfies the universal mapping property for homomorphisms from $\mathcal{F}$ to a sheaf. Construct $\operatorname{sh}(\mathcal{F})$ and prove the universal mapping property. The idea is to consider the stalks $\mathcal{F}_P$, and set

$$\Gamma(U, \mathcal{F}) = \text{good maps } P \mapsto s_P \in \mathcal{F}_P \text{ for all } P \in U,$$

where "good" means that all the s_Q for Q in some small neighbourhood V_P of P are the restrictions of some $s \in \Gamma(V_P, \mathcal{F})$. If you have trouble with this question, refer to [**H1**], Chap. II or one of the books on sheaf theory.

28. If $\mathcal{F} \subset \mathcal{G}$ is a subsheaf, construct the quotient sheaf $\mathcal{G}/\mathcal{F}$ as the associated sheaf of the presheaf $U \mapsto \Gamma(U, \mathcal{G})/\Gamma(U, \mathcal{F})$, and prove that it has the universal mapping property for maps from $\mathcal{G}$ to a sheaf killing $\mathcal{F}$. Prove also that its stalks are $\mathcal{G}_P/\mathcal{F}_P$, so that the sequence $0 \to \mathcal{F} \to \mathcal{G} \to \mathcal{G}/\mathcal{F} \to 0$ is exact.

29. If $f\colon X \to Y$ is a continuous map of topological spaces and $\mathcal{F}$ is a sheaf on Y, construct the sheaf theoretic pullback $f^{-1}\mathcal{F}$, whose stalk at $P \in X$ is $\mathcal{F}_{f(P)}$. Prove that it has the universal mapping property for sheaves $\mathcal{G}$ on X such that there exists a sheaf homomorphism $\mathcal{F} \to f_*\mathcal{G}$.

Incidentally, you mustn't write f^* for f^{-1}, because f^* is usually reserved for the pull back of sheaves of $\mathcal{O}_X$-module, given by $f^{-1}\mathcal{F} \otimes_{f^{-1}\mathcal{O}_Y} \mathcal{O}_X$.

CHAPTER C
Guide to the Classification of Surfaces

The classification of surfaces goes back to Castelnuovo and Enriques in the early years of the 20th century. It divides algebraic surfaces into 4 big classes. There will be lots to say about surfaces in each of the 4 classes, and the characteristic methods and results for each class are different. The subject matter thus divides naturally into the 4 separate classes, together with the logical division into cases, or the *proof of classification.*

This chapter introduces the classification of surfaces in overall terms. It is primarily intended as a guide to results that lots of people need to use, and most of the results are given without proof. However, to give a veneer of modernity I cast the classification within the logical structure of Mori theory: the dichotomy K_X nef or otherwise, and the numerical dimension of K_X form the main logical framework, taking precedence over criteria for rationality and ruledness and Kodaira dimension. The underlying material is of course essentially the same as that of Castelnuovo, Enriques and Kodaira. I return to the proof of the classification in Chapters D–E below.

I assume that X is a nonsingular projective surface over $\mathbb{C}$ to make life easier for myself and the reader. It is of course often reasonable to weaken these conditions: a topologist or analyst might want compact complex surfaces without the assumption of algebraicity, a number theorist might want singular projective surfaces over a finite field or a number field, students interested in higher dimensional geometry will need to know how the results for surfaces work in several other contexts. Getting the right level of generality is also important for the internal development of the subject.

C.1. Invariants

C.1.1. Numerical invariants

The main numerical invariants are

1. the Betti numbers B_i with the Euler number $e(X)$;
2. the signature (B_2^+, B_2^-) of the quadratic form Q_X on $H^2(X,\mathbb{Z})/\operatorname{Tors}$;
3. the Hodge numbers $h^{p,q}$ with the Poincaré characteristic $\chi(\mathcal{O}_X)$;
4. the Chern classes $c_1 = [-K_X] \in H^2(X,\mathbb{Z})$ and $c_2 \in H^4(X,\mathbb{Z}) = \mathbb{Z}$, and the Chern numbers c_1^2 and c_2.

Fortunately there are lots of relations between these numbers: only two or three of them are independent. There are also three other discrete invariants that are not simply numbers, but are closely related to the numerical invariants: the quadratic form Q_X on $H^2(X,\mathbb{Z})/\operatorname{Tors} X$ (up to isomorphism), the torsion subgroup $\operatorname{Tors} X = \operatorname{Tors} H^2(X,\mathbb{Z})$ and the fundamental group $\pi_1(X)$. The intersection form on $H^2(X,\mathbb{Z})$ is completely determined from a knowledge of the numerical invariants (including $c_1 \in H^2(X,\mathbb{Z})$), and in turn, for simply connected surfaces, it is known that this determines the homotopy type and the homeomorphism type of X.

C.1.2. Topological invariants

A complex surface is a compact 4-manifold with a chosen orientation. Accordingly, its Betti numbers $B_i = \operatorname{rank} H^i(X,\mathbb{Z})$ satisfy Poincaré duality

$$B_0 = B_4 = 1, \qquad B_1 = \operatorname{rank} H^1(X,\mathbb{Z}) = B_3.$$

Hence the Euler number is given by

$$e(X) = \sum (-1)^i B_i = 2 - 2B_1 + B_2.$$

(Hirzebruch's convention is to write $e(X)$ for the Euler characteristic, rather than something like $\chi_{\mathrm{top}}(X,\mathbb{Z})$, to distinguish it from Euler–Poincaré characteristics such as $\chi(\mathcal{O}_X)$ in sheaf cohomology; I follow this throughout, and urge you to do likewise.) In addition, Poincaré duality says that the symmetric bilinear form defined by cup product

$$\cup\colon H^2(X,\mathbb{Z}) \times H^2(X,\mathbb{Z}) \to H^4(X,\mathbb{Z}) = \mathbb{Z}$$

is a perfect pairing on $H^2(X,\mathbb{Z})/\operatorname{Tors} X = \operatorname{Hom}(H_2(X,\mathbb{Z}),\mathbb{Z})$. This means, equivalently, that $\cup$ induces an isomorphism $H^2(X,\mathbb{Z})/\operatorname{Tors} X \to \operatorname{Hom}_{\mathbb{Z}}(H^2(X,\mathbb{Z}),\mathbb{Z})$, or that in any $\mathbb{Z}$-basis of $H^2(X,\mathbb{Z})/\operatorname{Tors} X$, the symmetric matrix representing the pairing is unimodular, that is, has determinant ± 1.

C.1.3. Analytic invariants

Most of the analytic invariants are defined in terms of the sheaves of differentials: recall that T_X is a rank 2 complex vector bundle over X, whose sections are vector fields on (open subsets of) X; the dual bundle corresponds to the sheaf of Kähler differentials $\Omega^1_X = \operatorname{Hom}_{\mathcal{O}_X}(T_X, \mathcal{O}_X)$. This is a locally free sheaf of $\mathcal{O}_X$ modules of rank 2, based locally by dx, dy where x, y are local coordinates. The sheaf of canonical differentials $\Omega^2_X = \mathcal{O}_X(K_X)$ is the sheaf of holomorphic 2-forms, defined as $\bigwedge^2 \Omega^1_X$, and is the locally free sheaf of $\mathcal{O}_X$ modules of rank 1, based locally by $dx \wedge dy$.

Global sections of these bundles give us important invariants: the *geometric genus*

$$p_g(X) = h^0(X, \Omega^2_X) = \dim \Gamma(X, \Omega^2_X),$$

that is, the number of global holomorphic canonical differentials, and the *irregularity* $q = h^0(X, \Omega^1_X)$. *Hodge theory* allows us to express some of the other invariants of X in terms of these: firstly, it defines a decomposition

$$H^1(X,\mathbb{Z}) \otimes_{\mathbb{Z}} \mathbb{C} = H^1(X,\mathbb{C}) = H^{0,1} \oplus H^{1,0},$$

where $H^{0,1} = H^1(X, \mathcal{O}_X)$ and $H^{1,0} = H^0(X, \Omega^1_X)$; in this decomposition $H^{0,1}$ is the complex conjugate of $H^{1,0}$, so that in particular

$$h^{0,1} = h^{1,0} = q \quad \text{and} \quad B_1 = \operatorname{rank} H^1(X, \mathbb{Z}) = \dim_{\mathbb{C}} H^1(X, \mathbb{C}) = 2q.$$

Next, it tells us the signature of the pairing on $H^2(X, \mathbb{R}) = H^2(X, \mathbb{Z}) \otimes \mathbb{R}$: diagonalising this quadratic form over $\mathbb{R}$ gives the pairing as a diagonal matrix with B_2^+ entries $+1$ and B_2^- entries -1; the pair (B_2^+, B_2^-) is the Sylvester inertia index of the form. Now Hodge theory tells us that $B_2^+ = 2p_g + 1$. In particular, p_g is an invariant of the (oriented) homotopy type of X.

More precisely, the Hodge decomposition of $H^2(X, \mathbb{C})$ is

$$H^2(X, \mathbb{C}) = H^{0,2} \oplus H^{1,1} \oplus H^{2,0},$$

where $H^{2,0} = H^0(X, \Omega^2_X)$ and $H^{0,2}$ is the complex conjugate of $H^{2,0}$, so that the vector subspace

$$V^+_{\mathbb{C}} = H^{2,0} \oplus H^{0,2} \subset H^2(X, \mathbb{C})$$

is invariant under complex conjugation, and corresponds to a $2p_g$-dimensional real subspace $V^+ \subset H^2(X, \mathbb{R})$. Then if $L \in H^2(X, \mathbb{Z})$ is the class of a hyperplane section of X (the first Chern class of the line bundle $\mathcal{O}_X(1)$ for some embedding $X \subset \mathbb{P}^N$), by Hodge theory $L \in H^{1,1}$, and the Hodge index theorem says that the intersection form is positive definite on $V^+ \oplus \mathbb{R} \cdot L$, and negative definite on the "primitive part" $H^{1,1}_0$, the orthogonal complement of L in $H^{1,1}$.

Remark. More generally, the set-up of Hodge theory on a Kähler manifold is

$$H^n(X, \mathbb{C}) = \bigoplus_{p+q=n} H^{p,q}(X), \qquad (*)$$

where $H^{p,q}(X) = H^q(X, \Omega^p_X)$. One proves that $H^{p,q}$ and $H^{q,p}$ are complex conjugates, that the Hodge decomposition $(*)$ determines the signature of the intersection pairing on the "primitive" cohomology, etc. What I said above is enough for complex projective surfaces.

C.1.4. Chern classes

A crude consequence of the presence of the complex structure is that the tangent bundle T_X has the structure of a complex vector bundle with fibre $\mathbb{C}^2$. This vector bundle has characteristic classes $c_1(T_X) \in H^2(X, \mathbb{Z})$ and $c_2(T_X) \in H^4(X, \mathbb{Z})$. Each of these objects has an alternative interpretation: since Ω^1_X is the dual of T_X and Ω^2_X is the determinant bundle $\bigwedge^2 \Omega^1_X$,

$$c_1(T_X) = -c_1(\Omega^1_X) = -c_1(\Omega^2_X) = -[K_X] \in H^2(X, \mathbb{Z}).$$

Also, a general result is that $c_2(T_X) = e(X)$: morally speaking, the top Chern class of T_X should be the number of zeros of a generic section, but the number of zeros of a generic vector field is the Euler number.

The final relation between the numerical invariants is Noether's formula:

$$\chi(\mathcal{O}_X) = 1 - q + p_g = \frac{1}{12}(c_1^2 + c_2) = \frac{1}{12}(K_X^2 + e(X));$$

here I interpret c_1^2 and $c_2 \in H^4(X, \mathbb{Z}) = \mathbb{Z}$ as integers. The formula is part of the set-up of Hirzebruch–RR. Since $c_2(X) = e(X) = 2 - 4q + B_2$, and since q and p_g are homotopy invariants, so is c_1^2.

C.1.5. Exercises

1. For regular surfaces (that is, surfaces with $q = 0$), show how to express all the numerical invariants in terms of K_X^2 and $\chi = \chi(\mathcal{O}_X)$, and verify the Hirzebruch signature formula
$$B_2^+ - B_2^- = 4\chi - e(X) = \frac{1}{3}(c_1^2 - 2c_2).$$
2. Assume that $c_1^2 \geq 0$, $c_2 \geq 0$ and $c_1^2 \leq 3c_2$. (Lots of surfaces satisfy these conditions.) Prove that $B_2^- = 0$ is only possible if X satisfies $q = 0$, $B_2 = 1$, $c_1^2 = 9$ and $c_2 = 3$ (that is, X has the numerical invariants of $\mathbb{P}^2$). This proves that the quadratic form on H^2 is either $\mathbf{1}$ or is indefinite.

Solutions

1. $q = 0$, $p_g = \chi - 1$, $e(X) = 12\chi - K_X^2$,
$B_2 = e(X) - 2$, $B_2^+ = 2p_g + 1 = 2\chi - 1$, $B_2^- = e(X) - 2\chi - 1$.

(It is easy to write out similar formulas for any q.)

2. $B_2^- = B_2 - 2p_g - 1$
$= e(X) - 2 + 4q - 2p_g - 1$
$= c_2 - \frac{1}{6}(c_1^2 + c_2) + 2q - 1$
$= \frac{1}{3}c_2 + \frac{1}{6}(3c_2 - c_1^2) + 2q - 1.$

So $B_2^- = 0$ gives $q = 0$ and $2c_2 + (3c_2 - c_1^2) = 6$, which has only one solution compatible with $c_1^2 + c_2 \equiv 0 \bmod 12$.

C.1.6. Quadratic form on $H^2(X, \mathbb{Z})$

Consider a quadratic form (H, Q) over $\mathbb{Z}$, that is, a free $\mathbb{Z}$-module (lattice) $H \cong \mathbb{Z}^n$, together with a quadratic map $Q\colon H \to \mathbb{Z}$. There is a whole theory of isomorphism classes of quadratic forms over $\mathbb{Z}$; let me start by going through the easy bits of this theory; a nondegenerate quadratic form (H, Q) has a *signature* (n^+, n^-), a *discriminant* $\det Q$, and a *parity*. Parity just means the following: if $Q(x) \in \mathbb{Z}$ is even for all $x \in H$ then the quadratic form is *even*, otherwise it's *odd*.

For $H^2(X, \mathbb{Z})$ of a complex surface, all these invariants are under control: the signature (B_2^+, B_2^-) is discussed in C.1.3 above; the quadratic form Q_X is unimodular (that is, $\det Q_X = 1$) by Poincaré duality; and the parity is determined by $c_1(X)$ or K_X:
$$Q_X(x) \equiv c_1 \cdot x \bmod 2 \quad \text{for all } x \in H^2(X, \mathbb{Z}).$$
Hence Q_X is even if and only if $c_1(X)$ maps to $0 \in H^2(X, \mathbb{Z}/2)$, or equivalently, K_X is divisible by 2 as a divisor class. A surface having this property is *even* or is a *spin surface.*

A unimodular quadratic form (H, Q), assumed to be indefinite, is determined up to isomorphism by its signature (n^+, n^-) and its parity. In fact if Q is odd then it is isomorphic to the diagonal form:
$$Q \text{ is odd} \implies (H, Q) \cong \mathbf{1}^{n^+} \oplus (\mathbf{-1})^{n^-},$$
and there is a similar standard form for an even indefinite form
$$Q \text{ is even} \implies (H, Q) \cong \begin{pmatrix} 0 & 1 \\ 1 & 0 \end{pmatrix}^a \oplus (\pm E_8)^b,$$

where $\begin{pmatrix} 0 & 1 \\ 1 & 0 \end{pmatrix}$ is the standard hyperbolic lattice, E_8 is the lattice based by 8 vectors e_i in bijection with the vertexes of the Dynkin diagram E_8

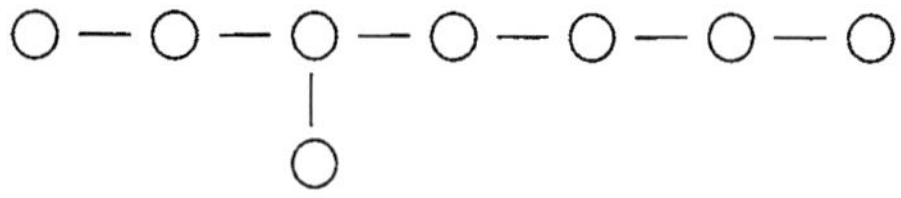

and quadratic form given by $e_i^2 = 2$, $e_ie_j = 0$ or -1 according as the corresponding vertexes are unjoined or joined, and $-E_8$ the same with the opposite signs $e_i^2 = -2$, $e_ie_j = 0$ or $+1$. That is, $\pm E_8 = \mathbb{Z}^8$ with the quadratic form

$$\pm\begin{pmatrix} -2 & 0 & 0 & 1 & 0 & 0 & 0 & 0 \\ 0 & -2 & 1 & 0 & 0 & 0 & 0 & 0 \\ 0 & 1 & -2 & 1 & 0 & 0 & 0 & 0 \\ 1 & 0 & 1 & -2 & 1 & 0 & 0 & 0 \\ 0 & 0 & 0 & 1 & -2 & 1 & 0 & 0 \\ 0 & 0 & 0 & 0 & 1 & -2 & 1 & 0 \\ 0 & 0 & 0 & 0 & 0 & 1 & -2 & 1 \\ 0 & 0 & 0 & 0 & 0 & 0 & 1 & -2 \end{pmatrix}$$

Definite quadratic forms over $\mathbb{Z}$ are of course much more complicated, but fortunately, we never have to deal with them (compare Exercise 1.5 above).

C.1.7. Exercise

All these invariants are usually easy enough to calculate if your surface is given fairly explicitly. For example, suppose that $X = X_d \subset \mathbb{P}^3_{\mathbb{C}}$ is a nonsingular hypersurface of degree d. It is simply connected by Lefschetz theory, and in particular $q = 0$. (See also Ex. 3.3.) The adjunction formula gives $K_X = (-4+d)H = \mathcal{O}_X(-4+d)$, so that

$$p_g = \begin{cases} 0 & \text{for } d \le 3 \\ \binom{d-1}{3} & \text{for } d \ge 4 \end{cases} \quad \text{and } K_X^2 = d(d-4)^2.$$

Finally, K_X is divisible by 2 in $\operatorname{Pic} X$ if and only if $d-4$ is even, that is, d is even. It's an easy exercise to determine all the invariants of X from this.

C.2. Birational versus biregular classification: blowups

There are several different points of view on what may be meant by the classification of algebraic varieties, among them the following three:

1. birational, that is, up to birational equivalence;
2. biregular, that is, up to isomorphism;
3. projective, that is, up to projective equivalence.

To say that X is a projective surface means that X is capable of being embedded as a closed subvariety of $\mathbb{P}^N$ for some N. The gap between (2) and (3) is the problem of finding a projective embedding of an abstract surface X, or separating off intrinsic properties of $X \subset \mathbb{P}^N$ from the properties of the ambient space and of the embedding. I leave this for the moment. Most of this section and the next is concerned with the gap between (1) and (2).

Notice that for nonsingular projective curves, (1) = (2): for C and C' any two nonsingular projective curves, any birational map from C to C' is an isomorphism. The blowup (see 1.10) shows that this is not the case for surfaces. The main point of this section is that blowups account for the whole of the gap between (1) and (2).

Remark. For surfaces having a minimal model with K_X nef, it turns out that the minimal model is unique, which restores (1) = (2). A birational map between two surfaces with K_X nef is necessarily an isomorphism.

C.2.1. Proposition. (a) Blowups. *Given a surface X and a point $P \in X$, there exists a surface X_1 and a morphism $\sigma: X_1 \to X$ such that $\sigma^{-1}P = E$ is a curve, and σ restricts to an isomorphism $X_1 \setminus E \to X \setminus P$.*

Here $E \cong \mathbb{P}^1$ and $E^2 = -1$, so that E is a -1-curve; if C is a curve of X then $\sigma^*C = C' + mE$, where $m = \operatorname{mult}_P C$, C' is the birational transform of C (that is, the closure in X_1 of $\sigma^{-1}(C \setminus P)$), and the points of C' over P correspond one-to-one to the distinct tangent lines to C.

(b) Castelnuovo's contractibility criterion. *Given a surface X and a -1-curve $E \subset X$, there exists a contraction $\sigma: X \to Y$ such that E maps to a point P of a (nonsingular projective) surface Y, and σ is the blowup of $P \in Y$.*

For (a), see 1.10. (b) is proved in Chapter 4, Contraction Theorem 4.15.

C.2.2. Theorem. (a) Resolution of indeterminacies. *Let $f: X \dashrightarrow \mathbb{P}^N$ be a rational map. Then there exists a commutative diagram*

$$\begin{array}{ccc} & Y & \\ {}^{g}\swarrow & & \searrow^{h} \\ X & \dashrightarrow & \mathbb{P}^N \end{array}$$

where g is a composite $Y = X_n \to \cdots \to X_1 \xrightarrow{\sigma_1} X$ of blowups, and h is a morphism.

(b) Factorisation of birational morphisms. *Let $h: X \to Y$ be a birational morphism between nonsingular projective surfaces. Then h is the composite of a chain of blowups $X \xrightarrow{\sigma_1} Y_1 \to \cdots \to Y_N = Y$.*

See for example [**Beauville**], Chapter 2 or [**Sh**], Chapter IV, §3.

C.2.3. Topological view of a blowup

For surfaces over $\mathbb{C}$, a blowup $\sigma: X_1 \to X$ corresponds to taking a connected sum $X\#Q$ of X with the 4-manifold $Q = \overline{\mathbb{P}^2_{\mathbb{C}}}$, that is, $\mathbb{P}^2_{\mathbb{C}}$ with the opposite orientation.

The idea is that it replaces a small disc around P by a tubular neighbourhood of the -1-curve E; it is not hard to see that this is diffeomorphic to $\mathbb{P}^2_{\mathbb{C}} \setminus (\text{disc})$, but with the opposite orientation (both of them are disc bundles over $\mathbb{P}^1_{\mathbb{C}} \cong S^2$, but the "core" copy of $\mathbb{P}^1_{\mathbb{C}}$ has selfintersection ± 1 in the two cases).

C.2.4. Fundamental asymmetry

It's important to note the asymmetry here: taking connected sum with Q is a blowup, so is an operation of algebraic geometry. But taking connected sum with $\mathbb{P}^2_{\mathbb{C}}$ itself cannot be an operation of algebraic geometry: it can never happen that X and $X\#\mathbb{P}^2_{\mathbb{C}}$ are both diffeomorphic to algebraic surfaces (with oriented connected sum and diffeomorphism preserving the orientation), if only because B_2^+ passes from odd to even.

At a much deeper level, the contrast can be seen in a very striking way from the following two results:

Theorem (Moishezon–Mandelbaum). *For most naturally occurring large classes of algebraic surfaces, the connected sum $M \# \mathbb{P}^2_{\mathbb{C}}$ has a decomposition*

$$M \# \mathbb{P}^2_{\mathbb{C}} \overset{\text{diffeo}}{\sim} (\mathbb{P}^2_{\mathbb{C}})^{\# n} \# Q^{\# m}.$$

Theorem (Donaldson). *If M is the 4-manifold underlying an algebraic surface, there does not exist any connected sum decomposition*

$$M \overset{\text{diffeo}}{\sim} M_1 \# M_2$$

with each of M_1, M_2 having $B_2^+ > 0$.

C.3. Results of minimal model theory

The following is a modern statement of minimal model theory. It is proved in Chapter D. In the classical theory, it would be a corollary of a whole chain of results. Recall that the *canonical divisor* K_X is the divisor corresponding to the invertible sheaf $\Omega^2_X = \mathcal{O}_X(K_X)$ and that a divisor D on a surface X is *nef* ("numerically eventually free") if $DC \geq 0$ for every curve $C \subset X$.

C.3.1. Theorem. *Let Y be any nonsingular projective surface. Then there is a chain*

$$Y \xrightarrow{\sigma_1} Y_1 \to \cdots \to Y_N = X$$

such that each $\sigma_i \colon Y_{i-1} \to Y_i$ is the contraction of a single -1-curve E_i, and X satisfies

either (i) K_X is nef;
or (ii) $X \cong \mathbb{P}^2$ or a $\mathbb{P}^1$-bundle over a curve.

More crudely, every surface is birational to a surface with K_X nef, or to a $\mathbb{P}^1$-bundle over a curve or $\mathbb{P}^2$. This form of the result is easy to remember, and is very useful. On the other hand, the information it contains can be analysed to give results that are more complicated, but much more general. Notice that as stated, the theorem refers to three different situations:

1. a contraction $\sigma \colon X \to X_1$ of a -1-curve E, with $E \cong \mathbb{P}^1$, $E^2 = -1$, $K_X E = -1$;
2. a $\mathbb{P}^1$-bundle $X \to C$ with fibre $F \cong \mathbb{P}^1$, $F^2 = 0$, $K_X F = -2$;
3. $X \cong \mathbb{P}^2$; here I can write the constant morphism $\mathbb{P}^2 \to$ pt., and consider a line $L \subset \mathbb{P}^2$ with $L \cong \mathbb{P}^1$, $L^2 = 1$, $K_X L = -3$.

All three of these situations can be classed together as a morphism $\varphi \colon X \to Y$ such that $-K_X C > 0$ for every curve C in a fibre of f. Thus Theorem C.3.1 is closely related to the following statement.

C.3.2. Theorem. *Let X be a surface, and suppose that K_X is not nef. Then there exists a morphism $\varphi \colon X \to Y$ (with $\dim Y = 0, 1$ or 2) contracting at least one curve of X to a point, and such that $-K_X C > 0$ for every curve C in a fibre of φ. There is no loss of generality in assuming $\varphi_* \mathcal{O}_X = \mathcal{O}_Y$, which means that Y is normal and the general fibre of φ is connected.*

Theorem C.3.2 is more primitive than Theorem C.3.1, but much more general in scope: it holds for surfaces over any field k, for a large class of singular surfaces, and for nonsingular varieties of any dimension over $\mathbb{C}$. The modern strategy for proving Theorem C.3.1 is first to prove Theorem C.3.2, then to analyse the morphism $\varphi \colon X \to Y$ for which $-K_X$ is ample.

C.3.3. Proposition. *Let $\varphi: X \to Y$ be as in the conclusion of Theorem C.3.2. Then one of the following 3 cases holds.*

$\dim Y = 2$. *Then Y is a nonsingular surface, and $\varphi: X \to Y$ is the blowdown of a number of disjoint -1-curves;*

$\dim Y = 1$. *Then Y is a nonsingular curve, and $\varphi: X \to Y$ is a conic bundle; the general fibre of φ is isomorphic to $\mathbb{P}^1$, whereas a special fibre is isomorphic to a line pair in $\mathbb{P}^2$;*

$Y = \text{pt}$. *In this case, the constant morphism $\varphi: X \to Y$ does not give any information, but $-K_X$ is an ample divisor.*

Surfaces in the final case are called *del Pezzo surfaces*, and they are similar in most of their properties to the cubic surfaces $X_3 \subset \mathbb{P}^3$ discussed in Chapter 1. The main step in deducing Theorem C.3.1 from Theorem C.3.2 is to prove that if X is a del Pezzo surface then either $B_2(X) \geq 2$ and X admits another contraction $X \to X'$ of one of the other two types; or $B_2(X) = 1$ and $X \cong \mathbb{P}^2$.

C.4. Ruled and rational surfaces, Tsen's theorem

I define a *ruled surface* to mean a nonsingular projective surface X together with a fixed morphism $\varphi: X \to B$ to a base curve B, whose fibre $F_{k(B)}$ over the generic point is a nonsingular curve of genus 0. The main theorem of this section, proved below, is Tsen's theorem: $F_{k(B)}$ is isomorphic over $k(B)$ to $\mathbb{P}^1_{k(B)}$. To discuss the results of minimal model theory, I assume this for the moment. That is, I assume that $\varphi: X \to B$ is a given morphism with generic fibre isomorphic to $\mathbb{P}^1_{k(B)}$.

There are many alternatives, going back to the ancients: you could require only that φ exists, without taking the responsibility for specifying it, or you could allow φ to be a rational map $\varphi: X \dashrightarrow B$. For the purposes of this introduction, call this "birationally ruled". Thus the plane $\mathbb{P}^2$ or any rational surface is birationally ruled. The $\kappa = -\infty$ part of the classification of surfaces (due to Castelnuovo and Enriques) can be stated in the form

$$P_{12}(X) = 0 \iff \kappa = -\infty \iff X \text{ is birationally ruled.}$$

Compare Corollary E.1.

C.4.1. Models

According to the definition, X has a morphism $X \to B$ with generic fibre $\mathbb{P}^1_{k(B)}$. Choosing an isomorphism of the generic fibre defines a birational map

$$\begin{array}{ccc} X & \dashrightarrow & B \times \mathbb{P}^1 \\ & \searrow \quad \swarrow & \\ & B & \end{array}$$

Since the definition of ruled surface only fixes the generic fibre, the general picture is as follows: X is isomorphic to $B \times \mathbb{P}^1$ over a Zariski open set of B, but over finitely many points $\{P\} \in B$ the fibre can have lots of components, say $\varphi^{-1}P = \bigcup \Gamma_i$. We know that the intersection matrix $(\Gamma_i\Gamma_j)_{ij}$ is negative semidefinite (see Theorem A.7 and Ex. A.17). It's an exercise to see from first principles that if the fibre is not $\mathbb{P}^1$ then at least one of the Γ_i is a -1-curve.

There are other conditions that can be added to a ruled surface that restrict the types of singular fibres. These conditions appear in a natural way in classical work, and more especially in Mori theory:

1. $-K_X$ ample characterises conic bundles. Every singular fibre of φ is a line pair, a union of two -1-curves $L_1 \cup L_2$ meeting transversally at one point. In fact $-K_X$ is relatively very ample, and embeds X into a $\mathbb{P}^2$-bundle over B.
2. $-K_X$ ample and $\rho(\varphi) = 1$ characterise $\mathbb{P}^1$-bundles. The condition $\rho(\varphi) = 1$ means that every curve in a fibre is numerically a multiple of the fibre. This implies that every fibre is nonsingular. Using Tsen's theorem, it's not hard to see that then $X = \mathbb{P}(E)$ is the projectivisation of a rank 2 vector bundle over B.

The case of conic bundle is important over an algebraically nonclosed field k: it can happen that $X \to B$ has $-K_X$ ample, so is geometrically a conic bundle, and has $\rho(\varphi) = 1$, so that there are no curves defined over k in the fibres of $X \to B$ that can be contracted out by a birational morphism; but $X \to B$ can still have singular fibres, line pairs $L_1 \cup L_2$ with L_1 and L_2 conjugate over k.

C.4.2. Theorem (Tsen's theorem). *Let k be an algebraically closed field and $k \subset F$ a function field in one variable over k (this means that $F = k(B)$, where B is an irreducible algebraic curve over k). Let X_F be a nonsingular projective curve over F, and assume that X_F is absolutely irreducible, in the sense that it remains irreducible over the algebraic closure of F. Then the following conditions are equivalent:*

1. *The curve X_F has $g = 0$ in RR;*
2. *X_F is isomorphic over F to a nonsingular plane conic;*
3. *X_F becomes isomorphic to $\mathbb{P}^1$ after some field extension $F \subset F'$, that is, $X_F \otimes_F F' \cong \mathbb{P}^1_{F'}$;*
4. *X_F is isomorphic over F to $\mathbb{P}^1_F$.*

Proof. $2 \implies 3 \implies 1$ and $4 \implies 1$ are obvious. The implication $1 \implies 2$ is standard use of RR: the anticanonical line bundle $-K_{X_F}$ has degree 2, and is very ample by the usual criterion: $\ell(-K_{X_F}) = 3$, and $\ell(-K_{X_F} - D) = 1$ for any divisor D of degree 2. All this works over an arbitrarily field F.

The point is thus to prove $2 \implies 4$. I explain how to use the assumptions that $F = k(B)$ is a function field in one variable over an algebraically closed field. The point is that $k(B)$ is the field of fractions of a ring $k[B]$ with "linear growth", namely the coordinate ring of an affine model of B. In more detail, $k[B] = \bigcup k[B]_d$ (polynomials of degree $\le d$) with multiplication satisfying $k[B]_{d_1} \cdot k[B]_{d_2} \subset k[B]_{d_1+d_2}$. Moreover, $k[B]_d$ is finite dimensional over k, with dimension growing as a linear function: for some constants a, b_1, b_2

$$ad - b_1 \le \dim k[B]_d \le ad + b_2 \quad \text{(same } a\text{!)}. \tag{1}$$

There are various ways of proving this, using RR on the curve B, or the commutative algebra definition of $\dim k[B] = 1$ in terms of Hilbert–Samuel functions, but in any case the fact that B is 1-dimensional is crucial.

Now X_F is isomorphic over $k(F)$ to a plane conic

$$X_F : q(x, y, z) = 0 \subset \mathbb{P}^2_F \tag{2}$$

defined by a quadratic form q in x, y, z with coefficients in F. Notice that since X_F is a plane conic, to prove that it is isomorphic to $\mathbb{P}^1_F$, it is enough to prove that is has a point defined over F (the argument is given in [**UAG**], (1.7) and Ex. 1.5).

The point is thus very simple: under the assumption that F is a function field in 1 variable, I can always solve (2). Multiplying through by denominators, I can assume that the coefficients of q are in $k[B]$, and even in $k[B]_c$ for some choice of c. The plan of the proof is to try to solve (2) with $x, y, z \in k[B]_d$ subject to $q(x, y, z) = 0 \in k[B]_{2d+c}$.

The punch line is that by (1) there are $\geq 3(ad - b_1)$ free variables in the choice of x, y, z, and $\leq a(2d + c) + b_2$ polynomial conditions on them in the equation $q(x, y, z) = 0 \in k[B]_{2d+c}$. To explain this gently, x, y, z are any elements of the vector space $k[B]_d$; if $e_1, \dots, e_N$ is a basis then

$$x = \sum u_i e_i, \quad y = \sum v_i e_i, \quad z = \sum w_i e_i$$

give $3N$ variables u_i, v_i, w_i. In a similar way, the equation $q(x, y, z) = 0$ can be written out as $\dim k[B]_{2d+c}$ quadratic equations in these variables.

Since for $d \gg 0$, there are more free variables than equations, over the algebraically closed k, it follows that u_i, v_i, w_i can be given values in k so that x, y, z satisfy $q(x, y, z) = 0$.

C.4.3. Birational maps

If X and Y are rational surfaces, the set of birational maps $X \dashrightarrow Y$, or the group $\operatorname{Bir} X$ of birational selfmaps is very big. This set (resp. group) is the obstruction to the uniqueness of the model of X as a $\mathbb{P}^1$-bundle over a curve or $\mathbb{P}^2$. For $X = C \times \mathbb{P}^1$, $\operatorname{Bir} X$ contains $\mathrm{PGL}(2, k(C))$. For $X = \mathbb{P}^2$, the group $\operatorname{Bir} X$ is the Cremona group, and is qualitatively something like the free group on a continuum of generators.

C.5. Classification assuming K_X nef

Aim

Assuming X a surface with K_X nef, classify into 3 cases according to the numerical properties of K_X (the 3 possibilities $\nu = 0, 1, 2$ for the numerical dimension of K_X), and according to the analytic properties of K_X (the 3 possibilities $\kappa = 0, 1, 2$, for the Kodaira dimension of K_X); statement of $\kappa = \nu$.

C.5.1. General fact. D nef $\implies D^2 \geq 0$

Recall that by definition, D is nef if and only if $DC \geq 0$ for all curves $C \subset X$, so that $D \cdot (\sum n_i C_i) \geq 0$ for all effective divisors $\sum n_i C_i$ with $n_i \geq 0$. See D.2 for a proof.

Morally, the reason C.5.1 holds is that

$$D \text{ nef} \implies D \text{ is ``close to being effective''} \implies D^2 \geq 0.$$

C.5.2. Definition of $\overset{\text{num}}{\sim}$

Two divisors D_1 and D_2 on a surface X are numerically equivalent (written $D_1 \overset{\text{num}}{\sim} D_2$) if $D_1 C = D_2 C$ for all curves $C \subset X$. Notice that

$$D_1 \overset{\text{lin}}{\sim} D_2 \iff \mathcal{O}_X(D_1) \cong \mathcal{O}_X(D_2) \implies D_1 \overset{\text{num}}{\sim} D_2$$

because $DC = \deg_C \mathcal{O}_X(D)$.

C.5.3. Division into cases according to ν

Assume K_X is nef, so that $K_X^2 \geq 0$; then the following 3 cases are all-inclusive and mutually exclusive:

$\nu = 0$. $K_X \overset{\text{num}}{\sim} 0$, that is, $K_X C = 0$ for all $C \subset X$;

$\nu = 1$. $K_X C > 0$ for some curve $C \subset X$, but $K_X^2 = 0$;

$\nu = 2$. $K_X^2 > 0$.

You should think of ν as being defined by this division into cases. Formally, one can write down the definition of ν as a formula as follows:

$$\nu = \max\{k \mid K_X^k \overset{\text{num}}{\not\sim} 0\},$$

although this has no other meaning than the case division just given.

C.5.4. Theorem (Weak form of main theorem). *There exists a morphism $\varphi\colon X \to Y$ with $\dim Y = \nu(X)$ and $\varphi_* \mathcal{O}_X = \mathcal{O}_Y$ (that is, Y is normal and the generic fibre of φ is connected), such that*

$$K_X C = 0 \iff \varphi(C) = \text{pt}.$$

There are 3 separate statements here:

$\nu = 0$. *The statement is vacuous: φ maps X to a point.*

$\nu = 1$. *This is the most substantial case; here $\varphi\colon X \to Y$ is a fibration of X over a curve, with $K_X(\text{fibre}) = 0$. A general fibre E is a nonsingular curve with $E^2 = 0$, so that $K_X E = 0$ implies that $g(E) = 1$, and thus φ is an elliptic fibration.*

$\nu = 2$. *In this case the morphism $\varphi\colon X \to Y$ is birational, and contracts at most a finite number of connected configurations of -2-curves to Du Val singularities; also, $\mathcal{O}_Y(K_Y)$ is an ample line bundle, and $K_X = \varphi^* K_Y$.*

C.5.5. Definition of Kodaira dimension

The vector spaces $\Gamma(X, (\Omega_X^2)^m) = \Gamma(X, \mathcal{O}_X(mK_X))$ provide further invariants of X and K_X. Define the *plurigenera* P_m of X by $P_m(X) = \dim \Gamma(X, \mathcal{O}_X(mK_X))$. It is formal to see that the P_m must satisfy one of the following:

Case $\kappa = -\infty$. $P_m = 0$ for all $m > 0$;

Case $\kappa = 0$. $P_m \leq 1$ for all $m > 0$, and $P_m = 1$ for some m;

Case $\kappa = 1$ or 2. There exist constants $a, b > 0$ such that $am^\kappa < P_m < bm^\kappa$ for all sufficiently large m.

It can be shown that if $\kappa \geq 0$ then $\kappa = \dim \varphi_{mK_X}(X)$, where if $P_m \neq 0$ then $\varphi_{mK_X}\colon X \dashrightarrow \mathbb{P}^{P_m - 1}$ is the rational map defined by $\mathcal{O}_X(mK_X)$.

Note that in contrast to ν, the invariants P_m and κ are analytic invariants of X: P_m is the number of global holomorphic m-times canonical differentials on X.

C.5.6. Statement of $\kappa = \nu$

The statement $\kappa = \nu$ is true, and is one of the central structural results of the theory. It means the following:

Case $\nu = 0$. $K_X \overset{\text{num}}{\sim} 0 \implies \exists m$ such that $mK_X \overset{\text{lin}}{\sim} 0$.

Case $\nu = 1$. $K_X \overset{\text{num}}{\not\sim} 0$ but $K_X^2 = 0 \implies \exists m$ such that $P_m \geq 2$.

Case $\nu = 2$. $K_X^2 > 0 \implies P_m$ grows quadratically in m. (This is easy.)

There is a more precise description of the case $\nu = 0$.

C.5.7. Theorem. *If $\nu = 0$ then one of the following 4 cases holds:*

1. $K_X = 0$, $H^1(\mathcal{O}_X) = 0$, *that is, X is a K3 surface (see Chapter 3);*
2. $K_X = 0$, $h^1(\mathcal{O}_X) = 2$, *and X is an Abelian surface.*
3. $K_X \overset{\text{lin}}{\not\sim} 0$ *but* $2K_X \overset{\text{lin}}{\sim} 0$ *and* $H^1(\mathcal{O}_X) = 0$; *these conditions are the definition of an Enriques surface.*
4. $K_X \overset{\text{lin}}{\not\sim} 0$ *but* $mK_X \overset{\text{lin}}{\sim} 0$ *for* $m = 2, 3, 4$ *or* 6, *and* $h^1(\mathcal{O}_X) = 1$. *Then X is a bielliptic surface, that is, $X = E_1 \times E_2/(\mathbb{Z}/m)$ is a quotient of a product of two elliptic curves by the cyclic group $\mathbb{Z}/m$ acting by translations on one factor and a group automorphism in the other.*

The results sketched in this chapter are taken up again with complete proofs in Chapters D–E below.

CHAPTER 3
K3s

Let X be a nonsingular projective surface. The formal definition of K3 surface is $K_X = 0$ and $H^1(\mathcal{O}_X) = 0$. As I discuss presently, the historic geometric definition of Enriques and Fano is closely related: X is a K3 if $X \subset \mathbb{P}^g$ has a hyperplane section that is a canonical curve $C \subset \mathbb{P}^{g-1}$. Linear systems on K3s are closely related to the geometry of canonical curves. For example, the dichotomy between hyperelliptic curves and canonically embedded curves carries over to K3s. There is a general principle that a "sufficiently special" linear system g^r_d on a curve $C \subset X$ is cut out by something on the surface. Exceptional classes of K3s (monogonal, hyperelliptic, trigonal, tetragonal, etc.) have natural models in terms of divisors in scrolls. Much of the material is taken from [**Saint-Donat**].

K3s occupy a special place in the classification of surfaces: a surface X with $K_X = 0$ is either a K3, a special kind of Enriques surface (in characteristic 2 only), or an Abelian surface (that is, X is a commutative algebraic group, with trivial tangent bundle $T_X \cong \mathbb{C}^2$, and hence $h^1(\mathcal{O}_X) = 2$); thus you could make your own original definition of K3 by taking $K_X = 0$, together with any other conditions that exclude Abelian surfaces (such as $\pi_1(X) = 0$).

K3s also occupy a special place in the curriculum for anyone trying to master algebraic surfaces. Although considerably simpler than surfaces of general type, they are a marvellous testing ground for your understanding of linear systems, cohomology, vanishing theorems, the structure sheaf $\mathcal{O}_D$ of a nonreduced divisor D, the relation between geometry of linear systems and the algebra of graded rings, singularities, intersection numbers of curves and quadratic forms, Hodge structures, moduli, and many other things.

Any treatment of K3s would be incomplete without a discussion of their moduli, both algebraic and analytic, and their Hodge theory and period map. Although outside the scope of these notes, I mention these topics briefly.

3.1. Restriction and adjunction

There are two memorable exact sequences associated with a nonsingular n-fold X and a codimension 1 subvariety $Y \subset X$. First

$$0 \to \mathcal{I}_{X,Y} \to \mathcal{O}_X \to \mathcal{O}_Y \to 0, \tag{1}$$

where $\mathcal{I}_Y = \mathcal{I}_{X,Y} = \mathcal{O}_X(-Y)$ is the ideal sheaf defining Y; it is an invertible sheaf (line bundle) on X. The conormal sheaf $(\mathcal{N}_{X|Y})^\vee = \mathcal{I}_Y/\mathcal{I}_Y^2$ is the restriction of this invertible sheaf to Y, also written $\mathcal{I}_{Y|Y} = \mathcal{I}_Y \otimes_{\mathcal{O}_X} \mathcal{O}_Y$ or $\mathcal{O}_Y(-Y)$.

Tensoring (1) with $\mathcal{O}_X(Y)$ gives

$$0 \to \mathcal{O}_X \to \mathcal{O}_X(Y) \to \mathcal{O}_Y(Y) \to 0, \tag{2}$$

where $\mathcal{O}_Y(Y) = \mathcal{O}_X(Y)_{|Y} = N_{X|Y}$ is the normal bundle.

Second,

$$0 \to T_Y \to T_{X|Y} \to N_{X|Y} \to 0. \tag{3}$$

Taking determinants gives

$$\det T_{X|Y} = \det T_Y \otimes N_{X|Y}, \qquad \text{that is,} \qquad K_Y = (K_X + Y)_{|Y}. \tag{4}$$

(4) is called the *adjunction formula*. Note that the formula

$$2g(C) - 2 = (K_X + C)C$$

for the genus of a nonsingular curve $C \subset X$ on a surface is a particular case. Alternative treatments of adjunction are given in Ex. 3.25–26.

Example 1. $K_{\mathbb{P}^n} = \mathcal{O}(-n-1)$. Therefore if $X \subset \mathbb{P}^n$ is a hypersurface of degree d then $K_{\mathbb{P}^n} = \mathcal{O}_{\mathbb{P}^n}(-(n+1))$ and $K_X = \mathcal{O}_X(d-(n+1))$. If X is the complete intersection of two hypersurfaces of degrees d_1, d_2 then $K_X = \mathcal{O}_X(d_1+d_2-(n+1))$, etc.

Example 2. Let $X = X_4 \subset \mathbb{P}^3$ be a nonsingular quartic surface. Then $K_X = 0$. Also, since $H^i(\mathbb{P}^n, \mathcal{O}_{\mathbb{P}^n}(k)) = 0$ for all $0 < i < n$, it follows from the cohomology exact sequence

$$0 \to \mathcal{O}_{\mathbb{P}^3}(-4) \to \mathcal{O}_{\mathbb{P}^3} \to \mathcal{O}_X \to 0$$

that $H^1(\mathcal{O}_X) = 0$. Thus X is a K3. Similarly the complete intersections $X_{2,3} \subset \mathbb{P}^4$ and $X_{2,2,2} \subset \mathbb{P}^5$ are K3s. See Ex. 3.1–8 for more details of the proof and more examples.

Example 3. Let $X = X_{2,3} \subset \mathbb{P}^1 \times \mathbb{P}^2$ be a nonsingular divisor with the indicated bihomogeneity. Then it is easy to see that $K_{\mathbb{P}^1\times\mathbb{P}^2} = \mathcal{O}(-X)$ so that $K_X = 0$; and moreover, $H^1(\mathcal{O}_{\mathbb{P}^1\times\mathbb{P}^2}) = H^2(\mathcal{O}_{\mathbb{P}^1\times\mathbb{P}^2}(-X)) = 0$ so that X is again a K3.

3.2. RR for curves and canonical curves

Let C be a nonsingular projective curve. Recall that a divisor $D = \sum n_i P_i$ is a formal sum of points with integer coefficients. The degree of D is $\deg D = \sum n_i$. The RR space or associated vector space $\mathcal{L}(D) = H^0(\mathcal{O}_C(D))$ has dimension $\ell(D) = \dim \mathcal{L}(D) = h^0(\mathcal{O}_X(D))$. The RR theorem states that C has a genus $g \geq 0$ and a canonical divisor class K_C such that for every divisor D on C,

$$\ell(D) - \ell(K_C - D) = 1 - g + \deg D. \tag{1}$$

It follows easily from (1) that $\deg K_C = 2g-2$ and $\ell(K_C) = g$. Here $\ell(D) = h^0(\mathcal{O}_C(D))$ and $\ell(K_C - D) = h^0(\mathcal{O}_C(K_C - D)) = h^1(\mathcal{O}_C(D))$, so that (1) is a formula for $\chi(\mathcal{O}_C(D))$. In this form, the theorem is a fairly trivial consequence of the theorems of coherent cohomology (see Ex. B.23–24).

A divisor class D on C is *special* if $\ell(K_C - D) \neq 0$, or equivalently, $\ell(D) > 1 - g + \deg D$. Thus K_C is the biggest special divisor class on C: a divisor D is special if and only if (up to linear equivalence) $D \leq K_C$.

Hyperelliptic curves are counterexamples to many results in the theory of curves. A curve C is *hyperelliptic* if it is birational to $y^2 = f_{2g+2}(x)$, or, equivalently, has a g^1_2, a divisor class D with $\deg D = 2$, $\ell(D) = 2$. It is not hard to see that the canonical class K_C is then composed of the g^1_2, that is, $K_C = (g-1)D$ and $\mathcal{L}(K_C) = S^{g-1}\mathcal{L}(D)$ (polynomials of degree $g-1$); in other words, if x_1, x_2 is a basis of $\mathcal{L}(D)$ then $x_1^{g-1}, x_1^{g-2}x_2, \dots, x_2^{g-1}$ is a basis of K_C. The canonical map $\varphi_{K_C}\colon C \to \mathbb{P}^{g-1}$ of a hyperelliptic curve is obtained by composing the double cover $C \to \mathbb{P}^1$ defined by the g^1_2 with the $(g-1)$st Veronese embedding $\mathbb{P}^1 \hookrightarrow \mathbb{P}^{g-1}$. A curve of genus 2 is automatically hyperelliptic, since K_C is a g^1_2.

The following result is classical:

Theorem (Max Noether). *If C is nonhyperelliptic of genus ≥ 3 then $\varphi_{K_C}\colon C \to \mathbb{P}^{g-1}$ is an embedding such that K_C is the hyperplane section divisor. Moreover, $C \subset \mathbb{P}^{g-1}$ is projectively normal, that is, $H^0(\mathbb{P}^{g-1}, \mathcal{O}(k)) \twoheadrightarrow H^0(C, \mathcal{O}_C(kK_C))$ for all $k \geq 1$.*

The image curve $C = \varphi_{K_C}(C) \subset \mathbb{P}^{g-1}$ is called a *canonical curve.*

Idea of proof. See [**4 authors**] for the proof. The idea is that $\ell(K_C) = g$; and if $P_1, P_2 \in C$ are any two points then $\ell(K_C - P_1 - P_2) = g - 2$, for otherwise RR gives $\ell(P_1 + P_2) = 2$, so $P_1 + P_2$ is a g^1_2, and C is hyperelliptic. Therefore functions in $\ell(K_C)$ distinguish P_1 and P_2.

The final sentence of the theorem is the first step in the Petri analysis [**4 authors**]. If $P_1, \dots, P_g \in C$ are g "sufficiently general" points, I choose coordinates $x_1, \dots, x_g$ of $\mathbb{P}^{g-1}$ such that $P_i = (0, \dots, 1, \dots, 0)$. Then $x_1, x_2 \in \mathcal{L}(K_C - P_3 - \dots - P_g)$ span a free pencil. It then follows from the "free pencil trick" that

$$\left.\begin{array}{r} x_1^2, x_1x_2, x_1x_3, \dots, x_1x_g \\ x_2^2, x_2x_3, \dots, x_2x_g \\ x_3^2, \dots, x_g^2 \end{array}\right\} \in \mathcal{L}(2K_C)$$

are $3g-3$ linearly independent elements. Passing up from kK_C for $k \geq 3$ is similar (actually, a little easier). □

The canonical embedding $C \subset \mathbb{P}^{g-1}$ of a nonhyperelliptic curve allows the following restatement of RR as a result in linear projective geometry. For D an effective divisor on C, write $\langle D \rangle \subset \mathbb{P}^{g-1}$ for the smallest linear subspace of $\mathbb{P}^{g-1}$ containing D as a subscheme of C. If $D = P_1 + \dots + P_d$ consists of distinct points, this is just their ordinary linear span $\langle P_1 + \dots + P_d \rangle$. Then $|K_C - D|$ is the linear system cut out residually to D by hyperplanes of $\mathbb{P}^{g-1}$ through $\langle D \rangle$, and therefore $\ell(K_C - D) = g - 1 - \dim \langle D \rangle$.

Theorem (Geometric form of RR). *Let $C \subset \mathbb{P}^{g-1}$ be a canonical curve, and D an effective divisor on X. Then $\ell(D) = \deg D - \dim \langle D \rangle$, in other words, $\ell(D) - 1 = \dim |D| =$ the number of linear dependence relations between the points of D.*

3.3. The historic definition of K3

A variety $X \subset \mathbb{P}^n$ is *linearly normal* or *embedded by a complete linear system* if $H^0(\mathbb{P}^n, \mathcal{O}(1)) \to H^0(X, \mathcal{O}(1))$ is surjective; in other words, X is not the linear projection of a variety spanning a higher dimensional space $X \subset \mathbb{P}^{n+1}$.

Theorem. *Let $X \subset \mathbb{P}^n$ be a nonsingular surface. Then X is a K3 embedded by a complete linear system if and only if one (every) nonsingular hyperplane section is a canonical curve.*

Proof. First suppose that X is a K3, and let C be a nonsingular hyperplane section. Then the adjunction formula and the assumption $K_X = 0$ give $K_C = \mathcal{O}_C(C)$. Also $H^0(X, \mathcal{O}_X(C)) \to H^0(C, K_C)$ is surjective from the cohomology exact sequence of

$$0 \to \mathcal{O}_X \to \mathcal{O}_X(C) \to \mathcal{O}_C(C) \to 0$$

and the assumption $H^1(\mathcal{O}_X) = 0$. Therefore C is a canonical curve.

Conversely, suppose that C is a canonical curve. Then $n = g$, and by Theorem 3.2, $H^0(\mathbb{P}^{g-1}, \mathcal{O}(k)) \to H^0(C, \mathcal{O}_C(kK_C))$ is surjective for all $k \geq 0$. But since $C \subset X \subset \mathbb{P}^g$, it follows from the commutative diagram of restriction maps

$$\begin{array}{ccc} H^0(\mathbb{P}^g, \mathcal{O}(k)) & \to & H^0(X, \mathcal{O}_X(k)) \\ \downarrow & & \downarrow \\ H^0(\mathbb{P}^{g-1}, \mathcal{O}(k)) & \to & H^0(C, \mathcal{O}_C(kK_C)) \end{array}$$

that $H^0(X, \mathcal{O}_X(kC)) \to H^0(C, \mathcal{O}_C(kK_C))$ is surjective for all $k \geq 0$.

Consider the cohomology exact sequence

$$\begin{aligned} \cdots \to H^0(\mathcal{O}_X(kC)) \to H^0(\mathcal{O}_C(kK_C)) \to \\ H^1(\mathcal{O}_X((k-1)C)) \to H^1(\mathcal{O}_X(kC)) \to H^1(\mathcal{O}_C(kK_C)) \to \\ H^2(\mathcal{O}_X((k-1)C)) \to \cdots \end{aligned}$$

Now since $H^0(X, \mathcal{O}_X(kC)) \to H^0(C, \mathcal{O}_C(kK_C))$ is surjective, it follows that

$$H^1(\mathcal{O}_X) \hookrightarrow H^1(\mathcal{O}_X(C)) \hookrightarrow \cdots \hookrightarrow H^1(\mathcal{O}_X(kC))$$

for all k. But $H^1(\mathcal{O}_X(kC)) = 0$ for $k \gg 0$ by Serre vanishing. Therefore $H^1(\mathcal{O}_X) = 0$.

Also, when $k = 1$, I get that $0 \neq H^1(\mathcal{O}_C(K_C)) \hookrightarrow H^2(\mathcal{O}_X)$. Hence $H^2(\mathcal{O}_X) \neq 0$, so that by duality $H^0(\mathcal{O}_X(K_X)) \neq 0$. Thus K_X is linearly equivalent to an effective divisor $D \geq 0$. But the adjunction formula $K_C = (K_X + C)_{|C}$ together with the assumption $K_C = C_{|C}$ implies that $D \cap C = \emptyset$, so that $D = 0$. Thus $K_X = 0$. Q.E.D.

Linear systems on K3s are closely related to the geometry of canonical curves. If $C \subset X$ and $A = P_1 + \cdots + P_d$ is a g^r_d on C, with d fairly small, then by the geometric form of RR,

$$\mathbb{P}^g \supset \langle P_1, \ldots, P_d \rangle = \Pi = \mathbb{P}^{d-1-r}.$$

This property of $\{P_1, \ldots, P_d\}$ does not depend on C through Π, so that all curves C through $\{P_1, \ldots, P_d\}$ also have a g^r_d. Thus, very roughly, one expects that all curves in a given linear system on X have the same "very special" linear systems.

3.4. RR on a surface

If X is a surface and D a divisor then RR takes the form

$$\chi(\mathcal{O}_X(D)) = \chi(\mathcal{O}_X) + \tfrac{1}{2}D(D - K_X). \tag{1}$$

"Proof". The idea is to use induction on the components of D, as in the proof of RR for curves (see Ex. B.23). I can get from 0 to $D = \sum n_i C_i$ by successively adding or subtracting an irreducible curve C; and (1) holds for $D = 0$, so that (1) for all D will follow by induction if I can prove the formula

$$\begin{aligned}\chi(\mathcal{O}_X(D)) - \chi(\mathcal{O}_X(D-C)) &= \big(\chi(\mathcal{O}_X) + \tfrac{1}{2}D(D-K_X)\big)\\ &\quad - \big(\chi(\mathcal{O}_X) + \tfrac{1}{2}(D-C)(D-C-K_X)\big) \qquad (2)\\ &= -\tfrac{1}{2}(K_X + C)C + DC.\end{aligned}$$

Suppose for simplicity that C is a nonsingular curve. Then the cohomology long exact sequence of $0 \to \mathcal{O}_X(D-C) \to \mathcal{O}_X(D) \to \mathcal{O}_C(D) \to 0$ gives

$$\chi(\mathcal{O}_X(D)) - \chi(\mathcal{O}_X(D-C)) = \chi(\mathcal{O}_C(D)). \tag{3}$$

Now the genus of C is given by the adjunction formula: $2g(C) - 2 = (K_X + C)C$ and $\mathcal{O}_C(D)$ is a line bundle on C of degree DC, so that $\chi(\mathcal{O}_C(D))$ by RR on C:

$$\chi(\mathcal{O}_C(D)) = 1 - g(C) + \deg(\mathcal{O}_C(D)) = -\tfrac{1}{2}(K_X + C)C + DC.$$

It is not too difficult to get around the assumption that C is nonsingular. This "proves" (2) and hence (1).

Recall from A.9 that one way of defining intersections numbers D_1D_2 is based on knowing that $\chi(\mathcal{O}_X(D))$ is a quadratic function of D. If this is your definition, then, on the face of it, the proof just given is circular. Rather than as a logical proof, it is better to think of it as a compatibility between all the ingredients in the formula: intersection numbers, the genus of a curve, the degree of a divisor on a curve, etc. I suggest accepting RR as an axiom for the present until you have time to learn the general form of Hirzebruch RR or Grothendieck RR or RR via the Atiyah–Singer index formula all at one go.

3.5. Corollary (the Hodge algebraic index theorem). *If H is ample on X then $HD = 0$ implies $D^2 \le 0$; moreover, if $D^2 = 0$ then $D \overset{\text{num}}{\sim} 0$.* Here *numerical equivalence $D \overset{\text{num}}{\sim} 0$* means that $D\Gamma = 0$ for every curve $\Gamma \subset X$.

Another way of stating this, which is useful in calculations, is that if D_1, D_2 are divisors and $(\lambda D_1 + \mu D_2)^2 > 0$ for some $\lambda, \mu \in \mathbb{R}$ then the determinant

$$\det \begin{vmatrix} D_1^2 & D_1D_2 \\ D_1D_2 & D_2^2 \end{vmatrix} \le 0,$$

with equality if and only if some nonzero rational linear combination is numerically equivalent to zero, that is $\alpha D_1 + \beta D_2 \overset{\text{num}}{\sim} 0$.

I leave the proof as an exercise; see Ex. 3.9–1.9 and Corollary D.2.2 below.

3.6. The arithmetic genus of a curve

I use RR to extend the definition of genus to singular curves, and even arbitrary effective divisors. If $C \subset X$ is an irreducible curve, it has a nonsingular model $\nu\colon C^\nu \to C$ given by normalisation, and the genus of C^ν is traditionally called the *geometric genus* of C; a better term might be *birational genus*.

Definition. Let D be an effective divisor on a surface X. The *arithmetic genus* p_aD of D is defined by

$$2p_aD - 2 = (K_X + D)D. \tag{1}$$

If $D = C$ is a nonsingular curve then the definition is just the adjunction formula, so that $p_aC = g(C)$. (See the discussion in 4.10–11 for more information.)

Consider the quotient sheaf $\mathcal{O}_D = \mathcal{O}_X/\mathcal{O}_X(-D)$, where $\mathcal{O}_X(-D)$ is the ideal sheaf of regular functions vanishing along D. This is a coherent sheaf on X by construction, and can be viewed as the structure sheaf of the subscheme $D \subset X$; see 3.10 below for a more detailed discussion. Be that as it may, the cohomology long exact sequence of $0 \to \mathcal{O}_X(-D) \to \mathcal{O}_X \to \mathcal{O}_D \to 0$ gives

$$\chi(\mathcal{O}_D) = \chi(\mathcal{O}_X) - \chi(\mathcal{O}_X(-D)),$$

and plugging in RR gives

$$\chi(\mathcal{O}_D) = 1 - p_aD. \tag{2}$$

Proposition. *Let C be an irreducible curve, possibly singular, on a surface X, and $\nu\colon C^\nu \to C$ its normalisation. Then*

$$p_aC = g(C^\nu) + \sum_{P_i \in \operatorname{Sing} C} \delta(P_i),$$

where $\delta(P_i) > 0$ are numerical invariants of the singularities of C.

In particular, if $p_aC = 0$ then $C \cong \mathbb{P}^1$.

Proof. There is an exact sequence of sheaves on C

$$0 \to \mathcal{O}_C \to \nu_*(\mathcal{O}_{C^\nu}) \to \mathcal{N} \to 0, \tag{3}$$

where ν_* is the sheaf theoretic pushforward (see B.6), and $\mathcal{N}$ the cokernel. Now I claim that $\mathcal{N}$ consists of finite dimensional vector spaces $\mathcal{N}_{P_i}$ of dimension $\delta(P_i)$ supported at the singular points $P_1, \dots, P_k$ of C.

To see this, note that $\nu_*(\mathcal{O}_{C^\nu})$ is a coherent sheaf of $\mathcal{O}_C$-modules by finiteness of normalisation; hence so is $\mathcal{N}$. Moreover, outside the singular points, ν is an isomorphism and $\mathcal{O}_C = \nu_*(\mathcal{O}_{C^\nu})$, so that $\mathcal{N}$ is supported at $\{P_1, \dots, P_k\} \subset C$. If $n_i \subset \mathcal{O}_{X,P_i}$ is the ideal annihilating $\mathcal{N}_{P_i}$ then $V(n_i) = \{P_i\}$. By the Nullstellensatz, n_i contains a power $m_{P_i}^k$ of the maximal ideal at P_i, so that $\mathcal{N}_{P_i}$ is a finitely generated module over $\mathcal{O}_{X,P_i}/m_{P_i}^k$ and is finite dimensional.

Now the cohomology long exact sequence of (3) gives $\chi(\nu_*(\mathcal{O}_{C^\nu})) = \chi(\mathcal{O}_C) + \sum \delta(P_i)$, that is,

$$p_aC = g(C^\nu) + \sum \delta(P_i), \tag{4}$$

where $\delta(P_i) = \dim_k \mathcal{N}_{P_i}$. Here I have used that $H^i(C, \nu_*(\mathcal{O}_{C^\nu})) = H^i(C^\nu, \mathcal{O}_{C^\nu})$; for $i = 0$ this comes from the definition of pushforward ν_*, for $i = 1$ because ν has only zero dimensional fibres. If P_i is singular then $C^\nu \not\cong C$ at P_i, so that $\delta(P_i) > 0$.

Notice that (4) expresses p_aC as a sum of terms which are ≥ 0. If $p_aC = 0$ it follows that $\sum \delta(P_i) = 0$, so that there are no P_i, and $C = C^\nu$ has $g(C) = 0$, that is, $C \cong \mathbb{P}^1$. Q.E.D.

3.7. RR on a K3

If X is a K3 then $\chi(\mathcal{O}_X) = 2$ and $K_X = 0$, so that RR takes the simpler form

$$\chi(\mathcal{O}_X(D)) = h^0(D) - h^1(D) + h^0(-D) = 2 + \tfrac{1}{2}D^2.$$

Corollary. *(i) $D^2 \geq -2$ implies $H^0(D) \neq 0$ or $H^0(-D) \neq 0$.*

(ii) $D^2 \geq 0$ implies $D \overset{\text{lin}}{\sim} 0$, or $h^0(D) \geq 2$, or $h^0(-D) \geq 2$.

(iii) If D is an effective divisor on X with $h^0(D) = 1$ then $D'^2 \leq -2$ for every divisor D' with $0 < D' \leq D$, and in particular D is a sum of -2-curves with $D^2 \leq -2$.

If $C \subset X$ is an irreducible curve on a K3 then C^2 is even, and the adjunction formula gives $2p_aC - 2 = C^2 \geq -2$, and $C^2 = -2$ implies that $C \cong \mathbb{P}^1$ by Proposition 3.6. A curve $C \cong \mathbb{P}^1$ with $C^2 = -2$ is called a *-2-curve*; these curves are important throughout the classification of surfaces. Note that any curve on a K3 has C^2 even, and $C^2 < 0$ only for -2-curves.

The bunch of -2-curves occurring as components of D in (iii) is quite restricted, but it's not necessarily true that it has negative definite intersection matrix. See Ex. 3.12.

3.8. Easy properties of linear systems

If D is a divisor on a variety X with $\mathcal{L}(D) \neq 0$, the projective space $|D| = \mathbb{P}_*(\mathcal{L}(D))$ parametrises effective divisors $D' \geq 0$ linearly equivalent to D. This is called a *complete linear system.* Example: a pencil $\lambda F + \mu G$ or net $\lambda_1 F_1 + \lambda_2 F_2 + \lambda_3 F_3$ of plane curves of degree d is an example of a linear system in the plane; the complete linear system $|\mathcal{O}_{\mathbb{P}^2}(d)|$ is the system of all curves of degree d, of dimension $\binom{d+2}{2} - 1$. In general, $|D|$ may have a *base locus* $\operatorname{Bs}|D|$, that is, a subscheme Σ contained in every divisor $D' \in |D|$. It is traditional to treat the codimension 1 part and the codimension ≥ 2 part of $\operatorname{Bs}|D|$ separately. The *fixed part* of D is the biggest divisor F such that $F \leq D'$ for every $D' \in |D|$; or in other words, the gcd of all $D' \in |D|$.

On a general variety it may be hard to predict what $\operatorname{Bs}|D|$ looks like. However, on a K3 things are very nice. Recall that a divisor D is *nef* (numerically eventually free) if $D\Gamma \geq 0$ for every curve $\Gamma \subset X$.

Theorem. *(a) If X is a K3 and D any effective divisor on X then I can subtract of an effective sum of -2-curves $F = \sum n_i\Gamma_i$ to get $M = D - F$ such that M is effective and nef (possibly zero), $M^2 \geq D^2$ and $H^0(X, \mathcal{O}_X(M)) = H^0(X, \mathcal{O}_X(D))$.*

(b) If $D > 0$ is nef and $D^2 = 0$ then $D = aE$, where $|E|$ is a free pencil.

(c) If D is nef and $D^2 > 0$ (that is, D is nef and big*) then $H^1(D) = 0$, so that $h^0(D) = 2 + \frac{1}{2}D^2$.*

(d) The first dichotomy: *if D is nef and big then either $|D|$ has no fixed part, or $D = aE + \Gamma$, where $|E|$ is a free pencil and Γ an irreducible -2-curve such that $E\Gamma = 1$. In this case D is* monogonal.

Confusion. I clear up some possible sources of confusion. The last condition in (a) means that $F \subset \operatorname{Bs}|D|$, in particular $H^0(F) = 0$, but F does not have to be the whole base locus: an example is the monogonal linear system in (d). The decomposition $D = M + F$ is not the same as the Zariski decomposition, which looks like $D = P + N$ as $\mathbb{Q}$-divisors, with P nef and N orthogonal to P; here $F \geq N$, and $F > N$ sometimes happens (see Ex. 3.13).

Proof. (a) If D is nef, I'm home setting $F = 0$. An easy point that is crucial in all work on surfaces: if D is effective and $D\Gamma < 0$ then Γ is a component of D and $\Gamma^2 < 0$. Because, I can certainly write $D = a\Gamma + D'$ with $a \geq 0$, $D' > 0$ and $\Gamma \not\subset D'$; then $D\Gamma = a\Gamma^2 + D'\Gamma$. Since $D'\Gamma \geq 0$ and $a \geq 0$, the only way this can be negative is if $\Gamma^2 < 0$ and $a > 0$.

If D is not nef then there is a Γ with $D\Gamma < 0$, necessarily in the fixed part of $|D|$. Set $D_1 = D - \Gamma$, so that $D_1^2 = D^2 - 2D\Gamma + \Gamma^2$. Now $\Gamma^2 = -2$, and $D\Gamma < 0$, and therefore $D_1^2 \geq D^2$. Also, obviously $H^0(D_1) = H^0(D)$. Continuing by induction proves (a).

(b) $D^2 = 0$ gives $h^0(D) \geq 2$, so that D moves in a nontrivial linear system $|D|$. Write $|D| = |M| + F$ with F the fixed part; then M is mobile, hence also nef. As in (a), this is an easy point that occurs again and again: for any curve $\Gamma \subset X$, since M is mobile, there is an effective divisor in $|M|$ not containing Γ, so $M\Gamma \geq 0$.

Thus $0 = D^2 = DM + DF$, so that $DM = DF = 0$; next, $M^2 + MF = 0$ so that $M^2 = MF = F^2 = 0$. Now if $F \neq 0$, I get $h^0(F) \geq 2$ from RR, contradicting F fixed. Therefore $|D| = |M|$ has no fixed part. But $D^2 = 0$ implies that D is free; because $D_1, D_2 \in |D|$ with no common components have $D_1 D_2 = 0$, therefore $D_1 \cap D_2 = \emptyset$.

I claim that every element of $|D|$ is made up of components of fibres of a morphism $f\colon X \to \mathbb{P}^1$ with connected general fibre. Indeed, $D^2 = 0$ implies that the morphism $\varphi_D\colon X \to \mathbb{P}^{h^0(D)-1} = \mathbb{P}^*(H^0(D))$ defined by $|D|$ has image a curve C. If $X \to \widetilde{C} \to C$ is the Stein factorisation of φ_D then a general fibre E of $X \to \widetilde{C}$ is connected. In fact necessarily $\widetilde{C}$ is isomorphic to $\mathbb{P}^1$, since $h^1(\mathcal{O}_X) = 0$, and $\widetilde{C} = C$ since $\widetilde{C} \to \mathbb{P}^{h^0(D)-1}$ is defined by a complete linear system. Now a general fibre of φ is an irreducible curve E (by the first Bertini theorem, see [**Sh**], Chapter II, Theorem 6.1). Also E is reduced, since if $E = nE'$ then also ${E'}^2 = 0$ and E' also moves in a linear system.

Then it is easy to see that $D \overset{\text{lin}}{\sim} aE$, where $h^0(D) = a+1$: because some element in $|D|$ certainly intersects E, and therefore contains it, and then I can apply the argument to $D - E$, $D - 2E$, etc. This proves (b).

(c) The statement that D nef and big implies $H^1(D) = 0$ is proved in 3.11–12 below, and I assume it for the moment.

(d) Write $|D| = |M| + F$, with $|M|$ mobile and F fixed; note that M is also effective and nef. There are two cases:

Case $M^2 > 0$. I prove that then $F = 0$. Indeed, M is also nef and big, so that $H^1(M) = 0$, and so $h^0(M) = h^0(D)$ together with RR implies $M^2 = D^2$. Now

$$D^2 = D(M + F) \geq DM = (M + F)M \geq M^2,$$

and equality implies that $DF = MF = 0$; hence also $F^2 = 0$. If $F \neq 0$ then $H^0(F) \geq 2$, contradicting that F is fixed.

Case $M^2 = 0$. Then $M = aE$ by (b), so that $h^0(M) = a+1$, and RR gives $h^0(D) \geq 2 + \frac{1}{2}D^2$. Thus

$$\begin{aligned} a+1 = h^0(M) = h^0(D) &\geq 2 + \tfrac{1}{2}D^2 \\ &= 2 + \tfrac{1}{2}(aE+F)^2 = 2 + aEF + \tfrac{1}{2}F^2 \\ &= 2 + \tfrac{1}{2}(aEF + F^2) + \tfrac{1}{2}aEF \geq 2 + \tfrac{1}{2}aEF. \end{aligned}$$

Now $EF \geq 2$ contradicts this inequality. Clearly $EF = 0$ is impossible, since then $DE = 0$, which would contradict the index theorem. Therefore $EF = 1$ and $F^2 \leq -2$.

There is a unique irreducible component Γ of F with $E\Gamma = 1$, and $F = \Gamma + F'$, say, with $E\Gamma = 0$. Then $D = aE + \Gamma + F'$; but it's easy to check that $D' = aE + \Gamma$ is also nef, and $D'^2 = 2a - 2 = D^2$. It follows as above that $(F')^2 = 0$, so that $F' = 0$. Q.E.D.

3.9. Numerically connected divisors on surfaces

Let X be a nonsingular projective surface. For an effective divisor D, the quotient sheaf $\mathcal{O}_D = \mathcal{O}_X/\mathcal{O}_X(-D)$ is a sheaf of rings, which can be viewed as the structure sheaf of a subscheme $D \subset X$. For a nonreduced divisor, $\mathcal{O}_D$ involves the nilpotents of scheme theory, but in a fairly mild way. The key to controlling the global sections $H^0(\mathcal{O}_D)$, and to the proof of Theorem 3.8, (c), is the notion of numerically connected divisor due to Franchetta and C. P. Ramanujam.

Definition. Let $k \in \mathbb{Z}$ (usually with $k \geq 0$). An effective divisor D is *numerically k-connected* if $D_1D_2 \geq k$ for every effective decomposition $D = D_1 + D_2$ with $D_1, D_2 > 0$. (Of course, an irreducible curve is k-connected for all k, since the condition is vacuous.)

Example. If E is irreducible with $E^2 = 0$, then $2E$ is numerically 0-connected but numerically 1-disconnected. If $|E|$ moves in a free pencil then it is the fibre of a morphism $f\colon X \to B$ to a base curve B, say $E = f^{-1}P$. Then the structure sheaf of the subscheme $aE \subset X$ is a module over $\mathcal{O}_{B,P}/m_P^a \cong k[\varepsilon]/\varepsilon^a$, and $H^0(\mathcal{O}_{aE}) = k[\varepsilon]/\varepsilon^a$; you can think of the nilpotent sections as (dual to) a normal vector field to X along E that goes out $a-1$ infinitesimal steps. See Ex. 3.20.

Example. If $\sigma\colon Y \to X$ is the blowup of a nonsingular point $P \in X$ and E the exceptional curve then E is a -1-curve, that is $E \cong \mathbb{P}^1$ and $E^2 = -1$. Obviously $2E$ is not 0-connected. It is easy to see that $H^0(\mathcal{O}_{2E}) = \mathcal{O}_{X,P}/m_P^2 \cong k[x,y]/(x,y)^2$. See Ex. 3.21.

3.10. Lemma (*Dévissage* of $\mathcal{O}_D$). *If $D = D' + D''$ with $D', D'' > 0$ then there is an exact sequence*

$$0 \to \mathcal{O}_{D''}(-D') \to \mathcal{O}_D \to \mathcal{O}_{D'} \to 0;$$

here $\mathcal{O}_{D''}(-D') = \mathcal{O}_X(-D') \otimes \mathcal{O}_{D''}$.

Proof. The exact sequence defining $\mathcal{O}_{D''}$ is

$$0 \to \mathcal{O}_X(-D'') \to \mathcal{O}_X \to \mathcal{O}_{D''} \to 0;$$

Tensoring by $\mathcal{O}_X(-D')$ gives

$$0 \to \mathcal{O}_X(-D) \to \mathcal{O}_X(-D') \to \mathcal{O}_{D''}(-D') \to 0.$$

The ideal sheaves satisfy $\mathcal{O}_X(-D) \subset \mathcal{O}_X(-D')$, so that $\mathcal{O}_D \to \mathcal{O}_{D'}$ is surjective, and the kernel can be calculated by the snake lemma applied to the commutative diagram

$$\begin{array}{ccccccccc} 0 & \to & \mathcal{O}_X(-D) & \to & \mathcal{O}_X & \to & \mathcal{O}_D & \to & 0 \\ & & \downarrow & & \| & & \downarrow & & \\ 0 & \to & \mathcal{O}_X(-D') & \to & \mathcal{O}_X & \to & \mathcal{O}_{D'} & \to & 0 \\ & & \downarrow & & & & & & \\ & & \mathcal{O}_{D''}(-D') & & & & & & \\ & & \downarrow & & & & & & \\ & & 0. & & & & & & \end{array}$$

Q.E.D.

3.11. Lemma. *Let X be a nonsingular projective surface and D an effective divisor.*

(i) D nef and $D^2 > 0$ imply that D is numerically 1*-connected.*

(ii) D numerically 1*-connected implies that* $H^0(\mathcal{O}_D) =$ constants.

(iii) Suppose that D is numerically 1*-connected, and let $\mathcal{L}$ be a line bundle on D with $\mathcal{L}\Gamma_i \le 0$ for every component Γ_i of D. Then*

$$H^0(\mathcal{L}) \neq 0 \implies \mathcal{L} \cong \mathcal{O}_D.$$

Proof. (i) Let $D = D_1 + D_2$ with $D_1, D_2 \geq 0$. Then D nef gives

$$D_1^2 + D_1D_2 = DD_1 \geq 0$$
$$D_1D_2 + D_2^2 = DD_2 \geq 0$$

Now if $D_1D_2 \leq 0$ then $D_1^2D_2^2 \geq (D_1D_2)^2 \geq 0$, which contradicts the Hodge algebraic index theorem (Corollary 3.5) unless D_1 or $D_2 = 0$.

(ii) $H^0(\mathcal{O}_D)$ is a finite dimensional algebra over k, so that if $h^0(\mathcal{O}_D) \neq 1$ then it contains either an idempotent $e \neq 0, 1$, or a nilpotent element $\varepsilon \neq 0$ with $\varepsilon^2 = 0$. If $e \neq 0, 1$ is an idempotent then $e \cdot \mathcal{O}_D$ and $(e-1) \cdot \mathcal{O}_D$ are ideals defining disjoint subschemes $D_1, D_2 \subset D$, so that D is disconnected.

If $0 \neq \varepsilon \in H^0(\mathcal{O}_D)$ is a nilpotent element and $D_1 \subset D$ is the greatest divisor on which it vanishes then the ideal sheaf $\mathcal{I}_{D_1} \subset \mathcal{O}_D$ is generated by ε outside a set of codimension ≥ 1. By Lemma 3.10, this kernel is $\mathcal{O}_{D_2}(-D_1)$; thus $\mathcal{O}_{D_2}(-D_1)$ is generated by ε at the generic point of each component, and it follows that $D_1\Gamma \leq 0$ for every component $\Gamma \subset D_2$. In particular $D_1D_2 \leq 0$, contradicting 1-connected.

I give a second proof which is more elementary: suppose that $D_1 \subset D$ is a divisor for which $h^0(\mathcal{O}_{D_1}) = 1$, for example an irreducible component of D. By the 1-connectedness assumption, if $D_1 < D$ then $D_1(D - D_1) \geq 1$, so there is a component Γ of $D - D_1$ such that $D_1\Gamma \geq 1$. Then I claim that also $h^0(\mathcal{O}_{D_1+\Gamma}) = 1$. Indeed, the sequence

$$H^0(\mathcal{O}_\Gamma(-D_1)) \to H^0(\mathcal{O}_{D_1+\Gamma}) \to H^0(\mathcal{O}_{D_1})$$

is exact, the first term is 0, and the third term 1-dimensional. Then $h^0(\mathcal{O}_D) = 1$ follows by induction.

(iii) is an exercise (see Ex. 3.27). Q.E.D.

3.12. Ramanujam vanishing

I start by completing the proof of Theorem 3.8, (c) on K3s. $H^0(D) \neq 0$, so that I can assume that D is effective. By assumption it is nef and big, so that $H^0(\mathcal{O}_D) =$ constants by Lemma 3.11. Now in the exact sequence

$$0 = H^0(\mathcal{O}_X(-D)) \to H^0(\mathcal{O}_X) \to H^0(\mathcal{O}_D) \to \\ \to H^1(\mathcal{O}_X(-D)) \to H^1(\mathcal{O}_X),$$

$H^1(\mathcal{O}_X) = 0$ by assumption (X is a K3), and $H^0(\mathcal{O}_X)$ also equals the constant functions, and hence maps onto $H^0(\mathcal{O}_D)$. Therefore $H^1(-D) = 0$, which is the dual of $H^1(D)$. Q.E.D.

On a surface X with $H^1(\mathcal{O}_X) = 0$, the same argument proves the vanishing $H^1(\mathcal{O}_X(-D)) = 0$ if D is effective and 1-connected, in particular if D is effective, nef and big. Dually, $H^1(\mathcal{O}_X(K_X + D)) = 0$. This is a weak form of Kodaira vanishing.

Discussion. Kodaira vanishing states that on a nonsingular n-fold over a field of characteristic zero,

$$H \text{ ample} \quad \Longrightarrow \quad H^i(\mathcal{O}_X(K_X + H)) = 0 \quad \text{for all } i > 0,$$

or dually, $H^j(\mathcal{O}_X(-H)) = 0$ for all $j < n$. Although the statement is purely in terms of coherent cohomology, the result is deeper than algebraic geometry in characteristic zero. There is no proof purely within coherent cohomology, and the result is false in characteristic p, already for surfaces. The two known proofs in all dimensions are Kodaira's (representing a cohomology class by a harmonic form, and integrating to give something which must be zero by Stokes' theorem, and at the same time strictly positive if the cohomology class is nonzero), and Deligne and Illusie's (by reducing modulo a sufficiently large prime p, and analysing the splitting of the characteristic p de Rham complex by Frobenius and Cartier operators). Generalisations of Kodaira's proof in a number of directions have been given, and this is a key ingredient in Mori theory and higher dimensional classification; compare the discussion in D.3 below.

A modern view of what vanishing *really is* has been given by Kollár: vanishing comes about when a coherent cohomology group admits a topological interpretation. Thus the argument given above works because the coherent cohomology group $H^0(\mathcal{O}_D)$ has an interpretation in terms of connectedness of D.

There are two different proofs of vanishing for algebraic surfaces in characteristic zero (without the assumption $H^1(\mathcal{O}_X) = 0$): the Bogomolov–Mumford proof based on rank 2 vector bundles, their Chern numbers and the inequality $c_1^2 < 4c_2$, and instability,[1] and C. P. Ramanujam's, which I now sketch.

3.13. Theorem. *If D is effective, nef and big then $H^1(\mathcal{O}_X(-D)) = 0$.*

---

[1]This proof uses essentially only the leading term in RR, and an argument on inseparable sections of scrolls. I hope to make this into a later chapter in these notes.

Sketch proof. We have already seen that D nef and big implies D 1-connected, and $H^0(\mathcal{O}_D) =$ constants. Then cohomology gives an exact sequence

$$0 \to H^1(\mathcal{O}_X(-D)) \to H^1(\mathcal{O}_X) \xrightarrow{\alpha(D)} H^1(\mathcal{O}_D).$$

Now C. P. Ramanujam proves that D effective and big also implies the injectivity $\alpha(D)\colon H^1(\mathcal{O}_X) \hookrightarrow H^1(\mathcal{O}_D)$.

Key lemma. *For any effective divisor D, write $\alpha(D)\colon H^1(\mathcal{O}_X) \to H^1(\mathcal{O}_D)$. Then, in characteristic zero,* $\ker\alpha(D)$ *depends only on the reduced divisor D_{red}.*

Corollary. *If D is effective and big then $\alpha(D)$ is injective.*

Proof of Corollary, Step 1. The condition D *big* means that $|nD|$ defines a birational map $\varphi_{|nD|}\colon X \dashrightarrow X' \subset \mathbb{P}^N$ for some $n > 0$; this is also popularly expressed in terms of the *Iitaka dimension* of D as $\kappa(X, D) = 2$. Now if $|nD|$ is a big linear system, I can write $nD = D' + F$ (equality of divisors) with $D', F \geq 0$, and $|D'|$ a big linear system without fixed part. The map $\alpha(D')$ factors as

$$H^1(\mathcal{O}_X) \to H^1(\mathcal{O}_{nD}) \to H^1(\mathcal{O}_{D'}),$$

so that obviously $\alpha(D')$ injective implies $\alpha(nD)$ injective, and by the Key Lemma, $\alpha(nD)$ injective implies $\alpha(D)$ injective. Thus it is enough to prove the corollary for D'. Hence from now on, replacing D by D', I assume D is nef and big.

Step 2. It follows from big that for any ample H, there exists an $m \gg 0$ such that $h^0(mD - H) \neq 0$, so that $mD \overset{\text{lin}}{\sim} H + E$ with E effective; indeed, just take a form in $\mathbb{P}^N$ vanishing on $\varphi(H)$. Taking a still larger multiple gives $lmD \overset{\text{lin}}{\sim} lH + lE = H' + lE$, where $H' \overset{\text{lin}}{\sim} lH$ is a very ample divisor. Then $H' + (lE)_{\mathrm{red}} \overset{\text{lin}}{\sim} lH + E_{\mathrm{red}}$, which is as ample as I like, so that, by Serre vanishing, $H^1(\mathcal{O}_X(-H' - E_{\mathrm{red}})) = 0$.

Step 3. Therefore, in particular, $\alpha(H' + E_{\mathrm{red}})$ is injective, so by using the Key Lemma, also $\ker\alpha(H' + lE) = 0$.

Step 4. Now $H' + lE \overset{\text{lin}}{\sim} lmD$ is nef and big, so that by Lemma 3.11, $h^0(\mathcal{O}_{H'+lE}) = 1$, and together with $\ker\alpha(H' + lE) = 0$ this implies that $H^1(\mathcal{O}_X(-H' - lE)) = 0$. Now $lmD \overset{\text{lin}}{\sim} H' + lE$ gives $\mathcal{O}_X(-lmD) \cong \mathcal{O}_X(-H' - lE)$, so also $H^1(\mathcal{O}_X(-lmD)) = 0$.

Step 5. Therefore $\ker\alpha(mD) = 0$. Hence by using the lemma again, since D and mD have the same reduced divisor, $\ker\alpha(D) = 0$.

3.14. Proof of Key Lemma 3.13

This is where we could hope to see Kollár's principle in action: we're talking about the map $H^1(\mathcal{O}_X) \to H^1(\mathcal{O}_D)$. If the group $H^1(\mathcal{O}_D)$ appearing here was topological in nature (say, homotopy invariant), then the group for D and for D_{red} should be the same, since they are the same topological space. Unfortunately, this is definitely not the case.

C. P. Ramanujam's idea is to view $H^1(\mathcal{O}_X)$ and $H^1(\mathcal{O}_D)$ as the Zariski tangent spaces to the Picard schemes $\operatorname{Pic}^0 X$ and $\operatorname{Pic}^0 D$. Assume we are over $\mathbb{C}$ and give X the complex topology. Then the Picard group $\operatorname{Pic} X = H^1(\mathcal{O}_X^*)$ fits into the exact sequence of sheaf cohomology

$$\begin{aligned} 0 \to H^1(X,\mathbb{Z}) \to H^1(X,\mathcal{O}_X) \to H^1(\mathcal{O}_X^*) \to \\ \to H^2(X,\mathbb{Z}) \end{aligned}$$

Thus $\operatorname{Pic}^0 X = \ker\{H^1(\mathcal{O}_X^*) \to H^2(X,\mathbb{Z})\}$ is the moduli space of topologically trivial line bundles. By Hodge theory, $H^1(X,\mathbb{Z}) \subset H^1(X,\mathcal{O}_X)$ is a full lattice (that is, $\mathbb{Z}^{2q} \subset \mathbb{C}^q$ as a discrete cocompact subgroup), so that the quotient $\operatorname{Pic}^0 X = H^1(X,\mathcal{O}_X)/H^1(X,\mathbb{Z})$ is naturally an Abelian variety.

Now $\operatorname{Pic}^0 D =$ the identity component of $H^1(\mathcal{O}_D^*)$ fits into a similar exact sequence, and can be made into a Lie group $\operatorname{Pic}^0 D = H^1(D,\mathcal{O}_D)/H^1(D,\mathbb{Z})$ in the same way. However, it can be shown that $H^1(D,\mathbb{Z}) \subset H^1(D,\mathcal{O}_X)$ is a full lattice if and only if D is a nonsingular curve. Indeed, if $\nu\colon C \to D_{\text{red}} \subset D$ is the resolution of singularities of the reduced subscheme of D, then there is a natural homomorphism $\operatorname{Pic}^0 D \to \operatorname{Pic}^0 C$ (the product of the Jacobians of the components of C), and the kernel comes from the nonreduced structure of D or the singularities of D_{red}. The nonreduced structure contributes only additive subgroups (a direct sum of copies of $\mathbb{C}^+$), and the singularities of D_{red} contribute either additive or multiplicative groups (a direct sum of copies of $\mathbb{C}^+$ and $\mathbb{C}^*$).

Now a morphism from an Abelian variety A to an additive or multiplicative group G is necessarily zero, for example because A is compact and G is affine. Therefore, there is no nonconstant map from $\operatorname{Pic}^0 X$ to $\ker\{\operatorname{Pic}^0 D \to \operatorname{Pic}^0 C\}$, and hence

$$\ker\{\operatorname{Pic}^0 X \to \operatorname{Pic}^0 D\} = \ker\{\operatorname{Pic}^0 X \to \operatorname{Pic}^0 D_{\text{red}}\} = \ker\{\operatorname{Pic}^0 X \to \operatorname{Pic}^0 C\}.$$

Since $\alpha(D)$ and $\alpha(D_{\text{red}})$ are the derivatives at 0 of these maps, this concludes the sketch proof of the Key Lemma. Q.E.D.

3.15. Final remarks on K3s

1. The 2nd dichotomy

Saint-Donat's theory of linear systems on K3s continues with a criterion for $\varphi_{|D|}$ to define an embedding modulo -2-curves: if D is nef and big, and not monogonal, then $|D|$ is free, and $\varphi_{|D|}$ is either 2-to-1 to its image (then D is *hyperelliptic*), or $\varphi_{|D|}$ is birational to a normal surface $\overline{X}$, and is an embedding, except that it contracts -2-curves Γ with $D\Gamma = 0$ to Du Val singularities. In the first case $X \to D$ is a double cover of $\mathbb{P}^2$, or the Veronese surface, or a rational normal scroll. The hyperelliptic case is characterised by the fact that X has $\varphi_{|D|}$ an elliptic pencil $|E|$ with $DE = 2$, or (in the Veronese case) a free linear system $|B|$ with $B^2 = 2$ and $D = 2B$.

2. The 3rd dichotomy

If D is nef and big, and $\varphi_D\colon X \to \overline{X} \subset \mathbb{P}^g$ is birational, then either $\overline{X}$ is contained in a 3-fold W contained in the intersection of all quadrics through $\overline{X}$, or $\overline{X}$ is an ideal theoretic intersection of quadrics. In the first case $X \to D$ is contained in the 3-fold W, which is either $\mathbb{P}^3$ or $Q \subset \mathbb{P}^4$ or the cone over the Veronese surface, or a rational normal scroll. In the final case D is *trigonal.* The trigonal case is characterised by the fact that X has $\varphi_{|D|}$ an elliptic pencil $|E|$ with $DE = 3$, or a free linear system $|B|$ with $B^2 = 2$ and $DB = 5$.

These dichotomies certainly continue for a while. In general, for the reason mentioned briefly at the end of 3.3, if one curve $C \in |D|$ on X has a very special linear system, then one expects all the other curves $C \in |D|$ to have a closely related linear system. For example, if C has a g^1_d and d is small compared to D^2 then C is cut out on X by a linear system. Lazarsfeld has shown that if C has a g^r_d which is special in the sense of Brill–Noether theory then $\operatorname{Pic} X$ must be strictly bigger than $\mathbb{Z} \cdot D$. Green and Lazarsfeld have proved that all $C \in |D|$ have the same Clifford index. It seems to be known that with a single well-known exception involving g^1_4s and g^2_6s, all $C \in |D|$ have the same gonality. I believe there are open research problems in this area.

3. Moduli and periods

If you fix a primitive sublattice L of rank $\rho \leq 20$ and signature $(1, \rho - 1)$ of the K3 lattice $H^2(X, \mathbb{Z}) \cong 2\begin{pmatrix} 0 & 1 \\ 1 & 0 \end{pmatrix} \oplus 3E_8$ then K3s having $\operatorname{Pic} X \supset L$ form a nonempty moduli space of dimension $20 - \rho$. In particular, there are countably many moduli spaces of algebraic K3s, each depending on 19 moduli, with (generically) $\operatorname{Pic} X = \mathbb{Z} \cdot D$ with $D^2 = 2g - 2$.

Analytic K3s depend on a single irreducible 20-dimensional moduli space. The Teichmüller space (parametrising K3s plus a basis of $H^2(X, \mathbb{Z})$) is just one half of a 20-dimensional quadric. To study K3s from this point of view, the essential result is the Torelli theorem, which says that a polarised K3 surface is uniquely determined by its Hodge structure. Every complex K3 has a Kähler metric.

A more leisurely discussion of these topics can be found in Beauville [**B2**]–[**B3**] or [**3 authors**], Chapter VIII.

Exercises to Chapter 3

1. Prove that $H^1(\mathbb{P}^1, \mathcal{O}(k)) = 0$ for all $k \geq -1$. [Hint: It's true if $k \gg 0$ by Serre vanishing; use the cohomology of $0 \to \mathcal{O}(k-1) \to \mathcal{O}(k) \to k_P \to 0$ for $P \in \mathbb{P}^1$, and induction.]

2. Prove that

$$\begin{aligned} H^0(\mathbb{P}^n, \mathcal{O}(k)) &= 0 && \text{for all } k < 0; \\ H^n(\mathbb{P}^n, \mathcal{O}(k)) &= 0 && \text{for all } k > -n-1; \\ H^i(\mathbb{P}^n, \mathcal{O}(k)) &= 0 && \text{for all } 0 < i < n \text{ and all } k \in \mathbb{Z}. \end{aligned}$$

[Hint: As in the preceding exercise, argue by induction on n using the cohomology of $0 \to \mathcal{O}_{\mathbb{P}^n}(k-1) \to \mathcal{O}_{\mathbb{P}^n}(k) \to \mathcal{O}_{\mathbb{P}^{n-1}}(k) \to 0$ for $\mathbb{P}^{n-1} \subset \mathbb{P}^n$.]

3. If $X = X_d \subset \mathbb{P}^n$ is a nonsingular hypersurface, prove that

$$\begin{aligned} H^0(X, \mathcal{O}(k)) &= 0 && \text{for all } k < 0; \\ H^{n-1}(X, \mathcal{O}(k)) &= 0 && \text{for all } k > d-n-1; \\ H^i(X, \mathcal{O}(k)) &= 0 && \text{for all } 0 < i < n-1 \text{ and all } k \in \mathbb{Z}. \end{aligned}$$

[Hint: Use the previous exercises, and argue on $0 \to \mathcal{O}_{\mathbb{P}^n}(k-d) \to \mathcal{O}_{\mathbb{P}^n}(k) \to \mathcal{O}_X(k) \to 0$.]

4. Let X be nonsingular and $Y \subset X$ a nonsingular subvariety of codimension c. Define the normal sheaf $N_{X|Y}$ as the dual of $\mathcal{I}_Y/\mathcal{I}_Y^2$ and prove that it is locally free of rank c. Prove that it fits into an exact sequence with T_Y and the restriction $T_{X|Y}$. State and prove the adjunction formula giving K_Y in terms of K_X.

5. Prove that every nonsingular complete intersection $X_4 \subset \mathbb{P}^3$, $X_{2,3} \subset \mathbb{P}^4$ or $X_{2,2,2} \subset \mathbb{P}^5$ is a K3.

6. Show that the list of K3 complete intersections in $\mathbb{P}^n$ (of hypersurfaces of degree ≥ 2) of the preceding question is complete.

7. Find all values of a_1, a_2, a_3 and e for which the general hypersurface $X_{3,e} \subset \mathbb{F}(a_1, a_2, a_3)$ is a nonsingular K3. [Hint: The canonical class of $\mathbb{F}$ is given in Proposition A.9. The criterion for $X \in |eL+3M|$ to be a nonsingular cubic curve generically over $\mathbb{P}^1$ is the Newton polygon argument of Worked Example 2.11. Note that the result of 2.11 does not give nonsingularity in general: when $|eL+3M|$ has a base locus, you have to check for isolated singularities along the base locus.]

8. Find all values of $a_1, \dots, a_4$ and d_1, d_2 for which the general codimension 2 complete intersection $Q_{d_1,2} \cap Q'_{d_2,2} \subset \mathbb{F}(a_1, \dots, a_4)$ is a nonsingular K3.

9. Prove the Hodge algebraic index theorem, Corollary 3.5. [Hint: If $D^2 > 0$, deduce from RR that either nD or $-nD$ is equivalent to an effective divisor for $n \gg 0$ (if $K_X - nD$ grows quadratically, you have to invent a restriction argument to show that also $-nD$ grows quadratically). Prove then that $HD \neq 0$. Now for the case of equality, if $HD = 0$, and $D^2 = 0$ but $D\Gamma \neq 0$ for some curve Γ, find a linear combination $D' = \alpha D + \beta A + \gamma H$ with $(D')^2 > 0$ but $DH = 0$.]

10. Prove the $\det \begin{vmatrix} D_1^2 & D_1D_2 \\ D_1D_2 & D_2^2 \end{vmatrix} \leq 0$ form of the Hodge algebraic index theorem given in 3.5.

11. Show that C^2 is even and ≥ -2 for every irreducible curve C on a K3, in particular, X does not contain any -1-curves. Prove Corollary 3.7 (the easy corollaries of RR on a K3).

12. Let $C \subset \Pi = \mathbb{P}^2 \subset \mathbb{P}^3$ be a nonsingular plane conic, and $X = X_4 \subset \mathbb{P}^3$ a quartic surface tangent to Π along C, but otherwise general. Prove that X has six nodes at points of C. [Hint: Take $C : (x_0 = q(x_1, x_2, x_3) = 0) \subset \mathbb{P}^3$ and $X : q^2 + x_0 f_3$.]

Let $f\colon Y \to X$ be the blowup of the 6 nodes, and $H = f^*\mathcal{O}_X(1)$, so that $\varphi_{|H|}\colon Y \to \mathbb{P}^3$ defines f. Show that $|H|$ contains a divisor of the form $2C + \sum_{i=1}^{6} E_i$, where C and the E_i are -2-curves, and $h^0(X, \mathcal{O}_X(C + \sum_{i=1}^{6} E_i)) = 1$. Show that the intersection matrix of C and the E_i is not negative definite.

13. Let D be an effective divisor on a surface X. The Zariski decomposition of D is an expression $D = P + N$ (pronounced "positive plus negative") with P a nef $\mathbb{Q}$-divisor, $N = \sum q_i \Gamma_i$ with $q_i \in \mathbb{Q}$, $q_i > 0$, the intersection matrix $\Gamma_i \Gamma_j$ is negative definite, and $P\Gamma_i = 0$ for all Γ_i. Show that this exists and is unique. [Hint: You can get uniqueness by a simple argument in quadratic forms. For existence: if D is nef there is nothing to prove. Otherwise $D\Gamma_1 < 0$, so $\Gamma_1^2 < 0$, and subtracting off a suitable multiple of Γ_1 I get $D_1 = D - a_1\Gamma_1$ with $D_1\Gamma_1 = 0$. Proceed in the same way, keeping D_i orthogonal to $\Gamma_1, \dots, \Gamma_i$.]

Let D be the divisor in Ex. 3.12. Find its Zariski decomposition $D = N + P$ and its fixed part F, and compare with the remark in Theorem 3.8.

14. Let C and D be irreducible curves. When is nC numerically k-connected? When is $nC + D$ numerically k-connected?

15. Suppose that $D = \sum_{i=1}^{k} n_i \Gamma_i$, with all $\Gamma_i^2 = 0$. When does D fail to be numerically 1-connected? For any n, give an example of such a divisor which is numerically n-connected but not numerically $n+1$-connected.

16. Suppose that C_1, C_2, C_3 form a triangle $D = C_1 + C_2 + C_3$, and each $C_i^2 < 0$. List all cases of nD numerically 2-connected.

17. Work out the exact sequence of Lemma 3.10 for a reduced divisor D. Understand why $h^0(\mathcal{O}_D) = 1$ if D is connected.

18. Ample certainly implies nef and big for a divisor on a surface. Deduce from Lemma 3.11 that a hyperplane section of an irreducible variety is a connected set.

19. Let $C : (x = 0)$ be the y-axis in $\mathbb{C}^2$ and let $D = 2C$. Write out explicitly the dévissage of the sheaf $\mathcal{O}_D$. The same question for $D = (x^2y^2 = 0)$.

20. Let $E = f^{-1}P$ be a nonsingular reduced fibre of a morphism $f\colon X \to B$ of a surface to a base curve. Prove that $H^0(\mathcal{O}_{aE}) = k[t]/(t^a)$. [Hint: Start with $a = 2$. You can do this in two different ways. First by carrying out the dévissage of Lemma 3.10 systematically, using the fact that the conormal bundle is trivial, $\mathcal{O}_E(-E) \cong \mathcal{O}_E$. Secondly, you can embed $k[t]/(t^a) = \mathcal{O}_{B,P}/m_P^a$ into $H^0(\mathcal{O}_{aE})$ using f^* and the fact that t^a defines exactly the subscheme aE; then show that $h^0(\mathcal{O}_{aE}) \leq a$ by restricting to a section of f.]

21. If $E = \sigma^{-1}P$ is a -1-curve, the exceptional curve of a blowup $\sigma\colon X \to Y$, prove that $H^0(\mathcal{O}_{aE}) = k[x, y]/(x, y)^a$.

22. If E is a -2-curve and $Q \in Y$ is the ordinary double point $xz = y^2$ then $H^0(\mathcal{O}_{aE}) = \left(k[x, y, z]/(xz - y^2)\right)/(x, y, z)^a$.

23. Let $D = 3L \subset \mathbb{P}^2$ be a plane cubic that happens to be a triple line. Calculate $\chi(\mathcal{O}_D)$ using dévissage, and compare with the formula for the arithmetic genus of D as a plane cubic.

24 (harder). Use the C. P. Ramanujam method to prove the following result (due essentially to P. Francia). Let X be a surface D an effective divisor on X and $P \in \operatorname{Sing} D$; write $\sigma \colon Y \to X$ for the blowup of P, E for the exceptional curve, $D' = \sigma^* D - E$ and $D'' = \sigma^* D - 2E$. Then P is a base point of $|D+K|$ if and only if $\sigma^* \colon H^0(\mathcal{O}_D) \to H^0(\mathcal{O}_{D''})$ is not surjective. [Hint: By cohomology, $P \in \operatorname{Bs}|D+K|$ if and only if $H^1(m_P \cdot \mathcal{O}_X(D+K)) \to H^1(\mathcal{O}_X(D+K))$ is not injective. Write this on Y in terms of $H^1(\mathcal{O}_Y(\sigma^*(D+K) - E))$, and use Serre duality to express in terms of $H^1(\mathcal{O}_Y) \to H^1(\mathcal{O}'_D)$ and $H^1(\mathcal{O}''_D)$. Then use the C. P. Ramunujam trick together with the fact that $\ker \colon \operatorname{Pic}^0 D' \to \operatorname{Pic}^0 D''$ does not contain any Abelian variety.]

25. The adjunction formula via Poincaré residue. X is a nonsingular variety and $Y \subset X$ a nonsingular codimension 1 subvariety. Suppose that $z_0, \dots, z_n$ are local coordinates on X, and Y is locally defined $f(z_0, \dots, z_n) = 0$, with $\partial f / \partial z_0 \neq 0$. Prove that there is a sheaf homomorphism

$$\operatorname{Res} \colon \Omega_X^{n+1}(Y) \to \Omega_Y^n \quad \text{taking the local basis} \quad \frac{dz_0 \wedge \cdots \wedge dz_n}{f} \mapsto \frac{dz_1 \wedge \cdots \wedge dz_n}{\partial f / \partial z_0}.$$

(The exercise is to show that the homomorphism is intrinsic, that is, independent of the choice of local coordinates and f). This is called *Poincaré residue.* It fits into an exact sequence of sheaves $0 \to \Omega_X^{n+1} \to \Omega_X^{n+1}(Y) \to \Omega_Y^n \to 0$.

26. The adjunction formula via Serre duality. The full form of Serre duality was mentioned in Chapter B, Rule vii: for X projective and Cohen–Macaulay (say nonsingular), there exists a duality pairing

$$H^i(X, \mathcal{F}) \times \operatorname{Ext}^{n-i}_{\mathcal{O}_X}(\mathcal{F}, \omega_X) \to H^n(X, \omega_X) = k$$

for any coherent sheaf $\mathcal{F}$. Let $Y \subset X$ be a codimension 1 subvariety which is a Cartier divisor. Now coherent sheaves on Y are particular cases of coherent sheaves on X via $\mathcal{O}_X \twoheadrightarrow \mathcal{O}_Y$, and it can be checked that $\operatorname{Ext}^{n-i}_{\mathcal{O}_X}(\mathcal{F}, \omega_X) = \operatorname{Ext}^{n-i-1}_{\mathcal{O}_Y}(\mathcal{F}, \omega_Y)$, where $\omega_Y = \mathcal{E}xt^1_{\mathcal{O}_X}(\mathcal{O}_Y, \omega_X)$.

Apply the cohomological δ-functor $\operatorname{Ext}^*_{\mathcal{O}_X}$ to the exact sequence

$$0 \to \mathcal{O}_X(-Y) \to \mathcal{O}_X \to \mathcal{O}_Y \to 0$$

and prove that ω_Y fits in an exact sequence $0 \to \omega_X \to \omega_X(Y) \to \omega_Y \to 0$. [Hint: $\omega_X(Y) = \mathcal{H}om_{\mathcal{O}_X}(\mathcal{O}_X(-Y), \omega_X)$ and $\mathcal{E}xt^1_{\mathcal{O}_X}(\mathcal{O}_X(-Y), \omega_X) = 0$.]

27. Prove Lemma 3.11, (iii). [Hint: For $s \in H^0(\mathcal{L})$, decompose $D = A + B$ so that A is the biggest divisor on which s vanishes. Then consider the exact sequence

$$0 \to \mathcal{L} \otimes \mathcal{O}_B(-A) \to \mathcal{L} \to \mathcal{L}_{|B} \to 0,$$

and make a numerical consequence from the fact that $\mathcal{L}(-A)$ is generically generated by the section s.]

CHAPTER 4
Singularities and Surfaces

This chapter turns to surface singularities. The main object of study is a normal (isolated) surface singularity $P \in X$ and a resolution $f\colon Y \to X$, with $f^{-1}P = \bigcup \Gamma_i$ a bunch of curves on a nonsingular surface Y. I introduce examples of three classes of singularities, the Du Val singularities, more general rational singularities and elliptic Gorenstein singularities; the latter two classes can be viewed as generalisations of the Du Val singularities. These singularities play an important role all over the classification of surfaces and 3-folds. At the same time, I discuss invariants of singularities, mainly concerned with quantifying the difference between X and Y in the resolution $f\colon Y \to X$. The results here are mostly taken from Artin [**A1–2**], Laufer [**L1–3**] and my unpublished manuscript [**R**], although much of it was known in some form to Du Val in the 1930s. Any area of math can be given a superior treatment using several decades' hindsight (and not least, Mori's notes [**Mori**]), and in this spirit, this chapter contains new proofs of the foundational results on rational and elliptic Gorenstein surface singularities.

Even for the reader only interested in nonsingular varieties or complex manifolds, there are several reasons for studying singularities. A philosophical point is that the study of varieties and the study of singularities each contains the other many times over. The techniques of this chapter mostly involve a bunch of curves $\{\Gamma_i\}$ on a surface Y, and don't really need the singularity $P \in X$ as such; they apply in other situations, notably to a fibre $f^{-1}P$ of a surface fibred over a curve $f\colon X \to C$. The main technical methods used here are those of the previous chapters, divisors on a nonsingular surface, their cohomology, and so on. As I said for K3s at the start of Chapter 3, this material provides excellent practice for calculations involving intersection numbers, coherent cohomology, and results based on these.

4.1. Example: the ordinary double point

As a first example, consider the ordinary quadratic cone

$$P = (0,0) \in X : (xz = y^2) \subset \mathbb{A}^3.$$

This singularity occurs throughout the theory of algebraic surfaces, and can be used to illustrate a whole catalogue of arguments. Because X is a cone with vertex P, it has the standard "cylinder" resolution $Y \to X$; that is, the cone is a union of generating lines through P, and Y is the disjoint union of the generating lines. In other words, Y is the blowup, or the correspondence between the cone and the base of the cone. This singularity and its resolution have already appeared in Chapter 2, where the surface scroll $\mathbb{F}_2 = \mathbb{F}(0,2)$ has a morphism to $\overline{\mathbb{F}}(0,2) \subset \mathbb{P}^3$ (the ordinary quadratic cone) contracting the negative section.

The exceptional curve $f^{-1}P = \Gamma$ of the resolution is a -2-*curve*, satisfying

$$\Gamma \cong \mathbb{P}^1, \quad \Gamma^2 = -2 \quad \text{(or equivalently, } K_Y\Gamma = 0\text{)}.$$

A particular case of Contraction Theorem 4.15 is that every -2-curve is obtained in this way by resolving an ordinary double point (that is, a double point with nondegenerate tangent cone). On minimal surfaces (where there are no -1-curves), -2-curves are characterised as the irreducible curves with $K_Y\Gamma \le 0$ and $\Gamma^2 < 0$, and they arise naturally for this reason in many important contexts of the classification of surfaces.

The ordinary double point also appears in Chapter 1 on cubic surfaces, and in the theory of projective embeddings of K3s in Chapter 3. As we saw in Chapter 1, a nonsingular cubic surface is isomorphic to $\mathbb{P}^2$ blown up in 6 "general" points $P_1, \dots, P_6$, embedded by the linear system of cubics through $\{P_1, \dots, P_6\}$. If P_1, P_2, P_3 are collinear (but the $\{P_i\}$ are otherwise "general") then on the blowup $X = \mathrm{Bl}_6(\mathbb{P}^2)$, the line through P_1, P_2, P_3 becomes a -2-curve, and is contracted by $X \to \overline{X} \subset \mathbb{P}^3$ to an ordinary double point of a cubic surface. Similarly, as described in 3.15.1, if X is a K3 and $|D|$ a free nonhyperelliptic linear system on X, the birational morphism $\varphi_D\colon X \to \overline{X} \subset \mathbb{P}^g$ contracts exactly curves Γ with $D\Gamma = 0$; any such curve is a -2-curve (because $K_X\Gamma = 0$ and $\Gamma^2 < 0$), and if there is exactly one then $\varphi_D(\Gamma) = P \in \overline{X}$ is an ordinary double point.

4.2. The Du Val singularities

Example: D_4**.** Consider the singularity

$$P = (0,0,0) \in X : (g = x^2 + y^3 + z^3 = 0) \subset \mathbb{A}^3.$$

The blow-up $X_1 \to X$ is covered by 3 affine pieces, of which I only write down one: consider $\mathbb{A}^3$ with coordinates x_1, y_1, z, and the morphism $\sigma\colon \mathbb{A}^3 \to \mathbb{A}^3$ defined by $x = x_1z, y = y_1z, z = z$. The inverse image of X under σ is defined by

$$g(x_1z, y_1z, z) = x_1^2z^2 + y_1^3z^3 + z^3 = z^2g_1, \qquad \text{where } g_1 = x_1^2 + (y_1^3 + 1)z.$$

Here the factor z^2 vanishes on the exceptional (x_1, y_1)-plane $\mathbb{A}^2 = \sigma^{-1}P \subset \mathbb{A}^3$, and the residual component $X_1 : (g_1 = 0) \subset \mathbb{A}^3$ is the birational transform of X. Now clearly the inverse image of P under $\sigma\colon X_1 \to X$ is the y_1-axis, and X_1 has ordinary double points at the 3 points where $y_1^3 + 1 = 0$. (Please check for yourselves that the other affine pieces of the blow-up have no further singular points.) The resolution Y is obtained on blowing up these three points $Y \to X_1 \to X$.

I claim that $f^{-1}P$ consists of -2-curves $\Gamma_0, \Gamma_1, \Gamma_2, \Gamma_3$ meeting as follows:

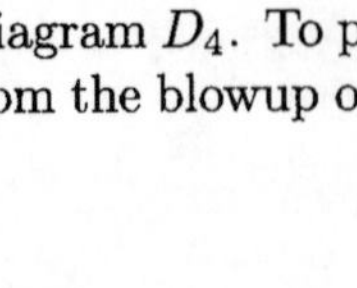

the configuration is the Dynkin diagram D_4. To prove this, it is clear that $\Gamma_1, \Gamma_2, \Gamma_3$ are -2-curves, since they arise from the blowup of ordinary double points. Also the fact that $\Gamma_0 \cong \mathbb{P}^1$, and Γ_0 meets each of Γ_i transversally in 1 point, can be verified directly from the coordinate description of Y. Finally, to see that $\Gamma_0^2 = -2$, note that y is a regular function on Y whose divisor is $\operatorname{div} y = 2\Gamma_0 + \Gamma_1 + \Gamma_2 + \Gamma_3 + C$, where C is the curve $y_1 = 0$ in Y, which also meets Γ_0 transversally in 1 point. Thus

$$0 = (\operatorname{div} y)\Gamma_0 = 2\Gamma_0^2 + \Gamma_0(\Gamma_1 + \Gamma_2 + \Gamma_3 + C) = 2\Gamma_0^2 + 4, \qquad \text{so that } \Gamma_0^2 = -2.$$

The remaining Du Val singularities can be resolved in a similar way (see Ex. 4.3). The equations are

$$\begin{aligned}
A_n: &\quad x^2 + y^2 + z^{n+1} = 0,\\
D_n: &\quad x^2 + y^2 z + z^{n-1} = 0,\\
E_6: &\quad x^2 + y^3 + z^4 = 0,\\
E_7: &\quad x^2 + y^3 + yz^3 = 0,\\
E_8: &\quad x^2 + y^3 + z^5 = 0,
\end{aligned}$$

and the resolution is a bunch of -2-curves whose configuration is given by the corresponding Dynkin diagram; for example, E_7 is the following:

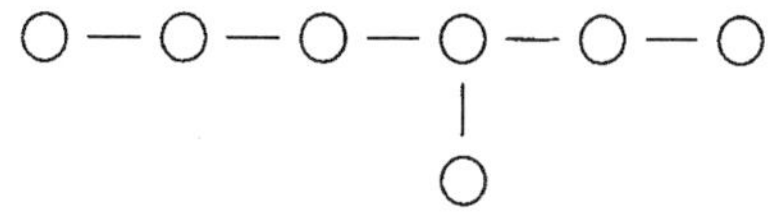

4.3. The general set-up

Let X be an affine surface and $P \in X$ a singular point of X. I want to talk about isolated surface singularities $P \in X$; old-fashioned singularity theorists always write (X, P), but my notation is shorter and self-documenting. It is reasonable to restrict attention to normal singularities, for reasons I will explain. So X is a surface, that is, a 2-dimensional algebraic variety (over an algebraically closed field k, for example, $k = \mathbb{C}$), and $P \in X$ an isolated singular point. I'm usually only interested in a neighbourhood of $P \in X$, so without loss of generality, I can choose X affine and $P \in X$ the only singularity. However, X is contained in a projective variety $\overline{X}$, and for some purposes it may be more convenient to think of X and Y as projective.

A *resolution* of X is a birational projective morphism $f\colon Y \to X$ from a nonsingular surface Y. It is an important and difficult theorem that a resolution always exists. I can assume that f gives $Y \setminus f^{-1}P \xrightarrow{\sim} X \setminus P$. By definition, f is *projective* if it factors as $Y \hookrightarrow X \times \mathbb{P}^N \to X$ with the first map a closed embedding. If X is projective, the condition on f is equivalent to Y also projective. In general, $f^{-1}P$ is a connected bunch of projective curves $\bigcup_{i=1}^n \Gamma_i$; it is traditional to say that f is a *good* resolution (or, more recently, a *log* resolution, meaning resolution in the category of *log varieties*) if $f^{-1}P$ is a union of nonsingular curves meeting transversally, but I do not necessarily assume this.

The surface X is normal if and only if $\mathcal{O}_X = f_*\mathcal{O}_Y$. Here $\mathcal{O}_Y$ is the structure sheaf of Y. The point is just that as a topological space, the normal variety X is the quotient space $Y/{\sim}$ where $\sim$ is the equivalence relation that collapses $f^{-1}P$ to a point; and X has structure sheaf

$$f_*\mathcal{O}_Y = \bigl\{\text{regular functions on } Y \text{ that are constant on } f^{-1}P\bigr\}.$$

Any other variety structure X' on this topological space (having a morphism $Y \to X'$) is obtained by taking a subring $\mathcal{O}_{X'} \subset \mathcal{O}_X$ of finite codimension, that is, $\mathcal{O}_{X'}$ is a subring containing m_P^n for some n, where $m_P \subset \mathcal{O}_{X,P}$ is the maximal ideal. Normal implies nonsingular in codimension 1, and the converse holds for hypersurfaces or complete intersections by the Serre criterion, compare [**YPG**], App. to §3. If my display of erudition confuses you, just concentrate on the case of isolated hypersurface singularities, which are automatically normal, and lots of fun in their own right.

4.4. Changing to the analytic category

When studying singularities, I do not spend much time worrying about which category of geometry I work in. A natural alternative to the language used here is to work up to local analytic equivalence: any ordinary double point is locally isomorphic to the singularity $y^2 = xz$ of 4.1 by a suitable analytic coordinate change. This does not make sense in the category of algebraic varieties: ordinary double points appear on singular cubic surfaces, or on (contracted) K3 surfaces, etc., and then, of course, no Zariski neighbourhood of $P \in X$ is isomorphic to the cone $y^2 = xz$ in the category of algebraic varieties. My justification for being sloppy here is that everything I need to know about an isolated singularity $P \in X$ of a variety depends not on the local ring $\mathcal{O}_{X,P}$ itself, but only on a quotient $\mathcal{O}_{X,P}/m_P^N$ for some large N.

To be able to use the ideas of singularity theory, especially local normal forms, I could assume that the base field is $\mathbb{C}$, and use local complex geometry (typically, the inverse function theorem). Very roughly, since I am mostly concerned with the invariants of coherent cohomology, affine corresponds to Stein. On the other hand, it is not difficult to work over any field using formal completions or the language of the etale topology. Beware that the normal forms of the singularities used here may not work at all in characteristic p, especially when p is small.

4.5. The numerical cycle of a resolution Z_{num}

Given a resolution $f\colon Y \to X$ of $P \in X$, any nonzero element $g \in m_P \subset \mathcal{O}_{X,P}$ vanishes on the set theoretic fibre $f^{-1}P$, so that $\operatorname{div} g = \sum a_i\Gamma_i + C'$ is as in the proof of Theorem A.7. As we saw there, the exceptional divisor $D = \sum a_i\Gamma_i$ satisfies $D\Gamma_i \le 0$ for every contracted curve Γ_i.

Proposition. *There exists a unique effective exceptional divisor* $Z = Z_{\text{num}} = \sum b_i\Gamma_i > 0$ *such that* $Z > 0$,

$$Z\Gamma_i \le 0 \qquad \textit{for every curve } \Gamma_i, \tag{1}$$

and Z *is minimal with this property.*

Proof. There certainly exists some cycle Z with this property, as I just said, so the only thing to prove is that if Z_1 and Z_2 both satisfy (1) then so does $Z = \gcd(Z_1, Z_2)$. This is more or less obvious: I can write $Z_1 = Z + A_1$ and $Z_2 = Z + A_2$, where $A_1, A_2 \ge 0$ have no common components; then any $\Gamma \not\subset A_1$ has $Z\Gamma = Z_1\Gamma - A_1\Gamma \le Z_1\Gamma \le 0$, and similarly for any $\Gamma \not\subset A_2$. Q.E.D.

Definition. Z_{num} is called the *numerical cycle* (or *numerical fundamental cycle*) of the bunch $\{\Gamma_i\}$. The discussion preceding the proposition shows that every $g \in m_P$ vanishes at least along Z_{num}, that is, $\operatorname{div} g \geq Z_{\text{num}}$.

The numerical cycle is often called the *fundamental cycle* in the literature, but lots of other cycles are equally fundamental. In 4.6, I introduce the *fibre cycle* Z_{f}, which is closely related to the scheme theoretic fibre, and later in this chapter the *cohomological cycle* Z_1, which carries the cohomology $H^1(Y, \mathcal{O}_Y)$ (see 4.8), and the *canonical cycle* Z_K for a Gorenstein singularity $P \in X$, which carries the canonical class of Y in a neighbourhood of $f^{-1}P$ (see 4.19).

Note that Z_{num} is very easy to calculate from the intersection matrix $\Gamma_i\Gamma_j$ (see Ex. 4.4).

4.6. Fibre cycle Z_{f} and multiplicity

I define the *fibre cycle* Z_{f} to be the greatest divisor contained in every $\operatorname{div} g$, that is,

$$Z_{\mathrm{f}} = \gcd\bigl\{\operatorname{div} g \bigm| \text{for all } g \in m_P \subset \mathcal{O}_{X,P}\bigr\}.$$

In other words, consider the ideal sheaf $\mathcal{I} = m_P\mathcal{O}_Y$ of $\mathcal{O}_Y$ generated by all $g \in m_P$. The subscheme of Y defined by $\mathcal{I}$ is the scheme theoretic fibre of the morphism f, that is, the subscheme of Y defined by $g = 0$ taken over all $g \in m_P$. By definition the fibre cycle Z_{f} is the maximal effective divisor contained in the scheme theoretic fibre, that is, such that $\mathcal{I} \subset \mathcal{O}_Y(-Z_{\mathrm{f}})$. In more geometric terms, take a finite dimensional vector subspace $V \subset \mathcal{O}_{X,P}$ which generates m_P as a $\mathcal{O}_{X,P}$-module, and for $g \in V$, consider the divisor H_g, a variable hyperplane through P. Then for fixed g, $\operatorname{div} g = f^*(H_g) = L + \sum b_i\Gamma_i$, where L has no components in common with $\{\Gamma_i\}$ and $\sum b_i\Gamma_i$ is an effective divisor. If $g \in V$ is "fairly general" then $\sum b_i\Gamma_i$ takes on a certain minimal value, which is Z_{f}.

It follows from the remarks in 4.5 that $Z_{\mathrm{f}} \geq Z_{\text{num}}$; equality holds in some important cases: see Theorem 4.17 for rational singularities and Theorem 4.23 for elliptic Gorenstein singularities. However, in more complicated cases, for example, when some of the curves Γ_i have large genus, Z_{f} is not determined by simple numerical properties of the configuration $\{\Gamma_i\}$, and it is not hard to find examples with $Z_{\mathrm{f}} > Z_{\text{num}}$ (see Ex. 4.13).

Definition. I take the *multiplicity* of a normal surface singularity $P \in X$ in the pedestrian sense: $\operatorname{mult}_P X$ is the dimension over k of $\mathcal{O}_{X,P}/(h_1, h_2)$ (also called its *length*), where $h_1, h_2 \in m_P$ are sufficiently general elements. In other words, this is the local intersection number at P of two general hyperplane sections through P. If $X \subset \mathbb{P}^N$ is normal and of degree d then any two hyperplane sections H_1, H_2 with no common components meet in d points counted with multiplicity, so that two general hyperplanes through P intersect in P counted $\operatorname{mult}_P X$ times plus $d - \operatorname{mult}_P X$ free intersection points outside P. Thus the multiplicity of P is the number of intersection points of H_1, H_2 absorbed[1] into P.

[1]This is proved exactly as the Bezout theorem in A.9. However, note that it is essential for this that $P \in X$ is normal, so that sufficiently general $h_1, h_2 \in m_P$ form a regular sequence, and the usual Koszul sequence $0 \to \mathcal{O}_{X,P} \to \mathcal{O}_{X,P} \oplus \mathcal{O}_{X,P} \to (h_1, h_2) \subset \mathcal{O}_{X,P}$ defined by h_1, h_2 is exact. In general, the right definition of the multiplicity $\operatorname{mult}_P X$ is in terms of the leading term of the Hilbert–Samuel function: $h(k) = \dim m_P^k/m_P^{k+1} = (\operatorname{mult}_P X)k + \text{const}$. For a nonnormal isolated surface singularity, the Koszul complex for any h_1, h_2 is *never* exact in the middle, and $\dim \mathcal{O}_{X,P}/(h_1, h_2)$ is strictly bigger than $\operatorname{mult}_P X$. See [**YPG**], App. to §3 for more discussion.

Lemma. *(i) Suppose that* $m_P\mathcal{O}_Y = \mathcal{O}_Y(-Z_{\mathfrak{f}})$ *with* $Z_{\mathfrak{f}}$ *the fibre cycle. Then* $\operatorname{mult}_P X = -Z_{\mathfrak{f}}^2$.

(ii) More generally, if $m_P\mathcal{O}_Y = \mathcal{I}\mathcal{O}_Y(-Z_{\mathfrak{f}})$ *with* $\mathcal{I}$ *an ideal defining a zero dimensional subscheme of* Y*, then* $\operatorname{mult}_P X \geq -Z_{\mathfrak{f}}^2 + \dim_k \mathcal{O}_Y/\mathcal{I}$.

Remark. $m_P\mathcal{O}_Y = \mathcal{O}_Y(-Z_{\mathfrak{f}})$ holds if and only if the resolution dominates the ordinary blowup of $P \in X$ (that is, the blowup of the maximal ideal $m_P \subset \mathcal{O}_X$). Because either condition says that the ratio between generators of m_P is a regular function on Y. Thus, by appropriate choice of resolution, I can always arrange that case (i) holds.

Proof. What's going on is the following: think of X as a projective surface $X \subset \mathbb{P}^N$, and consider the linear system $|m_P\mathcal{O}_X(1)|$ of hyperplane sections $H \ni P$. Taking f^* of these divisors defines a linear system $f^*|m_P\mathcal{O}_X(1)|$ on Y with some base locus, which, following tradition, I break up as the divisorial part $Z_{\mathfrak{f}}$ and the codimension ≥ 2 part. The intersection of two elements $H_1, H_2 \in |m_P\mathcal{O}_X(1)|$ consists of P with multiplicity $\operatorname{mult}_P X$ plus a free part; I can calculate the free part just as well on Y, and I get the intersection outside $f^{-1}P$ of two elements of $|f^*H - Z_{\mathfrak{f}}|$.

In case (i), $|f^*H - Z_{\mathfrak{f}}|$ is free. If $h_1, h_2 \in \mathcal{O}_{X,P}$ are two general elements, and $H_1, H_2 \subset X$ the corresponding divisors, then $f^*(H_1) = M_1 + Z_{\mathfrak{f}}$ and $f^*(H_2) = M_2 + Z_{\mathfrak{f}}$ with M_1 and M_2 disjoint over P, that is, $M_1 \cap M_2 \cap f^{-1}P = \emptyset$. Then M_1M_2 is the free intersection of H_1, H_2 outside P.

Then since $(f^*H)Z_{\mathfrak{f}} = 0$, I get

$$\begin{aligned} H^2 = f^*(H_1)f^*(H_2) &= f^*(H_1)(M_2 + Z_{\mathfrak{f}}) = f^*(H_1)M_2 \\ &= (M_1 + Z_{\mathfrak{f}})M_2 = M_1M_2 + Z_{\mathfrak{f}}(f^*(H_2) - Z_{\mathfrak{f}}) \\ &= M_1M_2 - Z_{\mathfrak{f}}^2. \end{aligned}$$

That is, the base locus $Z_{\mathfrak{f}}$ of $f^*|m_P\mathcal{O}_X(1)|$ absorbs $-Z_{\mathfrak{f}}^2$ points of intersection, as required.

(ii) is similar; you have to prove that two general curves through a zero dimensional base scheme at $P \in Y$ have local intersection number $\geq \dim_k \mathcal{O}_{Y,P}/\mathcal{I}_P$. It is in any case obvious that any points of a zero dimensional base locus make a strictly positive contribution. Q.E.D.

4.7. $R^1f_*\mathcal{O}_Y$, and how to calculate it

Here I'm interested in the sheaf $R^1f_*\mathcal{O}_Y$ for a resolution $f\colon Y \to X$. Assuming that X is affine, we will see that this is the finite dimensional vector space $H^1(Y, \mathcal{O}_Y)$ at P.

Proposition. *Let* $P \in X$ *be a surface singularity,* $f\colon Y \to X$ *its resolution and* $P \in U \subset X$ *an affine neighbourhood of* P*. Then for any coherent sheaf* $\mathcal{F}$ *on* Y*,*

1. $H^p(f^{-1}U, \mathcal{F}) = 0$ *for* $p \geq 2$*, and*
2. $H^1(f^{-1}U, \mathcal{F})$ *is independent of* U*.*

Definition. $R^1f_*\mathcal{F}$ is defined to be $H^1(f^{-1}U, \mathcal{F})$, viewed as a sheaf supported at the point $P \in X$. Note that $R^1f_*\mathcal{F} = \varinjlim_U H^1(f^{-1}U, \mathcal{F})$, because by (2), the injective limit is constant for affine U. Since $H^1(f^{-1}U, \mathcal{F})$ is a $\Gamma(U, \mathcal{O}_X)$-module, it is clear that $R^1 f_*\mathcal{F}$ is a module over $\mathcal{O}_{X,P}$. I will prove shortly that it is a finite dimensional vector space.

Proof. $f^{-1}U$ is covered by 2 affine open sets V_1, V_2, so that the first assertion follows at once by calculating $H^p(f^{-1}U, \mathcal{F})$ by the Čech complex. If $U = \operatorname{Spec} A$ then arbitrarily small neighbourhoods of P are given by $U_g = \operatorname{Spec} A[g^{-1}]$ for $g \in A$; the Čech complex of the corresponding cover $V_i \cap f^{-1}(U_g)$ is obtained by applying the exact functor $\otimes_A A[g^{-1}]$, and therefore

$$H^1(f^{-1}(U_g), \mathcal{F}) = H^1(f^{-1}U, \mathcal{F}) \otimes_A A[g^{-1}] = H^1(f^{-1}U, \mathcal{F}). \quad \text{Q.E.D.}$$

Discussion. My treatment here is self-contained, but I give a brief discussion of some background in sheaf theory for the interested reader. Quite generally, for a morphism $f\colon Y \to X$ and sheaf $\mathcal{F}$ on Y, the sheaf $R^p f_*\mathcal{F}$ is defined as the associated sheaf of the presheaf $U \mapsto H^p(f^{-1}U, \mathcal{F})$. This means that $R^p f_*\mathcal{F}$ is the sheaf on X whose stalk at a point $P \in X$ is $(R^p f_*\mathcal{F})_P = \varinjlim_U H^p(f^{-1}U, \mathcal{F})$, where the $\varinjlim$ runs over opens $U \ni P$.

Note that $R^p f_*\mathcal{F}$ is defined in terms of cohomology of thinner and thinner open neighbourhoods of the fibre $f^{-1}P$. In contrast, the holomorphic functions theorem of Zariski and Grothendieck interprets $(R^p f_*\mathcal{F})_P$ for a coherent sheaf $\mathcal{F}$ in terms of the fibre $f^{-1}P$ and sheaves supported on it:

$$(R^p f_*\mathcal{F})_P = \varprojlim_Z H^p(f^{-1}P, \mathcal{F} \otimes_{\mathcal{O}_Y} \mathcal{O}_Z),$$

where the limit is taken over subschemes Z with support in $f^{-1}P$, which in practice means a limit over fatter and fatter infinitesimal neighbourhoods of the fibre. I prove this in the simple case I need in the next section (the general case is done in [**EGA III**$_1$], 4.1.5 or [**H1**], III.11).

The etymology comes from the idea that an element of the left-hand side is something like a Taylor series expansion of a cohomology class around the fibre, and the point is to prove that it converges to a "holomorphic" cohomology class on a neighbourhood of the fibre.

4.8. Cohomological cycle

Let $f^{-1}P = \bigcup_{i=1}^n \Gamma_i$; as in 3.10, given a divisor $D = \sum n_i\Gamma_i$ with $n_i \in \mathbb{Z}$, $n_i \geq 0$, write $\mathcal{O}_Y(-D) \subset \mathcal{O}_Y$ for the ideal sheaf of regular functions on Y vanishing along D, and $\mathcal{O}_D = \mathcal{O}_Y/\mathcal{O}_Y(-D)$. This is a sheaf of rings on $f^{-1}P$, the structure sheaf of a subscheme $D \subset Y$. If $D_1 \leq D$ is another divisor supported on $\bigcup \Gamma_i$ then $D = D_1 + D_2$ with $D_1, D_2 > 0$, so by Lemma 3.10, the inclusion $\mathcal{O}_Y(-D) \subset \mathcal{O}_Y(-D_1)$ induces a short exact sequence

$$0 \to \mathcal{O}_{D_2}(-D_1) \to \mathcal{O}_D \to \mathcal{O}_{D_1} \to 0 \tag{2}$$

Theorem. *(a) For every $D_1 \leq D$ the induced map $H^1(\mathcal{O}_D) \to H^1(\mathcal{O}_{D_1})$ is surjective; the dimension $h^1(\mathcal{O}_D)$ is bounded above.*

(b) There is a unique smallest Z_1 such that $h^1(\mathcal{O}_{Z_1})$ takes the maximal value. Then for every $D \geq Z_1$, the map of (a) is an isomorphism $H^1(\mathcal{O}_D) = H^1(\mathcal{O}_{Z_1})$, and $h^1(\mathcal{O}_D) < h^1(\mathcal{O}_{Z_1})$ if $D \not\geq Z_1$.

(c) The exact sequence of sheaves $0 \to \mathcal{O}_Y(-D) \to \mathcal{O}_Y \to \mathcal{O}_D \to 0$ *induces a surjective homomorphism* $R^1 f_* \mathcal{O}_Y \to H^1(\mathcal{O}_D)$*, which is an isomorphism for any* $D \geq Z_1$.

Definition. The divisor Z_1 is called the *cohomological cycle* of $P \in X$.

The point of the theorem is that $R^1 f_* \mathcal{O}_Y$, defined in 4.7 in terms of open neighbourhoods $f^{-1}U$ of the fibre, is equal to the cohomology group $H^1(\mathcal{O}_{Z_1})$ of an exceptional divisor. This is the same thing as $\varprojlim_D H^1(f^{-1}P, \mathcal{O}_D)$, where the limit runs over fatter and fatter exceptional divisors D, as discussed in 4.7. Moreover, the limit is achieved at Z_1. The theorem is due to M. Artin in modern form, although it seems to have been known in substance to P. Du Val some 30 years earlier.

Proof. (a) The surjectivity of $H^1(\mathcal{O}_D) \to H^1(\mathcal{O}_{D_1})$ follows from (2) since by dimension

$$H^2(\mathcal{O}_{D_2}(-D_1)) = 0.$$

To prove boundedness, note that there exists an effective exceptional divisor $Z = Z_{\text{ample}} = \sum a_i \Gamma_i$ such that $-Z\Gamma_i > 0$ for every Γ_i. This follows from negative definiteness. In other words, $-Z$ is relatively ample for f.

It follows that there exists some exceptional divisor D such that $-D\Gamma_i \geq K_Y \Gamma_i$ for every i (for example, take D to be a large multiple of Z_{ample}). Then for every $D' > 0$,

$$H^1(\mathcal{O}_{D'}(-D)) \stackrel{\text{d}}{=} H^0(\mathcal{O}_{D'}(K_Y + D' + D))$$

(by Serre duality on D', see 4.10. The relation $\stackrel{\text{d}}{=}$ is pronounced "equal to the dual of"). But you can see that $H^0(\mathcal{O}_{D'}(D') \otimes \mathcal{L}) = 0$ if $\deg_{\Gamma_i} \mathcal{L} \leq 0$ for all i as an easy exercise (compare Ex. 4.14). Therefore $H^1(\mathcal{O}_{D+D'}) \to H^1(\mathcal{O}_D)$ is also injective for every $D' \geq 0$.

(b) I need to prove that if D' and D'' are divisors for which $h^1(\mathcal{O}_{D'}) = h^1(\mathcal{O}_{D''})$ both take the maximum value, then $A = \gcd(D', D'')$ has the same property, that is, $h^1(\mathcal{O}_A) = h^1(\mathcal{O}_{D'}) = h^1(\mathcal{O}_{D''})$.

So let $D' = A + B'$ and $D'' = A + B''$, where B' and B'' have no common components, and $C = A + B' + B'' = \operatorname{lcm}(D', D'')$. Then I claim that there exists an exact sequence

$$H^1(\mathcal{O}_C) \to H^1(\mathcal{O}_{D'}) \oplus H^1(\mathcal{O}_{D''}) \to H^1(\mathcal{O}_A) \to 0. \qquad (3)$$

This proves what I want, since C, D', D'' all have the same value of h^1, so

$$h^1(\mathcal{O}_A) \geq h^1(\mathcal{O}_{D'}) + h^1(\mathcal{O}_{D''}) - h^1(\mathcal{O}_C).$$

(3) comes by chasing the diagram

$$\begin{array}{ccccccc}
 & & H^1(\mathcal{O}_{B''}(-D')) & \to & H^1(\mathcal{O}_{B''}(-A)) & \to & 0 \\
 & & \downarrow & & \downarrow & & \\
H^1(\mathcal{O}_{B'}(-D'')) & \to & H^1(\mathcal{O}_C) & \to & H^1(\mathcal{O}_{D''}) & \to & 0 \\
\downarrow & & \downarrow & & \downarrow & & \\
H^1(\mathcal{O}_{B'}(-A)) & \to & H^1(\mathcal{O}_{D'}) & \to & H^1(\mathcal{O}_A) & \to & 0 \\
\downarrow & & \downarrow & & \downarrow & & \\
0 & & 0 & & 0 & &
\end{array}$$

here B' and B'' are without common components, so that $\mathcal{O}_{B'\cap B''}$ has support the finite set $B'\cap B''$, and $H^1(\mathcal{O}_{B'\cap B''})=0$ together with the exact sequence

$$0\to\mathcal{O}_{B''}(-D')\to\mathcal{O}_{B''}(-A)\to\mathcal{O}_{B'\cap B''}\to 0$$

provides the exactness of the top row and the left-hand column. I leave you the pleasure of the diagram chase.

(c) The surjectivity of $H^1(\mathcal{O}_Y)\to H^1(\mathcal{O}_D)$ follows from the assertion $H^2=0$ of Proposition 4.7. The point is to prove that it is injective for some D, so that in particular $H^1(\mathcal{O}_Y)$ is finite dimensional. If $Z=Z_{\text{ample}}$ is the divisor constructed in (a) then the sheaf $\mathcal{O}_Y(-Z)$ is relatively ample on Y (see Ex. 4.16, essentially by the Nakai–Moishezon ampleness criterion[2]). From this, $H^1(\mathcal{O}_Y(-nZ))=0$ for $n\gg 0$ by Serre vanishing, and the result follows. Q.E.D.

4.9. Corollary.

1. *$R^1f_*\mathcal{O}_Y=0$ if and only if $H^1(\mathcal{O}_D)=0$ for every D.*
2. *$R^1f_*\mathcal{O}_Y$ is a 1-dimensional vector space if and only if $H^1(\mathcal{O}_D)\le 1$ for every D and $=1$ for some D.*

If $P\in X$ is an affine neighbourhood, then $R^1f_*\mathcal{O}_Y$ is the sheaf consisting of $H^1(Y,\mathcal{O}_Y)$ supported at P, so that the conditions in (1) and (2) characterise respectively $H^1(Y,\mathcal{O}_Y)=0$ and $h^1(Y,\mathcal{O}_Y)=1$.

4.10. Serre duality

Let Y be a projective nonsingular surface and D an effective divisor on Y. I want to use the fact that the line bundle $\mathcal{O}_D(K_Y+D)$ has the properties of a dualising sheaf for D (for example, in the above proof of Theorem 4.8). I don't need the full strength.

Proposition. *If L is a divisor on Y then there is a natural duality*

$$H^i(\mathcal{O}_D(L))\overset{\mathrm{d}}{=}H^{1-i}(\mathcal{O}_D(K_Y+D-L))\qquad \textit{for } i=0,1.$$

Compare Ex. 3.26, or see **[H1]**, III.7 and **[R1]**, Theorem 2.12 for a more general treatment of the dualising sheaf ω_D and the adjunction formula

$$\omega_D=\mathcal{O}_D(K_D)=\mathcal{O}_D(K_Y+D)=\omega_Y(D)_{|D}.$$

Proof. This follows formally from duality on Y: the short exact sequence

$$0\to\mathcal{O}_Y(L-D)\to\mathcal{O}_Y(L)\to\mathcal{O}_D(L)\to 0$$

leads to

$$\cdots\to H^i(\mathcal{O}_Y(L-D))\to H^i(\mathcal{O}_Y(L))\to H^i(\mathcal{O}_D(L))$$
$$\to H^{i+1}(\mathcal{O}_Y(L-D))\to H^{i+1}(\mathcal{O}_Y(L))\to\cdots$$

[2]This is the central point of the proof. Nakai–Moishezon says that if H is a Cartier divisor on an n-dimensional complete scheme, then $H^rZ>0$ for every r-dimensional irreducible subvariety Z implies that H is ample. This is completely elementary if you know coherent cohomology: by a dévissage argument, you can see that the numerical condition implies that $H^i(\mathcal{F}\otimes\mathcal{O}(nH))=0$ for every $i>0$, for every coherent sheaf $\mathcal{F}$, and for $n\gg 0$ (see **[H1]**, Chapter V, 1.10). I should have included this as a guided exercise in the section on cohomology.

which by duality on Y is dual term by term to

$$\cdots \leftarrow H^{2-i}(\mathcal{O}_Y(K_Y + D - L)) \leftarrow H^{2-i}(\mathcal{O}_Y(K_Y - L)) \leftarrow \cdots$$
$$\cdots \leftarrow H^{1-i}(\mathcal{O}_Y(K_Y + D - L)) \leftarrow H^{1-i}(\mathcal{O}_Y(K_Y - L)) \leftarrow \cdots$$

But this can be identified with the cohomology long exact sequence of

$$0 \to \mathcal{O}_Y(K_Y - L) \to \mathcal{O}_Y(K_Y + D - L) \to \mathcal{O}_D(K_Y + D - L) \to 0,$$

which identifies $H^{1-i}(\mathcal{O}_D(K_Y + D - L))$ with the dual of $H^i(\mathcal{O}_D(L))$, as required. Q.E.D.

4.11. RR on a divisor D

In the same notation, define $\deg_D L = DL = \sum_i n_i \deg_{\Gamma_i} L$. Then

$$\chi(\mathcal{O}_D) = -\tfrac{1}{2}(D^2 + DK_Y),$$

and

$$\chi(\mathcal{O}_D(L)) = \chi(\mathcal{O}_D) + \deg_D L.$$

The proof is an exercise in same style as 4.10, using RR on Y. Notice that the formulas discussed in 3.6 involving the arithmetic genus p_aD,

$$\chi(\mathcal{O}_D) = 1 - p_aD \quad \text{and} \quad 2p_aD - 2 = (D + K_Y)D$$

are a particular case. If $D = C$ is a nonsingular curve, then $g(C) = p_aC = H^1(\mathcal{O}_C)$, and the formula here reduces to the usual RR formula $\chi(\mathcal{O}_C(L)) = 1 - g + \deg L$.

4.12. Rational singularities

Definition. Let $f\colon Y \to X$ be a resolution of a normal surface singularity $P \in X$. Then $P \in X$ is a *rational singularity* if $R^1f_*\mathcal{O}_Y = 0$. It is an *elliptic singularity* if $R^1f_*\mathcal{O}_Y$ is 1-dimensional. Elliptic Gorenstein singularities are treated in 4.21–23 below. Without the Gorenstein condition, elliptic is too weak a condition to be very interesting.

Proposition (Numerical characterisation of rational singularities).

1. *$P \in X$ is rational if and only if*

$$\chi(\mathcal{O}_D) \geq 1 \quad \textit{for every effective divisor } D \textit{ supported on } f^{-1}P.$$

2. *For a rational singularity, the numerical cycle Z_{num} is the maximal divisor with $\chi(\mathcal{O}_D) = 1$. In other words,*

$$h^0(\mathcal{O}_D) \geq 2 \quad \textit{for every } D > Z_{\mathrm{num}}.$$

3. *Z_{num} is numerically 1-connected (see Definition 3.9).*

Remark. By the adjunction formula and RR,

$$(K_Y + D)D = 2p_aD - 2 = -2\chi(\mathcal{O}_D),$$

so that the condition in (1) is equivalent to $p_aD \leq 0$ or $(K_Y + D)D < 0$.

Proof. (1) The implication $\Longrightarrow$ is clear: $R^1 f_* \mathcal{O}_Y = 0$ means that $H^1(\mathcal{O}_D) = 0$ for every D, and $H^0(\mathcal{O}_D) \neq 0$ (because it contains the constant functions $k \subset H^0(\mathcal{O}_Y) \to H^0(\mathcal{O}_D)$), so $\chi(\mathcal{O}_D) \geq 1$.

For $\Longleftarrow$, if $R^1 f_* \mathcal{O}_Y \neq 0$, there exists a divisor $D > 0$ such that $H^1(\mathcal{O}_D) \neq 0$, and therefore there exists a minimal one; that is $H^1(\mathcal{O}_D) \neq 0$, but $H^1(\mathcal{O}_{D-\Gamma}) = 0$ for every $\Gamma \subset D$.

By considering the exact sequence

$$H^1(\mathcal{O}_\Gamma(-D+\Gamma)) \to H^1(\mathcal{O}_D) \to H^1(\mathcal{O}_{D-\Gamma}) = 0,$$

we see that $H^1(\mathcal{O}_\Gamma(-D+\Gamma)) \neq 0$ for every $\Gamma \subset D$. It follows by Serre duality that $H^0(\mathcal{O}_\Gamma(K_Y + \Gamma + D - \Gamma)) \neq 0$, that is, $H^0(\mathcal{O}_\Gamma(K_Y + D)) \neq 0$. Now Γ is an irreducible curve, and $\mathcal{O}_\Gamma(K_Y + D)$ a line bundle over it, so that the existence of a section implies deg ≥ 0, that is $\Gamma(K_Y + D) \geq 0$ for every $\Gamma \subset D$. Summing over all components of D gives $D(K_Y + D) \geq 0$, that is $p_a D \geq 1$.

(2) and (3) are easy exercises (see Ex. 4.11–12). Q.E.D.

Example. Consider the bunch of $\mathbb{P}^1$s consisting of a central -3-curve $\times$ meeting 3 simple chains of -2-curves of length $p-1, q-1, r-1$ respectively:

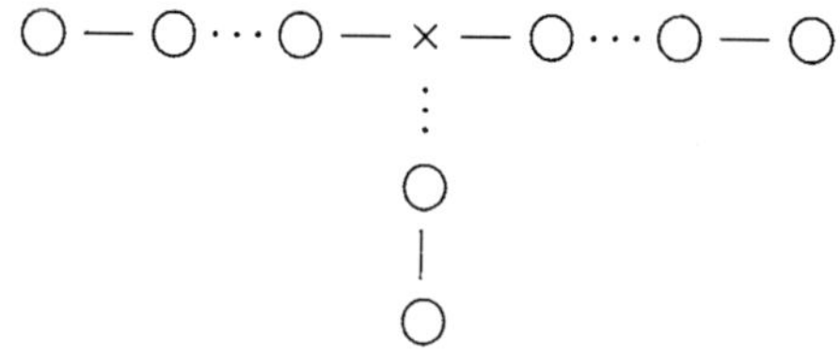

Check that Z_{num} is the reduced sum of all the curves, and that the configuration is rational. It's a fun calculation to obtain this configuration by resolving the singular surface $X \subset \mathbb{C}^4$ defined by

$$\begin{aligned} xy + xt^q + t^{p+q} &= 0 \\ yz + yt^r + t^{q+r} &= 0 \\ xz + zt^p + t^{p+r} &= 0 \end{aligned} \qquad \text{or} \qquad \operatorname{rank}\begin{pmatrix} z + t^r & -x & -t^q \\ -t^r & x + t^p & -y \\ -z & -t^p & y + t^q \end{pmatrix} \leq 1.$$

[Hint: Check that if $p = q = r = 1$, you get the cone over the rational normal curve of degree 3. Then check that if $p = 0$, X has two Du Val singularities of type A_{q-1}, A_{r-1} on the x-axis. Finally, if $p, q, r > 1$, the blowup ($x = x_1 t$, etc.) gives 3 coordinate axes as exceptional locus, through a point of the same type with numbers $p-1, q-1, r-1$, and you continue inductively.]

4.13. Structure of Pic D

Let $\{\Gamma_i\}_{i=1}^k$ be a bunch of curves on Y and $D = \sum n_i \Gamma_i$ an effective divisor with $n_i > 0$ for each i. $\mathcal{O}_D^*$ denotes the sheaf of units of $\mathcal{O}_D$, so that $\mathcal{O}_Y^* \to \mathcal{O}_D^*$ is surjective (sheaves for the Zariski topology). Then

$$H^1(\mathcal{O}_D^*) = \{\text{isomorphism classes of invertible sheaves over } \mathcal{O}_D\} = \operatorname{Pic} D.$$

For any curve $\Gamma = \Gamma_i$, there is a natural group homomorphism

$$\rho_\Gamma \colon H^1(\mathcal{O}_D^*) \to H^1(\mathcal{O}_\Gamma^*) = \operatorname{Pic} \Gamma.$$

Geometrically, this takes an invertible sheaf L over D to $L_{|\Gamma} = L \otimes \mathcal{O}_\Gamma$. Now compose with the degree map $\operatorname{Pic}\Gamma \to \mathbb{Z}$ to get

$$\deg_\Gamma \colon H^1(\mathcal{O}_D^*) = \operatorname{Pic} D \to \mathbb{Z}.$$

Putting together this map for each component Γ_i gives a map

$$\deg \colon H^1(\mathcal{O}_D^*) \to \mathbb{Z}^k$$

defined by $L \mapsto \deg_{\Gamma_i} L$. It is easy to see that deg is surjective: choose a general point Q on any component $\Gamma \subset D$, and construct a Cartier divisor with support Q and degree 1 on Γ, whose local equation $s \in \mathcal{O}_{D,Q}$ restricts to a local equation of Q in $\mathcal{O}_{\Gamma,Q}$. (This operation is important in what follows, and is called a *transverse cut* at Q.)

4.14. Proposition (Artin). *(i) Write $D' = D_{\mathrm{red}}$. Then the two groups*

$$K = \ker\{H^1(\mathcal{O}_D) \to H^1(\mathcal{O}_{D'})\}$$

and

$$K^* = \ker\{H^1(\mathcal{O}_D^*) \to H^1(\mathcal{O}_{D'}^*)\}$$

have filtrations with isomorphic quotients.

(ii) $\deg \colon H^1(\mathcal{O}_D^*) \to \mathbb{Z}^k$ *is an isomorphism if and only if* $H^1(\mathcal{O}_D) = 0$.

Proof. (i) Suppose C is a divisor such that $2C \subset D$; then define the subsheaves $J \subset \mathcal{O}_D$ and $J^* \subset \mathcal{O}_D^*$ to be the kernels in the exact sequences

$$0 \to J \to \mathcal{O}_D \to \mathcal{O}_{D-C} \to 0$$
$$0 \to J^* \to \mathcal{O}_D^* \to \mathcal{O}_{D-C}^* \to 0$$

(here of course $J = \mathcal{O}_C(-D+C)$ by Lemma 3.10); then J and J^* are isomorphic as sheaves of groups, because $J^2 = 0$, so that

$$\varepsilon \colon \Gamma(U, J) \xrightarrow{\sim} \Gamma(U, J^*) \quad \text{defined by } x \mapsto 1 + x$$

gives an isomorphism from the additive to the multiplicative group, a kind of "first order exponential". Also it is clear that $H^0(\mathcal{O}_{D-C}^*) \subset H^0(\mathcal{O}_{D-C})$ (as a set), and contains a Zariski neighbourhood of 1. A trivial cocycle calculation based on this proves that

$$H^0(\mathcal{O}_{D-C}) \to H^1(J) \quad \text{and} \quad H^0(\mathcal{O}_{D-C}^*) \to H^1(J^*) = \varepsilon(H^1(J))$$

have the same images. Hence

$$\ker\bigl\{H^1(\mathcal{O}_D) \to H^1(\mathcal{O}_{D-C})\bigr\} \cong \ker\bigl\{H^1(\mathcal{O}_D^*) \to H^1(\mathcal{O}_{D-C}^*)\bigr\}.$$

(Of course, the right-hand group is a priori not a k-vector space.) The argument continues by induction from D down to D'; $\ker\{H^1(\mathcal{O}_D^*) \to H^1(\mathcal{O}_{D'}^*)\}$ and $\ker\{H^1(\mathcal{O}_D) \to H^1(\mathcal{O}_{D'})\}$ thus have filtrations by subgroups with isomorphic quotients.

(ii) For a reduced curve D', it can be shown easily that

$$\operatorname{Pic} D' \cong \mathbb{Z}^k \iff H^1(\mathcal{O}_{D'}) = 0.$$

$\Longleftarrow$ By RR, a line bundle L on D' with $\deg_\Gamma L = 0$ for every $\Gamma \subset D'$ has

$$h^0(L) \ge 1 - h^1(\mathcal{O}_{D'}),$$

so $H^1(\mathcal{O}_{D'}) = 0 \implies H^0(L) \neq 0$. Then it is not hard to see that the section s has the property that $\mathcal{O}_{D'} \cdot s = L$. (See Lemma 3.11, (iii) and Ex. 3.27.)

$\Longrightarrow$ If L is a line bundle with sufficiently high degree on every component of D' then $H^1(L) = 0$, so by RR, $h^0(L) = 1 - h^1(\mathcal{O}_{D'}) + \deg_{D'} L$. If I pick $\deg_{\Gamma_i} L$ "general" points on each component Γ_i then obviously the resulting set of points $\{P_i\}$ gives a Cartier divisor $\sum P_i$ on D' with the same degrees as L, so $L_0 = L(-\sum P_i)$ is a line bundle of degree 0 on each component. But a simple argument shows that these points can be chosen one at a time to strictly decrease the dimension $h^0(L(-\sum P_i))$ until we get to zero, so that if $H^1(\mathcal{O}_{D'}) \neq 0$, I get a line bundle L_0 of degree zero on each component, but $h^0(L_0) \neq 0$, so that $L_0 \cong \mathcal{O}_{D'}$. Q.E.D.

Remark. For a projective variety X, the group $\operatorname{Pic} X$ is an extension of a discrete group by the group $\operatorname{Pic}^0 X$ of divisors algebraically equivalent to zero modulo divisors linearly equivalent to zero. The elements of $\operatorname{Pic}^0 X$ are naturally parametrised (in the sense of moduli problems) by an algebraic group or algebraic group scheme, the *Picard variety* of X. If X is nonsingular in characteristic zero, then $\operatorname{Pic}^0 X$ is an Abelian variety. The Jacobian of a curve is a well-known example. For singular, or nonreduced X, the algebraic group $\operatorname{Pic}^0 X$ is in general more complicated; however, it has the structure of an extension of an Abelian variety by an affine group (that is, an extension of additive and multiplicative groups $\mathbb{G}_a$s and $\mathbb{G}_m$s), and the Abelian variety comes entirely from the resolution of reduced components of X. Compare the discussion in 3.14.

4.15. Contractibility of a rational bunch of curves

Let Y be a nonsingular quasiprojective surface and $\{\Gamma_i\}$ a bunch of curves on Y. By Propositions 4.12 and 4.14, we know that the following 3 conditions are equivalent:

1. $H^1(\mathcal{O}_D) = 0$ for every divisor D supported on $\bigcup \Gamma_i$;
2. $p_a D \leq 0$ for every D;
3. $\deg\colon H^1(\mathcal{O}_D^*) = \operatorname{Pic} D \xrightarrow{\sim} \mathbb{Z}^k$ for every D, where k is the number of components of D.

If $f\colon Y \to X$ is a resolution of a normal surface singularity $P \in X$ with exceptional locus $\{\Gamma_i\} = f^{-1}P \subset Y$, these conditions are necessary and sufficient for $P \in X$ to be rational.

Contraction Theorem (Artin). *A connected bunch of curves $\{\Gamma_i\}$ on Y is the exceptional locus of a resolution of a rational singularity $P \in X$ if and only if*

(a) the intersection matrix $(\Gamma_i\Gamma_j)$ is negative definite;

(b) $p_a D \leq 0$ for every D.

Examples. (i) A -1-*curve* on a surface Y is a curve L satisfying $L \cong \mathbb{P}^1$, $L^2 = -1$ and $K_Y L = -1$. This satisfies the rationality condition, so can be contracted.

Let's check the rationality condition: obviously if $L = \mathbb{P}^1$ then $H^1(\mathcal{O}_L) = 0$, and for $n \geq 2$,

$$0 \to \mathcal{O}_L(-(n-1)L) \to \mathcal{O}_{nL} \to \mathcal{O}_{(n-1)L} \to 0$$

gives

$$0 = H^1(\mathbb{P}^1, \mathcal{O}(n-1)) \to H^1(\mathcal{O}_{nL}) \to H^1(\mathcal{O}_{(n-1)L}) \to 0.$$

So $H^1(\mathcal{O}_{nL}) = 0$ for all n.

This case has the special feature that after contraction, the point $P \in X$ is a nonsingular point, and $Y \to X$ is the usual blowup of P (this is proved in Theorem 4.17). In this case the contraction theorem is Castelnuovo's contractibility criterion.

(ii) Similarly, for any $d > 0$, a curve $L \cong \mathbb{P}^1$, with $L^2 = -d$ satisfies the rationality condition. If $d = 1, 2$ we get -1-curve, -2-curve respectively.

(iii) A bunch of -2-curves $\{\Gamma_i\}$ has negative definite intersection matrix if and only if it forms one of the configurations A_n, D_n, E_6, E_7, E_8 (see Ex. 4.5); if this happens then $p_a D \le 0$ for every effective divisor D supported on Γ_i. This can be seen as follows: $K_Y\Gamma_i = 0$ by the adjunction formula, and so $K_Y D = 0$, so $K_Y D + D^2 < 0$, and $p_a D = \frac{1}{2}(K_Y + D)D + 1 \le 0$. In this case, Theorem 4.17 proves that the image $P \in X$ is a hypersurface double point.

4.16. Proof of Theorem

The conditions are clearly necessary (compare Theorem A.7 and Proposition 4.12). I now assume (a) and (b), and prove that the contraction exists. Suppose that H is ample on Y. Then $H\Gamma_i > 0$ for every curve Γ_i. Since the matrix $(\Gamma_i\Gamma_j)$ is nonsingular, there exists rational numbers q_i such that $(H + \sum q_i\Gamma_i)\Gamma_j = 0$ for every j. By taking n to be any common denominators of the q_i, I get a divisor $D = \sum a_i\Gamma_i$ such that $(nH + D)\Gamma_i = 0$ for every Γ_i. Take n to be sufficiently large, in a sense to be specified below. This means in particular that $D\Gamma_i < 0$ for every i.

I claim that $a_i > 0$: indeed, write $D = A - B$ with A, B effective and with no common components. Then $DB = AB - B^2 \ge 0$. So $B = 0$, or in other words $D > 0$; but then the only way that $D\Gamma_i < 0$ is possible is for $\Gamma_i \subset D$, so $a_i > 0$.

Consider the invertible sheaf $\mathcal{L} = \mathcal{O}_Y(nH + D)$; by construction, $\deg_{\Gamma_i} \mathcal{L} = 0$ for each i. So by Proposition 4.14, (ii), $\mathcal{L}_{D'} \cong \mathcal{O}_{D'}$ for any divisor D' supported on $\{\Gamma_i\}$.

Consider the restriction exact sequence

$$0 \to \mathcal{O}_Y(nH) \to \mathcal{O}_Y(nH + D) \to \mathcal{O}_D \to 0.$$

Now $\mathcal{O}_Y(H)$ is ample, so that by Serre vanishing,

$$H^1(\mathcal{O}_Y(nH)) = 0 \quad \text{for every } n \gg 0,$$

and so $H^0(\mathcal{O}_Y(nH + D)) \to H^0(\mathcal{O}_D) \to 0$ is exact.

I now claim that a suitable space of sections of $\mathcal{O}_Y(nH + D)$ defines the required contraction $f\colon Y \to X'$, at least set theoretically. Let n and $s_0, \dots, s_N \in H^0(\mathcal{O}_Y(nH))$ be chosen so that $\varphi\colon Y \to \mathbb{P}^N$ defined by $(s_0 : \cdots : s_N)$ is an embedding; and let $s_{N+1} \in H^0(\mathcal{O}_Y(nH + D))$ be any element which maps to the identity section $1 \in H^0(\mathcal{O}_D)$. Consider the sections $(s_0, \dots, s_{N+1}) \in H^0(Y, \mathcal{L})$, and the map $f'\colon Y \to \mathbb{P}^{N+1}$ by $(s_0 : \cdots : s_{N+1})$.

Then (a) at every point $Q \in Y$, some $s_i \in \mathcal{L}$ is a local generator; (take s_i if $Q \notin D$, and s_{N+1} if $Q \in D$).

(b) $f'(D) = P = (0, \dots, 0, 1)$.

(c) $f'_{|(Y \setminus D)}$ is an embedding (since already the ratios $(s_0 : \cdots : s_N)$ give an embedding). Write $X' = f'(Y)$. Then there is a unique way to factorise the map f' as $Y \xrightarrow{f} X \to X'$ where X is normal, and this is the required contraction. X is projective since X' is, and the normalisation $X \to X'$ is finite. Q.E.D.

4.17. Multiplicity and embedding dimension

To understand what singularities arise on making contractions such as those in Theorem 4.15, I need to be able to determine invariants of a singularity $P \in X$ in terms of its resolution Y and the bunch of curves $\{\Gamma_i\}$. Multiplicity and embedding dimension are two simple invariants, which give an easy characterisation of the nonsingular surface point $P \in X$ in Castelnuovo's criterion, and hypersurface double points arising on contracting a bunch of -2-curves.

Multiplicity was defined in 4.6 above. The *embedding dimension* of $P \in X$ is the dimension of the Zariski tangent space $\operatorname{emb}\dim(P \in X) = \dim m_P/m_P^2$. This is clearly the dimension of the smallest ambient space in which an analytic neighbourhood $P \in X$ can be embedded.

Theorem (Artin). *For a rational singularity $P \in X$, let $Z = Z_{\text{num}}$ be the numerical cycle (see Definition 4.5), and set $d = -Z^2$. Then*

1. $m_P\mathcal{O}_Y = \mathcal{O}_Y(-Z)$, *and therefore* $\operatorname{mult}(P \in X) = d$;
2. $m_P/m_P^2 = H^0(\mathcal{O}_Z(-Z))$, *and* $\operatorname{emb}\dim(P \in X) = d + 1$.

In particular, $Z^2 = -1$ implies $P \in X$ is nonsingular, and $Z^2 = -2$ implies $P \in X$ is a hypersurface double point.

Remarks. (a) The proof of (2) gives at once that $m_P^k/m_P^{k+1} = H^0(\mathcal{O}_Z(-kZ))$ and $\dim m_P^k/m_P^{k+1} = kd+1$ for every $k \geq 1$, so that the knowledgeable reader who so prefers can use the more professional definition of multiplicity in terms of the Hilbert–Samuel function (compare the footnote to 4.6).

(b) The example of the cone over the rational normal curve helps remember the numbers, but don't be misled into thinking that rational surface singularities are all that simple.

Lemma. *Let $f: Y \to X$ be a resolution of a rational surface singularity $P \in X$. Suppose that X is affine, and let $\mathcal{L}$ be a line bundle on Y which is nef on $f^{-1}P$ (that is, $\mathcal{L}\Gamma_i \geq 0$ for each i).*

Then $H^1(Y, \mathcal{L}) = 0$ and $|\mathcal{L}|$ is free near $f^{-1}P$.

Proof. Given any exceptional divisor D, I can construct a divisor A on D as a sum of local transverse cuts through general points of the Γ_i, so that A is an effective Cartier divisor on D with the same degree as $\mathcal{L}$; then by Proposition 4.14, (ii), $\mathcal{L}_{|D} \cong \mathcal{O}_D(A)$. Now the exact sequence of sheaves

$$0 \to \mathcal{O}_D \to \mathcal{O}_D(A) \to \mathcal{O}_A(A) \to 0$$

shows that $H^1(\mathcal{O}_D(A)) = 0$. Now $H^1(Y, \mathcal{L}) = R^1 f_*\mathcal{L}$ can be calculated by the arguments of 4.8 as an inverse limit of the groups $H^1(\mathcal{L}_{|D})$, all of which are zero (compare Ex. 4.15).

The transverse cuts making up A can be chosen freely, so that the linear system $|\mathcal{L}|_D = |A|_D$ is free on any D. Apply this to $D = Z$. Then $\mathcal{L}(-Z)$ is nef, so that by the first part, $H^1(\mathcal{L}(-Z)) = 0$, and therefore

$$H^0(Y, \mathcal{L}) \twoheadrightarrow H^0(Z, \mathcal{L}_{|Z}).$$

Thus $|\mathcal{L}|$ on Y cuts out the complete linear system $|\mathcal{L}|_Z$ on Z, and I just said that this is free. Q.E.D.

Proof of Theorem 4.17, (1). As discussed above, every function $g \in m_P \subset \mathcal{O}_{X,P}$ vanishes on $Z = Z_{\text{num}}$, so that $f_*\mathcal{O}_Y(-Z) = m_P$. By the lemma, $|\mathcal{O}_Y(-Z)| = |-Z|$ is free, which means that $m_P\mathcal{O}_Y = \mathcal{O}_Y(-Z)$ (compare the discussion in 4.6). This means $Z = Z_{\mathfrak{f}}$ and the assumption of Lemma 4.6, (i) holds, so $\operatorname{mult}(P \in X) = -Z^2$ follows.

4.18. Graded rings over divisors

Before proving (2), I make some general remarks on the graded ring

$$R(Z, \mathcal{L}) = \bigoplus_{k \geq 0} H^0(Z, \mathcal{L}^{\otimes k})$$

corresponding to a divisor $Z = \sum n_i\Gamma_i$ on a nonsingular surface and a nef line bundle $\mathcal{L} \in \operatorname{Pic} Z$ on Z with $d = \deg \mathcal{L} > 0$.

I add a brief word of explanation in case you haven't seen this kind of thing before. There is a graded ring $R(V, \mathcal{L})$ corresponding to any line bundle over a variety (or scheme) V: by definition of tensor product of sheaves, there are local multiplication maps $\mathcal{L} \times \mathcal{L} \to \mathcal{L}^{\otimes 2}$ (more generally, $\mathcal{L}^{\otimes a} \times \mathcal{L}^{\otimes b} \to \mathcal{L}^{\otimes a+b}$). The graded ring $R(V, \mathcal{L})$ puts together all global sections of all powers of $\mathcal{L}$, with multiplication between the H^0 induced by the local maps of sheaves.

The case to bear in mind is when $\mathcal{L}$ is very ample, so that $V \subset \mathbb{P}^n$ with $H^0(\mathcal{L}) = \langle x_0, \dots, x_n \rangle$, the vector space of homogeneous coordinates of $\mathbb{P}^n$. Then the multiplication $H^0(\mathcal{L}) \times \dots \times H^0(\mathcal{L}) \to H^0(\mathcal{L}^{\otimes k})$ just views monomials in the x_i as restrictions to V of homogeneous forms in $H^0(\mathbb{P}^n, \mathcal{O}(k))$. In this case, $R(V, \mathcal{L})$ generated in degree 1 means

$$H^0(\mathbb{P}^n, \mathcal{O}(k)) \twoheadrightarrow H^0(V, \mathcal{L}^{\otimes k}) \quad \text{for all } k.$$

If V itself is a normal variety, this condition is called *projective normality*, and is equivalent to saying that the affine cone over V is normal. See 3.2 for an important example.

In the immediate application, Z is the numerical cycle of a rational singularity or bunch of curves, and $\mathcal{L} = \mathcal{O}_Z(-Z)$. Thus Z satisfies the assumptions of the next theorem.

Theorem. *Let $Z = \sum n_i\Gamma_i$ be a divisor on a nonsingular surface, and assume that Z is numerically 1-connected and $H^1(\mathcal{O}_Z) = 0$. Let $\mathcal{L} \in \operatorname{Pic} Z$ be a nef line bundle on Z with $d = \deg \mathcal{L} > 0$. Then the graded ring $R(Z, \mathcal{L}) = \bigoplus_{k \geq 0} H^0(Z, \mathcal{L}^{\otimes k})$ is generated in degree 1. In particular, the multiplication map*

$$H^0(\mathcal{L}) \times H^0(\mathcal{L}) \to H^0(\mathcal{L}^{\otimes 2})$$

is onto.

Proof. The proof is completely elementary: apart from the language, it is just the simplest case of Castelnuovo's free pencil trick.

Write $\deg_{\Gamma_i} \mathcal{L} = d_i$ (so that $d = \sum n_i d_i$), and choose d_i disjoint transverse cuts A_{ij} through Γ_i (for $j = 1, \dots, d_i$). As a scheme, $A_{ij} \cong \operatorname{Spec} k[x]/x^{n_i}$. Their sum $A = \sum A_{ij}$ is a Cartier divisor A with the same degree as $\mathcal{L}$ on every component of Z, so that Proposition 4.14 gives $\mathcal{L} \cong \mathcal{O}_Z(A)$. In other words, $\mathcal{L}$ has a section s_0 with divisor of zeros equal to A. As in Lemma 4.17, the cuts making up A can be chosen arbitrarily, so that $|\mathcal{L}| = |A|$ is a free linear system on Z.

Consider two general sections $s, s_0 \in H^0(\mathcal{L})$ (the eponymous free pencil), and write $A = \operatorname{div}(s_0)$; I assume that s is a local basis of $\mathcal{L}$ at every point of $\operatorname{Supp} A$. Then s_0 defines the short exact sequence

$$0 \to \mathcal{O}_Z \to \mathcal{L} \to \mathcal{O}_A \to 0.$$

Here I use the second section s to identify $\mathcal{L} = \mathcal{O}_Z(A)$ with $\mathcal{O}_Z$ near A, which simplifies the notation $\mathcal{L}_{|A} = \mathcal{O}_A(A)$ to $\mathcal{O}_A$.

Now the assumption $H^1(\mathcal{O}_Z) = 0$ obviously gives $H^1(\mathcal{L}) = 0$ and $H^0(\mathcal{L}) = d+1$. Let $s_1, \dots, s_d \in H^0(\mathcal{L})$ map to a basis of $H^0(\mathcal{O}_A)$. Then the theorem follows from the more precise claim: for any $k \geq 0$, the vector space $H^0(\mathcal{L}^{\otimes k})$ is based by the $kd+1$ monomials

$$s_0^k, \quad \text{and} \quad s_0^{k-a} s^{a-1} s_i \quad \text{for } a = 1, \dots, k \text{ and } i = 1, \dots d.$$

Indeed, $H^0(\mathcal{O}_Z)$ is just the constants (because Z is 1-connected), which gives the assertion for $k = 0$; for $k \geq 1$, it follows by induction from the exact sequence

$$H^0(\mathcal{L}^{\otimes k-1}) \hookrightarrow H^0(\mathcal{L}^{\otimes k}) \twoheadrightarrow H^0(\mathcal{O}_A). \quad \text{Q.E.D.}$$

The proof amounts to saying that $\mathcal{L}$ embeds the Artinian scheme A into $\mathbb{P}^{d-1}$ so that $H^0(\mathbb{P}^{d-1}, \mathcal{O}(1)) \to \mathcal{O}_A$ is onto. That is, the image $A \subset \mathbb{P}^{d-1}$ behaves like d points in linearly general position.

Proof of Theorem 4.17, (2). Under the identification $m_P = H^0(\mathcal{O}_Y(-Z))$, it is clear that m_P^2 maps to $H^0(\mathcal{O}_Y(-2Z))$, which gives a map

$$m_P/m_P^2 \to H^0(\mathcal{O}_Y(-Z))/H^0(\mathcal{O}_Y(-2Z)) = H^0(\mathcal{O}_Z(-Z)) \tag{4}$$

which is well defined and surjective. The point is to prove that it is also injective, so that $m_P/m_P^2 \cong H^0(\mathcal{O}_Z(-Z))$. This completes the proof, since by RR,

$$h^0(\mathcal{O}_Z(-Z)) = 1 - p_a Z + \deg(\mathcal{O}_Z(-Z)) = 1 - Z^2.$$

To see that (4) is injective, it is enough to prove that

$$H^0(Y, \mathcal{O}_Y(-2Z)) = m_P^2;$$

in words, a function vanishing along $2Z$ is a linear combination of products of two functions vanishing along Z. I prove this as a corollary of Theorem 4.18. Let $g \in H^0(Y, \mathcal{O}_Y(-2Z))$; its restriction to Z is a section $\overline{g} \in H^0(\mathcal{O}_Z(-2Z))$, so by the theorem, it is in the image of $H^0(\mathcal{O}_Z(-Z)) \otimes H^0(\mathcal{O}_Z(-Z))$; choosing an expression $\overline{g} = \sum \overline{x}_i \overline{y}_i$ and lifting to $x_i, y_i \in H^0(\mathcal{O}_Y(-Z))$ gives

$$g - f_2 \in H^0(\mathcal{O}_Y(-3Z)) \quad \text{with } f_2 = \sum x_i y_i \in m_P^2$$

Proceeding in the same way with $H^0(\mathcal{O}_Z(-3Z))$, etc., gives for any n an expression

$$g - f_2 - f_3 - \cdots - f_{n-1} \in H^0(\mathcal{O}_Y(-nZ)) \quad \text{with } f_i \in m_P^i. \tag{5}$$

On the other hand, I claim that $H^0(\mathcal{O}_Y(-nZ)) \subset m_P^2$ for some $n \gg 0$. Since also $m_P^i \subset m_P^2$, (5) implies that $g \in m_P^2$.

Note that the graded algebra $\bigoplus_{k \geq 0} H^0(\mathcal{O}_Y(-kZ))$ is finitely generated as a $k[X]$-algebra: for $\mathcal{O}_Y(-Z)$ is the pullback of the relatively ample line bundle $\mathcal{O}(1)$ under the morphism $\varphi_{|-Z|} \colon Y \to X \times \mathbb{P}^N$ defined by the free linear system $|-Z|$; (the image is the blowup of m_P). The required finite generation follows by projective normalisation. Each $H^0(\mathcal{O}_Y(-kZ))$ is contained in m_P, so that for n greater than the degrees of all the generators, the group $H^0(\mathcal{O}_Y(-nZ))$ is contained in a sum of products of at least two of these, therefore in m_P^2. Q.E.D.

I'm sorry this proof was so involved. Lots of proofs in algebraic geometry depend on finiteness arguments of this type, and projective normalisation is among the more elementary, provided it's in your vocabulary. A posteriori, the graded algebra is generated in degree 1.

An alternative is to go from (5) to the conclusion $g \in m_P^2$ using some kind of completion; [**A2**] uses Henselisation. If you think of Y as just a thin tubular neighbourhood of $f^{-1}P$ in the complex topology, any general divisor in $|-2Z|$ consists of a number of transverse cuts through general points of components of $f^{-1}P$, with each transverse cut a connected component of the divisor in the complex topology. By considering degrees, it is easy to see that this can be grouped as a sum of two divisors in $|-Z|$, which proves that $H^0(\mathcal{O}_Y(-2Z)) = m_P^2$.

4.19. The Gorenstein condition and the canonical cycle Z_K

The class of Gorenstein singularities $P \in X$ is a mild generalisation of hypersurface singularities; although I sketch a correct treatment of the definition and simple properties, you could for many practical purposes take the easy way out and assume throughout that $P \in X$ is a hypersurface singularity. Let $P \in X$ be a normal surface singularity; at some future point you need to know that $P \in X$ is automatically Cohen–Macaulay, as discussed in [**YPG**], App. to §3. Let $f\colon Y \to X$ be a resolution of singularities, subsequently usually assumed to be *minimal* (no -1-curves in the fibre).

Proposition–Definition. *Equivalent conditions:*

1. *The canonical class of Y can be written as an exceptional divisor plus a divisor disjoint from the exceptional locus:*
$$K_Y \overset{\text{lin}}{\sim} -Z_K + M \quad \textit{with} \quad M \cap f^{-1}P = \emptyset.$$
2. *There exists a rational canonical differential $s \in \mathcal{O}_X(K_X)$ on X which is regular and has no zeros on a punctured neighbourhood of $P \in X$.*
3. *The line bundle $\Omega^2_{X^0}$ on the nonsingular locus $X^0 = X \setminus P$ extends over X as a line bundle ω_X.*
4. *The canonical Weil divisor class K_X is Cartier (locally principal) at P.*
5. *The Serre–Grothendieck dualising sheaf ω_X is invertible at P.*
6. *The local ring $\mathcal{O}_{X,P}$ is Gorenstein in the sense of commutative algebra textbooks (for example* [**Matsumura**]*, Chapter 6, §18).*

If these conditions hold, $P \in X$ is Gorenstein*; the exceptional divisor Z_K is called the* canonical cycle *of the resolution (or the* antidiscrepancy*).*

Discussion of proof. The canonical divisor K_X of a normal n-fold is the Weil divisor class div s where $s \in \Omega^n_{k(X)}$ is a rational canonical differential. This was discussed in A.10 under the additional assumption that X is nonsingular, so that s could be compared with a volume form $dx_1 \wedge \cdots \wedge dx_n$ at every point of X. However, a rational canonical differential $s \in \Omega^n_{k(X)}$ is a birational notion, and to define its divisor div s as a Weil divisor, we only need to know X at each prime divisor.

It follows from this discussion that, essentially by definition, (1), (3) and (4) are merely restatements of (2). The (pre-) dualising sheaf ω_X is defined, and its fundamental dualising property proved in [**H1**], III.7. The equivalence of (4) and (5) comes from the fact that the (pre-) dualising sheaf ω_X equals the divisorial sheaf $\mathcal{O}_X(K_X)$, as discussed in [**C-3f**], App. to §1.

The point of the equivalence between (5) and (6) is to pass between global duality (properties involving ω_X and the category of quasicoherent sheaves on X) and local duality for the local ring $A = \mathcal{O}_{X,P}$. To do this categorically should be an exercise, but one which I can't do convincingly, partly because the literature on local duality leaves me speechless. A practical alternative uses the calculation of ω_X as an $\mathcal{E}xt$ sheaf: assume that $X \subset \mathbb{P}$ is an embedding into a nonsingular variety; then X Gorenstein at P means that $\mathcal{E}xt^{N-n}_{\mathcal{O}_\mathbb{P}}(\mathcal{O}_X, \omega_\mathbb{P}) = \omega_X$ is invertible at P and $\mathcal{E}xt^i = 0$ for $i \neq N-n$. Localising and playing with injective resolutions, it's not hard to see that this is equivalent to $\operatorname{Ext}^n_A(k, A) = k$ and $\operatorname{Ext}^i = 0$ for $i \neq n$, where $k = A/m_P$. This is one of the many equivalent definitions of a Gorenstein local ring, compare [**Matsumura**], Theorem 18.1.

Example. If $P \in X \subset \mathbb{C}^3$ is an isolated hypersurface singularity defined by $f = 0$, a basis of $\mathcal{O}_X(K_X)$ near P is given by

$$s = \operatorname{Res}_X \frac{\mathrm{d}x \wedge \mathrm{d}y \wedge \mathrm{d}z}{f} = \frac{\mathrm{d}x \wedge \mathrm{d}y}{\partial f/\partial z} = \text{etc.}$$

Then s is a rational canonical differential, and its divisor on Y is $\operatorname{div}_Y(s) = -Z_K$. That is, s has Z_K as its divisor of poles on Y. Compare the introductory discussion in [**YPG**], §1.

4.20. The canonical and cohomological cycles, $Z_1 = Z_K$

From now on I assume that the resolution Y is minimal, that is, K_Y is nef on the exceptional curves Γ_i. Then $-Z_K$ is nef, so that either $Z_K = 0$ or $Z_K \geq Z_{\text{num}}$, where Z_{num} is the numerical cycle of 4.2. In particular $Z_K \geq 0$, so I can view it as a subscheme $Z_K \subset Y$. It is easy to see that $Z_K = 0$ if and only if $Y \to X$ is a minimal resolution of a Du Val singularity.

Theorem. *If $P \in X$ is an isolated Gorenstein surface singularity then $Z_1 = Z_K$. Here Z_1 is the cohomological cycle of Theorem 4.8, and Z_K the canonical cycle defined in 4.19.*

Proof. First, if $D \geq Z_K$ then $H^1(\mathcal{O}_D) \to H^1(\mathcal{O}_{Z_K})$ is an isomorphism: indeed, it is clearly surjective, with kernel a quotient of

$$H^1(\mathcal{O}_{D'}(-Z_K)) \overset{\mathrm{d}}{=} H^0(\mathcal{O}_{D'}(D')), \quad \text{where } D' = D - Z_K.$$

But it is an easy exercise to see that $H^0(\mathcal{O}_{D'}(D')) = 0$ (see Ex. 4.14). Therefore Z_K achieves the maximum value of $H^1(\mathcal{O}_D)$, and so $Z_1 \leq Z_K$ by definition of Z_1 (see Theorem 4.8, (b)).

Next, if $Z_1 < Z_K$, then I claim that $h^1(\mathcal{O}_{Z_1}) < h^1(\mathcal{O}_{Z_K})$, which contradicts the definition of Z_1. The duals of these H^1s are the groups

$$H^1(\mathcal{O}_{Z_K}) \overset{\mathrm{d}}{=} H^0(\omega_{Z_K}) = H^0(Z_K, \mathcal{O}_{Z_K}(K_Y + Z_K)) = H^0(Z_K, \mathcal{O}_{Z_K})$$
$$H^1(\mathcal{O}_{Z_1}) \overset{\mathrm{d}}{=} H^0(\omega_{Z_1}) = H^0(Z_1, \mathcal{O}_{Z_1}(K_Y + Z_1)) = H^0(Z_1, \mathcal{O}_{Z_1}(Z_1 - Z_K))$$

which fit in the exact sequence

$$0 \to H^0(Z_1, \mathcal{O}_{Z_1}(Z_1 - Z_K)) \to H^0(Z_K, \mathcal{O}_{Z_K}) \to H^0(\mathcal{O}_{D''}), \tag{6}$$

where $D'' = Z_K - Z_1$. But if $D'' \neq 0$, the last map in (6) is certainly nonzero, because it is the identity on the constant functions. This completes the contradiction, and proves the theorem. Q.E.D.

4.21. Elliptic Gorenstein surface singularities

Definition. A normal surface singularity $P \in X$ is an *elliptic Gorenstein singularity* if it is Gorenstein (Proposition–Definition 4.19) and elliptic (Definition 4.12).

The treatment of elliptic Gorenstein singularities in what follows parallels closely that of rational singularities in 4.12–18.

Theorem (Numerical characterisation). *Let $P \in X$ be an elliptic Gorenstein surface singularity, $f\colon Y \to X$ its minimal resolution, and $Z = Z_K$ the canonical cycle. Then Z is the unique effective cycle with $p_a Z = 1$, and $p_a D \leq 0$ for $D \neq Z$. Moreover, Z is numerically 2-connected and $Z = Z_{\text{num}} = Z_1 = Z_K$.*

Conversely, suppose that the minimal resolution $Y \to X \ni P$ of a normal surface singularity has an effective cycle Z satisfying $p_a Z = 1$, and $p_a D \leq 0$ for $D \neq Z$. Then $P \in X$ is elliptic Gorenstein.

Proof. Note that by the adjunction formula 4.10, the canonical cycle Z_K of any Gorenstein singularity automatically satisfies $\omega_Z = \mathcal{O}_Z(K_Y + Z_K) \cong \mathcal{O}_Z$, so that $\chi(\mathcal{O}_{Z_K}) = 0$ and $p_a(Z_K) = 1$.

Suppose first that $P \in X$ is elliptic Gorenstein, and set $Z = Z_1 = Z_K$. Then by Theorem 4.8, (b), any effective exceptional divisor D satisfies $H^1(\mathcal{O}_D) = 0$ if $D \not\geq Z = Z_1$, and so $p_a D \leq 0$. If $D > Z = Z_K$ then

$$2p_a D - 2 = D(D - Z) = (D - Z)^2 + Z(D - Z) \leq (D - Z)^2 < 0, \tag{7}$$

($K_Y = -Z$ is nef on exceptional curves, because Y is minimal), so again $p_a D \leq 0$.

Claim. *If an effective divisor Z on a nonsingular surface satisfies $p_a Z = 1$ and $p_a D \leq 0$ for all $D < Z$ then $D(Z - D) \geq 2$, that is, Z is numerically 2-connected.*

Proof. Try it and see. □

This proves the "moreover" part: $-Z$ is nef (because $-Z \overset{\text{num}}{\sim} K_Y$), and for any divisor D with $0 < D < Z$, the claim implies that $D(Z - D) \geq 2$, so $-D$ is not nef. Therefore $Z = Z_{\text{num}}$.

The point of the converse is to prove that $K_Y + Z \overset{\text{lin}}{\sim} 0$ near $f^{-1}P$, so that $P \in X$ is Gorenstein by Proposition 4.19, (1). I divide this up into easy steps.

Step 1. $Z = Z_{\text{num}}$, *and in particular* $f^{-1}P = \operatorname{Supp} Z$.

Proof. As in (7), applying the assumption $p_a \leq 0$ to $Z + \Gamma_i$ implies that $-Z\Gamma_i \geq 0$ for any exceptional Γ_i:

$$\begin{aligned} -2 \geq 2p_a(Z + \Gamma) - 2 &= (K_Y + Z + \Gamma)(Z + \Gamma) \\ &= (K_Y + Z)Z + (K_Y + \Gamma)\Gamma + 2Z\Gamma \\ &\geq -2 + 2Z\Gamma. \end{aligned} \tag{8}$$

As in the claim, if $0 < D < Z$ then $-D$ is not nef, so that $Z = Z_{\text{num}}$.

Step 2. $K_Y + Z \overset{\text{num}}{\sim} 0$ *on exceptional curves.*

Proof. Since $(K_Y + Z)Z = 0$, there is nothing more to prove if Z is irreducible. If $Z = \sum n_i\Gamma_i$ then $\sum n_i(K_Y + Z)\Gamma_i = 0$, and the assumption $p_a(\Gamma_i) = 0$ together with the recent claim gives

$$(K_Y + Z)\Gamma_i = (K_Y + \Gamma_i)\Gamma_i + (Z - \Gamma_i)\Gamma_i = -2 + (Z - \Gamma_i)\Gamma_i \geq 0.$$

Thus $(K_Y + Z)\Gamma_i = 0$ for each i, as required.

Step 3. $\mathcal{O}_Z(K_Y + Z) \cong \mathcal{O}_Z$.

Proof. We know that $\mathcal{O}_Z(K_Y + Z)$ is a numerically trivial line bundle on Z, and that Z is numerically connected, so that by Lemma 3.11, (iii), it is enough to observe that it has a section:

$$H^0(\mathcal{O}_Z(K_Y + Z)) \overset{\text{d}}{=} H^1(\mathcal{O}_Z) \neq 0.$$

Step 4. $H^0(\mathcal{O}_Y(K_Y + Z)) \to H^0(\mathcal{O}_Z(K_Y + Z))$ *is surjective.*

Proof. As in the proof of Theorem 4.8, the cokernel $H^1(\mathcal{O}_Y(-Z))$ can be calculated as an inverse limit $\varprojlim H^1(\mathcal{O}_D(-Z))$, and each of these groups is zero (see Ex. 4.15).

It follows from Steps 3–4 that $\mathcal{O}_Y(K_Y + Z)$ has a section which is a basis at every point of $f^{-1}P$. Therefore $K_Y + Z \overset{\text{lin}}{\sim} 0$ in a neighbourhood of $f^{-1}P$. This completes the proof that $P \in X$ is Gorenstein with canonical cycle $Z_K = Z$. Clearly by Theorems 4.8 and 4.20, $R^1 f_*\mathcal{O}_Y = H^1(\mathcal{O}_Z)$ is 1-dimensional, so that $P \in X$ is elliptic Gorenstein. Q.E.D.

4.22. Examples

Just as the affine cone over a rational normal curve of degree d serves as a model example of a rational singularity of multiplicity d (at least for mnemonic purposes), the affine cone over a normally embedded elliptic curve $E_d \subset \mathbb{P}^{d-1}$ of degree $d \geq 3$ is the model example of an elliptic Gorenstein singularity; in singularity theory these cones are called *simple elliptic singularities.* Note that the fact that the elliptic curve E_d has a modulus (the j-invariant) means that the analytic type of the singularity also depends on a modulus. The typical example is the ordinary hypersurface triple point

$$0 \in X : (x^3 + y^3 + z^3 + \lambda xyz = 0) \subset \mathbb{C}^3$$

These admit many degenerations (for example, $xyz + x^p + y^q + z^r$ for $p, q, r \geq 4$, known to singularity theorists as $T_{p,q,r}$). This class of singularities includes cases that are very complicated from the point of view of classification of singularities.

Elliptic Gorenstein singularities with $d = 1$ or 2 are hypersurface double points, again typified by the simple elliptic singularities

$$x^2 + y^3 + z^6 + \lambda xyz \quad \text{and} \quad x^2 + y^4 + z^4 + \lambda xyz,$$

which this time are cones in a weighted sense: they are the respective affine cones on the curves in weighted projective planes $C_6 \subset \mathbb{P}(3,2,1)$ and $C_4 \subset \mathbb{P}(2,1,1)$. The statements of the results are slightly different, but the theory is broadly similar to that for $d \geq 3$. They are alternative explicit methods to treat them. Compare Ex. 4.18.

Both [**L2**] and [**R**] contain systematic lists of examples of elliptic Gorenstein singularities with $d = 1, 2, 3$.

4.23. The main theorem

Set $d = -Z^2$, defined to be the *degree* of $P \in X$.

Main Theorem ([R], [L2], [Mori]).
(I) If $d \geq 2$ then $m_P\mathcal{O}_Y = \mathcal{O}_Y(-Z)$; in particular $Z = Z_{\mathrm{f}}$ and $|-Z|$ is a free linear system, so that $\operatorname{mult}_P X = d$.

(II) If $d \geq 3$, then $f_*\mathcal{O}_Y(-2Z) = m_P^2$, and $\operatorname{emb\,dim}_P X = d$.

Moreover, $f_*\mathcal{O}_Y(-kZ) = m_P^k$ for all k; that is, the $\mathcal{O}_X$-algebra

$$\bigoplus_{k\geq 0} f_*\mathcal{O}_Y(-kZ) = \bigoplus_{k\geq 0} m_P^k \tag{9}$$

is generated by its component m_P in degree 1.

The results for $d = 1$ or 2 are slightly longer to state, and are deferred to 4.25 for the sake of clarity; however, the proofs are closely related, and are given together when appropriate.

Discussion. (II) says that the ordinary blowup $X_1 \to X$ of X at P coincides with the relative canonical model of $Y \to X$. Indeed, the blowup is defined to be Proj of the algebra $\bigoplus m_P^k$ on the right-hand side of (9). On the other hand, by definition of the canonical cycle, $\mathcal{O}_Y(-Z) \cong \mathcal{O}_Y(K_Y)$ in a neighbourhood of $f^{-1}P$, so that the left-hand side of (9) is the relative canonical algebra $\mathcal{R}(Y, K_Y)$ of $Y \to X$. It is easy to see that $\operatorname{Proj}\mathcal{R}(Y, K_Y)$ is the normal surface $\overline{Y}$ with only Du Val singularities and ample $K_{\overline{Y}}$ obtained from Y by contracting exactly the -2-curves of $f^{-1}P$.

Proof of (I). Most of the proof takes place on Z. First note that the restriction $H^0(Y, \mathcal{O}_Y(-Z)) \to H^0(Z, \mathcal{O}_Z(-Z))$ is surjective, by the usual argument: the cokernel is $H^1(Y, \mathcal{O}_Y(-2Z))$, which is zero by Ex. 4.15. For $d = -Z^2 \geq 2$, the following result applied to $\mathcal{L} = \mathcal{O}_Z(-Z)$ gives at once that $|-Z|$ is free on Y, which proves (I).

Lemma. *Let $\mathcal{L}$ be a line bundle on Z which is nef and of degree $d > 0$. Assume if possible that $Q \in Z$ is a closed point at which $|\mathcal{L}|$ is not free. Then $d = 1$, $Q \in Z$ is a nonsingular point (therefore a Cartier divisor), and $\mathcal{L} \cong \mathcal{O}_Z(Q)$. Then $H^0(\mathcal{L})$ consists of the identity inclusion $s\colon \mathcal{O}_Z \hookrightarrow \mathcal{O}_Z(Q)$ and its scalar multiples.*

If $\deg\mathcal{L} = 1$ *then $h^0(\mathcal{L}) = 1$, so that $|\mathcal{L}|$ is not free, and all these assertions hold for some point $Q \in Z$.*

Proof. The assumption is that the evaluation map $H^0(Z, \mathcal{L}) \to k_Q$ is not surjective, so its cokernel $H^1(Z, m_Q\mathcal{L})$ is nonzero. I sidestep the difficulty that this H^1 might come from restriction to a smaller divisor by taking D to be a divisor with

$0 < D \le Z$ which is minimal with $H^1(D, m_Q\mathcal{L}_{|D}) \neq 0$. Minimal means:

$$H^1(D - \Gamma_i, m_Q\mathcal{L}_{|D - \Gamma_i}) = 0 \quad \text{for every component } \Gamma_i < D,$$

so that the nonzero H^1 comes from the sheaf kernel of the restriction map

$$m_Q\mathcal{L} \to m_Q\mathcal{L}_{|D - \Gamma_i}.$$

Outside Q, the kernel is $\mathcal{L} \otimes \mathcal{O}_{\Gamma_i}(-D+\Gamma_i)$ by the dévissage exact sequence of 3.10. To understand what happens at Q, note there are two cases: either $Q \in D - \Gamma_i$, and the kernel is $\mathcal{O}_{\Gamma_i}(\mathcal{L} - D + \Gamma_i)$, since everything vanishing on $D - \Gamma_i$ is automatically in m_Q; or $Q \notin D - \Gamma_i$ (which means that Γ_i is a reduced component of D, and is the unique component of D containing Q, so that the restriction is an isomorphism near Q), and the kernel is $m_Q\mathcal{O}_{\Gamma_i}(\mathcal{L} - D + \Gamma_i)$.

Therefore

$$\begin{aligned} &H^1(\Gamma_i, \mathcal{O}_{\Gamma_i}(\mathcal{L} - D + \Gamma_i)) \neq 0 \quad \text{if } Q \in D - \Gamma_i; \\ \text{and}\quad &H^1(\Gamma_i, m_Q\mathcal{O}_{\Gamma_i}(\mathcal{L} - D + \Gamma_i)) \neq 0 \quad \text{if } Q \notin D - \Gamma_i. \end{aligned}$$

Now quite generally, for a torsion free sheaf $\mathcal{F}$ of rank 1 on an irreducible curve Γ,

$$H^1(\Gamma, \mathcal{F}) \neq 0 \implies \deg_\Gamma \mathcal{F} \le 2p_a\Gamma - 2. \tag{10}$$

Write $D = \sum a_i\Gamma_i$. Since $K_Y = -Z$ locally, $2p_a\Gamma_i - 2 = (-Z + \Gamma_i)\Gamma_i$. Summing (10) over all components Γ_i of D, and noticing that replacing $\mathcal{L}$ by $m_Q\mathcal{L}$ decreases the degree by 1 for at most one component Γ_i, I get

$$\sum a_i\Gamma_i(\mathcal{L} - D + \Gamma_i) \le 1 + \sum a_i\Gamma_i(-Z + \Gamma_i).$$

The quadratic terms Γ_i^2 on each side cancel, giving

$$D\mathcal{L} + D(Z - D) \le 1. \tag{11}$$

But now Z is 2-connected, so that (11) implies that $D = Z$, $\deg \mathcal{L} = 1$. Thus $h^0(\mathcal{L}) = 1$. Moreover, $Q \in \Gamma_i$ holds for a unique irreducible component Γ_i, and the section $\mathcal{O}_Z \to \mathcal{L}$ vanishes at Q only, so that Q supports an effective Cartier divisor of degree 1, which means it is a nonsingular point of Z. This proves the lemma and completes the proof of (I). Q.E.D.

Remark. A minor modification of the same argument proves the following result (see Ex. 4.22): if Z is a 2-connected divisor on a nonsingular surface and $\mathcal{L}$ a line bundle with $\mathcal{L} \otimes \omega_Z^{-1}$ nef and of degree ≥ 1 then $|\mathcal{L}|$ is free unless $\mathcal{L} \cong \omega_Z(Q)$ with $Q \in Z$ a nonsingular point. Catanese and others **[CFHR]** contains a dual version of the argument, using Serre duality more systematically: $0 \neq H^1(m_Q\mathcal{L}) \overset{\mathrm{d}}{=} \operatorname{Hom}(m_Q\mathcal{L}, \omega_Z)$, so choose a nonzero element $s \in \operatorname{Hom}(m_Q\mathcal{L}, \omega_Z)$, and let $Z = A + B$ be the decomposition such that A is the maximal divisor on which s vanishes. Then s defines an inclusion $m_Q\mathcal{L}_{|B} \hookrightarrow \omega_Z(-A) = \omega_B$. A simple numerical argument based on 2-connectedness and $\mathcal{L} > K_Z$ proves that $A = 0$, and s defines an isomorphism $m_Q\mathcal{L} \cong \omega_Z$, so that $Q \in Z$ is a nonsingular point and $\mathcal{L} \cong \omega_Z(Q)$.

4.24. Proof of (II)

The main object of study of this section, as in 4.18, is the graded ring

$$R(Z,\mathcal{L}) = \bigoplus_{k\geq 0} H^0(Z,\mathcal{L}^{\otimes k}) \tag{12}$$

corresponding to a nef line bundle of degree $d>0$ over a divisor Z. For $d\geq 3$, the result is as follows.

Theorem. *Let Z be a 2-connected divisor on a nonsingular surface Y for which $\mathcal{O}_Z(K_Y+Z)\cong\mathcal{O}_Z$, and let $\mathcal{L}$ be a nef line bundle on Z with $d=\deg\mathcal{L}\geq 3$. Then the graded ring $R(Z,\mathcal{L})$ is generated by elements of degree $k=1$. In particular,*

$$H^0(\mathcal{L})\otimes H^0(\mathcal{L})\twoheadrightarrow H^0(\mathcal{L}^{\otimes 2})$$

is surjective.

Proof. Because $\mathcal{L}$ is free, a general section $s_1\in H^0(\mathcal{L})$ does not vanish along any component of Z, so is regular at every point (that is, locally defined by a non-zerodivisor); thus it defines an effective Cartier divisor A of degree d on Z, such that $\mathcal{L}\cong\mathcal{O}_Z(A)$. As in 4.18, I choose a second general section s and use it to identify $\mathcal{L}$ with $\mathcal{O}_Z$ in a neighbourhood of every point of A, and consider the exact sequence

$$0\to\mathcal{O}_Z\xrightarrow{s_1}\mathcal{L}\to\mathcal{O}_A\to 0. \tag{13}$$

Now A is an Artinian scheme of degree d, and since $h^0(\mathcal{O}_Z)=h^1(\mathcal{O}_Z)=1$ it follows that $H^0(\mathcal{L})\to\mathcal{O}_A$ has rank $d-1$. (I write $\mathcal{O}_A=H^0(\mathcal{O}_A)$ out of laziness.)

Claim 1. *$H^0(\mathcal{L})\twoheadrightarrow\mathcal{O}_{A'}$ is surjective for every subscheme $A'\subset A$ of length $d-1$. Geometrically, this means that $\varphi_{\mathcal{L}}$ embeds A in $\mathbb{P}^{d-2}$ as a subscheme of degree d with the property that any subscheme of degree $d-1$ spans $\mathbb{P}^{d-2}$, like d points forming a projective frame of reference.*

To prove this, write $\mathcal{I}_{A'}\subset\mathcal{O}_Z$ for the ideal sheaf of A', so that $\mathcal{I}_A\subset\mathcal{I}_{A'}$ and $\mathcal{I}_{A'}/\mathcal{I}_A\cong k_Q$ for some point $Q\in Z$. Then $\mathcal{F}=\mathcal{I}_{A'}\mathcal{O}_Z(A)$ is an intermediate torsion free sheaf

$$\mathcal{O}_Z\subset\mathcal{F}\subset\mathcal{O}_Z(A),$$

with $\mathcal{F}/\mathcal{O}_Z\cong k_Q$. Now by Serre duality $H^1(\mathcal{F})\stackrel{\mathrm{d}}{=}\operatorname{Hom}(\mathcal{F},\mathcal{O}_Z)$. But this group is zero; for a nonzero homomorphism $\mathcal{F}\to\mathcal{O}_Z$ must restrict to a nonzero map $\mathcal{O}_Z\to\mathcal{O}_Z$, necessarily an isomorphism by 2-connectedness, and this would split the exact sequence $\mathcal{O}_Z\to\mathcal{F}\to k_Q$, contradicting $\mathcal{F}$ torsion free. Therefore $H^1(\mathcal{I}_{A'}\mathcal{L})=0$, which gives the required surjectivity.

Claim 2. *A subscheme $A_d\subset\mathbb{P}^{d-2}$ with the property stated in Claim 1 imposes linearly independent conditions on quadrics. (This means that the restriction map $H^0(\mathbb{P}^{d-2},\mathcal{O}(2))\to\mathcal{O}_{A_d}$ is surjective.)*

The cases $d=3$ and $d=4$ of Claim 2 are very easy, see Ex. 4.23. Although the proof involves various bits of mess, in substance it is similar to, and rather easier than [**UAG**], §1.

Indeed, any Artinian scheme A has a filtration $A_1\subset\cdots\subset A_{d-1}\subset A$ by subschemes A_i of degree i, and by assumption, each A_i has scheme theoretic linear span $\langle A_i\rangle=\mathbb{P}^{i-1}$. Clearly each A_i imposes i conditions on quadrics (since it does on hyperplanes), and I prove inductively that A_i is a scheme theoretic intersection of quadrics. It follows from this that A_{d-1} imposes $d-1$ conditions on quadrics, and that there is a quadric through A_{d-1} not containing A_d, so that A_d imposes one further condition, which proves Claim 2.

The inductive step is as follows: write

$$\langle A_i \rangle = \mathbb{P}^{i-1} : (x_i = 0) \subset \langle A_{i+1} \rangle = \mathbb{P}^i$$

and $\mathcal{I}_i$ for the ideal sheaf of A_i. Suppose that A_i and A_{i+1} differ only at a point Q. Then it is easy to check that there is an exact sequence

$$0 \to m_Q \mathcal{O}_{\mathbb{P}^i}(1) \xrightarrow{x_i} \mathcal{I}_{i+1}\mathcal{O}_{\mathbb{P}^i}(2) \to \mathcal{I}_i \mathcal{O}_{\mathbb{P}^{i-1}}(2) \to 0. \tag{14}$$

(See Ex. 4.24.) Clearly, to say that A_i is a scheme theoretic intersection of quadrics means exactly that $\mathcal{I}_i\mathcal{O}_{\mathbb{P}^{i-1}}(2)$ is generated by its H^0. Now the right-hand end of (14) is generated by its H^0 by the inductive hypothesis, and the left-hand end obviously so. Therefore so is the middle term. This proves Claim 2.

The proof of Theorem 4.24 and Theorem 4.23, (II) is now straightforward. Let $s_1, \dots, s_d$ be a basis of $H^0(Z, \mathcal{L})$, with s_1 defining an effective Cartier divisor A as in (13). Since $s_2, \dots, s_d$ are the homogeneous coordinates in the $\mathbb{P}^{d-2}$ containing A_d, by Claim 2, I can choose a set $\{t_1, \dots, t_d\}$ of quadratic monomials in $s_2, \dots, s_d$ that, viewed as sections of $H^0(Z, \mathcal{O}_Z(2A))$, map to a basis of $\mathcal{O}_A$. For $k \geq 2$, consider the exact sequence

$$0 \to H^0(Z, \mathcal{O}_Z((k-1)A) \xrightarrow{s_1} H^0(Z, \mathcal{O}_Z(kA)) \to \mathcal{O}_A \to 0.$$

Recall that I chose a second section $s \in H^0(Z, \mathcal{L})$ that is a local basis of $\mathcal{L}$ at every point of $\operatorname{Supp} A$, and which I use to identify $\mathcal{L}_{|A}$ with $\mathcal{O}_A$. Then the products $s^{k-2} \cdot \{t_1, \dots, t_d\}$ form a complimentary basis to $s_1 \cdot H^0(\mathcal{O}_Z((k-1)A)$. This proves that the ring $R(Z, \mathcal{L})$ is generated in degree 1.

The final part of the proof of (II) is the same as in 4.18. Since

$$0 \to f_*\mathcal{O}_Y(-(k+1)Z) \to f_*\mathcal{O}_Y(-kZ) \to H^0(Z, \mathcal{O}_Z(-kZ)) \to 0$$

is exact, any $g \in f_*\mathcal{O}_Y(-kZ)$ can be written modulo $f_*\mathcal{O}_Y(-(k+1)Z)$ as a polynomial of degree k in elements of m_P. By induction,

$$f_*\mathcal{O}_Y(-kZ) \subset m_P^k + f_*\mathcal{O}_Y(-lZ) \quad \text{for any } l > k.$$

But since the $\mathcal{O}_X$-algebra $\mathcal{R}(Y, \mathcal{O}_Y(-Z))$ is finitely generated, the argument at the end of 4.18 proves that $f_*\mathcal{O}_Y(-lZ) \subset m^k$ for any fixed k and $l \gg 0$. Therefore $f_*\mathcal{O}_Y(-kZ) = m_P^k$, as requested. Q.E.D.

4.25. The special cases $d = 2, 1$

The results for $d = \deg \mathcal{L} = 2$ or 1 are similar in spirit, but the statements must involve weighted homogeneous rings because the ring $R(Z, \mathcal{L})$ is not generated in degree 1.

Theorem. *For $d = 2$:*

$$R(Z, \mathcal{L}) = k[x_1, x_2, y]/f, \quad \textit{where } \deg x_1, x_2, y = 1, 1, 2,$$

and f is homogeneous of degree 4, of the form $f = y^2 +$ terms involving x_1, x_2.

For $d=1$:

$$R(Z,\mathcal{L}) = k[x,y,z]/f, \quad \textit{where } \deg x,y,z = 1,2,3,$$

and f is homogeneous of degree 6, of the form $f = z^2+y^3+$ terms involving x.

Proof. In either case, there exists a section $x \in H^0(Z,\mathcal{L})$ defining a Cartier divisor A of degree d, and therefore an exact sequence

$$0 \to \mathcal{O}_Z \to \mathcal{L} \to \mathcal{O}_A \to 0.$$

When $d=2$, I set $x = x_1$, and I can find a second section x_2 which is a local basis of $\mathcal{L}$ near $\operatorname{Supp} A$. Now choose an element $y \in H^0(Z,\mathcal{L}^{\otimes 2})$ so that y, x_2^2 map to a basis of $\mathcal{O}_A$. Arguing as usual on the restriction exact sequence

$$0 \to H^0(Z,\mathcal{O}_Z((k-1)A)) \to H^0(Z,\mathcal{O}(kA)) \to \mathcal{O}_A \to 0,$$

we see that $x_2^k, x_2^{k-2}y$ forms a complimentary basis to $x_1 \cdot H^0(Z,\mathcal{O}_Z((k-1)A))$ in every degree $k \ge 2$. It's very easy to see from this that

$$S^k(x_1,x_2), S^{k-2}(x_1,x_2)y \quad \text{bases} \quad H^0(Z,\mathcal{L}^{\otimes k}).$$

In particular, x_1, x_2, y generate the ring, and there is a unique relation expressing y^2 in terms of the stated basis.

The case $d=1$ is similar, and is left as an easy exercise. [Hint: You just have to pick elements y, z in degrees 2, 3 that map to a basis of $\mathcal{O}_A = k_Q$.] Q.E.D.

Corollary. *If $d=2$ then $\operatorname{mult}_P X = 2$ and $\operatorname{emb\,dim}_P X = 3$, that is, $P \in X$ is a hypersurface double point. The linear system $|-Z|$ on Y is free; that is, $Z = Z_{\mathrm{f}}$, and $m_P\mathcal{O}_Y = \mathcal{O}_Y(-Z)$. Moreover, $f_*\mathcal{O}_Y(-2Z) = m_P^2 + (y)$ for some element $y \in m_P \setminus m_P^2$, and the $\mathcal{O}_X$-algebra $\mathcal{R}(Y,K_Y) = \bigoplus_{k\ge 0} f_*\mathcal{O}_Y(-kZ)$ is generated by m_P in degree 1 and y in degree 2 (together with m_P^2).*

If $d=1$ then $\operatorname{mult}_P X = 2$ and $\operatorname{emb\,dim}_P X = 3$, that is, $P \in X$ is a hypersurface double point. The linear system $|-Z|$ on Y has a single base point $Q \in Z$, which is a nonsingular point of Z; that is, $Z = Z_{\mathrm{f}}$, and $m_P\mathcal{O}_Y = m_Q\mathcal{O}_Y(-Z)$. The algebra $\mathcal{R}(Y,K_Y) = \bigoplus_{k\ge 0} f_*\mathcal{O}_Y(-kZ)$ is generated by m_P in degree 1, and elements $y \in m_P \setminus m_P^2$ in degree 2 and $z \in m_P \setminus (m_P^2, y)$ in degree 3.

Remark. In either case, the system of ideals

$$I_k = f_*\mathcal{O}_Y(-kZ) \subset \mathcal{O}_X \quad \text{for } k \ge 0$$

defines a filtration of $\mathcal{O}_X$, and the relative canonical model $Y \to \overline{Y} \to X$, defined as $\overline{Y} = \operatorname{Proj} \mathcal{R}(Y,K_Y)$, is obtained by blowing up X along this filtration. In either case, the ideals I_k can be defined by a weighting of the ambient space $\mathbb{A}^3$. That is, give

$$y, x_1, x_2 \text{ weights } (2,1,1) \text{ in case } d=2,$$
$$z, y, x \text{ weights } (3,2,1) \text{ in case } d=1.$$

then I_k is the ideal generated by monomials of weighted degree $\ge k$. Taking $\operatorname{Proj} \bigoplus_{k\ge 0} I_k$ can also be described in toric geometry as the weighted blowup.

When $d=2$, the relative canonical model can also be described as the blowup of X at P followed by normalisation.

Exercises to Chapter 4

1. Let $\mathbb{Z}/2$ act on $\mathbb{A}^2$ by $(u,v) \mapsto (-u,-v)$. Prove that the subring of $k[u,v]$ of invariant polynomials is generated by $x = u^2, y = uv, z = v^2$, and that the morphism $\mathbb{A}^2 \to \mathbb{A}^3$ given by $(u,v) \mapsto (x,y,z)$ defines an embedding of the orbit space $\mathbb{A}^2/(\mathbb{Z}/2)$ as the ordinary quadratic cone $X : (xz = y^2) \subset \mathbb{A}^3$ of 4.1.

2. Let $V \to \mathbb{A}^2$ be the blowup of the origin; show that the action of $\mathbb{Z}/2$ of Ex. 4.1 extends to an action on V which fixes the -1-curve pointwise, and that the quotient space $V/(\mathbb{Z}/2)$ is isomorphic to the blowup $Y \to X$, and fits into a commutative diagram

$$\begin{array}{ccc} V & \to & \mathbb{A}^2 \\ \downarrow & & \downarrow \\ Y & \to & X. \end{array}$$

Calculate the canonical class K_V and its relation with $K_{\mathbb{A}^2}$ and K_Y.

3. Carry out blowups to resolve the Du Val singularities

$$\begin{array}{ll} A_n : & x^2 + y^2 + z^{n+1} = 0, \\ D_n : & x^2 + y^2 z + z^{n-1} = 0, \\ E_6 : & x^2 + y^3 + z^4 = 0, \\ E_7 : & x^2 + y^3 + yz^3 = 0, \\ E_8 : & x^2 + y^3 + z^5 = 0 \end{array}$$

in the spirit of 4.2. [Hint: In 4.2, I only calculated one affine piece of the blowup and stated that nothing interesting happens on the others. In general, one has to consider more than one affine piece.] Check that the exceptional locus is a bunch of -2-curves with intersections given by the Dynkin diagram.

4. Calculate the numerical cycle Z_{num} for each of the Du Val singularities. Check that $Z_{\text{num}}^2 = -2$. [Hint: Start from $Z_0 = \sum \Gamma_i$, and successively increase the coefficients of Γ_i only if $Z_j \Gamma_i > 0$.]

5. If $\bigcup \Gamma_i$ is a connected bunch of -2-curves with negative definite intersection matrix $(\Gamma_i \Gamma_j)$, check that the configuration is given by one of the Dynkin diagrams A_n, D_n, E_6, E_7, E_8. [Hint: If the configuration is more complicated, you can write down a combination $D = \sum a_i \Gamma_i$ with $D^2 \geq 0$. All the completed Dynkin diagrams $\widetilde{A}_n, \ldots, \widetilde{E}_8$ appear as logical ends of the argument. For example, on the graph

$\widetilde{D}_4$:

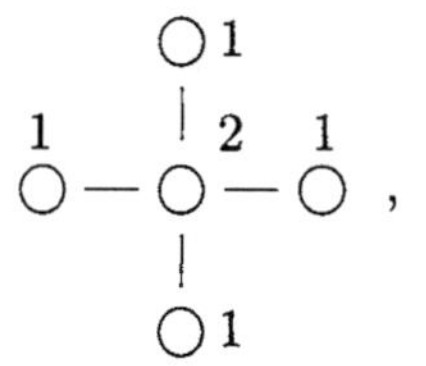

$D = \sum a_i \Gamma_i$ (with a_i as indicated) has $D^2 = 0$, which cannot happen in a negative definite graph G, so that G must have valency ≤ 3. You have to prove that G does not have a loop, or 2 vertexes of valency 3, or 1 vertex of valency 3 and very long arms out of it, etc.]

6. Let $\mathbb{Z}/3$ act on $\mathbb{A}^2$ by $(u,v) \mapsto (\varepsilon u, \varepsilon^2 v)$, where ε is a primitive 3rd root of 1. Study the quotient

$$\mathbb{A}^2/(\mathbb{Z}/3) = X : (xz = y^3) \subset \mathbb{A}^3$$

as in Ex. 4.1 in terms of the quotient map $\mathbb{A}^2 \to \mathbb{A}^3$ defined by

$$x = u^3, y = uv, z = v^3.$$

Let $Y \to X$ be the minimal resolution. Show how to construct a $\mathbb{Z}/3$ cover $V \to Y$ by first blowing up the origin in $\mathbb{A}^2$, then blowing up the two points in which the -1-curve meets the coordinate axes.

The point of the question is that the minimal resolution Y does not have a finite $\mathbb{Z}/3$ cover by a nonsingular blowup of $\mathbb{A}^2$.

7. Do the same as in Ex. 4.6 for $\mathbb{Z}/4$ acting by $(u,v) \mapsto (iu, i^3v)$, where $i^2 = -1$.

8. Let ε be a primitive 5th root of 1. Make $\mathbb{Z}/5$ act on $\mathbb{A}^2$ by

$$(u,v) \mapsto (\varepsilon u, \varepsilon^2 v).$$

Show that the subring of $k[u,v]$ of invariant polynomials is generated by 4 monomials, so that the quotient space $\mathbb{A}^2/(\mathbb{Z}/5) \cong X \subset \mathbb{A}^4$. (This is the quotient singularity of type $(5,2)$ in Hirzebruch's terminology, or of type $\frac{1}{5}(1,2)$ in that of [**YPG**].) Show that the ideal of relations holding between the invariant monomials is generated by 3 relations that can be written as the maximal minors of a 2×3 matrix. Show how to resolve the singularity by an explicit blowup, and describe the exceptional curves.

9. Let ε be a primitive $2n$th root of 1 and consider the two matrixes

$$A = \begin{pmatrix} \varepsilon & 0 \\ 0 & \varepsilon^{-1} \end{pmatrix} \quad \text{and} \quad B = \begin{pmatrix} 0 & -1 \\ 1 & 0 \end{pmatrix}.$$

Show that they generate a binary dihedral subgroup $\mathrm{BD}_n \subset \mathrm{SL}(2,\mathbb{C})$ of order $4n$. Find the ring of invariants of BD_n acting on $\mathbb{C}^2$ and prove that $\mathbb{C}^2/\mathrm{BD}_n$ is isomorphic to the Du Val singularity D_{n+2}. [Hint: Find first the ring of invariants of $\langle A \rangle \cong \mathbb{Z}/2n$, and show that B acts by an involution on it.]

10. In Ex. 4.9 you saw that the inclusion $\mathbb{Z}/2n \lhd \mathrm{BD}_n$ defines a double cover $A_{2n-1} \to D_{n+2}$ between Du Val singularities. Show also that $\mathrm{BD}_n \subset \mathrm{BD}_{2n}$ is a normal subgroup of index 2 and that it defines a double cover $D_{n+2} \to D_{2n+2}$ between Du Val singularities.

11. Prove Proposition 4.12, (2). [Hint: Several methods are possible. For example, you can prove that $h^0(\mathcal{O}_D(-Z_{\text{num}})) \neq 0$, so that $h^0(\mathcal{O}_{D+Z}) \geq 2$ for every $D > 0$. Or you can calculate $\chi(\mathcal{O}_{D+Z})$ using the numerical games of (1).]

12. Prove Proposition 4.12, (3). [Hint: If $Z_{\text{num}} = D_1 + D_2$, write out $p_a(Z_{\text{num}}) = 0$, $p_a(D_1), p_a(D_2) \leq 0$ in terms of the adjunction formula 3.6 or 4.10. Compare the method of 4.21, (7–8)]

13. Consider the codimension 2 singularity $P \in X \subset \mathbb{C}^4$ defined by the two equations $x_1^2 = y_1^3 - y_2^3$ and $x_2^2 = y_1^3 + y_2^3$. Prove that $P \in X$ has a resolution $f\colon Y \to X$ such that $f^{-1}P$ is a nonsingular curve C of genus 2 and $C^2 = -1$, but with $H^0(\mathcal{O}_C(-C)) = 0$. Prove that x_1, x_2 vanish along C with multiplicity 2 and y_1, y_2 with multiplicity 3. Hence $Z_{\mathfrak{f}} = 2C$, although obviously $Z_{\text{num}} = C$.

14. If $(\Gamma_i\Gamma_j)$ is negative definite, prove that $H^0(\mathcal{O}_D(D)) = 0$ for every effective divisor supported on Γ_i. More generally, if $\mathcal{L}$ is a line bundle on D such that $\deg_{\Gamma_i} \mathcal{L} \leq 0$ for all i then $H^0(\mathcal{O}_D(D) \otimes \mathcal{L}) = 0$. [Hint: Since $D^2 < 0$, any section $s \in H^0(\mathcal{O}_D(D))$ must vanish on some components of D. Now use the argument of Lemma 3.10 and Ex. 3.27.]

15. Let $Y \to X$ be a resolution of a normal surface singularity $P \in X$, and $\mathcal{L}$ a line bundle on Y such that $\mathcal{L}\Gamma_i \geq K_Y\Gamma_i$ for each exceptional curve Γ_i. Prove that $H^1(D, \mathcal{L}_{|D}) = 0$ for every exceptional divisor D, and deduce that $H^1(Y, \mathcal{L}) = R^1 f_*\mathcal{L} = 0$. This is related to the vanishing theorems of Kodaira and Grauert–Riemenschneider, but works in all characteristic.

16. Suppose that X is projective with ample H, and $f\colon Y \to X$ as usual. Let $Z = \sum a_i\Gamma_i$ be an effective exceptional divisor with $Z\Gamma_i < 0$. Prove that $nf^*H - Z$ is ample on Y for $n \gg 0$. [Hint: Some section of nH on X vanishes on Z, so that $nf^*H - Z$ is effective for some n. Now prove that by taking a larger n if necessary, $(nf^*H - Z)C > 0$ for every curve $C \subset Y$ and $(nf^*H - Z)^2 > 0$. The result followed by the Nakai–Moishezon ampleness criterion, compare footnote to 4.8 and [**H1**], Chapter V, 1.10.]

17. Show how to resolve the elliptic Gorenstein singularity

$$T_{p,q,r} \quad \text{defined by} \quad xyz + x^p + y^q + z^r \quad \text{for} \quad \frac{1}{p} + \frac{1}{q} + \frac{1}{r} \leq 1,$$

and describe the configuration of exceptional curves. [Hint: Start with the cases $p, q, r > 3$. Blowing up the maximal ideal gives a surface X_1 meeting the exceptional $\mathbb{P}^2$ in the tangent cone, which is the triangle $xyz = 0$, and having Du Val singularities at the 3 corners.]

18. Show how to resolve the elliptic Gorenstein singularities of degree 1 given by

$$x^2 + y^3 + z^k = 0 \quad \text{for } k = 6, 7, \ldots, 11, \quad \text{and} \quad x^2 + y^3 + yz^l = 0 \quad \text{for } l = 5, 7.$$

[Hint: Use the weighted blowup with weights 3, 2, 1, followed by the resolution of Du Val singularities.]

19. Ex. 4.8 is the special case $p = 2, q = r = 1$ of the rational triple points of Example 4.12. If you enjoyed it and want more of the same, carry out the calculations to verify the assertions made in Example 4.12.

20 (harder). The possible configurations of exceptional curves of a resolution $Y \to X$ of a rational triple point $P \in X$ are listed in [**A2**], p. 135. Verify this list following the hint given there. Find equations for some of these following the ideas of Example 4.12.

21 (harder). Rational surface singularities include the quotient singularities $\mathbb{C}^2/G$ where $G \subset \mathrm{GL}(2, \mathbb{C})$ is a finite group not containing any quasireflections. Prove that $P \in X$ is a quotient singularity if and only if it is a rational singularity, and the minimal resolution $Y \to X$ has $-K_Y \overset{\text{num}}{\sim} \sum a_i\Gamma_i$ with $a_i \in \mathbb{Q}$, $a_i < 1$. List all the cases. [Hint: Because $P \in X$ is rational, some multiple of K_X is a Cartier divisor. Then $P \in X$ has a cyclic cover which is a Du Val singularity. The lists can be found in Alexeev [**Utah2**], Chapter 3.]

22. Let Z be a 2-connected divisor on a nonsingular surface and $\mathcal{L}$ a line bundle with $\mathcal{L} \otimes \omega_Z^{-1}$ nef and of degree ≥ 1. Modify the proof of Lemma 4.23 to prove that if $|\mathcal{L}|$ is not free at Q then $Q \in Z$ is a nonsingular point, and $\mathcal{L} \cong \omega_Z(Q)$.

23. State and prove 4.24, Claim 2 for $d = 3$ and $d = 4$. [Hint: For $d = 4$, compare [**UAG**], §1.]

24. Check the exact sequence 4.24, (14). [Hint: Step 1. There is a restriction map $\mathcal{I}_{i+1} \to \mathcal{I}_i$. The kernel is obviously contained in $\mathcal{O}_{\mathbb{P}^i}(-1)$. Step 2. Because

$A_{i+1} \not\subset \mathbb{P}^{i-1}$, at Q, the kernel is equal to $m_Q\mathcal{O}_{\mathbb{P}^i}(-1)$. Step 3. It's tricky to see that $\mathcal{I}_{i+1} \to \mathcal{I}_i$ is surjective by local considerations at Q in the crucial case $Q \in \operatorname{Supp} A_i$; however, there is a very easy proof by calculating χ of the 3 sheaves in the exact sequence

$$0 \to m_Q\mathcal{O}_{\mathbb{P}^i}(-1) \to \mathcal{I}_{i+1} \to \mathcal{I}_i.]$$

CHAPTER D
Minimal Models of Surfaces

The final two chapters explain the proof of the classification of surfaces. The treatment of the classification of surfaces in these chapters is nonclassical, with emphasis on a new point of view, motivated in part by recent developments in 3-folds. For classical treatments, see [**Beauville**], or [**3 authors**].

D.0. Introduction and preview

Let X be a projective nonsingular surface defined over k; it's my responsibility to make everything work over any algebraically closed field k, but the student who is so inclined may suppose that $k = \mathbb{C}$ to make things easier. The classification of surfaces breaks up into two parts. (For the definition of K_X and of a nef divisor, see the end of D.1 below, or various places earlier in these notes.)

Case K_X not nef

This corresponds to studying the choice of birational model of X, the theory of minimal models and the characterisation of ruled and rational surfaces. Very roughly, the main result is that every surface is birational either to a surface X with K_X nef or to $\mathbb{P}^2$ or $\mathbb{P}^1 \times C$ with C a curve.

The main point to note is the dichotomy: after choosing a suitable birational model, either K_X is nef, or $-K_X$ is ample, or relatively ample on a nontrivial fibration $X \to C$. In this form, the result remains true over a nonclosed field, and is capable of extension to higher dimensions, although of course everything becomes harder.

Case K_X nef

This part is concerned with dividing surfaces into classes according to the numerical properties of K_X, and then proving that this coincides with the classification by the Kodaira dimension $\kappa(X)$. Very roughly, the main result is that X has "sufficiently many" holomorphic differential forms. More precisely:

(i) if $K_X \overset{\text{num}}{\not\sim} 0$ (numerical equivalence) then $h^0(mK_X) \to \infty$ with m;

(ii) if $K_X \overset{\text{num}}{\sim} 0$ then $mK_X \overset{\text{lin}}{\sim} 0$ for some m. In other words, the topological condition $c_1(X) = 0 \in H^2(X, \mathbb{R})$ implies that $\mathcal{O}_X(mK_X)$ is analytically isomorphic to $\mathcal{O}_X$.

Remark. In both parts, the conditions can be relaxed to allow any field k, and X to have various singularities; it can also make sense to study a *log surface*, that is, a surface X together with a marked divisor D (say reduced, normal crossing), and classify the set-up X with D, replacing K_X by $K_X + D$.

D.1. Preliminaries: 1-cycles versus codimension 1 cycles

Let X be any projective scheme. Write $\operatorname{Pic} X = H^1(X, \mathcal{O}_X^*)$ for the group of invertible sheaves (or divisor classes) on X, and $\{\text{1-cycles}(X)\}$ for the group of 1-cycles on X, that is, the free group on all curves $\Gamma \subset X$. Then there is a pairing

$$\operatorname{Pic} X \times \{\text{1-cycles}(X)\} \to \mathbb{Z}$$

defined by $(\mathcal{L}, \Gamma) \mapsto \mathcal{L}\Gamma = \deg_\Gamma \mathcal{L}_{|\Gamma}$, and extended by linearity.

D.1.1. Definition of numerical equivalence

Two invertible sheaves $\mathcal{L}_1, \mathcal{L}_2 \in \operatorname{Pic} X$ are *numerically equivalent* if $\mathcal{L}_1\Gamma = \mathcal{L}_2\Gamma$ for every curve $\Gamma \subset X$; we write $\mathcal{L}_1 \overset{\text{num}}{\sim} \mathcal{L}_2$. Similarly, two 1-cycles C_1, C_2 are *numerically equivalent* (written $C_1 \overset{\text{num}}{\sim} C_2$) if $\mathcal{L}C_1 = \mathcal{L}C_2$ for every $\mathcal{L} \in \operatorname{Pic} X$. Define

$$N^1X = (\operatorname{Pic} X/ \overset{\text{num}}{\sim}) \otimes \mathbb{R} \quad \text{and} \quad N_1X = (\text{1-cycles} \otimes \mathbb{R})/ \overset{\text{num}}{\sim};$$

obviously, by definition N^1X and N_1X are dual $\mathbb{R}$-vector spaces, and $\overset{\text{num}}{\sim}$ is the smallest equivalence relation for which this holds. Clearly, N^1X contains $\mathbb{Z}$- and $\mathbb{Q}$-forms, denoted by $N^1_{\mathbb{Z}}X$, $N^1_{\mathbb{Q}}X$ respectively. The fact that $\rho = \dim_{\mathbb{R}} N^1X$ is finite is the Néron–Severi theorem.

Notice that for a variety defined over $\mathbb{C}$, it is natural to consider N_1X as a subspace of $H_2(X, \mathbb{R})$ and N^1X as the corresponding quotient of $H^2(X, \mathbb{R})$ (in fact both are direct summands). Over $\mathbb{C}$, the finite dimensionality of N^1X and N_1X follows at once from this (see App. to D.1 below); thus the content of the Néron–Severi theorem is a generalisation of this to the purely algebraic set-up. Notice also that N^1X is the same thing as $\mathrm{NS}(X) \otimes \mathbb{R}$, where $\mathrm{NS}(X)$ is the Néron–Severi group of X (that is, Cartier divisors modulo algebraic equivalence).

D.1.2. The case of a surface

I have deliberately avoided the following point: if X is a nonsingular surface then, of course, 1-cycles are the same as codimension 1 cycles, so that $N^1X = N_1X$. This can be seen in two ways:

(1) every curve $\Gamma \subset X$ is a Cartier divisor (that is, locally defined by one equation) and

$$\{\text{1-cycles of } X\} = \operatorname{Div} X \to \operatorname{Pic} X = \operatorname{Div} X/ \overset{\text{lin}}{\sim}$$

is defined by

$$\sum n_iC_i \mapsto \mathcal{O}_X\left(\sum n_iC_i\right).$$

(2) $\operatorname{Pic} X$ has an intersection pairing defined by

$$\mathcal{L}_1\mathcal{L}_2 = \chi(\mathcal{O}_X) - \chi(\mathcal{L}_1) - \chi(\mathcal{L}_2) + \chi(\mathcal{L}_1 \otimes \mathcal{L}_2),$$

which becomes nondegenerate on N^1X.

Despite the canonical equality $N^1X = N_1X$, it is nevertheless convenient to distinguish between the two spaces. This is useful for dealing with the higher dimensional case, and also for singular surfaces, when it's no longer true that all divisors are Cartier.

Although very simple, this is one of the key ideas of Mori theory, and came as a surprise to anyone who knew the theory of surfaces before 1980: the quadratic intersection form for curves on a nonsingular surface can for most purposes be replaced by the bilinear pairing between N^1 and N_1, and in this form generalises to singular varieties and to higher dimension.

D.1.3. Definition of nef

An element $D \in N^1X$ is *numerically eventually free* or *nef* if $D\Gamma \geq 0$ for every curve $\Gamma \subset X$. This is related to the notion of a free linear system $|D|$: if $|mD|$ is free for some $m > 0$ ("eventually free") then D is nef.

The relation between nef and ample divisors is the content of the Kleiman criterion (see D.2).

D.1.4. The canonical class

If X is a nonsingular n-fold then $\mathcal{O}_X(K_X) = \omega_X = \Omega^n_X$ is a line bundle, locally generated by $dx_1 \wedge \cdots \wedge dx_n$. (Compare A.10.) This is the only nontrivial line bundle which can be intrinsically associated with any variety.

Appendix to D.1. Cycle class groups and homology

D.1.5. Proposition. *(i) If X is a variety over $\mathbb{C}$ then $N_1X \subset H_2(X,\mathbb{R})$ and $H^2(X,\mathbb{R}) \twoheadrightarrow N^1X$ is the dual quotient.*

(ii) In terms of Hodge theory,

$$N^1X = H^2(X,\mathbb{R}) \cap H^{1,1} = H^2(X,\mathbb{R})/\big(H^{2,0} \oplus H^{0,2} \cap H^2(X,\mathbb{R})\big).$$

(iii) Algebraic equivalence of 1-cycles implies numerical equivalence.

Sketch Proofs. (i) The inclusion takes a curve C into $[C]$ the 2-cycle obtained by triangulating C. The degree of a line bundle $\mathcal{L}$ on a curve is the same thing as the first Chern class $c_1(\mathcal{L}) \in H^2(X,\mathbb{R}) \twoheadrightarrow N^1X$ evaluated on the 2-cycle $[C] \in H_2(X,\mathbb{R})$.

(ii) Write down the exponential sequence of sheaves.

(iii) This follows easily from the fact that the degree of a line bundle is invariant in a flat family of curves. □

D.2. Easy consequences of RR

Let X be a nonsingular surface. Then RR takes the form

$$\chi(\mathcal{O}_X(D)) = \chi(\mathcal{O}_X) + \tfrac{1}{2}D(D - K_X).$$

The fact that this is a quadratic function of D has an immediate corollary.

D.2.1. Proposition. $D^2 > 0$ *implies either $h^0(nD) \neq 0$ or $h^0(-nD) \neq 0$ for $n \gg 0$. The sign of HD (for any ample H) distinguishes the two cases.*

Proof. If $D^2 > 0$ then

$$h^0(nD) + h^2(nD) \geq \chi(\mathcal{O}_X(nD)) \sim n^2D^2/2,$$

so that either $h^0(nD)$ or $h^2(nD)$ goes to infinity with n. Now by Serre duality,

$$h^2(nD) = h^0(K_X - nD).$$

I can write down the same argument with D replaced by $-D$, and conclude that either $h^0(-nD)$ or $h^0(K_X + nD)$ goes to infinity with n.

Now obviously, $h^0(K_X - nD)$ and $h^0(K_X + nD)$ cannot both go to infinity with n; indeed, if $h^0(K_X - nD) \neq 0$, multiplying by a nonzero section $s \in H^0(K_X - nD)$ gives an inclusion $H^0(K_X + nD) \hookrightarrow H^0(2K_X)$ so that $h^0(K_X + nD) \leq h^0(2K_X)$. Therefore either $h^0(nD)$ or $h^0(-nD)$ grows quadratically with n.

Here is an alternative way of concluding the argument: it is easy to see that for any divisor L on X,

$$|h^0(-nD) - h^0(L - nD)| < \text{const}.n.$$

(Just write $L = A - B$ with A, B curves and consider restriction maps.) Hence if $h^2(nD) = h^0(K_X - nD)$ grows quadratically with n, so does $h^0(-nD)$. Q.E.D.

D.2.2. Corollary (the Index Theorem). *If H is ample on X then $HD = 0$ implies $D^2 \leq 0$. Moreover if $D^2 = 0$ then $D \overset{\text{num}}{\sim} 0$.*

In other words, the intersection pairing on N^1X has signature $(+1, -(\rho - 1))$. The cone $D^2 > 0$ looks like the "light cone": the "past" and "future" components of the cone are separated by the hyperplane $H^\perp$. Another way of stating this, which is very useful in calculations, is that if D_1, D_2 are divisors and $(\lambda D_1 + \mu D_2)^2 > 0$ for some $\lambda, \mu \in \mathbb{R}$ then the determinant

$$\begin{vmatrix} D_1^2 & D_1D_2 \\ D_1D_2 & D_2^2 \end{vmatrix} \leq 0,$$

with equality if and only if D_1, D_2 are $\mathbb{Q}$-linearly dependent in N^1X.

Proof. If $D^2 > 0$ then either nD or $-nD$ is equivalent to a nonzero effective divisor for $n \gg 0$. Since H is ample, either of these conditions implies $HD \neq 0$. This proves the first statement.

The final part is left as an exercise. [Hint: If $DA \neq 0$ for some curve A then replace A by $B = A - \lambda H$ so that $HB = 0$; now if $DA \neq 0$ also $DB \neq 0$, and some linear combination of D and B has $(D + \alpha B)^2 > 0$, contradicting the first part.] Q.E.D.

D.2.3. Corollary (weak form of the Kleiman criterion). *D is nef implies*

1. *$D^2 \geq 0$ and*
2. *$D + \varepsilon H$ is ample for any $\varepsilon \in \mathbb{Q}$, $\varepsilon > 0$.*

Proof. Introduce the quadratic polynomial $p(t) = (D + tH)^2$. Then $p(t)$ is a continuous increasing function of $t \in \mathbb{Q}$ for $t > 0$, and $p(t) > 0$ for sufficiently large t. The following assertion then obviously implies (1):

D.2.4. Claim. *Let $t \in \mathbb{Q}$, $t > 0$; then $p(t) > 0$ implies that also $p(t/2) > 0$.*

Proof. For $(D+tH)^2 > 0$, together with $H(D+tH) > 0$ implies that $n(D+tH)$ is equivalent to an effective divisor for suitable $n \gg 0$. By the assumption that D is nef, $D(D+tH) \geq 0$, and hence $(D+(t/2)H)^2 = D(D+tH) + (t/2)^2H^2 > 0$.

For (2) it is enough to notice that $(D+\varepsilon H)\Gamma > 0$ for every curve $\Gamma \subset X$ and $(D+\varepsilon H)^2 > 0$, so that $D+\varepsilon H$ satisfies the conditions of the Nakai–Moishezon criterion (see for example [**H2**], Chapter I or [**H1**], Chapter V, 1.10). Q.E.D.

D.2.5. The Kleiman criterion

Corollary D.2.3 can be thought of as saying that nef divisors are limits of ample divisors. It is not hard to get from this to (a weak form of) Kleiman's ampleness criterion, which asserts that conversely, the cone of ample divisors is the interior of the nef cone in N^1X. For this make the following definitions: let $\mathrm{NE}\,X \subset N_1X$ be the cone of effective 1-cycles, that is

$$\mathrm{NE}\,X = \left\{C \in N_1X \;\middle|\; C = \sum r_i\Gamma_i \text{ with } r_i \in \mathbb{R}, r_i \geq 0\right\}.$$

Let $\overline{\mathrm{NE}}\,X$ be the closure of $\mathrm{NE}\,X$ in the real topology of N_1X. This is called the *Kleiman–Mori cone.*

One effect of taking the closure is the following trivial observation, which has many important uses in applications: if $H \in N^1X$ is positive on $\overline{\mathrm{NE}}\,X \setminus 0$, then the section $(Hz = 1) \cap \overline{\mathrm{NE}}\,X$ is compact. Indeed, the projectivisation of the closed cone $\overline{\mathrm{NE}}\,X$ is a closed subset of $\mathbb{P}^{\rho-1} = \mathbb{P}(N_1X)$, and therefore compact, and the section $(Hz = 1)$ projects homeomorphically to it. The same also holds for any face or closed subcone of $\overline{\mathrm{NE}}\,X$.

Theorem. *For $D \in \operatorname{Pic} X$, view the class of D in N^1X as a linear form on N_1X. Then*

$$D \text{ is ample} \iff DC > 0 \text{ for all } C \in \overline{\mathrm{NE}}\,X \setminus 0.$$

Note that it is not true that $DC > 0$ for every curve $C \subset X$ implies D ample. The necessary and sufficient condition $DC > 0$ for every $C \in \overline{\mathrm{NE}}\,X \setminus 0$ is just a tiny bit stronger than that. You can interpret it as saying that DC is "reliably bigger than $\|C\|$".

Sketch proof. The implication $\Longleftarrow$ comes at once from Corollary D.2.3, (2), since it is easy to see by a compactness argument that for H ample and sufficiently small $\varepsilon > 0$, $(D-\varepsilon H)C > 0$ for all $C \in \overline{\mathrm{NE}}\,X \setminus 0$, so $D = (D-\varepsilon H) + \varepsilon H$ is ample by Corollary D.2.3, (2). For $\Longrightarrow$ you need to use a compactness argument to prove that if a norm $\|\ \|$ is chosen on N_1X then for an ample divisor D there exists $\varepsilon > 0$ such that

$$DC > \varepsilon\|C\| \quad \text{for all } C \in \mathrm{NE}\,X$$

(see for example [**H2**], Chapter I).

Remarks. (1) This is only a weak form of Kleiman's criterion, since X is a priori assumed to be projective. The full strength of Kleiman's criterion gives a necessary and sufficient condition for ampleness in terms of the geometry of $\overline{\mathrm{NE}}\,X$. In particular, it can be used to prove that a variety is projective (without assuming it!). See [**Kleiman**].

(2) An analogous statement, the Kleiman criterion (weak form), holds for an arbitrary projective k-scheme: if D is a nef Cartier divisor then $D^rZ \geq 0$ for every

irreducible r-dimensional subvariety Z, and D is ample if and only if it is positive on $\overline{\mathrm{NE}}\,X \setminus 0$. The proof is more or less the same as (D.2.3–5), although you have to use cohomology and carry out an induction on the dimension. It goes something like this.

Step 1. (A version of the Nakai–Moishezon criterion.) If $D^r Z > 0$ for every r and every irreducible r-dimensional subvariety Z then

$$H^j(\mathcal{F} \otimes \mathcal{O}_X(nD)) = 0.$$

for every coherent sheaf $\mathcal{F}$, every $j > 0$, and all $n \gg 0$. (Just use induction on $\dim \operatorname{Supp} \mathcal{F}$.)

Step 2. If D is nef and H an ample Cartier divisor, then using RR, $(D+tH)^r Z > 0$ implies that $H^0(\mathcal{O}_Z(N(D+tH))) \neq 0$, so that $(D+tH)_{|Z}$ is an effective $\mathbb{Q}$-divisor. Then as in Claim D.2.4, using a further induction, one proves that for $t > 0$,

$$(D + tH)^r Z > 0 \implies (D + (t/r)H)^r Z > 0.$$

D.3. The rationality lemma and cone theorem

Another easy consequence of the form of RR is the following:

D.3.1. Rationality lemma. *Suppose that H is ample and K_X is not nef. Define the* nef threshold $t_0 = t_0(H)$ *by*

$$t_0 = \sup\{t \mid H + tK_X \text{ is nef}\},$$

so that $t_0 \in \mathbb{R}$, $t_0 > 0$. Then (i) $t_0 \in \mathbb{Q}$, and (ii) its denominator is ≤ 3.

Note that the condition K_X not nef means exactly that $\overline{\mathrm{NE}}\,X$ meets the half-space $(K_X z < 0)$ of $N^1 X$. Please draw a picture of the hyperplane $H + tK_X$ rotating from its initial position at $t = 0$ outside $\overline{\mathrm{NE}}\,X$ (since H is ample) to its asymptotic position for $t \gg 0$, when it cuts into $\overline{\mathrm{NE}}\,X$; t_0 is the threshold value at which it first hits $\overline{\mathrm{NE}}\,X$.

Discussion. The underlying reason behind the cone theorem and the contraction theorem, and all their generalisations to higher dimensional singular and log varieties is *vanishing*. Although the arguments get technically quite complicated, the main mechanism is very simple: when $t = t_0 + \varepsilon$, just after the critical value, $H + tK_X$ is not nef, but vanishing applies to it, because $H + (t-1)K_X$ is ample.

The rationality lemma is proved in much greater generality in [**Kawamata**] and [**Kollár**]. For a good introduction to the problem, see [**Utah1**].

Step 1. If $n(H + t_1 K_X) = D_1$ is effective for some $n > 0$, with $t_1 \in \mathbb{Q}$ and $t_1 > t_0$ (so that D_1 is not nef), then t_0 is determined by

$$t_0 = \min_{\Gamma \subset D_1} \left\{ \frac{H\Gamma}{-K_X\Gamma} \right\},$$

where the minimum runs over the irreducible components Γ of D_1 such that $K_X\Gamma < 0$. In fact for $0 < t < t_1$ the divisor $H + tK_X$ is a positive combination of H and D_1, so that it fails to be nef if and only if $(H + tK_X)\Gamma < 0$ for a component Γ of D_1.

Step 2. Now if $t_0 \notin \mathbb{Q}$ then for $n, m \in \mathbb{Z}$ with $n < mt_0 < n+1$, it follows that $mH + nK_X$ is ample but $mH + (n+1)K_X$ is not nef. Since $mH + nK_X$ is ample, by Kodaira vanishing $h^0(mH + (n+1)K_X)$ is given by RR, and will sometimes be positive; this gives either a proof by Step 1 or a contradiction. More precisely, for $m > 0$, set $mt_0 = n + \alpha$ with $0 \le \alpha < 1$, and write $D_0 = H + t_0K_X \in N^1X$. Then RR gives

$$\begin{aligned} H^0(mH + (n+1)K_X) &\ge \chi(\mathcal{O}_X) + \tfrac{1}{2}(mH + (n+1)K_X)(mH + nK_X) \\ &= \chi(\mathcal{O}_X) + \tfrac{1}{2}(m^2D_0^2 + m(1-2\alpha)D_0K_X - \alpha(1-\alpha)K_X^2). \end{aligned}$$

(Here ≥ 0 comes from the vanishing of H^2, which is trivial in any characteristic; if $\operatorname{char} k = 0$, the vanishing of H^1 gives equality.)

Hence if $D_0^2 > 0$ then $H^0 \ne 0$ for all large m; if $D_0^2 = 0$ and $D_0 \overset{\text{num}}{\not\sim} 0$ then necessarily $D_0K_X < 0$ (because $D_0(H + t_0K_X) = 0$ and $D_0H > 0$), and $H^0 \ne 0$ if m is large and $1 - 2\alpha$ is bounded away from 0. If $D_0^2 = D_0K_X = 0$ then also $D_0H = 0$, so that $D_0 \overset{\text{num}}{\sim} 0$, and this implies that $K_X \overset{\text{num}}{\sim} -(1/t_0)H$; in particular $-K_X$ is ample.

This proves (i).

The proof of (ii) is an exercise. [Hint: If $D_0^2 > 0$ then there must exist a curve L with $D_0L = 0$, since otherwise D_0 would be ample; it is easy to see that $D_0L = 0$ implies that L is a -1-curve, so that $t_0 \in \mathbb{Z}$. If $D_0^2 = 0$, one sees that $H+tK_X$ cannot be effective for any $t > t_0$. If $D_0 \ne 0$ and $2t_0 \notin \mathbb{Z}$, the above inequality gives a contradiction by taking $m \gg 0$ with $(1-2\alpha)$ negative and bounded away from zero. Finally if $D_0 \overset{\text{num}}{\sim} 0$ then $-K_X$ is ample, and it is easy to see (in any characteristic, see D.4 or especially E.2) that $\chi(\mathcal{O}_X) = 1$. Now since $H + t_0K_X \overset{\text{num}}{\sim} 0$, it follows that the denominator of t_0 divides K_X in $\operatorname{Pic} X$. It is now not hard to check that if $-K_X = dL$ with L ample, then $d \le 3$: indeed, a simple argument of projective geometry shows that $h^0(L) \le 2 + L^2$, whereas RR gives $h^0(L) \ge 1 + (1/2)(d-1)L^2$. (ii) also follows easily by the method of proof of the contraction theorem.]

The rationality lemma implies the Mori cone theorem by an argument that is pure convex body theory, taken from [**Utah1**], Lecture 11. (This result is not logically necessary for the proof of classification.)

D.3.2. Cone Theorem. *Set*

$$\overline{\text{NE}}_{K_X} X = \{C \in \overline{\text{NE}}\, X \mid K_XC \ge 0\}.$$

Then

$$\overline{\text{NE}}\, X = \overline{\text{NE}}_{K_X} X + \sum R_i, \tag{$*$}$$

where the R_i are extremal rays of $\overline{\text{NE}}\, X$ contained in $(K_Xz < 0)$. Moreover, for any ample Cartier divisor $H > 0$ and any given $\varepsilon > 0$, there are only finitely many extremal rays R_i with $(K_X + \varepsilon H)R_i \le 0$.

In other words, away from the halfspace $K_X \ge 0$, the cone is locally polyhedral. (Draw a picture for God's sake! I'm imprisoned in a word-processor.) It follows by the result of D.4 that each ray R_i is spanned by the class of an irreducible curve.

Idea of the proof. Suppose that $F_L = L^\perp \cap \overline{\mathrm{NE}}\,X$ is a face of the cone $\overline{\mathrm{NE}}\,X$ contained in the region $(K_X < 0)$ of the cone; if by bad luck F_L is positive dimensional, I can wiggle L slightly to make it a ray. This shows that each face of the cone contained in $(K_X < 0)$ is spanned by rays. In turn, I then show that each ray is defined by a set of linear equations with bounded denominators, so that the rays are discrete in $(K_X < 0)$.

To start on the formal proof, fix once and for all a basis of N^1X of the form $K_X, H_1, \dots, H_{\rho-1}$, where the H_i are ample and $\rho = \operatorname{rank} N^1$. As in the rationality lemma, for any nef element $L \in N^1X$, set

$$t_0(L) = \max\{t \mid L + tK_X \text{ is nef}\}.$$

If $L \in \operatorname{Pic} X$ is nef and the corresponding face $F_L = L^\perp \cap \overline{\mathrm{NE}}\,X$ is contained in $(K_X z < 0)$ (for example, if L is ample) then the rationality lemma gives $rt_0(L) \in \mathbb{Z}$ for the fixed integer $r = 6$.

D.3.3. Main Claim. *Let $L \in \operatorname{Pic} X$ be a nef divisor which supports a face F_L of $\overline{\mathrm{NE}}\,X$ contained in $(K_X z < 0)$. Consider $\nu L + H_i$ for all i and for $\nu \gg 0$.*

1. *$t_0(\nu L + H_i)$ in an increasing function of ν, is bounded above, and attains its bound.*
2. *Let ν_0 be any point after $t_0(\nu L + H_i)$ has attained its upper bound, and suppose $\nu > \nu_0$. Set*

$$L' = 6\big(\nu L + H_i + t_0(\nu L + H_i)K_X\big),$$

(multiplying by 6 is simply to ensure that $L' \in \operatorname{Pic} X$ by the rationality lemma). Then L' supports a face $F_{L'} \subset F_L$.

3. *If $\dim F_L \ge 2$ then there exists i and $\nu \gg 0$ such that*

$$L'_i = 6\big(\nu L + H_i + t_0(\nu L + H_i)K_X\big)$$

supports a strictly smaller face $F_{L'} \subsetneq F_L$.

4. *In particular, F_L contains an extremal ray R of $\overline{\mathrm{NE}}\,X$.*
5. *If $F_L = R$ is an extremal ray and $z \in R$ is a nonzero element then $6H_i z/K_X z \in \mathbb{Z}$.*
6. *The extremal rays of $\overline{\mathrm{NE}}\,X$ are discrete in $(K_X z < 0)$.*

Proof. (1) is almost obvious: $t_0(\nu L + H_i)$ is an increasing function of ν by construction. It is bounded above, because any point $z \in F_L \setminus 0$ obviously gives $Lz = 0$ and $(H_i + tK_X)z < 0$ for $t \gg 0$. It attains its bounds because t_0 varies in the discrete set $(1/6)\mathbb{Z}$.

(2) Suppose that $t_0 = t_0(\nu L + H_i)$ does not change with $\nu \ge \nu_0$, and let F_0 be the face orthogonal to the nef element $\nu_0 L + H_i + t_0 K_X$. Then for $\nu > \nu_0$, any $z \in F_{L'_i}$ satisfies

$$0 = (\nu L + H_i + t_0 K_X)z = (\nu - \nu_0)Lz + (\nu L + H_i + t_0 K_X)z,$$

and therefore $Lz = 0$.

For (3), consider $\nu \gg 0$, and

$$L'_i = 6\big(\nu L + H_i + t_0(\nu L + H_i)K_X\big)$$

for each i. Since $\nu \gg 0$, these are "small wiggles" of L in $\rho - 1$ linearly independent directions. Each $F_{L'_i} \subset F_L$ is a face of F_L. The intersection of all the $F_{L'_i}$ is contained in the set defined by

$$(\nu L + H_i + t_0(\nu L + H_i)K_X)z = 0,$$

which are $\rho - 1$ linearly independent conditions on z. Therefore at least one of the $F_{L'_i}$ is strictly smaller than F_L. (4) follows obviously from (3).

(5) follows from (2) and the rationality lemma. Indeed, since $F_L = R$ is a ray, (2) implies that $F_{L'_i} = F_L = R$. That is, R is orthogonal to $H_i + t_0K_X$, and $6t_0 \in \mathbb{Z}$.

Finally, (6) follows from (5), since in $(K_Xz < 0)$, every ray contains a unique element z with $K_Xz = -1$ and $H_iz \in (1/6)\mathbb{Z}$. Q.E.D.

Proof of Theorem D.3.2. Write $B = \overline{\mathrm{NE}}_{K_X} X + \sum R_i$ for the right-hand side of $(*)$. Note that $B \subset \overline{\mathrm{NE}}\, X$ is a closed convex cone: indeed, by (6) the R_i can only have accumulation points in $(K_Xz \geq 0)$, that is, in $\overline{\mathrm{NE}}_{K_X} X$.

Suppose that $B \subsetneq \overline{\mathrm{NE}}\, X$. Then there exists an element $M \in N^1X$ which is nef, and supports a nonzero face F_M of $\overline{\mathrm{NE}}\, X$ disjoint from B, necessarily contained in $(K_Xz < 0)$. By the usual compactness argument, for sufficiently small $\varepsilon > 0$, $M - \varepsilon K_X$ is ample, and $M + \varepsilon K_X$ is not nef, but is positive on B. These are all open conditions on M, and any open neighbourhood of M in N^1X contains a rational element M, so that by passing to this and taking a multiple, I can assume that $M \in \operatorname{Pic} X$. But now the rationality lemma and Claim D.3.3, (4) imply that there is a ray R of $\overline{\mathrm{NE}}\, X$ with $MR < 0$, so that $R \not\subset B$. This contradicts the choice of B, and proves $(*)$. Q.E.D.

D.4. Contraction theorem and minimal models of surfaces

A consequence of the rationality lemma is that if K_X is not nef then there exists a divisor D on X which is nef but not ample, and such that $D - \varepsilon K_X$ is ample for $\varepsilon > 0$ (namely the divisor $D = n(H + t_0K_X)$ appearing in the rationality lemma). Similarly, it is clear from the statement of the cone theorem that for each extremal ray $R = R_i$ of $\overline{\mathrm{NE}}\, X$ there is a divisor class D on X such that $D \geq 0$ on $\overline{\mathrm{NE}}\, X$ and $\overline{\mathrm{NE}}\, X \cap D^\perp = R$, and this divisor satisfies the same condition.

D.4.1. Contraction theorem. *Let D be a divisor which is nef but not ample, and such that $D - \varepsilon K_X$ is ample for some $\varepsilon > 0$. Then $|mD|$ is free for all $m \gg 0$.*

Equivalently, there exists a morphism $\varphi\colon X \to Y$ to a projective variety Y such that $D = \varphi^*(\text{ample})$. Of course, this means that for a curve $C \subset X$,

$$\varphi(C) = \mathrm{pt.} \iff DC = 0.$$

By taking normalisation, I can assume that $\mathcal{O}_Y = \varphi_*\mathcal{O}_X$. In the 3 possible cases, this morphism is

Case $D^2 > 0$: φ is the contraction of a finite number of disjoint -1-curves L_i to nonsingular points of a surface Y.

Case $D^2 = 0$ but $D \overset{\text{num}}{\not\sim} 0$: φ is a conic bundle structure $X \to$ (curve).

Case $D \overset{\text{num}}{\sim} 0$: X is a surface with $-K_X$ ample and $p_g = q = 0$ (del Pezzo surface).

Proof. If $D^2 > 0$ then by the Nakai–Moishezon ampleness criterion there must be curves L with $DL = 0$. Now any such curve is a -1-curve, that is, $L \cong \mathbb{P}^1$ and $L^2 = -1$; for by the index theorem $L^2 < 0$, and since $D - \varepsilon K_X$ is ample $K_X L < 0$, and therefore by the adjunction formula $K_X L = L^2 = -1$ and $p_a L = 0$. The index theorem implies that any two of these are disjoint. The standard classical proof of Castelnuovo's contractibility criterion then gives the result.

If $D^2 = 0$ but $D \overset{\text{num}}{\not\sim} 0$ then since $D - \varepsilon K_X$ is ample, $K_X D < 0$, and RR gives $h^0(mD) \sim (-K_X D/2)m$, so in particular $h^0(mD) \to \infty$. The mobile part of $|mD|$ then gives a conic bundle structure $X \to$ (curve), and using this it is not hard to see that $|mD|$ itself is free.

Finally, if $D \overset{\text{num}}{\sim} 0$ then the condition on $D - \varepsilon K_X$ implies that $-K_X$ is ample, and the assertion is just that $D \overset{\text{lin}}{\sim} 0$. The key to this is to prove that $\chi(\mathcal{O}_X) = 1$, therefore $\chi(\mathcal{O}_X(D)) = \chi(\mathcal{O}_X) = 1$, and $H^0(\mathcal{O}_X(D)) \neq 0$, giving $D \overset{\text{lin}}{\sim} 0$ as required. If $\operatorname{char} k = 0$ then by Kodaira vanishing $H^i(\mathcal{O}_X) = 0$ for $i = 1, 2$, and so $\chi(\mathcal{O}_X) = 1$. If $\operatorname{char} k = p$ then $p_g = 0$ and $K_X^2 > 0$, so that the following alternative argument implies that $q = 0$ also in the case $\operatorname{char} k = p$. (See E.2 for how to get rid of the reference to Mumford.)

D.4.2. Traditional argument

Because $p_g = 0$, the Picard scheme is reduced (see E.2, Remark (c), or compare [**M1**], Lecture 27), and $B_1 = B_3 = 2q$. Hence (in etale cohomology), $c_2 = 2 + B_2 - 4q$, so as $B_2 \geq 1$ and $K_X^2 > 0$ I get

$$12\chi(\mathcal{O}_X) = 12(1-q) = c_1^2 + c_2 \geq 4 - 4q;$$

thus $q \leq 1$, and $q = 1$ is only possible if $B_2 = 1$. But if $q = 1$ then the fibre and section of the Albanese morphism imply $B_2 \geq 2$, so a contradiction. Therefore $\chi(\mathcal{O}_X) = 1$. Q.E.D.

D.4.3. Corollary. *Let R be an extremal ray of $\overline{\mathrm{NE}}\, X$; then there exists a morphism $\varphi\colon X \to Y$ to a projective variety Y such that $-K_X$ is relatively ample for φ, $\mathcal{O}_Y = \varphi_*\mathcal{O}_X$, and for all curves $C \subset X$,*

$$\varphi(C) = \text{pt.} \iff DC = 0.$$

The corresponding classification of extremal rays gives

1. contraction of a -1-curve;
2. $\mathbb{P}^1$-bundle $\varphi\colon X \to C$ over a curve C;
3. $\mathbb{P}^2$.

Note that in (2), the assertion that $X = \mathbb{P}(\mathcal{E})$ for a rank 2 vector bundle is nontrivial: you need to know that $X \to C$ has a rational section, either by Tsen's theorem (see Theorem C.4.2 or [**Sh**], I.6), or an argument in topology using Poincaré duality, and the fact that because $p_g = 0$ the cohomology $H^2(X, \mathbb{Z})$ is spanned by curves on X (see for example [**Beauville**], Theorem III.4). Similarly, to know that $X \cong \mathbb{P}^2$ in (3), the main point is to prove that $-K_X$ is divisible by 3 in $\operatorname{Pic} X$, which again follows from $K_X^2 = 9$ (Noether's formula) by Poincaré duality.

D.4.4. Corollary (Main Theorem of minimal model theory). *Let X be any surface. Then there exists a chain of contractions of -1-curves $X \to X_1 \cdots \to X_N = Y$ such that either K_Y is nef, or Y is $\mathbb{P}^2$ or a $\mathbb{P}^1$-bundle over a curve.*

CHAPTER E
Classification of Surfaces with K nef

E.1. Statement of the main result

The first part of the classification of surfaces, the theory of minimal models, was concerned with the case K_X not nef. Thus I assume throughout this chapter that K_X is nef. It is convenient to summarise the main theorem of the classification of surfaces in the following form:

Table–Main Theorem. *A nonsingular projective surface X with K_X nef belongs to just one of the following cases.*

Name of Case:	$\nu = 2$	$\nu = 1$	$\nu = 0$
Definition:	$K_X^2 > 0$	$K_X^2 = 0, K_X \overset{\text{num}}{\not\sim} 0$	$K_X \overset{\text{num}}{\sim} 0$
P_m and κ :	$P_m \sim m^2, \kappa = 2$	$P_m \sim m, \kappa = 1$	$P_m = 1 \ \exists m, \kappa = 0$
Effective results:	$P_m \geq 2$ for all $m \geq 2$	$P_m \geq 1$ for some $m \in \{1, 2, 3, 4, 6\}$ $P_m \geq 2$ for some $m \leq 42$	$mK_X \overset{\text{lin}}{\sim} 0$ for some $m \in \{1, 2, 3, 4, 6\}$
Structure result:	$\exists$ canonical model $\varphi \colon X \to Y$	$\exists$ elliptic fibre space $\varphi \colon X \to C$ (or more special structure in char. 2 or 3)	either Abelian or $K3$ or etale quotient by $\mathbb{Z}/m$ (same m)

In this table, the first two rows define the invariant $\nu(X) = \nu(K_X) = \kappa_{\text{num}}(X)$, the *numerical Kodaira dimension* of X: the three possible cases for K_X^2 and K_X enumerated in the second row are exclusive and cover all surfaces with K_X nef, and ν is defined to be 2, 1 or 0 accordingly; more generally, for any nef divisor on a projective variety X, the *characteristic dimension* of D is defined by

$$\nu(X, D) = \max\{k \mid D^k \overset{\text{num}}{\not\sim} 0\}.$$

The third row states the theorem that $\nu(X)$ is equal to the Kodaira dimension $\kappa(X)$, which, as everyone knows, is defined by

$$\begin{aligned}\kappa = 2 &\iff P_m \sim m^2 \text{ for } m \gg 0,\\ \kappa = 1 &\iff P_m \sim m \text{ for } m \gg 0,\\ \kappa = 0 &\iff P_m \le 1 \text{ for all } m, \text{ and } = 1 \text{ for some } m.\end{aligned}$$

The fourth row is a slightly more precise version of the same result, giving effective values of m for which P_m grows. The last row summarises the structure results in each case.

The special structures referred to in the final line are as follows: a surface with $\kappa = 1$ has in any case a fibre space structure $\varphi\colon X \to C$ with fibres of arithmetic genus 1, but in characteristic 2 and 3 there is an extra possible case, a "quasielliptic" fibre space: it can happen that every geometric fibre is a singular curve isomorphic to a cuspidal plane cubic. A surface with $\kappa = 0$ is either as stated in the table, or (in characteristic 2 or 3 only) a quotient of an Abelian surface by a nonreduced finite group scheme, or "quasi-bielliptic" (see E.8.4), or an Enriques surface in characteristic 2. I do not have time in these notes to treat these fascinating cases in the detail they deserve.

Crude corollary. *If K_X is nef then there exists a morphism $\varphi\colon X \to Y$ to a projective variety Y such that*

$$K_X \overset{\text{num}}{\sim} \varphi^*(\text{ample } \mathbb{Q}\text{-divisor on } Y).$$

In higher dimension this statement is called the "abundance conjecture", and remains at the time of writing a hard problem. The 3-fold case has recently been settled in several papers of Miyaoka and Kawamata (see Kollár and others [**Utah2**], especially Chapters 13–14 for the latest information). For surfaces, the nontrivial assertion is

$$\nu = 1 \implies \exists \text{ an elliptic fibre space.}$$

This is proved in E.6, and is really the main aim of this chapter. The remainder of the proof of the main theorem can be viewed simply as tidying up after this, using basically similar arguments.

Note that there is essentially nothing to prove in the theorem if $p_g \ge 2$: the case $\nu = 0$ is excluded, and $\nu = 1 \iff K_X^2 = 0$, in which case all the conclusions in the $\nu = 1$ column are satisfied rather trivially.

Corollary (Enriques' criterion). *Let X be a surface, without the condition that K_X nef. Then equivalent conditions*

1. *X is birationally ruled;*
2. *$\nexists$ birational model of X with K_X nef;*
3. *$P_m = 0$ for all $m \in \{1, 2, 3, 4, 6\}$;*
4. *$\kappa(X) = -\infty$;*
5. *adjunction terminates;*

etc.

Corollary (Castelnuovo's criterion). *X is rational if and only if $p_g = P_2 = 0$ and $q = 0$.*

In the current view of minimal models of surfaces via Mori theory, ruled surfaces are characterised in the first instance by the numerical property of having no model with K_X nef; hence the proof of Enriques' criterion, in the form

$$\exists \text{ model with } K_X \text{ nef} \implies P_m \neq 0 \text{ for some } m \in \{1, 2, 3, 4, 6\}$$

becomes a problem in the classification of surfaces with K_X nef. Thus minimal model theory is concerned (as a matter of ideology) purely with the numerical properties of K_X, leaving analytic questions such as the dimension of $H^0(mK_X)$ to the second stage of classification theory. The classical arguments of Castelnuovo and Enriques work with $P_2 = 0$ via "adjunction terminates", mixing up analytic constructions such as the Albanese map together with numerical considerations of components of an effective divisor D with $H^0(D + K_X) = 0$. See [**Beauville**], Chapter 6 for a readable account.

E.2. Traditional numerical game

It turns out to be useful to tabulate the possibilities for the numerical invariants of surfaces with $p_g \leq 1$ (compare Bombieri and Mumford [**B–M2**], Introduction). For this, use Max Noether's formula

$$({c_1}^2 + c_2)(X) = 12\chi(\mathcal{O}_X),$$

and interpret the terms on the two sides in terms of other invariants of X.

In characteristic 0 this is very easy (refer to Chapter C for a discussion of the invariants): on the right-hand side $\chi(\mathcal{O}_X) = 1 - q + p_g$, and on the left-hand side ${c_1}^2(X) = {K_X}^2$ and $c_2(X) = E(X) = 2 - 4q + 2p_g + h^{1,1}$.

In characteristic p, (or for non-Kähler complex surfaces) a little care is needed. Write $q' = h^1(\mathcal{O}_X)$, and $q = \dim \operatorname{Pic}^0 X = \dim \operatorname{Alb} X$.

Proposition (Igusa). *(i) $q' \geq q$; (ii) $c_2(X) = 2 - 4q(X) + B_2$ where $B_2 \geq \rho(X)$.*

Discussion of proof. (i) The connected group scheme $\operatorname{Pic}^0 X$ contains an Abelian variety $A = (\operatorname{Pic}^0 X)_{\mathrm{red}}$ such that $(\operatorname{Pic}^0 X)/A$ is a finite group scheme with only one point (possibly nonreduced). Now $q' = h^1(\mathcal{O}_X)$ is the dimension of the Zariski tangent space to $\operatorname{Pic}^0 X$, and so $q' \geq \dim \operatorname{Pic}^0 X = \dim A$; the Albanese variety of X is the dual Abelian variety to A, so that $q = \dim A$.

(ii) The second Chern class $c_2(X) \in H^4_{\mathrm{et}}(X, \mathbb{Q}_\ell)$ equals the Euler characteristic of X in etale cohomology,

$$c_2(X) = B_0 - B_1 + B_2 - B_3 + B_4,$$

where $B_i = B_i(X) = \dim_{\mathbb{Q}_\ell} H^i_{\mathrm{et}}(X, \mathbb{Q}_\ell)$ and ℓ is a prime $\neq p$. Now $B_0 = B_4 = 1$ by Poincaré duality, and

$$B_1 = B_3 = 2q,$$

since H^1_{et} is related to π_1 (more precisely, their ℓ-primary parts), and $\pi_1(X)$ to $\pi_1(\operatorname{Alb} X)$ essentially as in the classical case.

Finally, under the cycle class map $\mathrm{NS}(X) \to H^2_{\mathrm{et}}(X, \mathbb{Q}_\ell)$, the intersection number of divisors goes over to the cup product in etale cohomology, which implies that $\mathrm{NS}(X) \hookrightarrow H^2_{\mathrm{et}}$, and hence $B_2 \geq \rho(X)$. Q.E.D.

Putting everything together gives

$$K_X^2 + 2 - 4q + B_2 = 12 - 12q' + 12p_g;$$

that is,

$$K_X^2 + 12(q' - q) + 8q + B_2 = 10 + 12p_g. \qquad (*)$$

Famous Table. *$p_g \le 1$ leads to the following table of cases:*

1	$p_g = 1$	$q' = q = 0$	$K_X^2 + B_2 = 22$	(if $K_X \overset{\text{num}}{\sim} 0$ then $K3$)
2	$p_g = 1$	$q' = q = 1$	$K_X^2 + B_2 = 14$	(doesn't exist with $K_X \overset{\text{num}}{\sim} 0$)
3	$p_g = 0$	$q' = q = 0$	$K_X^2 + B_2 = 10$	(if $K_X \overset{\text{num}}{\sim} 0$ then Enriques)
3′	$p_g = 1$	$q' = 1, q = 0$	$K_X^2 + B_2 = 10$	(if $K_X \overset{\text{num}}{\sim} 0$ then Enriques, char 2)
4	$p_g = 1$	$q' = q = 2$	$K_X^2 + B_2 = 6$	(if $K_X \overset{\text{num}}{\sim} 0$ then Abelian)
5	$p_g = 0$	$q' = q = 1$	$K_X^2 = 0, B_2 = 2$	(if $K_X \overset{\text{num}}{\sim} 0$ then bielliptic)
5′	$p_g = 1$	$q' = 2, q = 1$	$K_X^2 = 0, B_2 = 2$	(if $K_X \overset{\text{num}}{\sim} 0$ then quasi-bielliptic)

Proof. If I restrict to surfaces with $p_g \le 1$ the right-hand side of $(*)$ is ≤ 22, and the left a sum of positive terms. It's easy to tabulate the possibilities. For the conclusion $K_X^2 = 0$ in (5) and (5′), note that if $q = 1$ then $B_2(X) \ge \rho(X) \ge 2$; indeed, then $\operatorname{Alb} X$ is a curve, and $\alpha\colon X \to \operatorname{Alb} X$ a nonconstant morphism, so that an ample divisor and the fibre of α provide two linearly independent classes in $\operatorname{NS}(X)$. Q.E.D.

Remarks. (a) The proof of abundance when $p_g \le 1$ appeals to Famous Table E.2 several times. However, it's not clear whether the numerical game is really central to the classification of surfaces; if so, then it seems to me that the abundance conjecture in higher dimension is in serious trouble.

(b) The last 3 cases have $\chi(\mathcal{O}_X) = 0$, and otherwise $\chi(\mathcal{O}_X) > 0$. A priori, surfaces with $\chi(\mathcal{O}_X) < 0$ may exist, but not under the assumption $p_g \le 1$, by what I've just said. Note that in characteristic 0, Bogomolov's inequality ${c_1}^2 \le 4c_2$ implies that $\chi(\mathcal{O}_X) \ge 0$ and $\chi(\mathcal{O}_X) > 0$ if $K_X^2 > 0$.

(c) Notice that one reads directly from (3) and (5) of the table that $p_g = 0$ implies $q = q'$, that is, that $\operatorname{Pic}^0 X$ is reduced. More generally, for any projective scheme X, in the same way that $H^1(\mathcal{O}_X)$ is the space of 1st order infinitesimal deformations of a line bundle, the vector space $H^2(\mathcal{O}_X)$ is the obstruction space for deformations, hence $H^2(\mathcal{O}_X) = 0$ implies that $\operatorname{Pic}^0 X$ is reduced, so that $q = q'$. Mumford [**M1**], Lecture 27 shows how the tangent space to $(\operatorname{Pic}^0 X)_{\text{red}}$ as a vector subspace of $H^1(\mathcal{O}_X)$ can be determined more generally. However, none of this fancy stuff is required for the current proof.

E.3. General type

For the proof of the main theorem, I start by separating off the case $\nu = 2$.

Theorem. *Suppose $K_X^2 > 0$. Then the curves (if any) with $K_X\Gamma = 0$ are all -2-curves (that is, $\Gamma \cong \mathbb{P}^1$ and $\Gamma^2 = -2$), and there is at most a finite set $\{\Gamma_i\}$ of them; each connected component of $\bigcup \Gamma_i$ supports only rational cycles, and can be contracted by a morphism $\varphi\colon X \to Y$, so that Y has only Du Val singularities, and K_Y is an ample Cartier divisor. Thus Y is the canonical model of X. Hence X is of general type, and its canonical ring $R(X, K_X) = R(Y, K_Y)$ is finitely generated over k.*

Moreover $P_m \geq 2$ *for all* $m \geq 2$.

Proof. The first part is standard and the proof is omitted; see Chapter 4 for Du Val singularities (especially Contraction Theorem 4.15), or **[M2]**. If $p_g \geq 2$ then $|K_X|$ is positive dimensional, hence also $|mK_X|$ for each $m \geq 1$. Thus to prove $P_m \geq 2$, I can assume $p_g \leq 1$. This is the first application of Famous Table E.2: I can deal separately with the different cases.

Cases (5) and (5′) are excluded since $K_X^2 > 0$. Cases (1), (2), (3) and (3′) all satisfy $\chi(\mathcal{O}_X) \geq 1$, so RR gives

$$h^0(mK_X) \geq \chi(\mathcal{O}_X(mK_X)) \geq 1 + \binom{m}{2} K_X^2 \geq 2 \qquad \text{for all } m \geq 2.$$

The only remaining case is (4). Then $\chi(\mathcal{O}_X) = 0$ and $c_2 = -K_X^2 < 0$; in characteristic zero, this is impossible, since surfaces of general type have $c_2 > 0$ (for example by Bogomolov's inequality ${c_1}^2 \leq 4c_2$). In any case,

$$h^0(mK_X) \geq \chi(\mathcal{O}_X(mK_X)) \geq \binom{m}{2} K_X^2 \geq 1 \qquad \text{for all } m \geq 2,$$

and ≥ 2 except possibly for $m = 2$ and $K_X^2 = 1$. Together with the assumptions of case (4), this gives

$$p_g = 1, \quad q = q' = 2 \quad \text{and} \quad K_X^2 = 1. \tag{**}$$

Claim (Shepherd-Barron). (**) *is impossible.*

Proof. Consider the Albanese morphism $X \to \operatorname{Alb} X$ and its Stein factorisation $\alpha\colon X \to A$. By **[Sh-Barron]**, Theorem 7, the image of the Albanese morphism $\alpha\colon X \to A \subset \operatorname{Alb} X$ is a curve, which must be of genus 2 since $q = 2$. (In **[Sh-Barron]**, it then follows that the fibres are geometrically rational curves.)

Let $C \in |K_X|$ be the unique curve. Then since $K_X^2 = 1$, C has a unique irreducible component with $K_X\Gamma = 1$, possibly together with some -2-curves. If C is reducible, or irreducible but singular, then Γ has geometric genus ≤ 1, so has no nonconstant morphism to A. It would follow that α contracts the whole of C to a point of A, which contradicts the index theorem since $C^2 = K_X^2 = 1$. Therefore C is irreducible and nonsingular.

The restriction of α to C is either an isomorphism, or has $\deg \alpha > 1$ (then α is necessarily inseparable, since $g(C) = g(A) = 2$). Either case leads to a contradiction: if $C \to A$ is an isomorphism, C meets any fibre F of α transversally in one point, so $CF = 1$; but then $K_XF = CF = 1$ contradicts $F^2 = 0$.

Now since $X \to A$ is its own Stein factorisation, the function field $k(A)$ is algebraically closed in $k(X)$. It follows that the differential of α is a nonzero map $s : \alpha^*(\Omega_A^1) \to \Omega_X^1$. Let $k \geq 0$ be the maximum number such that s vanishes k times along C, so that $s\colon \alpha^*(\Omega_A^1) \to \Omega_X^1(-kC)$ has nonzero restriction to C. Then the diagram

$$\begin{array}{ccccccccc} & & & & \alpha^*(\Omega_A^1) & & & & \\ & & & & \downarrow & & & & \\ & & & & \Omega_X^1(-kC) & & & & \\ & & & & \downarrow & & & & \\ 0 & \to & \mathcal{O}_C(-(k+1)C) & \to & \Omega_X^1(-kC)_{|C} & \to & \Omega_C^1(-kC) & \to & 0 \end{array}$$

gives a nonzero map of $\alpha^*(\Omega^1_A)$ to one of

$$\Omega^1_C(-kC) \quad \text{or} \quad \mathcal{O}_C(-(k+1)C).$$

This is a contradiction, since $\alpha^*(\Omega^1_A)$ has degree $2\deg\alpha > 2$, whereas

$$\deg\Omega^1_C(-kC) = 2-k \quad \text{and} \quad \deg\mathcal{O}_C(-(k+1)C) = -(k+1). \quad \text{Q.E.D.}$$

E.4. The cases with $\chi(\mathcal{O}_X) > 0$

The following result deals with all surfaces with $\chi(\mathcal{O}_X) > 0$ and $p_g \le 1$ (that is, cases (1), (2), (3), (3′) of Famous Table E.2).

Theorem. *Assume that* $K_X^2 = 0$ *and* $\chi(\mathcal{O}_X) = 1 - q' + p_g > 0$. *Then*

either $K_X \overset{\text{lin}}{\sim} 0$ *and* $q' = 0$ *(K3 surface);*

or $2K_X \overset{\text{lin}}{\sim} 0$ *and* $p_g = q' = 0$ *(classical Enriques surface);*

or char $k = 2$, $K_X \overset{\text{lin}}{\sim} 0$ *and* $p_g = q' = 1$, $q = 0$ *(nonclassical Enriques surface);*

or $P_6 \ge 2$.

Proof. The idea of the proof is to plug the assumption $\chi(\mathcal{O}_X) = 1$ or 2 into RR, giving $\chi(\mathcal{O}_X(mK_X)) \ge 1$. For $m \ge 2$, note that

$$h^0(mK_X) + h^2(mK_X) \ge \chi(\mathcal{O}_X(mK_X)) \ge 1.$$

But $H^2(mK_X) \overset{\text{d}}{=} H^0(-(m-1)K_X)$, so that $H^2 \ne 0$ implies that $-(m-1)K_X$ is effective, hence $(m-1)K_X \overset{\text{lin}}{\sim} 0$, since K_X is nef. The theorem follows easily from a small case division based on this. In more detail:

Case (1): $q' = 0$, $p_g = 1$, so that $\chi = 2$. Then $\chi(\mathcal{O}_X(2K_X)) = 2$, and hence either $P_2 = h^0(2K_X) \ge 2$ or $H^2(2K_X) \ne 0$. The second case gives $H^0(-K_X) \ne 0$, so that $K_X \overset{\text{lin}}{\sim} 0$. Note that this leads directly to the definition of K3s as surfaces with $K_X \overset{\text{lin}}{\sim} 0$ and $q' = 0$.

Case (2): $p_g = 1$, $q' = q = 1$; then $\chi(\mathcal{O}_X) = 1$, so RR gives

$$\chi(\mathcal{O}_X(mK_X + \sigma)) = 1 \quad \text{for every } m \text{ and every } \sigma \in \operatorname{Pic}^0 X.$$

Hence there is an effective divisor $D_\sigma \in |K_X + \sigma|$ for every $\sigma \in \operatorname{Pic}^0 X$; divisors of the form $D_\sigma + D_{-\sigma}$ provide infinitely many distinct elements of $|2K_X|$, and therefore $P_2 \ge 2$. In particular $\kappa(X) = 1$.

Case (3): $p_g = q' = 0$; then $\chi(\mathcal{O}_X(2K_X)) = 1$. Now as explained at the start of the proof, $p_g = 0$ implies that $H^2(2K_X) = 0$, so that $H^0(2K_X) \ne 0$. Let $0 \le D_2 \overset{\text{lin}}{\sim} 2K_X$ be an effective divisor. If $D_2 = 0$ then I have the second conclusion of the theorem (a classical Enriques surface). Otherwise $D_2 > 0$ implies that in turn $H^2(3K_X) = 0$, so again by RR $H^0(3K_X) \ne 0$. Choose an effective divisor $0 < D_3 \overset{\text{lin}}{\sim} 3K_X$; now obviously $3D_2 = 2D_3$ would imply that $D_2 = 2D_1$ and $D_3 = 3D_1$, with $0 < D_1 \overset{\text{lin}}{\sim} K_X$, contradicting $p_g = 0$. Hence $|6K_X|$ contains two distinct elements $3D_2$ and $2D_3$, which gives $P_6 \ge 2$.

Case (3′): $p_g = q' = 1$, $q = 0$. Suppose first that $K_X \overset{\text{lin}}{\sim} 0$. The argument of [**B–M2**], p. 39 shows that in this case char $k = 2$. (Roughly, if char $k = p$, a simple subgroup scheme μ_p or α_p of the group scheme $\operatorname{Pic}^0 X$ corresponds to a cover $Y \to X$ which is respectively etale and Galois with group $\mathbb{Z}/p$, or a torsor under α_p; then $\chi(\mathcal{O}_Y) = p \cdot \chi(\mathcal{O}_X) = p$, but on the other hand one sees that Y is a reduced irreducible surface with $K_Y \overset{\text{lin}}{\sim} 0$, so that $\chi(\mathcal{O}_Y) \le 2$. Thus $\operatorname{char} k = p \le 2$.) In **[B–M2]**, p. 26, Enriques surfaces are defined by $K_X \overset{\text{num}}{\sim} 0$ and $B_2 = 10$, so that cases (3) and (3′) of Famous Table E.2 are classical and nonclassical Enriques surfaces by definition.

If $K_X \overset{\text{lin}}{\sim} D > 0$, you can prove that P_2, P_3 or $P_4 \ge 2$ as a very instructive exercise in the techniques of E.6, using the assumption $q' = 1$. [Hint: Using the terminology and basic results of E.6.1, write $K_X \overset{\text{lin}}{\sim} D = rE + D_1$, with E a 0-curve, $r \ge 1$ and D_1 disjoint from E. If $D_1 \ne 0$ then D is disconnected, and $h^0(\mathcal{O}_D) \ge 2$; by RR and duality, this implies that $h^0(\omega_D) = h^0(\mathcal{O}_D(K_X + D)) \ge 2$, from which by cohomology $P_2 = h^0(\mathcal{O}_X(K_X + D)) \ge 2$. Otherwise, $D_1 = 0$, so that $K_X \overset{\text{lin}}{\sim} D = rE$, and by adjunction $\mathcal{O}_E((r+1)E) \cong \omega_E \cong \mathcal{O}_E$. Then the exact sequence

$$0 \to \mathcal{O}_E(-(r+1)E) \to \mathcal{O}_{(r+2)E} \to \mathcal{O}_{(r+1)E} \to 0$$

implies that $h^0(\mathcal{O}_{(r+2)E}) \ge 2$. Thus also $h^0(\mathcal{O}_{(r+2)E}(K_X + (r+2)E)) \ge 2$. From this, using the assumption $q' = 1$ and cohomology, $|K_X + (r+2)E| = |(2r+2)E|$ moves, so that $P_m \ge 2$ for all $m \ge 4$ (and for all $m \ge 3$ if $r > 1$).] Q.E.D.

Example

In every case except (3) and (3′) I've proved that $P_2 \ge 2$. An elliptic surface $f\colon X \to \mathbb{P}^1$ with a double fibre $F_2 = 2E_2$ and a triple fibre $F_3 = 3E_3$, and whose Jacobian fibration is a rational elliptic surface has $\chi(\mathcal{O}_X) = 1$ and

$$K_X = f^*\mathcal{O}(-1) + E_2 + 2E_3 = -F + \frac{1}{2}F_2 + \frac{2}{3}F_3;$$

it's easy to see that X has $p_g = 0$, $P_2 = \cdots = P_5 = 1$, $P_6 = 2$. (Compare with Champion E.7.4.)

E.5. Digression: the case $\chi(\mathcal{O}_X) = 0$ in characteristic 0

This section is not needed for the main proof, and is intended just to put the difficulty in its place.

In characteristic 0, $c_2 \ge 0$ (for example by Bogomolov's inequality), so also $\chi(\mathcal{O}_X) = (1/12)c_2 \ge 0$. The arguments given above in the case $\chi(\mathcal{O}_X) > 0$ are rather simple, so treating the remaining cases $K_X^2 = c_2 = 0$ should be considered as the heart of the classification of surfaces. The thing that makes the proof awkward in these cases is that RR gives $\chi(\mathcal{F}) = 0$ for any naturally occurring sheaf on X (compare the discussion in E.9.3). Curiously, although in this case the proof of classification is quite involved, the answer itself is very simple (see E.8.6 for the proof):

E.5.1. Theorem. *If X is an algebraic surface over $\mathbb{C}$ such that $K_X^2 = 0$ and $c_2 = 0$ then X is either an Abelian surface, or a surface of the form $X = (C \times E)/G$, where C is a curve of genus ≥ 1, E a curve of genus 1, and G is a finite Abelian group acting anyhow on C, and on E by translations, in such a way that the diagonal action on $C \times E$ is free.*

E.5.2. Definition. I propose to call the surfaces $(C\times E)/G$ *sesquielliptic surfaces.* When $K_X \overset{\text{num}}{\sim} 0$ then $g(C)=1$, so that the surfaces here are the *bielliptic surfaces* which occupy the position $p_g=0$, $q=1$ in the classification of surfaces with $K_X \overset{\text{num}}{\sim} 0$ (see [**Beauville**], VI.19–20 and [**B–M2**], §3). The name is good, since they are characterised by having two elliptic fibrations, to $\mathbb{P}^1$ and $\operatorname{Alb} X$. A sesquielliptic surface also has two projections to C/G and $E/G = \operatorname{Alb} X$, the first of which is an elliptic pencil; (according to the dictionary, the prefix "sesqui-", meaning $1\frac{1}{2}$, is a contraction of Latin *semi* + *que* (and)).

E.5.3. Theorem. *A surface X with $p_g=0, q=1$ is sesquielliptic. A sesquielliptic surface $(C\times E)/G$ has $\nu=0$ if $g(C)=1$, and $\nu=1$ if $g(C)\geq 2$.*

Note that the sesquielliptic conclusion can be interpreted as saying that X is a quotient of $\mathbb{C}\times\mathbb{C}$ or $\mathbb{C}\times\mathcal{H}$ (where $\mathcal{H}$ is the complex upper half-plane). Thus Theorem E.5.1 says that $c_1^2(X)=c_2(X)=0$ implies that X has a locally homogeneous differential geometric structure. It would be reasonable to look for a differential geometric proof of this fact. See E.9.4 for a wild discussion.

E.6. The existence of an elliptic pencil

I now give the proof of abundance: if $\nu(X)=1$ then X has an elliptic pencil. The proof breaks up into two big steps: (I) X has a 0-curve E (an effective divisor having the numerical properties of a nonmultiple fibre of an elliptic pencil); and (II) if K_X is nef and E a 0-curve then a multiple of E moves in an elliptic pencil.

It's not necessary to assume $\operatorname{char} k=0$, or anything about the values of c_2 or p_g, although if $p_g\geq 2$ or $c_2>0$ everything has been proved in the preceding section.

E.6.1. 0-curves

I write down some preliminary material, leaving the easy proofs to the reader. For details, see [**Beauville**], Chapter VIII, or [**3 authors**], or [**M3**], p. 332 (or work them out for yourself).

(a) A divisor *of elliptic fibre type* is a nef divisor $D \overset{\text{num}}{\not\sim} 0$ with $D^2=DK_X=0$. It follows by the Index Theorem that $K_X \overset{\text{num}}{\sim} aD$ with $a\in\mathbb{Q}$ and $a\geq 0$. If $D=\sum n_iC_i$ is effective then $DC_i=K_XC_i=0$ for every component C_i; it is easy to see from this that the intersection matrix (C_iC_j) is negative semidefinite (see Theorem A.7, (2)). In other words, D has the numerical properties of a sum of fibres of an elliptic fibre space.

(b) A 0-*curve* is an effective divisor $E=\sum n_iC_i$ of elliptic fibre type for which $\operatorname{Supp} E$ is connected and $\operatorname{hcf}(n_i)=1$. Then E is a Gorenstein curve with $\omega_E\cong\mathcal{O}_E$ and $H^0(\mathcal{O}_E)=1$; E behaves throughout in every respect as if it were an irreducible nonsingular elliptic curve; the crucial property is that for any line bundle $\mathcal{L}$ on E,

$$\deg_{C_i}\mathcal{L}=0 \text{ for all } i \text{ and } H^0(\mathcal{L})\neq 0 \quad\Longrightarrow\quad \mathcal{L}\cong\mathcal{O}_E.$$

(c) If $E=\sum n_iC_i$ is a 0-curve and D an effective divisor on X with $DC_i=0$ for all components of E then

$$D=rE+D',$$

with $r\geq 0$ and D' disjoint from E; in particular the line bundle $\mathcal{O}_X(D')$ is trivial in a neighbourhood of E, so that $\mathcal{O}_{aE}(D')\cong\mathcal{O}_{aE}$ for every $a>0$.

(d) Any effective divisor D of elliptic fibre type is a sum of 0-curves $D = \sum n_i E_i$.

Mumford's rather regrettable terminology is "canonical type" for "elliptic fibre type" and "indecomposable divisor of canonical type" for "0-curve".

E.6.2. Remark

Miyaoka's inequality for log surfaces [**Miyaoka**] implies that in characteristic 0, a surface with K_X nef, $K_X^2 = 0$ and $c_2 = 0$ does not contain -2-curves or nodal or cuspidal elliptic curves. So as far as the main case $\operatorname{char} k = c_2 = 0$ is concerned, a 0-curve E actually is a nonsingular elliptic curve.

E.6.3. Step I, the existence of a 0-curve

Since $\nu(X) = 1$, it is enough to prove that there exists an effective divisor numerically equivalent to mK_X with $m > 0$, and then apply (d). If $p_g \neq 0$ this is trivial. If $p_g = 0$ then by (3) and (5) of Famous Table E.2, $q' = q \leq 1$; as above, if $q' = 0$ and $K_X \overset{\text{num}}{\not\sim} 0$ then $P_2 > 0$ by RR. This leaves only the case $p_g = 0$, $q' = q = 1$. Then $A = \operatorname{Pic}^0 X$ is reduced, so an elliptic curve.

E.6.4. Proposition. *$H^0(K_X + \sigma) \neq 0$ for some $\sigma \in \operatorname{Pic}^0 X$.*

Proof. The Albanese variety is the dual Abelian variety $\operatorname{Alb} X = A^\vee = A$; write $0 \in A$ for the zero of the group law. On the product $A \times A$, consider the divisor

$$L = \Delta_A - A \times 0 - 0 \times A.$$

L can be thought of as the universal family parametrising line bundles of degree 0 on A, since for each $a \in A$, restricting $\mathcal{O}_{A\times A}(L)$ to $A \times a$ gives $\mathcal{O}_A(a - 0)$.

Consider the Albanese map $\alpha \colon X \to A$ and the pullback $L' = (\alpha \times \mathrm{id})^* L$ via

$$\alpha \times \mathrm{id} \colon X \times A \to A \times A.$$

This is the universal line bundle on $X \times \operatorname{Pic}^0 X$, essentially by definition of the Albanese morphism. Note that $L^2 = -2$, so that $(L')^2$ on $X \times A$ works out as $-2F$, where F is a fibre of α.

E.6.5. Base change

Let p, q denote the projections of $X \times \operatorname{Pic}^0 X$ to its factors. The idea of the proof is to compute $\chi = \chi(X \times A, p^*K_X + L')$ in two different ways: first, by RR on a 3-fold and the following easy calculation

$$\begin{aligned}\chi(X \times A, p^*K_X + L') &= \operatorname{ch}(p^*K_X + L') \cdot \mathrm{Td}_{X\times A} \\ &= \frac{1}{6}(p^*K_X + L')^3 - \frac{1}{4}K_X(p^*K_X + L')^2 \\ &= \frac{1}{4}K_X(L')^2 = 1 - f\end{aligned}$$

(the other terms all vanish for simple reasons), one sees that $\chi = 1 - f \leq 0$, where $f = p_a F$ is the genus of a fibre of α. Secondly, by the base change theorem, there is a complex

$$M^\cdot : 0 \to M^0 \xrightarrow{a} M^1 \xrightarrow{b} M^2 \to 0$$

of vector bundles on $A = \operatorname{Pic}^0 X$ such that $R^i q_*(p^*K_X + L')$ is the homology of $M^\cdot$. (To construct $M^\cdot$, take a Čech complex that computes $R^i q_*(p^*K_X + L')$, then use the fact that any bounded complex with coherent cohomology is quasi-isomorphic to a complex of vector bundles; you can truncate the complex so that $M^i = 0$ whenever $H^i = 0$ just by taking ker and coker of maps of constant rank. See [**M4**], Chapter 2, §5 or [**H1**], Chapter III.)

Now a and b are maps between locally free sheaves, and a is injective since $H^0(K_X + \sigma) = 0$ for general $\sigma \in A$. The rank of a drops at $\sigma \in A$ if and only if $H^0(K_X + \sigma) \neq 0$, and that of b if and only if $H^2(K_X + \sigma) \neq 0$; either of these only happens at finitely many points $\sigma \in A$, so that $R^i q_*(p^*K_X + L')$ is of finite length; write $r^i(\sigma)$ for the length of the stalk of the sheaf $R^i q_*(p^*K_X + L')$ at σ. Then the Leray spectral sequence gives

$$\chi = \sum_{\sigma \in A} -r^1(\sigma) + r^2(\sigma).$$

Now $r^2(0) \neq 0$ (since $h^2(K_X) = h^0(\mathcal{O}_X) = 1$), and so $\chi \leq 0$ implies $r^1(\sigma) \neq 0$ for some $\sigma \in A$; then the rank of a drops there, as required. Q.E.D.

E.6.6. Enriques' argument

If the Albanese map $\alpha\colon X \to \operatorname{Alb} X$ has fibre genus $f = 1$ then the existence of an elliptic pencil is established. At this point Enriques (see [**M3**], p. 331) argues that if $f \geq 2$ and $F = \alpha^{-1}P$ is a general fibre then the linear system $|2K_X + F|$ contains an element vanishing on some other fibre F'; hence $|2K_X + \sigma| \neq \emptyset$ where $\sigma = F - F' \in \operatorname{Pic}^0 X$. Enriques' argument is more geometric and much more picturesque, but it depends on one further dichotomy, and the result it proves directly is slightly weaker than Proposition E.6.4.

Step II, a 0-curve moves

E.6.7. Lemma. *Let $E = \sum n_i C_i$ be a 0-curve and D a divisor with $DC_i = 0$ for all i; suppose that n, $b > 0$. Then the exact sequence*

$$0 \to \mathcal{O}_{(n+b-1)E}(D + (b-1)E) \to \mathcal{O}_{(n+b)E}(D + bE) \to \mathcal{O}_E(D + bE) \to 0$$

gives rise to an inclusion $H^0(\mathcal{O}_{(n+b-1)E}(D + (b-1)E)) \subset H^0(\mathcal{O}_{(n+b)E}(D + bE))$, and

$$\text{this inclusion is strict} \iff \mathcal{O}_{(n+b)E}(D + bE) \cong \mathcal{O}_{(n+b)E}.$$

Proof. If there exists a section s having nonzero restriction to E, then s is a global basis of $\mathcal{O}_{(n+b)E}(D + bE)$ by E.6.1, (b). Q.E.D.

E.6.8. Corollary. *If the inclusion $H^0(\mathcal{O}_{nE}(D)) \subsetneq H^0(\mathcal{O}_{(n+b)E}(D+bE))$ is strict then there exists b' with $0 < b' \leq b$ such that*

$$\mathcal{O}_{(n+b')E}(D + b'E) \cong \mathcal{O}_{(n+b')E}.$$

Proof. At least one of the inclusions in the chain

$$H^0(\mathcal{O}_{nE}(D)) \subset \cdots \subset H^0(\mathcal{O}_{(n'-1)E}(D + (n' - n - 1)E)) \\ \subset H^0(\mathcal{O}_{n'E}(D + (n' - n)E)) \subset \cdots$$

must be strict, so that this follows from Lemma E.6.7. Q.E.D.

E.6.9. Theorem. *Let X be a surface with K_X nef and E a 0-curve on X.*
(i) For every $\mu \geq 2$,

$$H^0(\mathcal{O}_X(\mu(K_X + E))) \neq 0.$$

(ii) There exist effective divisors D_1, D_2 disjoint from E such that

$$\mu K_X + \nu E \overset{\text{lin}}{\sim} D_1 - D_2.$$

with $\mu, \nu \in \mathbb{Z}$ and $\mu > \nu$, $\mu > 0$. In particular

$$\mathcal{O}_{nE}(\mu K_X + \nu E) \cong \mathcal{O}_{nE} \quad \text{for all } n \geq 1.$$

(iii) The main point: (ii) implies that

$$H^0(\mathcal{O}_{nE}(K_X + nE)) \to \infty \quad \text{as} \quad n \to \infty;$$

therefore eventually $h^0(\mathcal{O}_{nE}(K_X + nE)) \geq h^1(\mathcal{O}_X) + 2$, and hence by cohomology

$$\dim |K_X + nE| \geq 1.$$

E.6.10. Proof. (i) Consider the exact sequence

$$0 \to \mathcal{O}_X(\mu K_X + (\mu - 1)E) \to \mathcal{O}_X(\mu(K_X + E)) \to \mathcal{O}_E \to 0$$

(where I substitute $\mathcal{O}_E(K_X + E) = \omega_E \cong \mathcal{O}_E$ in the last term). Since $H^1(\mathcal{O}_E) \neq 0$ and $H^2(\mathcal{O}_X(\mu K_X + (\mu - 1)E)) = 0$ for $\mu \geq 2$ (by duality, using the fact that K_X is nef), it follows that $h^1(\mathcal{O}_X(\mu(K_X + E))) \geq 1$, and therefore

$$h^0(\mathcal{O}_X(\mu(K_X + E))) = \chi(\mathcal{O}_X) + h^1(\mathcal{O}_X(\mu(K_X + E))) \geq \chi(\mathcal{O}_X) + 1.$$

Now if $p_g \neq 0$ then $\mu(K_X + E) = \mu K_X + \mu E$ is obviously effective for $\mu \geq 1$, and there is no problem. If $p_g = 0$ then by (3) and (5) of Famous Table E.2, $\chi(\mathcal{O}_X) \geq 0$. Thus in any case $|\mu(K_X + E)| \neq \emptyset$. Q.E.D.

(ii) Let $D \in |\mu(K_X + E)|$; then by E.6.1, (c),

$$D = rE + D_1,$$

with $r \geq 0$ and D_1 disjoint from E. Now I have a dichotomy:
Case $r > 0$. Then (ii) is satisfied with $\nu = \mu - r$.
Case $r = 0$. Then $D_1 = D \neq 0$, so that it contains a 0-curve E_1 disjoint from E; in this case apply (i) to E_1 to get $D' \in |\mu(K_X + E_1)|$; if

$$D' = sE + D_2$$

with $s \geq 0$ and D_2 disjoint from E then

$$\mu K_X - sE \overset{\text{lin}}{\sim} D_2 - \mu E_1,$$

which proves (ii) in this case. Q.E.D.

E.6.11. Proof of (ii) $\implies$ (iii). Let E be a 0-curve. By Lemma E.6.7, in the chain

$$0 \subset H^0(\mathcal{O}_E(K_X+E)) \subset \cdots \subset H^0(\mathcal{O}_{(n-1)E}(K_X+(n-1)E)) \\ \subset H^0(\mathcal{O}_{nE}(K_X+nE)) \subset \cdots,$$

the inclusion from $(n-1)$ up to n is strict if and only if $\mathcal{O}_{nE}(K_X+nE) \cong \mathcal{O}_{nE}$; thus (iii) says exactly that this coincidence occurs infinitely often. It first occurs when $n=1$, since $\omega_E = \mathcal{O}_E(K_X+E) \cong \mathcal{O}_E$. It's clearly enough to prove the following:

E.6.12. Claim. *Suppose that $\mathcal{O}_{nE}(K_X+nE) \cong \mathcal{O}_{nE}$ for some $n \geq 1$. Then*

$$\mathcal{O}_{n'E}(K_X+n'E) \cong \mathcal{O}_{n'E}$$

for some $n' > n$.

To prove this, note that by (ii), also $\mathcal{O}_{nE}(\mu K_X + \nu E) \cong \mathcal{O}_{nE}$. Thus

$$\mathcal{O}_{nE} \cong \mathcal{O}_{nE}(\mu(K_X+nE) - (\mu K_X + \nu E)) \cong \mathcal{O}_{nE}(bE), \quad \text{where } b = \mu n - \nu > 0.$$

(notice that $\mu > \nu$ is needed for the case $n=1$). The claim therefore follows from the next lemma.

E.6.13. Lemma. *Let E be a 0-curve and $n, b > 0$. Suppose that $\mathcal{O}_{nE}(bE) \cong \mathcal{O}_{nE}$; then there exists n' with $n < n' \leq n+b$ such that*

$$\omega_{n'E} = \mathcal{O}_{n'E}(K_X+n'E) \cong \mathcal{O}_{n'E}.$$

Proof. The restriction map

$$H^0(\mathcal{O}_{(n+b)E}) \to H^0(\mathcal{O}_{bE})$$

is nonzero (because of the constant sections). Therefore I have a strict inclusion

$$H^0(\mathcal{O}_{nE}(-bE)) \subsetneq H^0(\mathcal{O}_{(n+b)E}),$$

and in particular $h^0(\mathcal{O}_{nE}(-bE)) < h^0(\mathcal{O}_{(n+b)E})$. By RR and duality,

$$h^0(\mathcal{O}_{nE}(-bE)) = h^1(\mathcal{O}_{nE}(-bE)) = h^0(\omega_{nE}(bE)) \\ h^0(\mathcal{O}_{(n+b)E}) = h^1(\mathcal{O}_{(n+b)E}) = h^0(\omega_{(n+b)E}),$$

so that the preceding inequality gives $h^0(\omega_{nE}(bE)) < h^0(\omega_{(n+b)E})$. Now using the assumption $\mathcal{O}_{nE}(bE) \cong \mathcal{O}_{nE}$, I conclude that the inclusion

$$H^0(\mathcal{O}_{nE}(K_X+nE)) \subset H^0(\mathcal{O}_{(n+b)E}(K_X+(n+b)E)$$

is also strict. The lemma therefore follows by Corollary E.6.8. Q.E.D.

This completes the proof of Theorem E.6.9.

E.6.14. Remarks

(i) Every step of the proof is absolutely trivial, so what's going on? The subtle point is the curious argument in E.6.13

$$\begin{aligned}&\text{strict inclusion of vector spaces}\\ &\qquad\Longrightarrow \text{strict numerical inequality}\\ &\qquad\Longrightarrow \text{(by RR and duality) another strict numerical inequality}\\ &\qquad\Longrightarrow \text{another strict inclusion of vector spaces;}\end{aligned}$$

thus the bald statement that $h^0(\mathcal{O}_{nE}(-bE)) = h^0(\omega_{nE}(bE))$ conceals the fact that both $H^0(\mathcal{O}_{nE}(-bE))$ and $H^0(\omega_{nE}(bE))$ are modules over $H^0(\mathcal{O}_{nE})$, and the module structures contain nontrivial information about the nilpotents of the scheme nE. Hence the essence of the proof is using the numerical fact given by RR and duality to relate the two different nilpotent structures of the sheaves $\mathcal{O}_{(n+b)E}$ and $\omega_{(n+b)E}$.

(ii) The above argument is adapted from [**M3**], pp. 334–5. Mumford uses only $\mu = 1$ and applies duality 4 or 6 times more than necessary (for example, he starts his argument by saying $H^2(\mathcal{O}_X) \to H^2(\mathcal{O}_X(rE))$ is not injective, which can more simply be read $|K_X| \ni D = rE + D'$ with D' disjoint from E).

E.7. Effective growth of plurigenera

The remaining issues in Main Theorem E.1 are the effective results on the P_m, and the finer structure results, especially those for $\nu = 0$. As remarked in E.4, if $p_g \geq 2$ or if $\chi(\mathcal{O}_X) \geq 1$ then everything has already been proved. The structure results are treated in E.8. The remaining assertions divide up as follows:

E.7.1. Theorem (see especially [**B–M2**]).

1. *Case* $\nu = 0$. *Then* $mK \overset{\text{lin}}{\sim} 0$ *for some* $m \in \{1, 2, 3, 4, 6\}$.
2. *Case* $\nu = 1$. *Then* $P_m \geq 1$ *for some* $m \in \{1, 2, 3, 4, 6\}$.
3. *Case* $\nu = 1$. $P_m \geq 2$ *for some* $m \leq 42$.

Discussion of proof of 1 and 2. There is nothing to prove if $p_g = 1$, so we are in Case (5) of Famous Table E.2, that is, $p_g = 0$, $q = q' = 1$. The proof works in either case by applying the adjunction formula (canonical bundle formula) to a suitably chosen elliptic fibre space $X \to B$. You are almost certainly guessing that I'm going to use the Albanese morphism $\alpha\colon X \to A$, which is a map to an elliptic curve; however, it is a striking observation that *this never works.* Indeed, in the case $K_X \overset{\text{lin}}{\sim} 0$ the adjunction formula gives $K_X = \alpha^* L$ with $\deg L = 0$, but at this stage of the argument there's no reason why L should be torsion, so there is no conclusion to be drawn. In the case $\nu = 1$, we don't know that the Albanese map is an elliptic fibre space (in fact it never is).

In the case $\nu = 1$, I use the elliptic fibre space $X \to B$ provided by abundance (E.6), and in the case $\nu = 0$, the following result, which is the first step in the analysis of bielliptic and quasi-bielliptic surfaces.

E.7.2. Theorem. *If* $K_X \overset{\text{num}}{\sim} 0$ *and* $p_g = 0$, $q = q' = 1$ *then* X *has an elliptic fibration* $\varphi\colon X \to \mathbb{P}^1$ different from the Albanese morphism.

Proof. Given $B_2 = 2$ and the structure of the inner product on $\mathrm{NS}(X)$, if F is the fibre of α, it's not hard to find a class E with $EF > 0$ and $E^2 = 0$. Then, because $\chi(\mathcal{O}_X) = EK_X = E^2$, arguing exactly as in E.6.4–5 gives that

$$\chi(X \times A, p^*E + L') = \frac{1}{6}(p^*E + L')^3 = \frac{1}{2}p^*E(L')^2 = -EF,$$

so that the Leray spectral sequence implies $H^0(X, \mathcal{O}_X(E+\sigma)) \neq 0$ for some element $\sigma \in \mathrm{Pic}^0 X$. Thus E is numerically equivalent to an effective divisor. Now some multiple of E moves in a pencil, by Theorem E.6.9. It necessarily maps to $\mathbb{P}^1$, since otherwise $q \geq 2$. Q.E.D.

E.7.3. Vertical divisors on X and fractional divisors on B

The proof of Theorem E.7.1, 1–2 is based on the adjunction formula for the canonical bundle of an elliptic fibration $\varphi\colon X \to B$. For clarity, I start with the *tame* case, which is defined by the condition that $R^1\varphi_*\mathcal{O}_X$ is torsion free. Then $R^1\varphi_*\mathcal{O}_X = L^{-1}$, where L is a line bundle with $\deg L = \chi(\mathcal{O}_X)$, and the adjunction formula for φ is simply Kodaira's canonical bundle formula

$$K_X = \varphi^*(K_B + L) + \sum_i (m_i - 1)E_i, \qquad (*)$$

where $F_i = m_iE_i$ are all the multiple fibres (see for example [**3 authors**], Theorem V.12.1). In all our cases, $\chi(\mathcal{O}_X) = 0$, so that $\deg L = 0$ in $(*)$.

All the calculations in what follows take place at the level of fractional divisors on the curve B. To explain this, I write a divisor on X of the form $f^*D + \sum a_iE_i$ as the pullback $f^*\Delta$ of the fractional divisor on B of the form $\Delta = D + \sum(a_i/m_i)Q_i$. If $B = \mathbb{P}^1$ and $\deg D = d$ then $D \overset{\text{lin}}{\sim} dQ$ for a chosen general point $Q \in \mathbb{P}^1$, so that I write $\Delta = dQ + \sum(a_i/m_i)Q_i$. Note that the divisorial sheaf associated with a $\mathbb{Q}$-divisor is defined by $\mathcal{O}_B(m\Delta) = \mathcal{O}_B([m\Delta])$, where $[\ ]$ is the integral part or round-down. On B, this says that rational functions cannot use up a fractional allocation of poles, and on X, it corresponds exactly to the fact that only multiples of m_iE_i can move in a linear system made up of fibres.

You'll be able to follow the argument much better if you work out the details of the following example. (Compare Katsura and Ueno [**K–U**].)

E.7.4. Champion!

The worst case, when the estimates $P_6 \geq 1$ and $P_{42} \geq 2$ are best possible, arises as follows: let $X \to \mathbb{P}^1$ be an elliptic surface having trivial Jacobian fibration (that is $J \cong E \times \mathbb{P}^1$ with E an elliptic curve), and having multiple fibres $F_2 = 2E_2$, $F_3 = 3E_3$, $F_7 = 7E_7$. Then just as in the example of E.4,

$$K_X = f^*\mathcal{O}_{\mathbb{P}^1}(-2) + E_2 + 2E_3 + 6E_7 = f^*\Delta,$$

where Δ is the $\mathbb{Q}$-divisor

$$\Delta = -2Q + (1/2)Q_2 + (2/3)Q_3 + (6/7)Q_7$$

on $\mathbb{P}^1$. Then $H^0(mK_X) = H^0(\mathcal{O}_{\mathbb{P}^1}(m\Delta))$, and the canonical ring of X is the graded ring $R(\mathbb{P}^1, D)$. An elementary (but fairly long) calculation shows that this ring is the graded ring $k[x, y, z]/(f)$, where x, y, z have weights 6, 14, 21 and $f = x^7 + y^3 + z^2$. In particular $P_i = 0$ for all $i \leq 5$, $P_i \leq 1$ for all $i \leq 41$. (Also

$P_{43} = 0$ and $P_{85} = 1$. This is responsible for the slightly obscure statement that $P_m \geq 2$ for all $m \geq 86$ that sometimes appears in the literature.)

Note that this example shows that the apparently ineffective aspect of the proof in E.6.9–13 is not actually very far from the truth.

E.7.5. Proof of 1 and 2, the tame case. For the proof of 1, I apply (*) to the alternative fibration $\varphi\colon X \to \mathbb{P}^1$ obtained in Theorem E.7.2. Then K_X is the pullback of the fractional divisor

$$\Delta = -2Q + \sum \frac{m_i - 1}{m_i} Q_i \quad \text{on } \mathbb{P}^1,$$

where the $Q_i \in \mathbb{P}^1$ are distinct points, and each $m_i \geq 2$. Clearly $K_X \overset{\text{num}}{\sim} 0$ if and only if the fractional divisor has degree 0, that is,

$$-2 + \sum_i \frac{m_i - 1}{m_i} = 0 \quad \text{or} \quad \sum \frac{1}{m_i} = (\text{number of } i) - 2$$

This obviously has exactly 4 solutions

$$(2,2,2,2), \quad (3,3,3), \quad (2,4,4), \quad (2,3,6).$$

Thus $mK_X \overset{\text{lin}}{\sim} 0$ for m respectively $2,3,4,6$. This proves 1.

To prove 2, I apply (*) to the elliptic or quasielliptic fibre space $X \to B$ provided by abundance, that is, the results of E.6. Thus K_X is the pullback of $\Delta = K_B + L + \sum((m_i - 1)/m_i)Q_i$ and from $\nu = 1$ it follows that $\deg \Delta > 0$. The assumptions of Case (5) imply immediately that $\deg L = \chi(\mathcal{O}_X) = 0$ and $g(B) \leq 1$.

Because K_X is nef and nonzero, if $g(B) = 1$ there must be at least one multiple fibre; then every term $2((m_i - 1)/m_i)Q_i$ makes a positive contribution to $[2\Delta]$, so that $\deg[2\Delta] > 0$, and RR on B gives $P_2 = H^0(\mathcal{O}_B([2\Delta]) \geq 1$. (Similarly, $P_3 > 0$ and $P_m \geq 2$ for every $m \geq 4$. Here and below, results like this leak out for free, and are used in the proof of Theorem E.7.1, 3.)

The other possibility is that $B \cong \mathbb{P}^1$. Then K_X is the pullback of $\Delta = -2Q + \sum(m_i - 1/m_i)Q_i$, and K_X nef and nonzero gives

$$-2 + \sum_i \frac{m_i - 1}{m_i} > 0, \quad \text{that is,} \quad \sum \frac{1}{m_i} < (\text{number of } i) - 2$$

Now every point Q_i makes a positive integral contribution to 2Δ, so that $P_2 \geq 2$ if there are at least 5 points Q_i. If there are 4 points, at least one must be ≥ 3, and $P_2 \geq 1$, $P_4 \geq 2$. The final part of the proof is a pleasurable exercise. [Hint: If all $m_i \geq 3$, the smallest case is $(3,3,4)$, giving

$$3K_X \overset{\text{lin}}{\sim} \varphi^*(-6Q + 2Q_1 + 2Q_2 + (9/4)Q_3) \overset{\text{lin}}{\sim} E_3 = (1/4)F_3,$$

so $P_3 = 1$ and $P_{12} = 2$. Otherwise, $m_1 = 2$, and if both $m_2, m_3 \geq 4$ the smallest case is $(2,4,5)$, giving

$$4K_X \overset{\text{lin}}{\sim} \varphi^*(-8Q + 2Q_1 + 3Q_2 + (16/5)Q_3) \overset{\text{lin}}{\sim} E_3 = (1/5)F_3,$$

so that $P_4 = 1$ and $P_{20} = 2$. Otherwise, $m_1 = 2, m_2 = 3$, and $m_3 \geq 7$. This leads to Champion E.7.4.]

E.7.6. Exercise. Prove that $P_{12} \le 1$ happens only in 7 cases:

$$(2,5,5) \quad (2,4,5) \quad \text{and} \quad (2,3,m) \text{ with } m \le 11.$$

E.7.7. Wild!

Once the terminology and basic properties are established, the proofs in the wild case are essentially the same combinatorial arguments as in E.7.3–6.

Let $\varphi\colon X \to B$ be an elliptic or quasielliptic fibration and $F = mE = \varphi^* P$ a multiple fibre. Following **[B–M2]**, I introduce the following three numerical characteristics of F: the *multiplicity* m, the *order* n of $\mathcal{O}_E(E)$ as a torsion element of $\operatorname{Pic}^0 E$, and the *local canonical invariant*, that is, the integer $a < n$ such that $\mathcal{O}_X(K_X) \cong \mathcal{O}_X(aE)$ in a neighbourhood of F.

In characteristic zero, automatically $a = m-1$ and $n = m$. Indeed, if t_P is a local parameter at P on B, then locally near E, I have $\operatorname{div}(t_P) = mE$ and $\operatorname{div}(dt_P) = (m-1)E$. Also taking the mth root of t_P and normalising gives rise to an etale cyclic cover of a neighbourhood of F such that the inverse image of F is connected. On the other hand, the restriction to E is the cyclic cover defined by $\mathcal{O}_E(E)$, hence this has order m.

Remark. The argument here is essentially topological, and can be viewed as a beautiful illustration of Kollár's philosophy (discussed in 3.12) that *vanishing* is when a coherent cohomology group has a topological interpretation: in this case, $H^0(\mathcal{O}_E(iE)) = 0$ for all $0 < i < m$ because the inverse image of E in a cyclic cover is connected.

In characteristic p, the same argument applies only to the part of m coprime to p: if $m = m'p^a$ with m' not divisible by p then the m'th root of t_P still defines an etale cyclic cover in which F remains connected; therefore the torsion order n is given by $n = m'p^b$ for some $b \le a$.

Quite generally, a multiple fibre of $X \to B$ can be reduced to a nonmultiple fibre by taking the normalised pullback by a suitable Galois separable ramified cover $B' \to B$ of the base curve, which induces an etale cover of a neighbourhood of $F \subset X$. However, in characteristic p, a ramified cover can have a complicated p-group as local Galois group (ramification group), and in any case, the typical cyclic cover is of the form $x^p + x = a$ (an *Artin–Schreier extension*).

Definition. The fibre F is *tame* if $n = m$ and *wild* if $n < m$. Obviously since both $\mathcal{O}_E(F) = \mathcal{O}_E(mE)$ and $\omega_E = \mathcal{O}_E((a+1)E)$ are trivial, it follows that n divides m and $a+1$.

If F is a wild fibre, then clearly $h^0(\mathcal{O}_F) \ge 2$, so that also

$$h^1(\mathcal{O}_F) > 1.$$

In particular, the stalk of $R^1\varphi_*\mathcal{O}_X$ at P needs ≥ 2 generators, and it follows that $R^1\varphi_*\mathcal{O}_X = T \oplus L^{-1}$, where T is a torsion sheaf, nonzero at each wild fibre, and L is a line bundle with $\deg L = \chi(\mathcal{O}_X) + \operatorname{length}(T)$. (This follows simply by RR on B.) The adjunction formula in the general case is thus

$$K_X \doteq \varphi^*(K_B + L) + \sum a_i E_i, \qquad (**)$$

where $F_i = m_i E_i$ are the multiple fibres, and a_i their local canonical invariants.

E.7.8. Lemma. *If $h^1(\mathcal{O}_X) = 1$ then $a = m - 1$ or $a = m - n - 1$.*

Proof. Since $\mathcal{O}_E(nE) = \mathcal{O}_E$, it follows from Lemma E.6.13 that

$$h^0(\mathcal{O}_{n'E}(K + n'E)) \geq 2 \quad \text{for some } n' \text{ with } 1 < n' \leq n + 1.$$

Then by cohomology, $H^0(K_X) \subsetneq H^0(K_X + n'E)$, so that some part of the divisor $(a+n')E$ must move off E, that is, $m \leq a + n' \leq a + n + 1$. Thus $n \geq m - (a+1)$, but on the other hand, n divides m and $a + 1$. Therefore, $m - (a+1) = 0$ or n, as asserted. Q.E.D.

E.7.9. Proof of 1 and 2 in the wild case. As before, to prove 1, I work on the alternative fibration $\varphi\colon X \to \mathbb{P}^1$ provided by Theorem E.7.2. In $(**)$ I have $\deg L = \chi(\mathcal{O}_X) + \operatorname{length}(T)$. If $\deg L \geq 2$ then $(**)$ would give $p_g \geq 1$, a contradiction, so it follows that $\deg L = 1$ and there is exactly one wild fibre with $\operatorname{length}(T) = 1$. Set $F_0 = m_0 E_0$ for the wild fibre, n_0 and a_0 for its period and local canonical invariant, and $F_i = m_i E_i$ for $i = 1, \dots, k$ for the tame multiple fibres. Now argue exactly as in E.7.4: K_X is the pullback of the fractional divisor

$$-Q + \frac{a_0}{m_0} Q_0 + \sum_{i=1}^{k} \frac{m_i - 1}{m_i} Q_i.$$

Moreover $a_0 = m_0 - 1$ or $m_0 - n_0 - 1$. In the first case it is an easy exercise to deduce that $K_X \overset{\text{num}}{\sim} 0$ gives $k = 1$ and $(m_0, m_1) = (2, 2)$ so that $P_2 = 1$.

For the second case, it's useful to note that the wild fibre behaves for numerical purposes exactly like two usual multiple fibres of multiplicity m_0 and m_0/n_0: to see this, it is enough to write the above $\mathbb{Q}$-divisor throughout in the form

$$-2Q + \frac{m_0 - 1}{m_0} Q_0 + \frac{m_0/n_0 - 1}{m_0/n_0} Q_0 + \sum_{i=1}^{k} \frac{m_i - 1}{m_i} Q_i.$$

Thus $K_X \overset{\text{num}}{\sim} 0$ gives

$$k = \frac{1}{m_0} + \frac{n_0}{m_0} + \sum_{i=1}^{k} \frac{1}{m_i}.$$

The right-hand side is a sum of reciprocal integers, and an obvious calculation shows that the set of integers $(m_0, m_0/n_0, \{m_i\}_{i=1}^k)$ is one of the usual list:

$$(2,2,2,2) \quad (3,3,3) \quad (2,4,4) \quad (2,3,6).$$

Thus $mK_X \overset{\text{lin}}{\sim} 0$ for $m = 2, 3, 4$ or 6. This proves 1.

For the purposes of this proof, I didn't need to figure out which of the integers is m_0/n_0. Note that since $n_0 = p^a m_0$, where $p = \operatorname{char} k$, the argument shows that for a surface with $\nu = 0$ in Case (5), a wild fibre can only happen in characteristic 2 or 3.

As before, for 2, I work on the pluricanonical elliptic fibration $X \to B$ provided by abundance. The argument is almost identical to previous work. In view of $p_g = 0$, $q = q' = 1$, the adjunction formula $(**)$ gives that $B = \mathbb{P}^1$, $\deg L = 1$

and there is exactly one wild fibre to which Lemma E.7.8 applies. Now K_X is the pullback of the fractional divisor

$$-Q+\frac{a_0}{m_0}Q_0+\sum_{i=1}^{k}\frac{m_i-1}{m_i}Q_i,$$

where $a_0=m_0-1$ or m_0-n_0-1. In the first case I have

$$-1+\frac{m_0-1}{m_0}+\sum_{i=1}^{k}\frac{m_i-1}{m_i}>0 \quad\text{that is,}\quad \sum_{i=0}^{k}\frac{1}{m_i}<k$$

Thus $k\geq 1$. If $k\geq 2$ then $P_2\geq 2$; otherwise $k=1$ and one of $m_0,m_1\geq 3$, so that $P_2\geq 1$, $P_3\geq 1$ and $P_6\geq 2$ (compare Example E.4). In the second case I get

$$-1+\frac{m_0-n_0-1}{m_0}+\sum_{i=1}^{k}\frac{m_i-1}{m_i}>0 \quad\text{that is,}\quad \frac{1}{m_0}+\frac{n_0}{m_0}+\sum_{i=1}^{k}\frac{1}{m_i}<k$$

Here again $k\geq 1$, and the pleasurable exercise at the end of E.7.5 can be repeated verbatim to prove that $P_m\geq 1$ for some $m\leq 6$ and $P_m\geq 2$ for some $m\leq 42$.

E.7.10. Proof of Theorem E.7.1, 3. If $p_g\geq 2$ there is nothing to prove. If $\chi(\mathcal{O}_X)\geq 1$ and $\nu=1$ then $P_6\geq 2$ was proved in Theorem E.4. The remaining case with $p_g=0$ is Case (5), and everything has been proved in E.7.5 in the tame case, and in E.7.9 in the wild case.

The remaining cases are (4), (5′), which have $p_g=1$ and $q'=2$. In this case, the proof consists of using the construction of E.6.9–13 as an *effective* method. Write $K_X=rE+D'$ with E a 0-curve, $r\geq 1$ and $D'>0$ a divisor disjoint from E. The arguments of E.6.9–13 give $H^0(\mathcal{O}_{(r+2)E})\geq 2$, and $H^0(\mathcal{O}_{(r+2)E+D'})\geq 3$ if $D'\neq 0$, so that by RR and duality $H^0(\mathcal{O}_{(r+2)E+D'}(K_X+(r+2)E+D'))\geq 3$, therefore, by the cohomology long exact sequence, $|K_X+(r+2)E+D'|$ moves. In the worst case $r=1$, this gives $4K_X>K_X+(r+2)E+D'$, so that $P_4\geq 2$. If $D'=0$, then I take the argument of Lemma E.6.13 one step further: then $H^0(\mathcal{O}_{bE}(r+b)E)\cong\mathcal{O}_{bE}$ for some $b\leq r+2$, so that $H^0(\mathcal{O}_{(r+2b)E})\geq 3$ and $|K_X+(r+2b)E|$ moves. Thus in the worst case $D'=0$, $r=1$ and $b=3$, I get that $|8E|$ moves, so that $P_m\geq 2$ for all $m\geq 8$. This completes the proof of Theorem E.7.1. Q.E.D.

E.7.11. Remark. I'm rather disappointed that the arguments of E.6.9–13 don't seem to give the effective bound $P_{42}\geq 2$ also in the cases $p_g=0$. The interest of the question is that if you could find a better argument here, you might be able to tidy up the main proof of E.6.9–13.

In this case $q=q'\leq 1$, and you could try writing $mK_X=rE+D'$ for some $m=\{1,2,3,4,6\}$, with D' effective and disjoint from E. Thus

$$\mathcal{O}_E(mK_X-rE)\cong\mathcal{O}_E\cong\mathcal{O}_E(K_X+E)\implies\mathcal{O}_E((m+r)E)\cong\mathcal{O}_E.$$

Now arguing as in Theorem E.6.9, (i), there exists $c\leq m+r+1$ such that $H^0(\mathcal{O}_{cE})\geq 2$ and $\mathcal{O}_{cE}(K_X+cE)\cong\mathcal{O}_{cE}$. If $D'\neq 0$, you get as before a curve $cE+D'$ with $H^0(\mathcal{O}_{cE+D'})\geq 3$, and $|K_X+cE+D'|$ moves by cohomology. In the worst case $m=6$, $r=1$, $c=8$, this gives $P_m\geq 2$ for $m\geq 49$. If $D'=0$, the next step is to say that $\mathcal{O}_{cE}((mc+r)E)\cong\mathcal{O}_{cE}$, therefore $H^0(\mathcal{O}_{((m+1)c+r)E})\geq 3$, so that $|K_X+((m+1)c+r)E|$ moves. In the worst case $m=6$, $r=1$, $c=8$, so that

$|K_X + 57E| = |343K_X|$ moves (notice that $343 = 7^3$). Of course, this is far from best possible. The way to improve it would be to show that $\mathcal{O}_{8E}(7E) \cong \mathcal{O}_{8E}$, that is, replace 49 by 7. Probably you can show that the kernel of $\operatorname{Pic}^0(8E) \to \operatorname{Pic}^0 E$ is a vector space by the same kind of arguments as in 4.13, so has no torsion (at least in characteristic $\neq 7$)?

E.8. Abelian, bielliptic and sesquielliptic surfaces

E.8.1. Theorem (characterisation of Abelian surfaces). *Let X be a surface with $K_X \overset{\text{lin}}{\sim} 0$, $p_g = 1$, $q = q' = 2$ (that is, $\nu = 0$ and Case (4) of Famous Table E.2). Then the Albanese map $\alpha\colon X \to \operatorname{Alb} X$ is an isomorphism, so that X is an Abelian surface.*

E.8.2. Proof. Write $\alpha\colon X \to A = \operatorname{Alb} X$ for the Albanese map. Recall that it has the universal mapping property (UMP) for morphisms of X to an Abelian variety. I exclude the possibility that α maps to a curve $C \subset A$ by saying that C has genus 2, and thus has etale covers of large genus; hence, by pullback, X has etale covers with $q > 2$, therefore $p_g > 1$, which contradicts $K_X = 0$.

Consider the differential $d\alpha\colon T_X \to \alpha^* T_A$. In characteristic zero, it is generically an isomorphism, so gives an injective map

$$T_X \hookrightarrow \alpha^* T_A \cong \mathcal{O}_X^{\oplus 2},$$

necessarily an isomorphism because $K_X \overset{\text{lin}}{\sim} 0$; thus α is etale. But an etale cover of an Abelian variety is itself Abelian, so that $X = A$ follows from the UMP of $\alpha\colon X \to A$.

In characteristic p, the same proof goes through unchanged if $d\alpha$ is generically injective. On the other hand, $d\alpha$ is not generically zero. Indeed, $d\alpha = 0$ means that $\alpha^* k(A) \subset k(X)^p$ (the subfield of pth powers of elements of $k(X)$), or equivalently, $k(A)^{1/p} \subset k(X)$. This means that α factors through the geometric Frobenius map $A^{(-1)} \to A$, which again contradicts its UMP.[1]

Therefore I can assume that $d\alpha$ has generic rank 1, and argue on foliations as in Rudakov and Shafarevich [**R–S**] and [**Ekedahl**]. The foliation $\ker d\alpha \subset T_X$ is a rank 1 subsheaf, and is necessarily saturated; write $\ker d\alpha = \mathcal{O}_X(D)$. Then there is an exact sequence

$$0 \to \mathcal{O}_X(D) \to T_X \to \mathcal{I}_Z \mathcal{O}_X(-D) \to 0,$$

where $\mathcal{I}_Z$ is the ideal sheaf of a zero dimensional subscheme, and $D \geq 0$, since by assumption there is a nonzero homomorphism $\mathcal{I}_Z \mathcal{O}_X(-D) \to \alpha^* T_A \cong \mathcal{O}_X^{\oplus 2}$.

Consider the subfield of the function field $k(X)$ generated by $k(X)^p$ and $k(A)$, that is, the composite field $K = k(A)k(X)^p$. Then since $[k(X) : k(X)^p] = p^2$ and $k(A) \not\subset k(X)^p$, I deduce that $K \subset k(X)$ is inseparable of degree p. It follows that α factorises as $X \to Y \to A$, where Y is the surface obtained as the normalisation of A in K, the morphism $\pi\colon X \to Y$ is inseparable of degree p, and $\ker d\pi =$

[1] Here $A^{(-1)}$ is the normalisation of A in the field $k(A)^{1/p}$; it is isomorphic as a scheme to A via the absolute Frobenius, so is an Abelian variety in its own right. The point of the funny notation is mainly to avoid offending experts: $A^{(-1)}$ and the morphism $A^{(-1)} \to A$ are defined over k, and $A^{(-1)}$ is conjugate to A by the Frobenius automorphism of k, but is not isomorphic to A over k.

$\mathcal{O}_X(D) \subset T_X$. Therefore $\mathcal{O}_X(D)$ is a p-closed foliation and $Y = X/\mathcal{O}_X(D)$ is the corresponding quotient. (See [**R–Sh**] for the terminology.)

Case $D > 0$. By the adjunction formula for $X \to Y$ of [**R–Sh**], §2, Proposition 2 (p. 1211 of translation), $\pi^* K_Y = -(p-1)D$, so that $K_{Y'}$ cannot be nef on any birational model Y' of Y, and Y is rational or ruled by the main result of minimal model theory (Corollary D.4.4). This contradicts $Y \to A$ generically finite.

Case $D = 0$. Then $Z = 0$, since $c_2(X) = 0$. By [**R–Sh**], §1, Corollary of Theorem 1 (p. 1208 of translation), the surface Y is nonsingular, and $K_Y \overset{\text{num}}{\sim} 0$ by the same adjunction formula. Thus $p_g(Y) \le 1$; but since $q(Y) \ge 2$, it follows that Y is also in Case (4) of Famous Table E.2, that is, Y satisfies all the same hypotheses as X. I can therefore assume by induction on $\deg(Y/A)$ that $Y = A$, that is, $\alpha\colon X \to A$ is inseparable of degree p.

Moreover, by the exact sequence $0 \to \mathcal{O}_X(D) \to T_X \to \mathcal{I}_Z\mathcal{O}_X(-D) \to 0$, the differential $\mathrm{d}\alpha\colon T_X \to \alpha^* T_A \cong \mathcal{O}_X^{\oplus 2}$ has image isomorphic to $\mathcal{O}_X$, which is thus a direct summand. Thus $T_A = L_1 \oplus L_2$ is a direct sum of two trivial line bundles $L_i \cong \mathcal{O}_A$, such that $\mathrm{d}\alpha$ maps surjectively to $\alpha^* L_1$.

Now $X \to A$ is inseparable of degree p, so that the inclusion $k(X)^p \subset k(A)$ defines a factorisation $F\colon X \to A \to X^{(1)}$ of the Frobenius. On the other hand, since $\mathrm{d}\alpha$ has image $\alpha^* L_1$, and $\mathrm{d}F = 0$, it follows that $\beta\colon A \to X^{(1)}$ has $\ker \mathrm{d}\beta = L_1$, and β is the quotient by L_1. Therefore L_1 is p-closed.

Now also $L_1 \cong \mathcal{O}_A$, and is a direct summand of T_A, so that L_1 is a p-Lie subalgebra of the algebra of tangent fields on A. The Frobenius map $F\colon A \to A^{(1)}$ is a homomorphism of algebraic group schemes, and the vector space $H^0(T_A)$ is in a natural way the p-Lie algebra corresponding to the finite subgroup scheme $\ker F$. From the fact that L_1 is p-closed, it follows that $H^0(L_1)$ is a p-Lie subalgebra, so that by the Lie correspondence ([**M4**], Chapter III, §14, Theorem), it is the Lie algebra of a subgroup scheme $G \subset \ker F$. Therefore the quotient by L_1 is the same thing as the quotient by G, and it follows that $X^{(1)}$ is itself an Abelian variety. Therefore X is also an Abelian variety, and the factorisation $X \to A$ contradicts the UMP of α as before. Q.E.D.

E.8.3. Remarks. (1) The proof in [**B–M2**], pp. 40–41 is an extremely weird reduction to a finite field. My proof (based largely on suggestions of Nick Shepherd-Barron) is an expanded and simplified version of the argument of [**Ekedahl**], Proposition 4.3.

(2) It's an exercise to generalise the treatment of the final case to prove the following theorem of Igusa and Serre ([**Serre**], §2.6, Theorem 4): if $f\colon X \to A$ is a generically finite surjective morphism of an n-fold to an Abelian variety, the two conditions (a) and (b) are equivalent:

(a) There exists a 1-form $\omega \in H^0(\Omega_A^1)$ such that $f^*\omega = 0 \in \Omega_X^1$.

(b) f factors via $B \to A$, an inseparable morphism of height 1 between Abelian varieties. (This means that $B \to A$ has degree > 1 and factors the Frobenius morphism $B \to A \to B^{(1)}$; or, more simply, $k(B)^p \subset k(A) \subsetneq k(B)$.)

It follows that if the Albanese map $\alpha\colon X \to \operatorname{Alb} X$ is surjective for any variety X, then its differential is an injective map $\alpha^*\colon H^0(\Omega_A^1) \hookrightarrow H^0(\Omega_X^1)$.

E.8.4. Theorem.

1. *Case $K_X \overset{\text{num}}{\sim} 0$, $p_g = 0$, $q = q' = 1$ (that is, $\nu = 0$ and (5)). Then X is bielliptic (see Definition E.5.2).*

2. *Case* $K_X \overset{\text{lin}}{\sim} 0$, $p_g = 1$, $q = 1$, $q' = 2$ *(that is,* $\nu = 0$ *and (5′)). Then* $\operatorname{char} k = 2$ *or* 3, *and* X *is quasi-bielliptic.*

Here quasi-bielliptic means that X is a quasielliptic fibre space $X \to A$ over an elliptic curve, which becomes a $\mathbb{P}^1$-bundle on making an inseparable cover of the base curve.

E.8.5. Sketch proof. In either case (5–5′), consider the Albanese map $\alpha\colon X \to A$, and the alternative fibration $\varphi\colon X \to \mathbb{P}^1$ of Theorem E.7.2. Any fibre E of φ is a curve with $p_a E = 1$ having a finite morphism $\pi\colon E \to A$. Thus π is necessarily an isogeny of elliptic curves (after choosing base points), so that E is a nonsingular elliptic curve, and $E \to A = E/G$ is the quotient by the finite subgroup scheme $G = \ker \pi \subset E$.

Suppose first that α is an elliptic fibration, that is, that the general geometric fibre is nonsingular. Then the adjunction formula (**) applied to α implies that $\deg L = 0$, and all $a_i = 0$, and therefore, every fibre F is the same nonsingular elliptic curve. Now pulling back $X \to A$ by the finite cover $E \to A$ gives an elliptic fibre space $X_E = X \times_A E \to E$ having a section E (the diagonal $\Delta_E \subset E \times_A E$), and with all geometric fibres the same curve F. Therefore X_E is isomorphic to the product $E \times F$ and $X_E \to X$ is a quotient by the subgroup scheme $G \subset E$, acting on E by translation, and acting somehow on F. However, since X has $mK_X \overset{\text{lin}}{\sim} 0$ for some $m \in \{1,2,3,4,6\}$ it follows that the action of G on $H^0(K_F)$ factors via the group scheme μ_m of mth roots of unity, where m is the order of K_X in $\operatorname{Pic} X$. The kernel of this action is a translation subgroup scheme of $E \times F$, and the corresponding quotient is an Abelian variety. Therefore X is isomorphic to the quotient of an Abelian variety by μ_m.

If $\operatorname{char} k \neq 2$ or 3, it follows that X is an etale quotient of an Abelian variety by a cyclic group of order m. In any case, $m = 1$ would give that X is an Abelian surface, and contradict the case assumption (5–5′). But all surfaces with $\nu = 0$ in Case (5′) have $p_g = 1$, therefore $K_X \overset{\text{lin}}{\sim} 0$, so that α cannot be an elliptic fibration for these surfaces.

Next suppose that $\alpha\colon X \to A$ is a quasielliptic fibre space. As before, $\deg L = 0$ and all $a_i = 0$ in the adjunction formula (**), so that every fibre F of α is a reduced irreducible curve with a cusp. Pulling back by the isogeny $E \to A$ gives a ruled surface with a section, so that the normalised pullback X_E is isomorphic to a $\mathbb{P}^1$-bundle over E, and $X_E \to X$ is again a quotient by a subgroup scheme $G \subset E$. Q.E.D.

E.8.6. Sketch proof of Theorem E.5.1 and E.5.3. Consider a surface X over $\mathbb{C}$ with K_X nef, $\nu = 1$ and $c_2 = 0$. We know that X is an elliptic surface $f\colon X \to B$ with $c_2 = 0$. Hence by the Euler number calculation, the only degenerate fibres are multiple nonsingular fibres, say $E_i = m_i e_i$. In particular the modular invariant of the fibre is bounded, and therefore constant. Hence the Jacobian fibration is trivial, $J = E \times B$. Now the canonical bundle formula gives $K_X = f^*K_B + \sum(m_i - 1)e_i$, and it's easy to see that $p_g(X) = g(B)$, $q(X) = g(B) + 1$. By the complete reducibility property of Abelian varieties, $\operatorname{Alb} X$ has a projection to an elliptic curve A which restricts to a nonconstant morphism from the general fibre E of $X \to B$. In particular, the Albanese morphism $\alpha\colon X \to \operatorname{Alb} X$ never equals the canonical elliptic fibre space $f\colon X \to B$. The normalised pullback $X \times_A E$ is then a product. But $E \to A$ must be Galois with Abelian group, so that X is obtained as stated in the theorem. Q.E.D.

E.9. Any questions or comments?

This chapter has given a detailed, complete, self-contained proof of the main results of the classification of surfaces, following Enriques' argument as rewritten by Mumford. My reason for writing this in this form is my impression that while hundreds of mathematicians (and nowadays also a growing number of theoretical physicists and specialists in science fiction) need to know and use the results of the classification, very few have ever been through a proof. Although the coincidence $\nu = \kappa$ may seem at first sight such an obvious and coarse result, its proof is logically quite intricate and beautiful. Moreover, the role of the bielliptic and sesquielliptic surfaces with $c_2 = 0$ as a logical bottleneck of the proof seems to me to deserve some emphasis.

E.9.1. Abundance as a logical bottleneck

As far as I know, there is only one complete proof of abundance in the literature that does not pass via Enriques' argument at the central point, namely that given in [**3 authors**]. They first give Ueno's proof of Iitaka's additivity conjecture $C_{2,1}$, which uses the theory of moduli of curves. Recall that if X has $p_g = 0$, $q = 1$ and $K_X^2 = 0$ then X has an Albanese morphism $\alpha\colon X \to \operatorname{Alb} X = A$ to an elliptic curve A, with fibres of genus 1 if $\nu = 0$, or genus ≥ 1 if $\nu = 1$. Then $C_{2,1}$ proves at once that $P_m \geq 1$ for some m, and $P_m > 1$ for some m if the fibres of α have genus ≥ 2, and with a bit more work also if $\nu = 1$.

E.9.2. Finding cohomology when $\chi(\mathcal{F}) = 0$

The central problem in proving abundance is to find nonzero cohomology on X. The point is that

$$\chi(\mathcal{O}_X(nK_X)) = \chi(\mathcal{O}_X) \quad \text{for all } n,$$

so that RR on its own does not imply that $H^0(nK_X) \to \infty$ as $n \to \infty$. But if $P_m \geq 2$ for some m, then $D \in |mK_X|$ will consist of many fibres of the elliptic fibration φ_{mK_X}, so $H^0(\mathcal{O}_D)$ is large. It's interesting to analyse the argument of E.6–7 to note the key points in Proposition E.6.4, Theorem E.6.9 and Proposition E.7 at which one has to work to squeeze out nonzero cohomology one drop at a time.

E.9.3. Finding curves C with $KC = 0$

Abundance can be stated as saying that a suitably defined "moduli space" of maps $\varphi\colon C \to X$ with $K_X\varphi(C) = 0$ has the right dimension. Our experience with Mori theory teaches us that there is essentially only one way of proving that a nonsingular projective variety X with K_X not nef contains rational curves, namely Mori's bending-and-breaking argument: very roughly, a curve $C \subset X$ with $K_XC < 0$ moves in a positive dimensional family (for easy reasons), and must break off a rational component at some point (for delicate easy reasons). Put this crudely, this suggests that a similar argument might be capable of locating the curves with $K_XC = 0$ on a n-fold with K_X nef. Unfortunately, it seems almost certain that this kind of approach cannot work.

The point is that Mori's bending-and-breaking argument actually works with the deformation theory of morphisms $\varphi\colon C \to X$ with C a fixed curve; if you allow the moduli of C to vary, you might gain some parameters, but you completely lose the ability to predict that the family breaks up. Although, if you believe abundance, the moduli space of curves in X with $K_X C = 0$ has the right dimension, it seems impossible to approximate them by curves with $K_X C$ "small".

In the rest of this chapter I discuss three other approaches to the proof of abundance, none of which has been completely worked out (to my knowledge). Assume $k = \mathbb{C}$ for a partial sense of security.

E.9.4. Kähler–Einstein

This is the question left at the end of E.5. Consider minimal surfaces with $K_X^2 = c_2 = 0$. By a theorem of Uhlenbeck–Yau and Donaldson, if we knew that T_X were stable (or a direct sum of stable bundles), it could be given a Hermitian–Einstein metric. Of course, a posteriori this happens only for Abelian surfaces. Nevertheless, it seems to me (as an outsider in the analytic side of differential geometry) quite likely that one can prove a priori that there exists a metric that makes the Kähler structure locally symmetric.

H. Tsuji seems to have some general results on this kind of question in higher dimensions, which I paraphrase, probably rather wildly. Try to construct a "Kähler pseudometric" on the line bundle K_X whose curvature form vanishes along the fibres of the canonical map $\varphi\colon X \to Y$, and is positive in directions transverse to the fibres. Of course, a priori we don't know φ or any of its fibres, but if $K_X^n = 0$ then the curvature of a Kähler pseudometric must vanish on a subspace of the tangent bundle. Then there is a kind of heat equation argument saying that a Kähler pseudometric should exist as a distribution. If you could prove that this exists as an analytic pseudometric then its kernel should give the fibres of φ, at least as a subbundle of T_X or as a complex foliation. Then to prove that the canonical fibration exists is equivalent to proving the regularity of the Kähler pseudometric distribution. (??)

E.9.5. Miyaoka's argument

Let X be a 3-fold with K_X nef and $\nu = 1$ (that is $K_X \overset{\text{num}}{\not\sim} 0$, $K_X^2 \overset{\text{num}}{\sim} 0$), and assume $\kappa \geq 0$ to avoid appealing to Miyaoka's previous hard theorems. Let E be a suitable divisor chosen from the components of a divisor $D \in |mK_X|$, with $E^2 \overset{\text{num}}{\sim} 0$ and $K_{X|E} \overset{\text{num}}{\sim} 0$. Miyaoka proves that some multiple of E moves in a pencil on X. The idea of the proof is to restrict X to a tubular neighbourhood of E; then some easy homotopy theory shows that it is possible to make cyclic covers of X branched along E, and then to pass to a minimal model $F \subset Y$ via Kulikov's results. This model has the following good properties: F is a degenerate Abelian or $K3$ surface (that is, global normal crossings, $K_F \overset{\text{lin}}{\sim} 0$, and all components of F are algebraic surfaces), Y is nonsingular, and $K_{Y|F} \overset{\text{lin}}{\sim} F_{|F} \overset{\text{lin}}{\sim} 0$.

Now F has a local deformation space as an abstract complex space; next, Miyaoka's idea is to prove that there are sufficiently many abstract infinitesimal deformations of F that can be mapped to Y. (Specialists know that Miyaoka's appeal to Friedman's results in deformation theory in the case of degenerate Abelian surfaces is not valid as written, but I understand that his proof has been fixed up by slightly different arguments. Miyaoka's proof contains in any case a whole string of important new ideas. See [**Utah2**] for the current status of this.)

It seems that in characteristic 0, this argument can be translated back to the surface case without much trouble, and can be used to replace Step II, (E.6.7–13). (??)

E.9.6. There must be a better way

It seems conceivable to me that we are just overlooking some much coarser and much more fundamental reason why abundance must be true for surfaces, an argument that doesn't resort to miserable cookery with the constant term in RR. Here is one attempt at such an argument.

Let X be a surface with $\nu = 1$. Consider the set of curves $E \subset X$ with $K_X E = 0$. By the Hodge index theorem, the intersection form on these is negative semidefinite. Thus either there are only finitely many, or at most a single 1-dimensional family plus finitely many; if there is a positive dimensional family, it must be an elliptic pencil, and everything's lovely. I say that there must be a way of proving directly that there are infinitely many curves E with $K_X E = 0$, or better, that there is a curve E with $K_X E = 0$ through every sufficiently general point $P \in X$.

Let H be a divisor on X with $d = K_X H > 0$. Note that replacing H by $H + nK_X$ does not change $K_X H$, so that there is no loss of generality in assuming that also $H^2 > 0$.

Now resolutely ignoring the constant term in RR gives

$$h^0(\mathcal{O}_X(H + nK_X)) \sim (n/2)d.$$

E.9.7. Conjecture. *There exist integers $n \gg 0$ and $m < \sqrt{(H + nK_X)^2} - 1$, and a decomposition*

$$H + nK_X = D + E \qquad \textit{with } D, E > 0 \textit{ and } DE \leq m.$$

Moreover, E can be chosen to pass through a sufficiently general point P of X.

The statement itself follows from abundance: just take E to be the fibre of φ_{mK} through P. The conjectural part is that this statement can be proved directly without appealing to abundance.

E.9.8. Lemma. *The conjecture implies abundance.*

Proof. The form $D^2E^2 \leq (DE)^2$ of the Hodge index theorem implies at once that D^2 or $E^2 \leq 0$, since otherwise

$$(H + nK_X)^2 = D^2 + 2DE + E^2 \leq \max_{1 \leq x \leq m^2} \{x + 2m + m^2/x\} = (1 + m)^2$$

contradicts the assumption on m. Exchanging D and E allows me to assume $E^2 \leq 0$. Then $D^2 > 0$.

Set $d_1 = K_X D$ and $d_2 = K_X E$, so that $d = d_1 + d_2$ and $d_1, d_2 \geq 0$ (since K_X is nef). Since $D^2 > 0$, the Hodge index theorem gives $d_1 > 0$.

If $d_2 \neq 0$ then $d_1 < d$, and I just replace $H \mapsto D$ and repeat the same argument with a smaller value of d. Since d can only decrease a finite number of times, eventually every decomposition $D+E$ as in the statement of the conjecture satisfies $K_X E = 0$. Q.E.D.

E.9.9. Special clusters

The suggestion for proving that $H + nK_X$ has a decomposition of the required type is in terms of Reider's method. I say that $Z \subset X$ is a *special cluster* of degree m for $|H + (n+1)K_X|$ if Z is a zero dimensional subscheme (for example $Z = P_1 + \cdots + P_m$ with distinct $P_i \in X$) which is special in the sense that it imposes dependent conditions on $|H + (n+1)K_X|$. In other words, the evaluation homomorphism at Z

$$H^0(H + (n+1)K_X) \to \mathcal{O}_Z \qquad (*)$$

is not surjective. Z behaves like an analog of a special divisor on a general curve $C \in |H + nK_X|$.

The cokernel of $(*)$ is the group $H^1(I_Z \cdot \mathcal{O}_X(H + (n+1)K_X))$, which has Serre dual corresponding to extension classes

$$0 \to \mathcal{O}_X \to \mathcal{E} \to I_Z \cdot \mathcal{O}_X(H + nK_X) \to 0;$$

in most cases one can arrange that $\mathcal{E}$ is locally free, so a rank 2 vector bundle.

Now $c_1{}^2(\mathcal{E}) = (H + nK_X)^2$ and $c_2(\mathcal{E}) = m$, so if I take m to be fairly small, $\mathcal{E}$ has the Bogomolov numerical instability property $(c_1{}^2 - 4c_2)(\mathcal{E}) > 0$. It's not hard to get from this to a decomposition as in the conjecture.

The hard problem is the existence of special clusters for suitable values of n and m. One can set up a formalism of vector bundles over the Hilbert scheme of clusters, and define locuses W_m^r of special clusters in analogy with the theory of special linear systems on curves. It's easy to do a dimension count in the style of Brill–Noether to prove that suitable locuses W_m^r have dimension > 0, but I don't know how to do the intersection theory to prove that they are nonempty. (??)

E.9.10

Although this proof runs into technical problems, it seems to me to provide at least a possible reason of principle why abundance may be true, not depending on frail numerical considerations: since K_X is not ample, there must be lots of arbitrarily large cohomology groups around; if you can find one large group, e.g. $H^1(I_Z \cdot \mathcal{O}_X(H + nK_X))$, then it should be possible to chase this back to $H^0(nK_X)$.

References

[4 authors] E. Arbarello, M. Cornalba, P. A. Griffiths and J. Harris, *Geometry of algebraic curves, I.*, Springer, 1984.

[A1] M. Artin, *Some numerical criteria for contractibility of curves on algebraic surfaces*, Amer. J. Math **84** (1962), 485–496.

[A2] M. Artin, *On isolated rational singularities of surfaces*, Amer. J. Math **88** (1966), 129–136.

[3 authors] W. Barth, C. Peters and A. Van de Ven, *Compact complex surfaces*, Springer, 1984.

[Beauville] A. Beauville, *Complex algebraic surfaces*, Astérisque **54** (1978) and C.U.P., 1983.

[B2] A. Beauville, *Surfaces K3*, Séminaire Bourbaki **609** (1982/83); in Astérisque **106**.

[B3] A. Beauville, *Géométrie des surfaces K3: modules et périodes*, Astérisque **126**.

[Bertini] E. Bertini, *Introduzione alla geometria proiettiva degli iperspazi*, Enrico Spoerri, Pisa, 1907.

[B–M2] E. Bombieri and D. Mumford, *Enriques' classification of surfaces in characteristic p, II.*, Complex Analysis and Algebraic Geometry, Iwanami, Tokyo, 1977, pp. 23–42.

[B–M3] E. Bombieri and D. Mumford, *Enriques' classification of surfaces in characteristic p, III.*, Invent. Math. **35** (1976), 197–232.

[CFHR] F. Catanese, M. Franciosi, K. Hulek and M. Reid, *Embedding of curves and surfaces*, in preparation (currently 25 pp.).

[Utah1] H. Clemens, J. Kollár and S. Mori, *Higher dimensional complex geometry*, Astérisque **166** (1988), (second edition in preparation).

[E–H] D. Eisenbud and J. Harris, *On varieties of minimal degree (A Hundred Years of Scrollitude)*, Algebraic geometry (Bowdoin 1985), Proc. of Symp. in Pure Math, vol. 46 Part I, A.M.S., 1987, pp. 3–13.

[Ekedahl] T. Ekedahl, *Foliations and inseparable morphisms*, Algebraic geometry (Bowdoin 1985), Proc. of Symp. in Pure Math, vol. 46 Part II, A.M.S., 1987, pp. 139–149.

[F1] W. Fulton, *Algebraic curves*, Addison–Wesley, 1989.

[F2] W. Fulton, *Introduction to intersection theory in algebraic geometry*, A.M.S. Regional conference series in mathematics (1984), no. 54, A.M.S.

[F3] W. Fulton, *Intersection theory*, Springer, 1984.

[Harris] J. Harris, *A bound on the geometric genus of projective varieties*, Ann. Scuola Norm. Sup. Pisa Cl. Sci. (4) **8** (1981)), 35–68.

[H1] R. Hartshorne, *Algebraic Geometry*, Springer, 1977.

[H2] R. Hartshorne, *Ample subvarieties of algebraic varieties*, LNM **160**.

[K–U] T. Katsura and K. Ueno, *On elliptic surfaces in characteristic p*, Math. Ann. **272** (1985), 291–330.

[Kawamata] Yu. Kawamata, *The cone of curves of algebraic varieties*, Ann. of Math **119** (1984), 603–633.

[Kleiman] S. Kleiman, *Towards a numerical theory of ampleness*, Ann. of Math **84** (1966), 293–344.

[Kollár] J. Kollár, *The cone theorem*, Ann. of Math. **120** (1984), 1–5.

[Utah2] J. Kollár and others, *Flips and abundance for algebraic 3-folds*, Astérisque **211** (1992).

[L1] H. B. Laufer, *Normal surface singularities*, Ann. of Math Studies, Princeton.

[L2] H. B. Laufer, *On minimally elliptic singularities*, Amer. J. Math **99** (1977), 1257–1295.

[L3] H. B. Laufer, *Weak simultaneous resolution for deformations of Gorenstein surface singularities*, Singularities (Arcata, Jul 1981), Proc. Symp. Pure Math, vol. 40 Part II, 1983, pp. 1–29.

[Manin] Yu. I. Manin, *Cubic forms: algebra, geometry, arithmetic*, Nauka (1972), Moscow; English translation North-Holland 1974, Amsterdam-London, second edition 1986.

[M–Ts] Yu. I. Manin and M. A. Tsfasman, *Rational varieties: algebra, geometry, arithmetic*, Uspekhi Mat. Nauk **41** (1986), no. 2, 43–94; English translation in Russian Math. Surveys **41** (1986), no. 2, 51–116.

[Matsumura] H. Matsumura, *Commutative ring theory, C.U.P.*, Cambridge, 1986.

[Miyaoka] Y. Miyaoka, *The maximal number of quotient singularities on surfaces with given numerical invariants*, Math. Ann. **268** (1984), 159–171.

[Mori] S. Mori, *Elliptic Gorenstein singularities*, Notes of lecture circulated at Park City, 1993 (to appear as a chapter in the revised edition of [**Utah1**]).

[M1] D. Mumford, *Lectures on curves on a surface*, Ann. of Math. Studies, Princeton.

[M2] D. Mumford, *The canonical ring of an alagebraic surface*, Ann. of Math. **76** (1962), 612–615.

[M3] D. Mumford, *Enriques' classification of surfaces in characteristic p. I.*, in Global Analysis, papers in honour of K. Kodaira, Univ. of Tokyo Press (1969), pp. 325–339.

[M4] D. Mumford, *Abelian varieties*, Oxford Univ. Press, 1974.

[Parikh] Carol-Ann Parikh, *The unreal life of Oscar Zariski*, Academic Press, Boston, 1991.

[UAG] M. Reid, *Undergraduate Algebraic Geometry, C.U.P.*, 1988.

[YPG] M. Reid, *Young person's guide to canonical singularities*, Algebraic geometry, (Bowdoin 1985), Proc. of Symp. in Pure Math, vol. 46, Part I, A.M.S., 1987, pp. 345–414.

[R] M. Reid, *Elliptic Gorenstein singularities of surfaces*, unpublished manuscript (1975).

[C3-f] M. Reid, *Canonical 3-folds*, Journées de géométrie algébrique (Angers 1979), Sijthoff and Nordhoff, Alphen, 1980.

[R1] M. Reid, *Nonnormal del Pezzo surfaces*, Publ. of RIMS, Kyoto **30** (1994), no. 5, 695–727.

[R–S] A. Rudakov and I. R. Shafarevich, *Inseparable morphisms of algebraic surfaces*, Izv. Akad. Nauk SSR, Ser. Mat. **40** (1976), 1269–1307; English translation in Math. USSR Izv. **10** (1976), 1205–1237.

[Saint-Donat] B. Saint-Donat, *Projective models of K3 surfaces*, Amer. J. Math. **96** (1974), 602–639.

[FAC] J.-P. Serre, *Faisceaux algébriques cohérents*, Ann. of Math. **61** (1955), 197–278.

[Serre] J.-P. Serre, *Quelques propriétés des variétés abéliennes en caractéristique p*, Amer. J. Math. **80** (1956), 715–739.

[Sh] I.R. Shafarevich, *Basic Algebraic Geometry, 2 Vols.*, Springer, 1994.

[Sh-Barron] N.I. Shepherd-Barron, *Geography for surfaces of general type in positive characteristic*, Invent. Math **106** (1991), 263–274.

Lectures on Linear Series

Robert Lazarsfeld

with the assistance of
Guillermo Fernández del Busto

IAS/Park City Mathematics Series
Volume 3, 1997

Lectures on Linear Series

Robert Lazarsfeld

with the assistance of
Guillermo Fernández del Busto

Introduction

Linear series have long played a central role in algebraic geometry, and the basic results and techniques from their study form an essential part of the field's culture. However the past decade has witnessed two important new developments. First, vector bundles have emerged as powerful tools for analyzing linear series on curves and surfaces. These vector bundle techniques — the most important being Reider's method involving Bogomolov instability on surfaces — have led to considerable simplifications and extensions of classical results. A second major influence has come from the flowering of higher dimensional geometry. One now has a conjectural picture of how the most familiar facts about linear series on curves should extend to arbitrary smooth projective varieties, and some encouraging partial results have been proved. On the technical side, the cohomological machinery developed to study higher dimensional varieties — notably vanishing theorems for $\mathbf{Q}$-divisors — is proving to have applications even to concrete questions on surfaces. These new tools promise to play an important role in the future, but they have to a certain extent remained embedded in the research literature, and this has limited somewhat their accessibility to the novice or non-expert.

The purpose of these lectures is to provide a down-to-earth and gentle introduction to some of these new ideas and techniques. While I hope that these notes may have something to say to seasoned geometers who wish to learn about recent work on linear series, I have particularly tried to gear the discussion to a novice audience. My intention was that with a little faith and effort, the material here should be accessible to anyone having finished the standard texts, e.g. [**H2**] or

[1]Department of Mathematics, UCLA, Los Angeles,CA 90024. Partially supported by NSF Grant DMS 94-00815
E-mail address: rkl@math.ucla.edu

[**GH1**].* The underlying theme is the search for higher dimensional generalizations of the most basic theorems about linear series on algebraic curves, but to keep things elementary we work more or less entirely in dimension two. The philosophy is to illustrate in the setting of surfaces the various methods that have been used to attack these questions in general, and we end up repeatedly proving variants of one central result, namely Reider's theorem. Sticking to the case of surfaces allows one to eliminate many technical complexities, and some of the underlying ideas become particularly transparent. I hope that parts of the present notes might therefore provide a useful first introduction to the powerful and important cohomological tools of higher dimensional geometry, as well as to an active area of current research. To this end, I have included many exercises which sketch further developments and applications of the material discussed in the text. An overview of the questions we consider, and an outline of the contents of the lectures, will be found in §1.

It was not my intention to produce a balanced survey of work on linear series, and I have ignored a number of topics that might have fit very nicely. I am particularly cogniscent of the fact that the complex analytic side of the story is woefully under-represented here. Starting with Demailly's ground-breaking paper [**De1**], there has been an intriguing and fruitful interplay between algebraic and analytic approaches to many of these questions. I can only hope that the one-sided slant of the present notes will motivate someone more able than I to give an introductory account of the analytic viewpoint.

I'm grateful to O. Küchle, V. Maşek and G. Xu for helpful suggestions, and to F. Schreyer for bailing me out of a problem in a preliminary version of these lectures. G. Fernández del Busto — who served as my course assistant at the Park City Institute — produced a preliminary version of these notes and also contributed useful advice. I've written an exposition on linear series once before [**L2**], but the viewpoint here is rather different. This evolution is partly the result of contact with a number of mathematicians to whom I owe thanks. To begin with, H. Esnault and E. Viehweg initially explained to me a few years ago the philosophy of how vanishing for **Q**-divisors could be used to produce sections of adjoint bundles. I've also greatly profitted from suggestions and encouragement from J. Kollár, M. Reid and Y.-T. Siu. But above all I wish to acknowledge my debt to Lawrence Ein, with whom I have spent the last few years working on the questions discussed in these notes. He deserves a large share of the credit for any originality or utility the present lectures may possess.

§0. Notations and Conventions.

(0.1). We work throughout over the complex numbers **C**. Varieties are assumed to be smooth and projective unless otherwise stated.

*The only pre-requisites not covered in [**H2**] that we use systematically are Chern classes of vector bundles, and vanishing theorems. However the latter might be taken on faith. Griffiths and Harris [**GH1**] provide more than enough background for everything that appears here, except that the reader might want to supplement their discussion of the classical Kodaira vanishing theorem with an account (e.g. [**CKM**], Lecture 8, or [**Kol3**], Chapters 9, 10, or [**EV2**]) of the generalization due to Kawamata and Viehweg.

(0.2). We write K_X for the canonical divisor (class) of a smooth variety X. If $Z \subset X$ is a subvariety, $\mathcal{I}_Z \subset \mathcal{O}_X$ denotes its ideal sheaf.

(0.3). Given a line bundle L on X, much of our focus will be on the adjoint bundle $K_X + L$, for which it is traditional to use additive notation. But then to indicate the sheaf of sections of this bundle vanishing at a point $x \in X$, it seems most natural to write $\mathcal{O}_X(K_X+L)\otimes\mathcal{I}_x$. It soon becomes difficult to maintain notational consistency. We've finally surrendered to a certain amount of chaos concerning notation for line bundles, divisors and the corresponding invertible sheaves. If L is a line bundle and D is a divisor on X, we write $L(D)$, $L + D$ and $\mathcal{O}_X(L + D)$ more or less interchangably. We hope that this will not lead to undue confusion or annoyance. The notation $|\mathcal{O}_X(L + D)|$ is reserved for the complete linear system of divisors of sections of $\mathcal{O}_X(L + D)$. We say that a line bundle (or a divisor, or a linear series) L is *free* (or *globally generated* or *basepoint-free*) at a point $x \in X$ if there is a section $s \in H^0(X, L)$ with $s(x) \neq 0$. L is *free* if it is free at every point of X

(0.4). If L is a line bundle on a smooth projective variety X of dimension n, we denote by L^n the top self-intersection $\int c_1(L)^n$ of the first Chern class of L. If $C \subset X$ is a curve in X, $L \cdot C$ denotes the intersection number of $c_1(L)$ with C. Numerical equivalence of divisors or line bundles is denoted by $\equiv$. Recall that L is *nef*, or *numerically effective* if $L \cdot C \geq 0$ for every irreducible curve $C \subset X$. By Kleiman's criterion (cf. [**H1**], Chapter 1), L is nef if and only it is in the closure of the cone of ample line bundles. Note that the pull-back of a nef line bundle under any morphism remains nef. A nef line bundle is *big* if $L^n > 0$. This is equivalent (for nef bundles) to saying that $h^0(X, kL)$ grows like k^n. (See [**CKM**] or [**Mori**] for alternative characterizations.)

§1. Background and Overview

The theme of these lectures is the search for higher dimensional generalizations of the most familiar and elementary facts about linear series on curves. In this section we introduce the basic questions, and give a brief overview of their history and current status. The detailed contents of the notes are summarized at the end of the section.

To set the stage, we start by recalling the story in dimension one:

Theorem 1.1. *Let C be a compact Riemann surface of genus g.*

(A). If $g \geq 2$ then the canonical bundle K_C is globally generated, and the pluri-canonical series $|mK_C|$ are very ample when $m \geq 3$.

(B). Let N be a line bundle on C, with $\deg(N) = d$. If $d \geq 2g$ then N is globally generated, and if $d \geq 2g + 1$ then N is very ample.

Proof. We focus on statement (B). Suppose that $d \geq 2g$, and let $P \in C$ be a fixed point. We need to show that there exists a section $s \in \Gamma(C, N)$ with $s(P) \neq 0$. Consider to this end the exact sequence of sheaves

$$0 \longrightarrow N(-P) \longrightarrow N \longrightarrow N \otimes \mathcal{O}_P \longrightarrow 0.$$

It induces in cohomology the exact sequence

$$H^0(C,N) \xrightarrow{e_P} H^0(C, N\otimes \mathcal{O}_P) \longrightarrow H^1(C, N(-P)) \,,$$

with e_P evaluation at P. Now $H^0(C, N\otimes\mathcal{O}_P) \cong \mathbf{C}$, and it suffices to show that e_P is surjective. But this follows from the fact that $\deg N(-P) \geq 2g-1$, since in this case we have that $H^1(C, N(-P)) = 0$. The proof of very ampleness is similar, as is statement (A). (cf. [**H2**], IV.3.2). □

The attempt to generalize this theorem to higher dimensions is a very fundamental and interesting problem. In setting (A), it is relatively clear what to look for. The higher dimensional analogue of a curve of genus ≥ 2 is a variety of general type. Taking into account the expectation that one might want to work on a particularly tractable birational model, we can state the question as

Problem A. *Study pluri-canonical series $|mK_X|$ on "nice" varieties X of general type.*

By contrast, the higher dimensional analogue of statement (B) has only recently come into focus. The key is to rephrase (1.1)(B) without explicitly bringing in the genus of the curve. To this end, note that if N is a line bundle of degree $2g$ on a curve C of genus g, then we can write N in the form

$$N = K_C + 2A$$

for some ample line bundle A on C. Similarly, if $\deg(N) = 2g+1$, then $N = K_C + 3A$. Hence the natural question generalizing statement (B) is

Problem B. *Let X be a smooth projective variety of dimension n. Study the adjoint linear series*

$$|K_X + L|$$

where L is a suitably positive line bundle on X.

For example, given an ample line bundle A on X, one might take $L = (n+1)A$ or $L = (n+2)A$. As a start, one would like to understand when the adjoint bundles in question are globally generated or very ample.

Pluricanonical mappings of surfaces of general type were studied by Kodaira [**Kod**] and Bombieri [**Bomb**] in the late 1960's and early 1970's. In 1988, many of their results emerged as special cases of a theorem of Reider [**Rdr**] concerning adjoint linear series. Precise statements appear in §2. Suffice it to say here that Problems A and B are by now quite well understood on surfaces.

In higher dimensions, naturally enough, much less is known. But in recent years two conjectures of Fujita have attracted a great deal of interest:

Fujita's Conjectures. [**Fuj1**] *Let X be a smooth projective variety of dimension n.*

(A). Assume that X is minimal and of general type, i.e. suppose that the canonical bundle K_X is nef and big. Then $\mathcal{O}_X(mK_X)$ is globally generated for $m \geq n+2$.

(B). Let A be an ample line bundle on X (which is now not assumed to be minimal or of general type). Then $K_X + (n+1)A$ is globally generated, and $K_X + (n+2)A$ is very ample.

For curves this is essentially the content of Theorem 1.1, and on surfaces these statements follow from Reider's theorem. But in general Fujita's conjectures remain open as of this writing. Existing results are of two sorts. First, there are effective statements in all dimensions due to Demailly [**De1**], Kollár [**Kol2**] and most recently Siu [**Siu2**], which however are exponential in the dimension n. In another direction, one can stick to the simplest case of global generation on threefolds, and ask for statements closer to the bounds predicted by Fujita. Theorems of this sort are given by Ein and Lazarsfeld in [**EL2**], with Maşek in [**ELM**], and by Fujita in [**Fuj2**]. We refer to Ein's paper [**E**] for an overview of some of this work. Siu's paper on Matsusaka's theorem [**Siu1**] develops some related ideas (cf. Exercise 7.7).*

It is probably well to stress from the outset why it is that Problems A and B, while completely elementary on curves, become more subtle in higher dimensions. The proof of Theorem 1.1 reduces the question to the vanishing of a cohomology group, and this goes through in the general setting without any problem:

Proposition 1.2. *Let L be an ample line bundle on a smooth projective variety X (of arbitrary dimension). Then $K_X + L$ is globally generated if and only if*

$$H^1(X, \mathcal{O}_X(K_X + L) \otimes \mathcal{I}_x) = 0$$

for all points $x \in X$, where $\mathcal{I}_x$ is the ideal sheaf of $\{x\}$. □

(There is of course an analogous criterion for $K_X + L$ to be very ample.) If X is a curve, then $\mathcal{O}_X(K_X + L) \otimes \mathcal{I}_x$ is locally free, and one deduces the required vanishing from general facts about the cohomology of line bundles. However when X is a variety of dimension ≥ 2, the sheaf $\mathcal{O}_X(K_X+L) \otimes \mathcal{I}_x$ is no longer invertible and it becomes considerably harder to control the group $H^1(X, \mathcal{O}_X(K_X+L) \otimes \mathcal{I}_x)$. So in higher dimensions the issue in effect is to prove a Kodaira-type vanishing theorem for certain non-invertible sheaves.

The "classical" approach to these questions on surfaces is to apply vanishing theorems involving numerical connectedness on a blow-up of X to control the required cohomology groups. This is discussed in Reid's lectures. During the 1980's some new techniques emerged for studying linear series on curves and surfaces, revolving around the geometry of vector bundles. Methods along these lines were used for example to study projective normality of curves in [**GL**], and the geometry of curves on K3 surfaces in [**L1**]. But certainly the most important development was Reider's application [**Rdr**] of Bogomolov's instability theorem to study adjoint series on surfaces. Reider's method – which has its antecedents in a proof by Mumford of Kodaira vanishing [**Reid**] – has largely superceeded the classical approach using numerical connectedness à la Ramanujam. We remark that a number of geometers have attempted to apply vector bundles in a similar manner to study linear

*See note added in proof at the end of the paper.

series on varieties of dimension three and higher, but so far these efforts have not met with success.

At about the same time that vector bundle techniques were developed to study linear series on surfaces, Kawamata, Reid and Shokurov (among others) introduced some powerful but subtle cohomological techniques to analyze the asymptotic behavior of pluricanonical and other linear series on varieties of all dimensions. These techniques – which we shall refer to as the KRS package – are based on the Kawamata-Viewheg vanishing theorem for **Q**-divisors, and form one of the central tools of contemporary higher dimensional geometry. A typical result here is a theorem of Kawamata and Shokurov to the effect that if X is a smooth minimal variety of general type, then $|mK_X|$ is free for all $m \gg 0$. We refer to [**CKM**] or [**KMM**] for references and an overview of this circle of ideas. Ein and the author remarked that one could use the KRS machine to recover (some of) Reider's results. This in turn opened the door to the higher dimensional theorems appearing for instance in [**Kol2**] and [**EL2**].

The goal of these lectures is to explain some of the new methods that have been introduced to study Problems A and B on surfaces and higher dimensional varieties. To keep things elementary, we will work more or less exclusively on surfaces, and the plan is to discuss one central result (viz. Reider's theorem) from many points of view. To begin with, the vector bundle techniques apply here, and we go through these in some detail. However we also present in the surface setting some of the methods developed for higher dimensions. Thus we devote considerable attention to the circle of ideas surrounding vanishing theorems for **Q**-divisors. We also explain – still in the case of surfaces – some algebro-geometric analogues of Demailly's analytic approach, as well as some local results from [**EL3**] and [**EKL**].

The detailed contents of these notes is as follows. In §2 we give the statement of Reider's theorem, and present some of its applications. The following two sections are devoted to the proof of this result via vector bundles: §3 discusses Serre's method (and related techniques) for constructing vector bundles, and in §4 Bogomolov's instability theorem is applied to prove Reider's statement. The focus then turns to vanishing theorems. We consider in §5 the questions involving Seshadri constants and local positivity that arise if one tries to apply vanishing theorems in the most naive possible way to study adjoint series. Section 6 revolves around vanishing theorems for **Q**-divisors and the Kawamata-Reid-Shokurov machine: we use these techniques to reprove (parts of) Reider's theorem, as well as the rank two case of Bogomolov's theorem. Finally, §7 is devoted to the algebro-geometric analogues of Demailly's approach to these questions. Further applications and extensions of the material are outlined in numerous exercises scattered throughout the notes.

§2. Reider's Theorem – Statement and Applications

In this section, we state Reider's theorem [**Rdr**], and present some simple applications. Reider's result forms the core of the present notes: most of the subsequent sections are devoted to various different proofs of the theorem or its corollaries. Here we try to convey some feeling for its power and scope.

The theorem in question gives a very precise geometric explanation for the failure of an adjoint bundle on a surface to be globally generated or very ample.

Theorem 2.1 ([**Rdr**]). *Let X be a smooth projective surface, and let L be a nef line bundle on X.*

(i). Assume that $L^2 \geq 5$, and that the adjoint series $|K_X + L|$ has a base-point at $x \in X$. Then there exists an effective divisor $D \subset X$ passing through x such that either

$$D \cdot L = 0 \quad \textit{and} \quad D^2 = -1; \quad \textit{or}$$
$$D \cdot L = 1 \quad \textit{and} \quad D^2 = 0. \tag{2.2}$$

(ii). If $L^2 \geq 10$, and if $x,\ y \in X$ are two points (possibly infinitely near) which fail to be separated by $|K_X + L|$, then there exists an effective divisor $D \subset X$ through x and y such that either

$$D \cdot L = 0 \quad \textit{and} \quad D^2 = -1 \ \textit{or} \ -2; \quad \textit{or}$$
$$D \cdot L = 1 \quad \textit{and} \quad D^2 = 0 \ \textit{or} \ -1; \quad \textit{or} \tag{2.3}$$
$$D \cdot L = 2 \quad \textit{and} \quad D^2 = 0$$

Note that we do not assume that the base point x in (i) is isolated. Hence subject to the numerical inequalities on the nef line bundle L, the statement applies as soon as $\mathcal{O}_X(K_X + L)$ fails to be globally generated. Similarly, (ii) serves as a criterion for the bundle in question to be very ample.

Example 2.4. Typical examples of a divisor satisfying (2.2) may be obtained by taking D to be an exceptional curve of the first kind (in the first case), or as the fibre of a ruled surface (in the second).

Exercise 2.5. (i). Show that if D is one of the divisors satisfying (2.2) described in Example 2.4, then D is necessarily in the base locus of $|K_X + L|$.

(ii). Find examples of divisors D satisfying the various possibilities in (2.3), for which $\mathcal{O}_X(K_X + L)$ indeed fails to be very ample.

(iii). Show that the numerical hypothesis on L^2 in Theorem 2.1 cannot in general be weakened. □

Reider's theorem leads to a simple numerical criterion for an adjoint bundle to be globally generated or very ample:

Corollary 2.6. *Let L be an ample line bundle on a smooth projective surface X. Assume that $L^2 \geq 5$, and that $L \cdot C \geq 2$ for all irreducible curves $C \subset X$. Then $\mathcal{O}_X(K_X + L)$ is globally generated. If $L^2 \geq 10$ and $L \cdot C \geq 3$ for all $C \subset X$, then $\mathcal{O}_X(K_X + L)$ is very ample.* □

This in turn implies the dimension $n = 2$ case of Fujita's conjecture:

Corollary 2.7. *If A is an ample line bundle on X, then $|K_X + 3A|$ is free, and $|K_X + 4A|$ is very ample.*

Proof. In fact, take $L = 3A$. Then $L^2 = 9A^2 \geq 9$, and $L \cdot C = 3A \cdot C \geq 3$ for any curve $C \subset X$. So Corollary 2.6 shows that $\mathcal{O}_X(K_X + L)$ is globally generated, and similarly $\mathcal{O}_X(K_X + 4A)$ is very ample. □

As we indicated in the previous section, attempts to generalize Reider's theorem to higher dimensions have mainly focused on extending the statements of Corollaries 2.6 and 2.7. At the moment, one doesn't even have any clear conjectures as to what might be the analogue of the more precise information contained in Theorem 2.1. However in the case of surfaces, these restrictions on the self-intersection of D are very powerful. Indeed, we will now see that they allow one to deduce from Reider's theorem many of the classical facts concerning pluricanonical models of surfaces of general type.

Let X be a surface of general type. It is very natural and important to try to understand the geometry of the pluricanonical rational mappings

$$\Phi_m = \Phi_{|mK|} : X \dashrightarrow \mathbf{P} = \mathbf{P}H^0(mK_X)$$

defined by multiples of the canonical bundle K_X. Here Reider's theorem leads to an extremely quick proof of some of the fundamental results of Kodaira and Bombieri:

Corolllary 2.8. ([**Kod**], [**Bomb**]) *Assume that X is minimal i.e. not the blowing up of some other smooth surface at a point. Then:*

(i). The bundle $\mathcal{O}_X(mK_X)$ is globally generated (i.e. Φ_m is a morphism) if $m \geq 4$, or if $m \geq 3$ and $K_X^2 \geq 2$.

(ii). Φ_m is an embedding away from (-2)-curves if $m \geq 5$, or if $m \geq 4$ and $K_X^2 \geq 2$, or if $m \geq 3$ and $K_X^2 \geq 3$.

Remark. By an embedding away from (-2)-curves, we mean a morphism which is one-to-one and unramified away from a divisor $Z \subseteq X$ consisting of smooth rational curves with self-intersection -2. Catanese and Reider [**C**] have used Reider's method to prove the more precise theorem of Bombieri that K_V is very ample on the canonical model V of X obtained by blowing down all the $A - D - E$ cycles Z to rational double points.

Proof of Corollary 2.8. We will only consider the global generation of $|mK_X|$, leaving the proof of the second assertion as an exercise.

Recall that on a minimal surface of general type X, the canonical bundle K_X is nef [**BPV**], III.2.3. Hence Theorem 2.1 applies to $L = (m-1)K_X$. The numerical hypotheses of (2.1)(i) are satisfied thanks to the conditions on m and K_X^2. Thus if $|mK_X|$ has a base point, then there exists an effective divisor $D \subseteq X$ such that either $(m-1)K_X \cdot D = 1$ or else

$$(m-1)K_X \cdot D = 0 \text{ and } D^2 = -1.$$

The first possibilty is excluded by the assumption that $(m-1) \geq 2$. As for the second, if $K_X \cdot D = 0$ then $D^2 = D \cdot (D + K_X) \equiv 0 \pmod 2$ by adjunction, contradicting $D^2 = -1$. □

We refer to [**Rdr**] for further applications of Reider's method to pluricanonical mappings. The following exercises present some other applications of Theorem 2.1, and the conference proceedings [**SBL**] contain a sampling of some more recent developments.

Exercise 2.9. (Embeddings of Abelian Surfaces.) Prove the following result of Ramanan [**Ram**]. Let X be an abelian surface which contains no elliptic curves. If L is an ample line bundle on X such that $L^2 \geq 10$, then L is very ample. (It follows for example that on a sufficiently general abelian surface, a polarization of type $(1, d)$ is very ample if $d \geq 5$. In particular, taking d = 5, there exist smooth abelian surfaces in $\mathbf{P}^4$.) See [**LB**] Chapt. 10, §4, for more precise statements, and the following exercise for a generalization. The corresponding questions on higher dimensional abelian varieties are considered in [**BLR**] and [**DHS**]. □

Exercise 2.10. (Linear Series on Minimal Surfaces of Kodaira Dimension Zero.) In this exercise, X denotes a smooth projective surface whose canonical bundle K_X is numerically trivial. This hypothesis is satisfied e.g. by Abelian, K3 and Enriques surfaces; see [**BPV**], IV.1, for the complete list. Fix an ample line bundle L on X.

(i). Assume that $L^2 \geq 5$. Prove that $\mathcal{O}_X(K_X + L)$ fails to be globally generated if and only if there exists an irreducible curve $E \subset X$ with $p_a(E) = 1$ and $E \cdot L = 1$.

(ii). Assume that $L^2 \geq 10$, and that $|K_X + L|$ is free. Prove that $\mathcal{O}_X(K_X + L)$ fails to be very ample if and only if there exists a reduced curve $E \subset X$, with $p_a(E) = 1$ and $E \cdot L = 2$. [Note that the intersection form on X is even, i.e. D^2 is even for every divisor D on X.]

See [**Rdr**], Proposition 5, for a more precise statement due to Beauville. □

We conclude this section with some interesting open problems of an algebraic nature that arise in connection with Reider's theorem. Returning for an instant to the one-dimensional case, let N be a line bundle of degree d on a smooth projective curve C of genus g. A classical theorem of Castelnuovo, Mattuck and Mumford asserts that if $d \geq 2g + 1$, then N is normally generated, i.e. the natural maps $Sym^m(H^0(N)) \longrightarrow H^0(mN)$ are surjective for all $m \geq 0$. Furthermore, if $d \geq 2g + 2$, then in the embedding $C \subset \mathbf{P}\mathrm{H}^0(N)$ defined by N, the homogeneous ideal I_C of C is generated by quadrics. The famous theorems of Noether and Petri (cf. [**Mfd**], [**ACGH**] Chapter III, [**GL**]) give analogous statements for the canonical bundle, and Green [**Grn**] has shown that at least conjecturally the whole picture extends to higher syzygies as well. (See also Exercise 3.5.)

The question then arises whether similar statements hold for adjoint and pluricanonical bundles on surfaces and higher dimensional varieties. In the two dimensional case, the natural thing to hope for here is the following:

Conjecture 2.11. (**Mukai.**) *Let A be an ample line bundle on a smooth projective surface X, and let P be any nef line bundle. Then $\mathcal{O}_X(K_X + 4A + P)$ is normally generated, and in the embedding $X \subset \mathbf{P}$ defined by $\mathcal{O}_X(K_X + 5A + P)$, the homogeneous ideal I_X of X is generated by quadrics.*

One would also like results in the spirit of (2.1) dealing with the adjoint bundles $K_X + L$ subject only to numerical conditions on L. For instance, one might hope that if L satisfies the hypotheses of (2.1)(ii), then $\mathcal{O}_X(K_X + L)$ is normally generated. There should also be analogous statements for higher syzygies.

When X is a ruled surface, Butler [**But**] proves that K_X+5A+P is projectively normal, and he also obtains results for generation by quadrics and higher syzygies. For "hyper-adjoint" bundles of the form $K+mB+P$ where B is *very ample*, results in arbitrary dimension are given by Ein and Lazarsfeld in [**EL1**]. In fact, if V is a smooth projective variety of dimension n, then $K_V + (n+1)B + P$ is normally generated, and $K_V+(n+2)B+P$ defines an embedding in which the homogeneous ideal of V is defined by quadrics. (Compare Exercise 5.15. [**EL1**] also contains analogous statements for higher syzygies.) However it seems that new ideas will be needed to tackle Conjecture 2.11. It would be already very interesting to get a result along the lines of (2.11) but with weaker numbers.

§3. Building Vector Bundles

This section is devoted to preparations for the first proof of Reider's theorem, via vector bundles. The strategy for using vector bundles to study linear series involves two steps:

- Encode the geometric data at hand in a vector bundle E; and
- Study the geometry of E (e.g. stability, or sub-bundles, or endomorphisms) to arrive at the desired conclusions.

Here we focus on the first point, and discuss techniques for constructing bundles. Specifically, after some warm-up with extentions of line bundles, we study a method introduced by Serre, which underlies Reider's argument. In the exercises, we sketch several other constructions and applications. A good reference for the general homological machinery appearing here is [**GH1**], Chapter 5, §3 and §4.

Extensions of Line Bundles.

To set the stage, we begin with the simplest technique for producing vector bundles, namely as extensions of invertible sheaves. Let X be an irreducible projective variety, and let L and M be line bundles on X. Recall that an *extension* of L by M is a short exact sequence of sheaves:

$$0 \longrightarrow M \longrightarrow E \longrightarrow L \longrightarrow 0, \tag{3.1}$$

so that E is a rank two vector bundle on X. Two such extentions are *equivalent* if there is a map between them inducing the identity on the outer terms. The set of equivalence classes of such extensions can be given the structure of a complex vector space, denoted $\operatorname{Ext}^1(L, M)$. The zero element of $\operatorname{Ext}^1(L, M)$ corresponds to the split sequence. (Analogous considerations apply when L and M are arbitrary coherent sheaves, although of course in this case the middle term in (3.1) isn't in general locally free.)

Assuming as we are that L and M are line bundles, one has the basic isomorphism:

$$\text{Ext}^1(L, M) = H^1(X, L^* \otimes M), \tag{3.2}$$

whose proof we outline in Exercise 3.3. For the present purposes, the importance of (3.2) is that the bundle E serves as a geometric realization of a class $e \in H^1(X, L^* \otimes M)$. This is illustrated in Exercise 3.5, where we indicate how ideas along these lines may be used to prove Noether's theorem and various generalizations concerning the projective normality of algebraic curves.

Exercise 3.3. **(Extension Classes** – Compare **[H2]** Ex. III.6.1, and **[GH1]**, pp. 722 - 725.) We keep the notation just introduced. In particular, L and M are locally free sheaves of rank one on the projective variety X.

(i). We start by defining the map

$$\text{Ext}^1(L, M) \longrightarrow H^1(X, L^* \otimes M) \tag{3.3.1}$$

required for (3.2). To this end, fix an extension of L by M. Prove that there exists a covering of X by open sets $\{U_i\}$ on which the restriction of (3.1) splits. Show that on the intersections $U_i \cap U_j$, the difference of two such local splittings determines homomorphisms $L(U_i \cap U_j) \longrightarrow M(U_i \cap U_j)$, which in turn yields a Čech cocycle in $Z^1(\{U_i\}, L^* \otimes M)$. Prove that the resulting cohomology class $e \in H^1(X, L^* \otimes M)$ — which is called the *extension class* of the extension (3.1) — is independent of the choices made, and vanishes if and only if (3.1) splits. This defines (3.3.1).

(ii). Show that the map (3.3.1) is an isomorphism. [It may be helpful to remark that one can take transition matrices for the bundle E appearing in (3.1) to be triangular, with the cocycle representing the extension class as the off-diagonal entry.]

(iii). Prove that map (3.3.1) can alternatively be defined as follows. Given (3.1), tensor through by L^* to get $0 \longrightarrow L^* \otimes M \longrightarrow L^* \otimes E \longrightarrow \mathcal{O}_X \longrightarrow 0$. Then the corresponding extension class is the image of the constant function $1_X \in H^0(X, \mathcal{O}_X)$ under the connecting homomorphism $\delta : H^0(X, \mathcal{O}_X) \longrightarrow H^1(X, L^* \otimes M)$.

(iv). Note that cup product determines a map $H^0(X, L) \otimes H^1(X, L^* \otimes M) \longrightarrow H^1(X, M)$, or equivalently a homomorphism

$$H^1(X, L^* \otimes M) \longrightarrow Hom(H^0(X, L), H^1(X, M)). \tag{3.3.2}$$

Verify that (3.3.2) may be interpreted as the map which takes an extension to the connecting homomorphism it determines. □

Exercise 3.4. Let E be a rank two vector bundle on $\mathbf{P}^n$ which admits a nowhere vanishing section. Prove that if $n \geq 2$, then E is a direct sum of line bundles. (The same statement is true when $n = 1$, but the proof is a little more delicate. In fact, a theorem of Grothendieck (cf. **[OSS]**, I.2.1) states that any vector bundle on $\mathbf{P}^1$ splits as a direct sum of line bundles.) □

Exercise 3.5. (**Noether's Theorem and Generalizations.**) Let C be a smooth projective curve of genus $g \geq 2$. A classical theorem of Noether states that if C is non-hyperelliptic, then the natural maps

$$\rho_m : Sym^m(H^0(C, K_C)) \longrightarrow H^0(C, mK_C)$$

are surjective for $m \geq 2$. In other words, every pluri-canonical differential form can be expressed as a polynomial in one-forms, or equivalently, C is projectively normal in its canonical embeding $C \subset \mathbf{P}^{g-1}$. In this exercise we will outline a proof of this statement via vector bundles. There are quicker approaches to Noether's theorem, but with only a little extra effort the present argument yields also some substantial generalizations of the classical results (see (v) below). These appear in [**GL**], from which this exercise is adapted. For simplicity we will focus on the surjectivity of ρ_2; the case of ρ_m for $m > 2$ is similar but more elementary.

(i). Let C be any smooth curve of genus g, and assume that ρ_2 is not surjective. Then the map $H^0(C, \mathcal{O}(2K_C))^* \longrightarrow H^0(C, \mathcal{O}(K_C))^* \otimes H^0(C, \mathcal{O}(K_C))^*$ dual to multiplication has a non-trivial kernel. Using Serre duality and Exercise 3.3.(iv), show that there exists a non-split extension

$$0 \longrightarrow \mathcal{O}_C \longrightarrow E \longrightarrow \mathcal{O}_C(K_C) \longrightarrow 0 \tag{3.5.1}$$

which is exact on global sections. In particular, $h^0(C, E) = g + 1$. The plan is to show that the existence of such a bundle forces C to be hyperelliptic.

(ii). Now fix a point $P \in C$. Show that there exists a non-zero section $s \in H^0(C, E)$ vanishing at P. Denoting by D the effective divisor (containing P) on which s vanishes, one has an exact sequence:

$$0 \longrightarrow \mathcal{O}_C(D) \longrightarrow E \longrightarrow \mathcal{O}_C(K_C - D) \longrightarrow 0. \tag{3.5.2}$$

Using the fact that (3.5.1) is non-split, show that $\mathcal{O}_C(D) \not\cong \mathcal{O}_C(K_C)$.

(iii). Compare the estimates on $h^0(C, E)$ obtained from (3.5.1) and (3.5.2) to prove that $deg(D) \leq 2r(D)$, where as usual $r(D) = h^0(C, \mathcal{O}_C(D)) - 1$. Finally, conclude from Clifford's theorem that C is hyperelliptic. This completes the proof of Noether's theorem.

(iv). Prove the theorem of Castelnouvo, Mattuck and Mumford that if L is a line bundle of degree $\geq 2g+1$ on a curve C of genus g, then L is normally generated, i.e. the maps $Sym^m(H^0(C, L)) \longrightarrow H^0(C, mL)$ are surjective for $m \geq 1$.

(v). A result going back to Segre (cf. [**MS**]) states that if F is a rank two vector bundle of degree d on C, then F contains a line sub-bundle $A \subset F$ with $deg(A) \geq [\frac{d-g+1}{2}]$. Using this, prove the theorem of [**GL**] that if L is a very ample line bundle with

$$deg(L) \geq 2g + 1 - 2h^1(L) - \text{Cliff}(C), \tag{3.5.3}$$

then L is normally generated. Here

$$\text{Cliff}(C) = \min\left\{deg(A) - 2r(A) \mid h^0(A) \geq 2, deg(A) \leq g - 1\right\}$$

denotes the Clifford index of C, which measures how general C is from the point of view of moduli. (For instance, $\text{Cliff}(C) = 0$ if and only if C is hyperelliptic; $\text{Cliff}(C) = 1$ iff C is either trigonal or a smooth plane quintic; and if C is a general curve of genus g, then $\mathrm{Cliff}(C) = [\frac{g-1}{2}]$.) Note that (3.5.3) contains the theorems of Noether and Castelnuovo-Mattuck-Mumford as special cases. □

The Serre Construction.

We henceforth assume that X is a smooth projective surface. The bundles on X that arise as extensions of line bundles are rather special, as Exercise 3.4 suggests. Reider's theorem requires a more general construction. It was introduced by Serre, and applied notably in the analysis of codimension two subvarieties of projective space (cf. [**OSS**], I.5). The case of surfaces, which has a somewhat different flavor, was studied by Griffiths and Harris [**GH2**].

Consider to begin with a rank two vector bundle E on X, with $\det(E) = L$. Suppose that $s \in \Gamma(X, E)$ is a section of E that vanishes on a finite set. Denote by

$$Z = Z(s) \subset X$$

the zero-scheme of s: locally one may view s as given by a vector $s = (s_1, s_2)$ of regular functions, and Z is defined locally by the vanishing of s_1 and s_2. Remark that the Koszul complex associated to (E, s) determines a short exact sequence

$$0 \longrightarrow \mathcal{O}_X \xrightarrow{\cdot s} E \longrightarrow L \otimes \mathcal{I}_Z \longrightarrow 0 \tag{3.6}$$

resolving $L \otimes \mathcal{I}_Z$, where $\mathcal{I}_Z$ is the ideal sheaf of Z. Somewhat abusively, we will refer to (3.6) as the Koszul complex arising from s.

Serre addresses the possibility of making the inverse construction. Specifically, fix a line bundle L on X, plus a finite scheme $Z \subset X$. We pose

Question 3.7. *When does there exist a rank two vector bundle E on X, with $det(E) = L$, plus a section $s \in \Gamma(X, E)$ such that $Z(s) = Z$?*

Clearly there are local obstructions to finding E: for instance, Z must be locally defined by two equations. But there are also some very interesting global conditions, and these are the key to Reider's method.

Serre's idea is that one should try to construct E via (3.6), as an extension of $L \otimes \mathcal{I}_Z$ by $\mathcal{O}_X$. As above, we consider the group

$$\mathrm{Ext}^1(L \otimes \mathcal{I}_Z, \mathcal{O}_X)$$

parametrizing all such. Given $e \in \mathrm{Ext}^1(L \otimes \mathcal{I}_Z, \mathcal{O}_X)$ we denote by $\mathcal{F}_e$ the sheaf arising from the corresponding extension:

$$0 \longrightarrow \mathcal{O}_X \longrightarrow \mathcal{F}_e \longrightarrow L \otimes \mathcal{I}_Z \longrightarrow 0. \tag{3.8}$$

Thus $\mathcal{F}_e$ is a torsion-free $\mathcal{O}_X$-module of rank two. Suppose it happens that $\mathcal{F}_e$ is actually locally free for some $e \in \mathrm{Ext}^1(L \otimes \mathcal{I}_Z, \mathcal{O}_X)$. Then the map $\mathcal{O}_X \longrightarrow \mathcal{F}_e$ determines a section $s \in \Gamma(X, \mathcal{F}_e)$ with (3.8) as the corresponding Koszul complex. In particular, $Z(s) = Z$. Therefore $(\mathcal{F}_e, s)$ gives the required solution to (3.7). Thus the essential point is to determine when there exists an element $e \in \mathrm{Ext}^1(L \otimes \mathcal{I}_Z, \mathcal{O}_X)$ such that $\mathcal{F}_e$ is a vector bundle.

We begin with a criterion for the failure of $\mathcal{F}_e$ to be locally free.

Proposition 3.9. *Given an element $e \in \operatorname{Ext}^1(L \otimes \mathcal{I}_Z, \mathcal{O}_X)$, the correponding sheaf $\mathcal{F}_e$ fails to be locally free if and only if there exists a proper subscheme $Z' \subsetneqq Z$ (possibly $Z' = \emptyset$) such that*

$$e \in Im\big\{\operatorname{Ext}^1(L \otimes \mathcal{I}_{Z'}, \mathcal{O}_X) \longrightarrow \operatorname{Ext}^1(L \otimes \mathcal{I}_Z, \mathcal{O}_X)\big\}. \tag{3.9.1}$$

Remark 3.10. Let us explicate the map in (3.9.1). If $Z' \subseteq Z$, then

$$L \otimes \mathcal{I}_Z \subseteq L \otimes \mathcal{I}_{Z'}.$$

Thus starting with an extension of $L \otimes \mathcal{I}_{Z'}$ by $\mathcal{O}_X$, one can pull it back in an evident way to get an extension of $L \otimes \mathcal{I}_Z$ by $\mathcal{O}_X$. This gives rise to the homomorphism appearing in (3.9.1).

We now give a proof of Proposition 3.9 drawing on some facts from local algebra. A more geometric approach to the main implication is sketched in Exercise 3.19.

Proof of Proposition 3.9. We start with some general remarks of a homological nature, referring to [**OSS**], II.1.1 (especially pp. 148-149), for details and proofs. Let $\mathcal{F}$ be a torsion-free sheaf on the smooth surface X. Then the set of points at which $\mathcal{F}$ fails to be locally free is finite or empty. Denote by $\mathcal{F}^{**}$ the double dual of $\mathcal{F}$. Then $\mathcal{F}^{**}$ is reflexive, hence a vector bundle. (In general, a reflexive sheaf on a smooth variety is locally free off a set of codimension ≥ 3.) Furthermore, the natural injection $\mu : \mathcal{F} \longrightarrow \mathcal{F}^{**}$ fails to be an isomorphism exactly over the set where $\mathcal{F}$ is not locally free. It is suggestive to think of $\mathcal{F}^{**}$ as a sort of "canonical desingularization" of $\mathcal{F}$.

Turning now to the situation of the Proposition, write $\mathcal{F} = \mathcal{F}_e$, and assume that $\mathcal{F}$ fails to be locally free. The extension defining $\mathcal{F}$, plus the map $\mu : \mathcal{F} \longrightarrow \mathcal{F}^{**}$, give rise to an exact commutative diagram:

$$\begin{array}{ccccccccc}
 & & & & 0 & & 0 & & \\
 & & & & \downarrow & & \downarrow & & \\
0 & \longrightarrow & \mathcal{O}_X & \longrightarrow & \mathcal{F} & \longrightarrow & L \otimes \mathcal{I}_Z & \longrightarrow & 0 \\
 & & \| & & \downarrow{\scriptstyle \mu} & & \downarrow & & \\
0 & \longrightarrow & \mathcal{O}_X & \longrightarrow & \mathcal{F}^{**} & \longrightarrow & L \otimes \mathcal{I}_{Z'} & \longrightarrow & 0. \\
 & & & & \downarrow & & \downarrow & & \\
 & & & & \tau & = & \tau & & \\
 & & & & \downarrow & & \downarrow & & \\
 & & & & 0 & & 0 & &
\end{array} \tag{3.11}$$

Here $\tau = \mathcal{F}^{**}/\mathcal{F}$ is a finite sheaf supported on the set where $\mathcal{F}$ fails to be locally free. The map $\mathcal{O}_X \longrightarrow \mathcal{F}^{**}$ determines a section $s' \in \Gamma(X, \mathcal{F}^{**})$ vanishing on a finite scheme $Z' =_{\text{def}} Z(s')$, and the second row of (3.11) is the corresponding

Koszul complex. It follows from Remark 3.10 that if $e' \in \text{Ext}^1(L \otimes I_{Z'}, \mathcal{O}_X)$ denotes the extension class of the middle row, then e' maps to the given extension class $e \in \text{Ext}^1(L \otimes I_Z, \mathcal{O}_X)$. So to prove the first implication in Proposition, it suffices to show that Z' is a proper subscheme of Z. But $\tau = \text{coker}(\mu) \neq 0$ since $\mathcal{F}$ is not locally free, and the right hand column of (3.11) then shows that $Z' \subsetneq Z$, as required.

Conversely, suppose that there exists a proper subscheme $Z' \subsetneq Z$ such that the given extension defining $\mathcal{F} = \mathcal{F}_e$ is induced from an extension $e' \in \text{Ext}^1(L \otimes \mathcal{I}_{Z'}, \mathcal{O}_X)$. Denote by $\mathcal{E}$ the double dual of the torsion-free sheaf $\mathcal{F}_{e'}$ determined by e', so that $\mathcal{E}$ is locally free of rank two. Arguing from a diagram much like (3.11), one finds that there exists an injective sheaf homomorphism $\nu : \mathcal{F} \longrightarrow \mathcal{E}$ with a non-trivial finite cokernel. But a generically injective map between two vector bundles of the same rank is either an isomorphism or drops rank along a divisor (locally defined as a determinant). Therefore $\mathcal{F}$ cannot be locally free. □

We assume henceforth for simplicity that Z is a *reduced* finite scheme, i.e. that all the points of Z appear with "multiplicity one". Then we have:

Corollary 3.12. *There is an extension class $e \in \text{Ext}^1(L \otimes I_Z, \mathcal{O}_X)$ with $\mathcal{F}_e$ locally free if and only if for every proper subset*

$$Z' \subsetneq Z \quad (\textit{including } Z' = \emptyset),$$

$\text{Ext}^1(L \otimes \mathcal{I}_{Z'}, \mathcal{O}_X)$ maps to a proper subspace of $\text{Ext}^1(L \otimes \mathcal{I}_Z, \mathcal{O}_X)$.

Proof. In fact, since Z is reduced there are only finitely many proper subschemes $Z' \subset Z$. □

The next ingredient we'll need is:

Serre-Grothendieck Duality. *If $\mathcal{G}$ is any coherent sheaf on the smooth projective surface X, then there is a natural isomorphism*

$$\text{Ext}^1(\mathcal{G}, \mathcal{O}_X(K_X)) \longrightarrow H^1(X, \mathcal{G})^*.$$

We refer to [**H2**], III.7, for the proof (cf. also [**GH1**], Chapter 5, §4). However let us at least define the map appearing in the statement. Represent a given element $e \in \text{Ext}^1(\mathcal{G}, \mathcal{O}_X(K_X))$ by an extension $0 \longrightarrow \mathcal{O}_X(K_X) \longrightarrow \mathcal{F}_e \longrightarrow \mathcal{G} \longrightarrow 0$. Then the connecting homomorphism defines a map

$$H^1(X, \mathcal{G}) \longrightarrow H^2(X, \mathcal{O}_X(K_X))$$

which, after fixing an identification $H^2(X, \mathcal{O}_X(K_X)) = \mathbf{C}$, may be viewed as an element of $H^1(X, \mathcal{G})^*$. (See also Exercise 3.19.(iv).)

Still assuming that Z is reduced, we may now state the answer to Question 3.7 as:

Theorem 3.13. [**GH2**] *There exists a rank two vector bundle E with $\det E = L$, plus a section $s \in \Gamma(E)$ with $Z(s) = Z$, if and only if every section of $\mathcal{O}_X(K_X + L)$ vanishing at all but one of the points of Z also vanishes at the remaining point.*

Remark. When $Z = \{x\}$ consists of a single (reduced) point, the criterion is simply that x be a base-point of $\mathcal{O}_X(K_X + L)$.

Proof of Theorem 3.13. As we observed above, the question is equivalent to the existence of an extension (3.8) with $\mathcal{F}_e$ locally free. Note to begin with the formal fact that it is enough to test the condition in Corollary 3.12 for sets $Z' \subset Z$ obtained by deleting one point from Z. Hence there exists a locally free extension of $L \otimes \mathcal{I}_Z$ by $\mathcal{O}_X$ if and only if for every point $x \in Z$, the map

$$(*) \qquad \mathrm{Ext}^1(L \otimes \mathcal{I}_{Z-\{x\}}, \mathcal{O}_X) \longrightarrow \mathrm{Ext}^1(L \otimes \mathcal{I}_Z, \mathcal{O}_X)$$

is non-surjective. Now

$$\mathrm{Ext}^1(L \otimes \mathcal{I}_Z, \mathcal{O}_X) = \mathrm{Ext}^1(\mathcal{O}_X(K_X + L) \otimes \mathcal{I}_Z, \mathcal{O}_X(K_X)),$$
$$\mathrm{Ext}^1(L \otimes \mathcal{I}_{Z-\{x\}}, \mathcal{O}_X) = \mathrm{Ext}^1(\mathcal{O}_X(K_X + L) \otimes \mathcal{I}_{Z-\{x\}}, \mathcal{O}_X(K_X)).$$

So by Duality, the existence of a locally free extension is equivalent to the non-injectivity of

$$(**) \qquad H^1(X, \mathcal{O}_X(K_X + L) \otimes \mathcal{I}_Z) \longrightarrow H^1(X, \mathcal{O}_X(K_X + L) \otimes \mathcal{I}_{Z-\{x\}})$$

for every $x \in Z$. But the map (**) sits in the long exact sequence on cohomology determined by the sheaf sequence

$$0 \longrightarrow \mathcal{O}_X(K_X+L) \otimes \mathcal{I}_Z \longrightarrow \mathcal{O}_X(K_X+L) \otimes \mathcal{I}_{Z-\{x\}} \longrightarrow \mathcal{O}_X(K_X+L) \otimes \mathcal{O}_{\{x\}} \longrightarrow 0.$$

In particular, the non-injectivity of (**) is equivalent to evaluation at x giving the zero homomorphism $H^0(\mathcal{O}_X(K_X+L) \otimes \mathcal{I}_{Z-\{x\}}) \longrightarrow H^0(\mathcal{O}_X(K_X+L) \otimes \mathcal{O}_{\{x\}}) = \mathbf{C}$. But this means exactly that every section of $\mathcal{O}_X(K_X + L)$ vanishing on $Z - \{x\}$ also vanishes at x, as claimed. □

In the situation of Reider's Theorem 2.1(i), we have now achieved the goal of encoding geometric hypotheses on the linear series $|K_X + L|$ into the existence of a (hopefully!) interesting vector bundle on X:

Corollary 3.14. *Let L be any line bundle on the smoooth surface X, and suppose that $x \in X$ is a point at which every section of $\mathcal{O}_X(K_X + L)$ vanishes. Then there exists a rank two vector bundle E on X, with $\det E = L$, which has a section $s \in \Gamma(X, E)$ vanishing precisely at x, i.e. $Z(s) = \{x\}$.* □

It remains to use this bundle to produce the required curve D. We take this up in the next section. We conclude the present discussion with several exercises, some of which outline other applications of vector bundles to study linear series.

Exercise 3.15. (Non-Reduced Zero-Schemes.) Generalize Theorem 3.13 to allow for the possibility that Z is not reduced. In fact, assuming that $Z \subset X$ is a finite local complete intersection subscheme, show that the statement of Corollary 3.12 remains true if one deals with proper subschemes $Z' \subset Z$, with an analogous modification in the statement of Theorem 3.13. Deduce that if L is a line bundle on X such that $\mathcal{O}_X(K_X + L)$ fails to be very ample, then there is a rank two vector bundle E with det $E = L$ with a section vanishing on a finite scheme $Z \subset X$ of length two. [The crucial point is that a local Gorenstein ring – in particular, a local ring of Z – has a unique minimal non-zero ideal of dimension one, to wit the socle (cf. [C]). Hence a finite local complete intersection scheme does not contain continuous families of maximal proper subschemes.] □

Exercise 3.16. **(Cayley-Bacharach Theorem, [GH2].)** Let X be a smooth surface, and let $C_1, C_2 \subset X$ be effective reduced divisors on X which meet transversely. Prove that if $D \in |K_X + C_1 + C_2|$ is a divisor passing through all but one of the points of $C_1 \cap C_2$, then D passes through the remaining one as well. Generalize to the case when the intersection of C_1 and C_2 is proper but possibly not transversal. □

Exercise 3.17. **(Elementary Transformations.)** Let X be a smooth surface, let V be a vector bundle of rank e on X, and let $C \subset X$ be a smooth curve. We suppose given a line bundle A on C of degree d, plus a surjective map $\lambda : V|C \longrightarrow A$ from the restriction of V to C onto A. From these data, we construct a new rank e bundle F on X, as follows. We may view the invertible $\mathcal{O}_C$-module A as a torsion $\mathcal{O}_X$ - module; i.e. via "extension by zero", A becomes a coherent sheaf on X. We denote this sheaf by $\mathcal{A}$ to emphasize that it is not localy free on X. The composition $V \longrightarrow V|C \longrightarrow A$ determines a surjection of $\mathcal{O}_X$-modules $\bar{\lambda} : V \longrightarrow \mathcal{A}$, and we set $F = \ker\ \bar{\lambda}$. Thus one has the basic exact sequence:

$$0 \longrightarrow F \xrightarrow{\mu} V \longrightarrow \mathcal{A} \longrightarrow 0 \tag{3.17.1}$$

of sheaves on X.

(i). Prove that F is locally free on X, of rank e. By analogy with a classical construction on ruled surfaces (cf. [**H2**], V.5.7.1), F is called the *elementary transformation* of V determined by $\bar{\lambda} : V \longrightarrow \mathcal{A}$.

(ii). Show that the Chern classes of F are given by

$$\begin{aligned} c_1(F) &= c_1(V) - [C] \\ c_2(F) &= c_2(V) - c_1(V) \cdot [C] + d, \end{aligned} \tag{3.17.2}$$

where as above d is the degree of A, considered as a line bundle on C. [For c_1 one can argue for instance that the map μ in (3.17.1) is a homomorphism between two vector bundles of the same rank which drops rank on C. The formula for c_2 takes more work – for example, one could deduce it by working backwards from Riemann-Roch.]

(iii). Show that the transpose μ^* of μ sits in the exact sequence

$$0 \longrightarrow V^* \xrightarrow{\mu^*} F^* \longrightarrow \mathcal{B} \longrightarrow 0, \tag{3.17.3}$$

where $\mathcal{B}$ is the extension by zero of the line bundle $B = N_{C/X} \otimes A^* = \mathcal{O}_C(C) \otimes A^*$ on C. [This amounts to the assertion that $\mathcal{E}xt^1_{\mathcal{O}_X}(\mathcal{A}, \mathcal{O}_X) = \mathcal{B}$. When $A = \mathcal{O}_C$ this is elementary, and in general one can argue that given a point $P \in C$, the statement is true for A if and only if it holds for $A(P)$.] □

Exercise 3.18. (Special Divisors on Curves on a K3 Surface.) Let X be a K3 surface, and let $C \subset X$ be a smooth curve of genus g. In this exercise we outline a proof of the following

Theorem [L1]. *Assume that every curve in the linear series $|C|$ is reduced and irreducible. Then for every line bundle A on C, one has*

$$g(C) - h^0(A) \cdot h^1(A) \geq 0.$$

To give some background for this statement, suppose for a moment that C is any smooth curve of genus g. It is important to understand under what conditions on r, d, and g one can find a line bundle A on C of degree d with $h^0(A) \geq r+1$. Classical parameter counts show that the family of all such line bundles has expected dimension $\rho = \rho(d, g, r) =_{def} g - (r+1)(g-d+r) \geq 0$. The existence of g^r_d's when $\rho \geq 0$ was established by Kempf and Kleiman-Laksov in 1972. Classically, it was conjectured that if $\rho < 0$, then a *general* curve C of genus g carries no line bundles of degree d and $h^0 \geq r+1$. By Riemann-Roch, it is equivalent to predict:

(*) If A is any line bundle on a general curve C of genus g, then

$$g - h^0(A) \cdot h^1(A) \geq 0.$$

Griffiths and Harris proved this (and more) in 1980 using a rather elaborate degenerational argument. We refer to [**Mfd**] for a quick introduction to this circle of ideas, and to [**ACGH**] for a comprehensive overview and references. Returning to the Theorem above, the hypothesis on the linear series $|C|$ is certainly satisfied if $Pic(X) = \mathbf{Z} \cdot [C]$. On the other hand, one knows from Hodge theory that for any integer $g \geq 2$, there exists a K3 surface X whose Picard group is generated by the class of a curve $C \subset X$ of genus g. Thus the Theorem gives a quick proof of (*), without degenerations.

(i). Returning to the situation of the Theorem, let A be a line bundle of degree d on C, with $h^0(A) = r+1$, such that both A and $K_C \otimes A^*$ are globally generated. Then A is a quotient of the trivial vector bundle $\mathcal{O}_C^{r+1}$ on C, so we can make an elementary transformation of $V = \mathcal{O}_X^{r+1}$ to create a vector bundle $F = F(C, A)$ on X, of rank $r+1$, sitting in the exact sequence $0 \longrightarrow F \longrightarrow \mathcal{O}_X^{r+1} \longrightarrow A \longrightarrow 0$. Set $E = F^*$. Prove that $H^1(E) = H^2(E) = 0$, and show that E is generated by its global sections.

(ii). Prove that the holomorphic Euler characteristic of $E \otimes E^*$ satisfies $\chi(E \otimes E^*) = 2h^0(E \otimes E^*) - h^1(E \otimes E^*) = 2 - 2\rho(A)$, where $\rho(A) = g(C) - h^0(A) \cdot h^1(A)$. [Use the Hirzebruch–Riemann–Roch formula $\chi(E \otimes E^*) = \int Td(X) \cdot ch(E \otimes E^*)$, together with the multiplicativity of the Chern character.]

(iii). Assume now that $\rho(A) < 0$. Then E has an endomorphism w which is not a multiple of the identity. Use this to construct a homomorphism $u : E \longrightarrow E$ which drops rank everywhere on X. [If λ is an eigenvalue of $w(x)$ for some $x \in X$, put $u = w - \lambda \cdot 1$.] Then consider the exact sequence $0 \longrightarrow \text{im } u \longrightarrow E \longrightarrow \text{coker } u \longrightarrow 0$. Using the fact that E is globally generated, show that in the Chow group $A_1(X) = Pic(X)$, $c_1(\text{im } u)$ and $c_1(\text{coker } u)$ are represented by non-zero effective curves. Then deduce that $|C|$ contains a reducible or multiple curve.

(iv). The previous step proves the Theorem when both A and $K_C \otimes A^*$ are globally generated. Show that the general case of the Theorem reduces to this one. □

Exercise 3.19. (Alternative Approach to Theorem 3.13). We indicate here a proof of the existence statement in Theorem 3.13 which avoids explicit use of the sheaf-theoretic considerations appearing in the proof of Proposition 3.9. As in the statement of the Theorem, let $Z \subset X$ be a reduced finite scheme, say $Z = \{x_1, \ldots, x_r\}$. Let

$$f : Y = \mathrm{Bl}_Z(X) \longrightarrow X$$

be the blowing-up of X along Z, and let $D \subset Y$ be the exceptional divisor. Thus $D = \sum_1^r D_i$, where $\mathbf{P}^1 \cong D_i \subset Y$ is the (-1)-curve lying over $x_i \in X$. The idea is that if the pair (E, s) sought in (3.7) exists, then since $f^*(s) \in \Gamma(Y, f^*E)$ vanishes on D, the pull-back of (3.6) determines an extension $0 \longrightarrow \mathcal{O}_Y(D) \longrightarrow f^*E \longrightarrow f^*L(-D) \longrightarrow 0$ on Y. This suggests that we consider on Y extensions of $f^*L(-D)$ by $\mathcal{O}_Y(D)$. So fix $e \in \mathrm{Ext}^1(f^*L(-D), \mathcal{O}_Y(D))$ corresponding to the exact sequence

$$(*) \qquad 0 \longrightarrow \mathcal{O}_Y(D) \longrightarrow V \longrightarrow f^*L(-D) \longrightarrow 0.$$

(i). Let $\mathbf{P}^1 \cong D_i \subset Y$ be one of the exceptional curves. Show that $V|D_i$ is an extension of $\mathcal{O}_{\mathbf{P}^1}(1)$ by $\mathcal{O}_{\mathbf{P}^1}(-1)$. Hence either $V|D_i \cong \mathcal{O}_{\mathbf{P}^1}^2$, or $V|D_i \cong \mathcal{O}_{\mathbf{P}^1}(-1) \oplus \mathcal{O}_{\mathbf{P}^1}(1)$.

(ii). Prove that if $V|D_i \cong \mathcal{O}_{\mathbf{P}^1}^2$ for every i, then $V = f^*E$, where E is a rank two vector bundle on X with a section vanishing on Z. So in this case we are done. [See **[OSS]**, I.2.2.6.]

(iii). Suppose that $V|D_i \cong \mathcal{O}_{\mathbf{P}^1}(-1) \oplus \mathcal{O}_{\mathbf{P}^1}(1)$ for some i. Then make an elementary transformation along the resulting map $\lambda : V \longrightarrow \mathcal{O}_{\mathbf{P}^1}(-1)$ to define a new vector bundle V' on Y. Using the fact that the composition $\mathcal{O}_Y(D) \longrightarrow V \longrightarrow \mathcal{O}_{\mathbf{P}^1}(-1)$ is surjective, show that V' can be realized as an extension of $f^*L(-D)$ by $\mathcal{O}_Y(D - D_i)$ from which (*) is induced. Thus the given extension class e lies in the image of $\mathrm{Ext}^1(f^*L(-D), \mathcal{O}_Y(D - D_i)) \longrightarrow \mathrm{Ext}^1(f^*L(-D), \mathcal{O}_Y(D))$. (This is the analogue of Proposition 3.9.)

(iv). Recalling that $K_Y = f^*K_X + D$, reprove the existence statement in Theorem 3.13 using Serre duality for line bundles on Y. □

§4. Reider's Theorem via Vector Bundles

In the previous section we constructed a vector bundle E encoding the failure of an adjoint linear series $|K_X + L|$ to be very ample or free. The next step is to study the geometry of E. Reider's basic tool to this end is Bogomolov's instability theorem. We start with a quick review of this fundamental result. Then we present the proof of Reider's theorem.

Bogomolov's Instability Theorem.

Let X be a smooth complex projective surface, and let E be a rank two vector bundle on X. It is always easy to construct very negative rank one subsheaves of E: in fact, if H is an ample divisor, then for all $n \gg 0$ there exist sheaf monomorphisms $\mathcal{O}_X(-nH) \hookrightarrow E$. The notion of instability refers to the exceptional situation in which E has an unusually positive subsheaf. Bogomolov's theorem is a numerical criterion for instability in terms of the Chern numbers of E.

We start with some formal definitions. Let $N(X)$ be the Néron-Severi vector space of X, i.e. the subspace of $H^2(X, \mathbf{R})$ generated by the classes of algebraic curves. The Hodge index theorem implies that the intersection form on $N(X)$ has type $(+, -, \ldots, -)$, so

$$C = \{\alpha \mid \alpha^2 > 0\} \subset N(X)$$

is a cone with two connected components. The *positive cone* $N(X)^+$ of X is by definition the component of C containing the classes of ample divisors.

Definition 4.1. Let E be a rank two bundle on X. One says that E is *Bogomolov unstable* if there exist a finite subscheme $Z \subset X$ (possibly empty), plus line bundles A and B on X sitting in an exact sequence

$$0 \longrightarrow A \longrightarrow E \longrightarrow B \otimes \mathcal{I}_Z \longrightarrow 0, \tag{4.1.1}$$

where

$$c_1(A) - c_1(B) \in N(X)^+. \tag{4.1.2}$$

Very concretely, (4.1.2) is equivalent to the conditions that $(A - B)^2 > 0$ and that $(A - B) \cdot H > 0$ for all ample divisors H. Note that the finite scheme Z in (4.1.1) is the scheme defined by the vanishing of the vector bundle map $A \longrightarrow E$. It is suggestive to think of (4.1.2) as meaning roughly speaking that "A is more positive than B".

Bogomolov's statement is the following:

Theorem 4.2. ([**Bog**] cf. also [**Reid**].) *Let E be a rank two vector bundle on a smooth projective surface X. If*

$$c_1(E)^2 - 4c_2(E) > 0, \tag{4.2.1}$$

then E is Bogomolov unstable.

We will outline a proof of Bogomolov's theorem from [**FdB1**] in §6. We refer to [**Bog**], [**Gies**] or [**Mka**] for other arguments, as well as the corresponding statement for bundles of higher rank. Shepherd-Barron [**S-B**] has studied the geometry of this result in positive characteristic, and Moriwaki [**Mwk**] has given some arithmetic analogues.

Exercise 4.3. If E is a rank two bundle sitting in the exact sequence (4.1.1), show that the Chern classes of E are given by:

$$c_1(E) = c_1(A) + c_1(B); \qquad c_2(E) = c_1(A) \cdot c_1(B) + \text{ length}(Z).$$

[For c_2, note that one can think of (4.1.1) as determining a map $\mathcal{O}_X \longrightarrow E \otimes A^*$ with zero-scheme Z. In particular, $c_2(E \otimes A^*) = \text{length}(Z)$.] □

Exercise 4.4. Remark that if E is a rank two vector bundle on X, and N is a line bundle, then E is Bogomolov unstable if and only if $E \otimes N$ is. Prove that up to scalar multiples, $c_1^2 - 4c_2$ is the only weight two polynomial in the Chern classes of E which is invariant under twisting by line bundles. Thus the essential content of Theorem 4.2 is that there is *some* numerical criterion for instability. □

Proof of Reider's Theorem.

We now turn to the proof of Reider's theorem. We follow the approach of [**BFS**], which simplifies somewhat Reider's original presentation.* We limit ourselves to the first statement of Reider's result, leaving the second as an exercise for the reader.

Proof of Theorem 2.1.(i). Fix a point $x \in X$, and suppose that L is a nef line bundle on X, with $L \cdot L \geq 5$, such that $\mathcal{O}_X(K_X + L)$ has a base-point at x. By Corollary 3.14, there exists a rank two vector bundle E sitting in the exact sequence

$$(*) \qquad 0 \longrightarrow \mathcal{O}_X \longrightarrow E \longrightarrow L \otimes \mathcal{I}_x \longrightarrow 0.$$

Note from (4.3) that det $E = L$ and $c_2(E) = 1$, and hence

$$c_1(E)^2 - 4c_2(E) = L \cdot L - 4 > 0.$$

Therefore Bogomolov's theorem applies, and one has an exact sequence

$$(**) \qquad 0 \longrightarrow A \longrightarrow E \longrightarrow B \otimes \mathcal{I}_Z \longrightarrow 0,$$

where Z is some finite subscheme of X, and $c_1(A) - c_1(B) \in N(X)^+$.

The plan now is to play off (*) against (**). Taking determinants, we find in the first place that $A + B = L$, whence $A - B = 2A - L$. Hence by definition of the positive cone:

$$(4.5) \qquad \begin{gathered} (2A - L)^2 > 0 \\ (2A - L) \cdot H > 0 \quad \forall \text{ ample divisors } H. \end{gathered}$$

Denote by

$$\alpha : A \longrightarrow L \otimes \mathcal{I}_x$$

the composition $A \hookrightarrow E \longrightarrow L \otimes \mathcal{I}_x$ determined by (**) and (*). We claim to begin with that $\alpha \neq 0$. In fact, in view of (*) it is enough to show that $\text{Hom}(A, \mathcal{O}_X) = 0$, and this follows from the nefness of L and the second equation in (4.5). Then α is given by multiplication by (the equation of) an effective divisor $D \subset X$, with

$$x \in D \qquad \text{and} \qquad A = L - D.$$

It remains to show that D satisfies the numerical conclusions of Theorem 2.1.

To this end, we collect various inequalities. First:

$$(4.6) \qquad (L - 2D) \cdot L > 0.$$

*We remark that the argument given in [**L2**] contains an error.

In fact, note that $L - 2D = 2A - L$. Since L is nef, and hence a limit of ample divisors, (4.5) implies that in any event $(L - 2D) \cdot L \geq 0$. But if $(L - 2D) \cdot L = 0$, then $(L - 2D)^2 < 0$ by Hodge Index, and this contradicts (4.5).

Next:

$$(L^2)(D^2) \leq (L \cdot D)^2 \tag{4.7}$$

$$(L - D) \cdot D \leq 1. \tag{4.8}$$

Indeed (4.7) is a consequence of Hodge Index. As for (4.8), computing from (**) via (4.3), one finds $c_2(E) = (L - D) \cdot D +$ length Z. But $c_2(E) = 1$ and length $Z \geq 0$, and this gives (4.8). Finally, we claim:

$$2D^2 < L \cdot D. \tag{4.9}$$

Here we argue in cases. If $D^2 > 0$, then $L \cdot D \neq 0$ by Hodge index. Moreover

$$2(L \cdot D)(D^2) \underset{(4.6)}{<} (L^2)(D^2) \underset{(4.7)}{\leq} (L \cdot D)^2,$$

and (4.9) follows. Next, say $D^2 = 0$. Then Hodge index rules out the possibility that $L \cdot D = 0$. Therefore $L \cdot D > 0$, so again (4.9) is verified. Finally, if $D^2 < 0$, then (4.9) is trivial since $L \cdot D \geq 0$.

Combining (4.8) and (4.9), one finds:

$$L \cdot D - 1 \leq D^2 < \frac{L \cdot D}{2}.$$

But this is only possible if

$$L \cdot D = 0\ ,\ D^2 = -1; \text{ or}$$
$$L \cdot D = 1\ ,\ D^2 = 0.$$

This completes the proof. □

Exercise 4.10. (Mumford's Proof of Vanishing.) Let X be a smooth projective surface, and let L be a nef line bundle on X such that $L^2 > 0$. Prove that then $H^1(X, \mathcal{O}_X(K_X + L)) = 0$. [If not, there exists a non-split extension of L by $\mathcal{O}_X$, to which one can apply Bogomolov's theorem. See [**Reid**] for details.] □

Exercise 4.11. (Higher Order Embeddings.) This exercise is concerned with the following result of Beltrametti and Sommese:

Theorem. ([**BS**]) *Let X be a smooth surface, let L a nef line bundle on X, and let d be a positive integer such that $L^2 > 4d$. Then either the restriction*

$$e_Z : \Gamma(X, \mathcal{O}_X(K_X + L)) \longrightarrow \Gamma(Z, \mathcal{O}_Z(K_X + L))$$

is surjective for every subscheme $Z \subset X$ of length d, or else there exists an effective divisor $D \subset X$ such that

$$L \cdot D - d \leq D^2 < \frac{1}{2}(L \cdot D). \tag{*}$$

(i). Assume that there exists a finite subscheme $Z \subset X$ of length d such that e_Z fails to be surjective. By induction, one can assume that $e_{Z'}$ is surjective for every proper subscheme $Z' \subset Z$. Show that then there exists a rank two vector bundle E on X, with $\det E = L$, having a section $s \in \Gamma(X, E)$ such that $Z(s) = Z$.

(ii). Arguing as in the proof of Reider's theorem, construct a divisor D satisfying (*).

(iii). Let L be an ample line bundle on X such that $L^2 > 4d$ and $L \cdot C \geq 2d$ for every irreducible curve $C \subset X$. Show that then the restriction e_Z is surjective for every finite subscheme $Z \subset X$ of length $\leq d$. □

Exercise 4.12. (**Gonality of Complete Intersection Curves.**) Reider-type methods can sometimes be used to study linear series on subvarieties of codimension ≥ 2 in an ambient manifold. To illustrate the approach, we consider here a very concrete question in classical curve theory: what is the least degree required to express a complete intersection curve $C \subset \mathbf{P}^r$ as a branched covering $C \longrightarrow \mathbf{P}^1$ of the Riemann sphere? The answer is given in the following

Theorem. *Let $C \subset \mathbf{P}^r$ be a smooth complete intersection of hypersurfaces of degrees $2 \leq a_1 \leq a_2 \leq \cdots \leq a_{r-1}$. Let A be a base-point free line bundle on C, of degree d, with $h^0(C, A) \geq 2$. Then $d \geq (a_1 - 1) \cdot a_2 \cdot \ldots \cdot a_{r-1}$.*

The idea of the argument is this: let $S \supset C$ be a general complete intersection surface of type $(a_2, \ldots, a_{r-1})$. As in Exercise 3.18, one can associate to A a rank two vector bundle F on S. One finds that if $d < (a_1 - 1)a_2 \cdots a_{r-1}$, then F is Bogomolov unstable. It is easy to get a contradiction provided one knows that the destabilizing subsheaf is of the form $\mathcal{O}_S(k)$, but this doesn't seem to be guaranteed. To remedy this, instead of working on a surface we work on a complete intersection threefold $X \supset C$, whose Picard group is controlled by the Lefschetz theorems. Related results, proved by more classical methods, appear in [**CL**] and [**Bas**], and the Theorem also connects with some of the conjectures in [**EGH**]. Paoletti [**Paol**] has extended the techniques of this exercise to deal with certain non-complete intersection curves. He proves the striking result that under suitable numerical hypotheses, the gonality of a space curve $C \subset \mathbf{P}^3$ is governed by its *Seshadri constant*, which roughly speaking measures how positive the hyperplane bundle $\mathcal{O}_{\mathbf{P}^3}(1)$ is in a neighborhood of C (see §5).

(i). Put $\gamma = a_3 a_4 \cdots a_{r-1}$, and let $X \supset C$ be a general complete intersection threefold of type $(a_3, \ldots, a_{r-1})$. [If $r = 3$ take $X = \mathbf{P}^3$ and $\gamma = 1$.] Let $f : Y \longrightarrow X$ be the blowing-up of C, and let $E \subset Y$ be the exceptional divisor, with $\pi : E \longrightarrow C$ the natural map. Consider on E the globally generated line bundle $\mathcal{A} = \pi^* A$. Choosing a base-point free pencil in $\Gamma(E, \mathcal{A})$, we define in the usual way a rank two vector bundle $\mathcal{F}$ on Y via the sequence $0 \longrightarrow \mathcal{F} \longrightarrow \mathcal{O}_Y^2 \longrightarrow \mathcal{A} \longrightarrow 0$. Compute the Chern classes of $\mathcal{F}$.

(ii). Denote by H the pull-back to Y of the hyperplane divisor on X, and for $0 \leq \epsilon \in \mathbf{Q}$ consider the $\mathbf{Q}$-divisor $D_\epsilon = (a_2 + \epsilon)H - E$. Show that $D =_{\text{def}} D_0 = a_2 H - E$ is globally generated and that D_ϵ is ample if $\epsilon > 0$. Now assume that $d < (a_1 - 1)a_2\gamma$. Prove that then for $0 < \epsilon \ll 1$:

$$(*) \qquad \left(c_1(\mathcal{F})^2 - 4c_2(\mathcal{F})\right) \cdot D_\epsilon = (a_1 - \epsilon)a_1 a_2 \gamma - 4d > 0.$$

(iii). Fixing ϵ for which (*) holds, an extension by Miyaoka [**Mka**] of Bogomolov's instability theorem implies that there exists a rank one subsheaf $\mathcal{L} \subset \mathcal{F}$ such that $(2c_1(\mathcal{L}) - c_1(\mathcal{F})) \cdot D_\epsilon \cdot D > 0$. Show that one can assume that $\mathcal{L}$ is locally free, and that the vector bundle map $\mathcal{L} \longrightarrow \mathcal{F}$ drops rank (if at all) on a codimension two subset $Z \subset Y$. Prove that $\mathcal{L} = \mathcal{O}_Y(-tH - \mu E)$ for some integers $t, \mu \in \mathbf{Z}$. [Recall that $Pic(X) = \mathbf{Z}$ thanks to the Lefschetz Hyperplane Theorem.]

(iv). Now let $S \in |a_2 H - E| = |D|$ be a general divisor, so that S is isomorphic to a complete intersection surface of type $(a_2, \ldots, a_{r-1})$ through C. Setting $F = \mathcal{F}|S$, show that $c_2(F) = d$, and that the restriction to S of the subsheaf $\mathcal{L} \hookrightarrow \mathcal{F}$ gives rise to an exact sequence $0 \longrightarrow \mathcal{O}_S(-sH) \longrightarrow F \longrightarrow \mathcal{O}_S((s - a_1)H) \otimes I_W \longrightarrow 0$, where $s = t + \mu a_1$, and $W \subset S$ is some finite subscheme. Use instability to prove that $2s < a_1$. Then estimate $c_2(F)$ to deduce that $a_1 < s + 1$. But $s > 0$ since $h^0(S, F) = 0$, and this gives a contradiction.

(v). Prove that the inequality in the Theorem is the best possible, in the sense that for any integers $2 \leq a_1 \leq \cdots \leq a_{r-1}$, there exists a complete intersection curve C that carries a base-point free pencil of degree $(a_1 - 1) \cdot a_2 \cdot \ldots \cdot a_{r-1}$. □

§5. Vanishing Theorems and Local Positivity

The vector bundle methods described above seem essentially limited to surfaces. In higher dimensions, vanishing theorems are the only tools that have so far achieved significant success, and they will be the focus of the rest of these lectures. In this section we discuss the questions that arise if one tries to use vanishing in the most naive way to produce pluricanonical or adjoint divisors. In a word, one is led to study the "local positivity" of ample line bundles. We will see that one cannot hope to recover completely the known results (e.g. Corollary 2.7) in this fashion. However it turns out somewhat surprisingly that there are bounds on local positivity which apply at a *generic* point of a smooth surface, and these give results which in some respects go beyond Reider's theorem. While we try to explain in detail how vanishing theorems come into the picture, we content ourselves with just a sketch of the statements on local positivity.

Let X be a smooth projective surface, and let L be an ample (or nef and big) line bundle on X. We remarked in Proposition 1.2 that $\mathcal{O}_X(K_X + L)$ is free at a point $x \in X$ if and only if $H^1(X, \mathcal{O}_X(K_X + L) \otimes \mathcal{I}_x) = 0$. Standard vanishing theorems can't directly apply here because the sheaf in question isn't locally free. The traditional first step around this problem is to blow up at x, which at least reduces the question to one involving only invertible sheaves.

So fix a point $x \in X$, let

$$f : Y = Bl_x(X) \longrightarrow X$$

be the blowing up of X at x, and denote by $E \subset Y$ the exceptional divisor. The first point to note is:

Lemma 5.1. *Let L be a line bundle on the smooth surface X, and let $r > 0$ be any positive integer. Then for all $i \geq 0$ there are isomorphisms*

$$H^i(X, \mathcal{O}_X(K_X + L) \otimes \mathcal{I}_x^r) = H^i(Y, \mathcal{O}_Y(K_Y + f^*L - (r+1)E)).$$

Proof. Observe first that $f_*(\mathcal{O}_Y(-rE)) = \mathcal{I}_x^r$, whereas $R^j f_*(\mathcal{O}_Y(-rE)) = 0$ for $j > 0$. In fact, via the inclusion $f_*(\mathcal{O}_Y(-rE)) \subseteq f_*(\mathcal{O}_Y) = \mathcal{O}_X$, we may identify the stalk of $f_*(\mathcal{O}_Y(-rE))$ at x as consisting of germs of functions on X whose pull-backs to Y vanish to order $\geq r$ along the exceptional divisor, which yields the first assertion. The second may be proven inductively by taking direct images of the sequence $0 \longrightarrow \mathcal{O}_Y(-rE) \longrightarrow \mathcal{O}_Y(-(r-1)E) \longrightarrow \mathcal{O}_E(-(r-1)E) \longrightarrow 0$. (Compare [**H2**], V.3.4.) Now recall that $K_Y = f^*K_X + E$. Using the Leray spectral sequence and the projection formula, one finds:

$$\begin{aligned} H^i(X, \mathcal{O}_X(K_X + L) \otimes \mathcal{I}_x^r) &= H^i(Y, f^*(\mathcal{O}_X(K_X + L)) \otimes \mathcal{O}_Y(-rE)) \\ &= H^i(Y, \mathcal{O}_Y(K_Y + f^*L - (r+1)E)), \end{aligned}$$

as claimed. □

Recall next the statement of the basic:

Vanishing Theorem 5.2. *Let V be a smooth projective variety. If L is any ample line bundle on V, then*

$$H^i(V, \mathcal{O}_V(K_V + L)) = 0 \quad \textit{for all } i > 0.$$

More generally, the same statement holds assuming only that L is nef and big. □

The first assertion is of course the classical Kodaira vanishing theorem. The fact that it is enough that L be nef and big was proven by Kawamata [**K1**] and Viehweg [**V**]. As we will see, it is very convenient in applications only to have to check this weaker condition. We refer to [**SS**], Chapter VII, or [**Kol1**] for nice introductions to this extension of Kodaira vanishing.

The plan is to try to apply vanishing to the line bundle $f^*L - (r+1)E$ on Y. This requires knowing something about its positivity. Demailly [**De2**] has introduced a useful accounting mechanism for keeping track of what one needs:

Definition 5.3. With notation as above, the *Seshadri constant* of a nef line bundle L at x is the real number

$$\epsilon(L, x) = \sup\{\epsilon \geq 0 \mid f^*L - \epsilon \cdot E \text{ is nef}\}.$$

Here $f^*L - \epsilon E$ is considered as an $\mathbf{R}$-divisor on Y, and to say that it is nef means simply that $f^*L \cdot C' \geq \epsilon(E \cdot C')$ for all irreducible curves $C' \subset Y$. Needless to say, one can make the analogous definition on a smooth projective variety of any dimension.

The connection with Seshadri's criterion for ampleness (cf. [**H1**], Chapt. 1) occurs via:

Lemma 5.4. *One has*

$$\epsilon(L,x) = \inf_{C \ni x} \left\{ \frac{L \cdot C}{\operatorname{mult}_x(C)} \right\},$$

where the infimum is taken over all reduced and irreducible curves $C \subset X$ passing through x.

Proof. In fact, let $C \subset X$ be a reduced and irreducible curve with multiplicity m at x, and let $C' \subset Y$ denote the proper transform of C. Then $C' \cdot E = m$, and consequently for any ϵ:

$$(f^*L - \epsilon E) \cdot C' = L \cdot C - \epsilon m.$$

Hence if $f^*L - \epsilon E$ is nef, then $\epsilon \leq \frac{L \cdot C}{m}$. The reverse inequality is similar. □

It is suggestive to think of the Seshadri constant $\epsilon(L,x)$ as measuring how positive L is locally near x. We present two exercises that convey this point, and refer to [**De2**], §6, for other interpretations.

Exercise 5.5. (i). Show that if L is a *very* ample line bundle on the surface X, then $\epsilon(L,x) \geq 1$ for all points $x \in X$.

(ii). Prove that the inequality in (i) holds assuming only that L is ample and globally generated.

(iii). Show that for any ample line bundle L on X, there exists a positive constant $\epsilon = \epsilon(L) > 0$ such that $\epsilon(L,x) \geq \epsilon$ for all $x \in X$. (This is the elementary half of Seshadri's criterion for ampleness.) □

For the next exercise, and subsequent discussion, we need a definition. Given a line bundle B on X, and an integer $s \geq 0$, we say that the linear series $|B|$ *generates s-jets* at x if the natural map

$$H^0(X,B) \longrightarrow H^0(X, B \otimes \mathcal{O}_X/\mathcal{I}_x^{s+1})$$

is surjective. In other words, we ask that we be able to find a global section of B with arbitrarily prescribed s-jet at x. For instance $|B|$ generates 0-jets at every point of X if and only if it is free, and $|B|$ generates all 1-jets if and only if the differential $d\phi_{|B|}$ of the corresponding map $\phi_{|B|} : X \longrightarrow \mathbf{P}$ is everywhere injective.

Exercise 5.6. ([**De2**], Theorem 6.4). Given a line bundle B on X, let $s(B,x)$ be the largest integer such that the global sections in $H^0(X,B)$ generate s-jets at $x \in X$. Returning to an ample line bundle L on X, put

$$\sigma(L,x) = \limsup_{k \to \infty} \frac{1}{k} s(kL, x).$$

Show that $\epsilon(L,x) = \sigma(L,x)$ for every $x \in X$. [Compare Proposition 5.10 below.]□

We now have:

Proposition 5.7. *Let L be an ample line bundle on the smooth surface X, and let $x \in X$ be a fixed point. If $\epsilon(L,x) > 2$, then $\mathcal{O}_X(K_X + L)$ is free at x. More generally, if $\epsilon(L,x) > s+2$ for some integer $s \geq 0$, then the linear series $|K_X + L|$ generates s-jets at x. The same statement holds if $\epsilon(L,x) = s+2$ provided that $L^2 > (s+2)^2$.*

Proof. To prove that $|K_X + L|$ generates s-jets at x, it is sufficient to establish the vanishing

$$(*) \qquad H^1(X, \mathcal{O}_X(K_X + L) \otimes \mathcal{I}_x^{s+1}) = 0.$$

As before, let $f : Y \longrightarrow X$ be the blowing up of X at x, and let $E \subset Y$ be the exceptional divisor. Setting $\epsilon = \epsilon(L,x)$, one has on Y the numerical equivalence of $\mathbf{R}$-divisor classes:

$$f^*L - (s+2)E \equiv \frac{s+2}{\epsilon}(f^*L - \epsilon E) + (1 - \frac{s+2}{\epsilon})f^*L.$$

The first term on the right is nef, and the second is nef and big since $\epsilon > s+2$. Hence the vanishing theorem (5.2) applies to $f^*L - (s+2)E$, and (*) then follows from Lemma 5.1. If $\epsilon = s+2$, then $f^*L - (s+2)E$ is nef, and the inequality in the statement of the Proposition implies that it is big. □

Corollary 5.8. *Let A be an ample line bundle on the surface X such that $\epsilon(A,x) \geq 1$ at some point $x \in X$. Then for all $s \geq 0$ the linear series $|K_X + (s+3)A|$ generates s-jets at x.* □

Exercise 5.9. Show that if $\epsilon(L,x) > 4$ for all $x \in X$, then $\mathcal{O}_X(K_X + L)$ is very ample. State and prove the analogues of (5.1), (5.7) and (5.8) on a smooth projective variety V of arbitrary dimension. □

As one might expect, it can happen that $\mathcal{O}_X(K_X + L)$ is free at x without this being accounted for by the bound $\epsilon(L,x) \geq 2$. (See Proposition 5.12 below.) However by a small variant of [**De2**], Theorem 6.4, statements such as (5.8) for jets of arbitrarily high order are *equivalent* to inequalities on the Seshadri constant:

Proposition 5.10. *Let A be an ample line bundle on X, $x \in X$ a fixed point, and $\epsilon > 0$ a real number. Suppose that for all $s \gg 0$, the linear series $|K_X + rA|$ generates s-jets as soon as $r > (s+2)/\epsilon$. Then $\epsilon(A,x) \geq \epsilon$.*

Proof. Let $C \ni x$ be a reduced and irreducible curve, with $\mathrm{mult}_x(C) = m$. Fix $s \gg 0$, and let $r = r(s)$ be the least integer $> (s+2)/\epsilon$. Then $|K_X + rA|$ generates s-jets, and consequently we can find a curve $D = D_s \in |K_X + rA|$, with $\mathrm{mult}_x(D) = s$, having a prescribed tangent cone at x. In particular, we can choose D so that the tangent cones to C and D at x meet properly in T_xX, and since C is irreducible it follows that C and D themselves meet properly. Therefore

$$C \cdot (K_X + rA) \geq \mathrm{mult}_x(C) \cdot \mathrm{mult}_x(D) = ms,$$

whence

$$\frac{C \cdot A}{m} \geq \frac{s}{r} - \frac{C \cdot K_X}{rm}.$$

But we can assume that $r \leq (\frac{s+2}{\epsilon}) + 1$, and the result follows from Lemma 5.4 upon letting $s \to \infty$. □

So far this discussion has been quite formal. It remains to say something about the actual behavior of the Seshadri constants $\epsilon(L,x)$. As measures of local positivity, these are in any event very interesting invariants, quite apart from the potential application to adjoint series. We may summarize the story on surfaces in the following two statements:

Theorem 5.11. [**EL3**]. *Let L be an ample line bundle on a smooth projective surface X. Then $\epsilon(L,x) \geq 1$ for all except perhaps countably many points $x \in X$, and moreover if $L^2 > 1$ then the set of exceptional points is finite. If $L^2 \geq 5$ and $L \cdot C \geq 2$ for all curves $C \subset X$, then $\epsilon(L,x) \geq 2$ for all but finitely many $x \in X$.*

Proposition 5.12. (**Miranda**). *Given $\epsilon > 0$, there exists a surface X, a point $x \in X$, and an ample line bundle L on X such that $\epsilon(L,x) \leq \epsilon$.*

It follows for instance from (5.7) and the Theorem that if $L^2 \geq 5$ and $L \cdot C \geq 2$ for all $C \subset X$ then $\mathcal{O}_X(K_X + L)$ is free off a finite set. But of course we know from Reider's theorem (Corollary 2.6) that in fact $\mathcal{O}_X(K_X + L)$ is *everywhere* globally generated. So from this point of view, one may think of Seshadri constants as giving local Reider-type results, which however are valid only at a general point of the surface X. On the other hand, the statements coming from (5.7) and (5.11) for higher order jets are necessarily stronger than the uniform results deduced from Reider's method. (See Exercise 5.14 for a summary.)

We start by outlining Miranda's construction of examples of small Seshadri constants.

Proof of Proposition 5.12. Let $D \subset \mathbf{P}^2$ be an irreducible plane curve of degree $d \gg 0$ with a point $x \in D$ of multiplicity m. Let D' be a second irreducible curve of degree d, meeting D transversely. Choosing D' generally, we may suppose that all the curves in the pencil spanned by D and D' are irreducible. Blow up the base-points of this pencil to obtain a surface X, admitting a map $\pi : X \longrightarrow \mathbf{P}^1$ with irreducible fibres, among them $D \subset X$. Observe that π has a section $S \subset X$ (viz. an exceptional curve of the blowing up $X \longrightarrow \mathbf{P}^2$) meeting D transversely at one point. Fix an integer $a \geq 2$. It follows from the Nakai criterion that the divisor $L = aD + S$ on X is ample. But $L \cdot D = 1$ whereas $\mathrm{mult}_x(D) = m$, so $\epsilon(L,x) \leq \frac{1}{m}$. Note that by taking suitable a we can even make L^2 arbitrary large, and by taking L to be a multiple of $aD + S$ we can arrange that the intersection numbers $L \cdot C$ of L with irreducible curves $C \subset X$ be bounded below by any preassigned integer. □

Finally, we sketch without full details the main idea of the proof of Theorem 5.11. The argument is very elementary, but is essentially limited to surfaces.

Idea of proof of Theorem 5.11. We focus on the first statement, and we use the characterization (5.4) of Seshadri constants. The main point, which was inspired by [**Xu**], is to view the question variationally. In fact, the set

$$\{(C,x) \mid C \subseteq X \text{ is a reduced, irreducible curve with } \mathrm{mult}_x(C) > L \cdot C\}$$

consists of at most countably many algebraic families. The first statement of the theorem will follow if we show that each of these families is discrete, i.e. that pairs (C,x) forcing $\epsilon(L,x) < 1$ are rigid.

Suppose to the contrary that (C_t, x_t) is a non-trivial one-parameter family of reduced and irreducible curves $C_t \subseteq X$ and points $x_t \in C_t$ with $\mathrm{mult}_{x_t}(C_t) > L \cdot C_t$. Let $C = C_{t^*}$ and $x = x_{t^*}$ for general t^*, and set $m = \mathrm{mult}_{x_{t^*}}(C_{t^*})$. A local computation involving deformation theory of singular curves shows that

$$C^2 \geq m(m-1).$$

(In brief, the given deformation of C determines a section $\rho(\frac{d}{dt}) \in H^0(\mathcal{O}_C(C))$, and one shows that since the deformation preserves the m-fold point of C, $\rho(\frac{d}{dt})$ vanishes to order $\geq m-1$ at x.) But $L^2 \cdot C^2 \leq (L \cdot C)^2$ by the Hodge index theorem, and since $L \cdot C \leq m-1$ by assumption, we find that

$$m(m-1) \leq L^2 \cdot C^2 \leq (C)^2 \leq (m-1)^2.$$

But this is a contradiction when $m > 1$, and the first statement follows. □

Remark 5.13. Theorem 5.11 has recently been extended to varieties of arbitrary dimension, although the numerical bound obtained is somewhat weaker. In fact, it is shown in [**EKL**] that if L is an ample line bundle on a smooth projective variety of dimension n, then $\epsilon(L, x) \geq \frac{1}{n}$ for all $x \in X$ outside a countable union of proper subvarieties. The idea is to choose a divisor $D = D_x \in |kL|$ $(k \gg 0)$ with large multiplicity at x, and to study in effect the higher order deformation theory of the pair (D_x, C_x), where $C_x \subset X$ is a Seshadri-exceptional curve at x. The argument is inspired on the one hand by some techniques that come up in the theory of diophantine approximation, and on the other hand by the methods used to prove boundedness of Fano varieties of given dimension.

Exercise 5.14. **(Higher Jets and Adjoint Series.)** It is interesting to summarize the statements on separation of jets that come out of these discussions. Let A be an ample line bundle on the smooth projectve surface X. Prove the following:

(i). The adjoint series $|K_X + (s+3)A|$ generates s-jets at a sufficiently general point $x \in X$.

(ii). There cannot exist a linear function $f(s)$ such that for all X, A and $s \gg 0$, $|K_X + f(s)A|$ generates s-jets at *every* point $x \in X$.

(iii). There exists a *quadratic* function $f(s)$ (independent of X and A) such that for $s \gg 0$, $|K_X + f(s)A|$ generates s-jets at every point $x \in X$. In fact, for $s \geq 1$ one can take $f(s) = (s+1)(s+2)$. [Use Exercise 4.11.(iii).] □

Exercise 5.15. **(Normal Generation of "Hyper-adjoint" series, [BEL],§3, [ABS].)** Let V be a smooth projective variety of dimension n, and let B be a *very ample* line bundle on V.

(i). Prove that $K_V + (n+1)B$ is globally generated, and that $K_V + (n+2)B$ is very ample. [For the first statement, consider a general divisor $W \in |B|$ and argue by induction on n.]

(ii). Show that $\mathcal{O}_V(K_V + (n+1)B)$ is normally generated, i.e. that the homomorphisms

$$Sym^m(H^0(\mathcal{O}_V(K_V + (n+1)B))) \longrightarrow H^0(\mathcal{O}_V(m(K_V + (n+1)B)))$$

are surjective for $m \geq 0$. [The case $m = 2$ is the essential one. Put $N = \mathcal{O}_V(K_V + (n+1)B)$, and consider on $V \times V$ the line bundle $F = pr_1^*(N) \otimes pr_2^*(N)$. It suffices to prove that $H^1(V \times V, F \otimes \mathcal{I}_\Delta) = 0$, where $\Delta \subset V \times V$ is the diagonal. Using the hypothesis that B is very ample, show that $pr_1^*(B) \otimes pr_2^*(B) \otimes \mathcal{I}_\Delta$ is globally generated. Now apply vanishing on the blow-up of $V \times V$ along Δ.] See [**EL1**] for some generalizations involving defining equations and higher syzygies, and [**BEL**] for some similar elementary applications of vanishing theorems to study the equations defining projective varieties. □

Exercise 5.16. (**Seshadri Constants Along Finite Sets.**) This exercise gives an outline of some unpublished work of **Geng Xu**, and we thank him for his permission to include it. Let X be a smooth surface, and let $Z \subset X$ be a finite set consisting of r points, say $Z = \{x_1, \ldots, x_r\}$. Given an effective divisor $D \subset X$, define $\operatorname{mult}_Z(D) = \sum \operatorname{mult}_{x_i}(D)$. If L is a nef line bundle on X, we define the Seshadri constant of L along Z to be the real number

$$\epsilon(L, Z) = \inf_{C \subset X} \left\{ \frac{L \cdot C}{\operatorname{mult}_Z(C)} \right\},$$

the infimum being taken over all reduced and irreducible curves $C \subset X$ (not necessarily passing through Z).

(i). Assume that L is nef and $L^2 > r$. Show that if Z consists of r sufficiently general points, then $\epsilon(L, Z) \geq 1$. [If not, let $C \subset X$ be a Seshadri exceptional curve which deforms with all local deformations of the x_i. Put $m_i = \operatorname{mult}_{x_i}(C) \geq 0$. We may assume that C passes through x_1, so $m_1 \geq 1$, and by hypothesis $L \cdot C \leq (m_1 - 1) + m_2 + \cdots + m_r$. As C moves in an algebraic family generically covering X, $L \cdot C \geq 1$. Consider the deformation obtained by letting x_1 move and fixing $x_2, \ldots, x_r$. Since C is reduced and irreducible, one finds as in [**EL3**], §1, the inequality:

$$C^2 \geq (m_1 - 1)m_1 + m_2^2 + \cdots + m_r^2 \geq (m_1 - 1)^2 + m_2^2 + \cdots + m_r^2.$$

Hodge index and the hypothesis $L^2 \geq r + 1$ then gives

$$(r+1)\left((m_1 - 1)^2 + m_2^2 + \cdots + m_r^2\right) \leq (m_1 - 1 + m_2 + \cdots + m_r)^2,$$

and by minimizing the the difference of the two sides one sees that this is impossible.]

(ii). In the situation of (i), suppose that $D \subset X$ is any effective divisor (possibly non-reduced or reducible). Show that then $\operatorname{mult}_Z(D) \leq L \cdot D$.

See Exercise 7.8 for an interesting application of this result. □

§6. Adjoint Series and Bogomolov Instability via Vanishing.

We saw in the previous section that one can't hope to prove known and expected results on linear series using only the most naive application of vanishing theorems for line bundles. Our purpose here is to show how more subtle vanishing theorems, for **Q**-divisors, do lead to (parts of) Reider's theorem. This argument appears in [**EL2**], §1, and follows the approach pioneered by Kawamata, Reid, Shokurov *et* *al.* in connection with the minimal model program. Many of the complexities of the general KRS machine disappear on surfaces, and the ideas become particularly transparent. We hope that the present discussion can serve as a low-key introduction to this important and powerful tool. Completing this circle of ideas, Fernández del Busto [**FdB1**] has shown that one can use the approach of [**EL2**] to give a new proof of Bogomolov's instability theorem. We sketch the argument at the end of the section, and in Exercises 6.23 and 6.24.

By way of motivation, let L be an ample line bundle on a smooth projective surface X, and consider the problem of constructing a section of $\mathcal{O}_X(K_X+L)$ which is non-vanishing at some point $x \in X$. The "classical" approach might be to prove something along the following lines:

Exercise 6.1. Assume that $L^2 \geq 5$, and suppose there exists a reduced irreducible divisor

$$D \in |L| \text{ such that } q = \text{ mult}_x(D) \geq 2.$$

Then $H^1(X, \mathcal{O}_X(K_X + L) \otimes \mathcal{I}_x) = 0$, and consequently $\mathcal{O}_X(K_X + L)$ is free at x. [Let $f : Y \longrightarrow X$ be the blowing up of X at x, with $E \subset X$ the exceptional divisor. Let $D' \subset Y$ be the proper transform of D, so that $f^*D - 2E \equiv D' + (q-2)E$ is effective and has positive self-intersection. Prove that $f^*D - 2E$ is numerically 1-connected, and deduce from Ramanujam vanishing ([**BPV**], IV.8.2) that $H^1(Y, \mathcal{O}_Y(K_Y + f^*L - 2E)) = 0$. Then use Lemma 5.1 to conclude.] □

The essential drawback to this statement is that in the situations of interest there is no reason even to suppose that $|L|$ is non-empty. The advantage of the KRS method is that it lets one make an *asymptotic* construction. Specifically, we will take $k \gg 0$ and work with a divisor

$$D \in |kL| \text{ such that } \text{mult}_x(D) > 2k.$$

There is no problem in producing D if $L^2 \geq 5$ and k is sufficiently large. Then one will want to "divide D by k", and this is where **Q**-divisors come into the picture.

Vanishing Theorems for Q-Divisors.

We start with some notation and definitions. Let V be a smooth projective variety of dimension n. A **Q**-*divisor* on V is simply a **Q**-linear combination of prime divisors:

$$M = \sum a_i D_i \quad (a_i \in \mathbf{Q}).$$

The multiplicity $\text{mult}_x M$ of M at a point $x \in X$ is taken to be $\sum a_i \text{mult}_x D_i$. Assuming that the D_i are distinct, we define the *round-up* of M to be the integral divisor $\ulcorner M \urcorner = \sum \ulcorner a_i \urcorner D_i$, where $\ulcorner a_i \urcorner$ denotes the least integer $\geq a_i$. The *integer part*, or *round-down* $[M]$ is defined similarly, and the *fractional part* $\{M\}$ of M is $\{M\} = M - [M]$. There is a **Q**-valued intersection theory involving **Q**-divisors, defined in the evident way by first clearing denominators, and one has the usual functorial operations such as pull-backs under morphisms. This gives rise to the notion of numerical equivalence of **Q**-divisors, which we continue to denote by $\equiv$. We say that M is *nef* if $M \cdot C \geq 0$ for all irreducible curves $C \subset V$, and M is *ample* if the statement of Nakai's criterion holds. Equivalently, M is nef or ample if mM

is so, for some positive integer $m > 0$ such that mM is an integral divisor. If M is nef, it is in addition *big* if $(M^n) > 0$.

The basic result is:

Kawamata-Viehweg Vanishing Theorem 6.2. *Let M be a nef and big* **Q**-*divisor on the smooth projective variety V. Assume that the fractional part $\{M\}$ of M is supported on a divisor with global normal crossings. Then*

$$H^i(V, \mathcal{O}_V(K_V + \ulcorner M \urcorner)) = 0 \quad \textit{for} \quad i > 0.$$

In other words, the Theorem gives a vanishing for (integer) divisors of the form:

$$K_V + (\text{ nef and big } \mathbf{Q}\text{-divisor }) + \Delta,$$

where Δ is an effective fractional divisor (i.e. $[\Delta] = 0$) with normal crossing support. The statement may appear non-intuitive at first blush, but we will see momentarily that this is precisely the tool needed to generalize the argument sketched in Exercise 6.1. We won't prove (6.2) in these notes. The original arguments of [**K1**] and [**V**] used covering constructions to deduce the result from vanishing for integer divisors. The normal crossing hypothesis is used to contol the singularities introduced upon passing to a covering. A number of direct proofs have since been given, one based on connections with logarithmic singularities [**EV1**], another on Hodge theory for twisted coefficient systems [**Kol1**], and a third involving singular metrics on line bundles [**De2**]. We refer to [**Kol1**], [**CKM**], Chapter 8, [**Kol3**], Chapters 9 - 10, and [**EV2**] for good discussions. It is worth emphasizing that Theorem 6.2 is by now not much harder to prove than the classical Kodaira vanishing theorem.

Remark 6.3. Since rounding does not in general respect linear equivalence, it is essential that the fractional part of the **Q**-divisor M appearing in (6.2) be defined as an actual divisor, and not merely as an element in $Pic(X) \otimes \mathbf{Q}$. However we will often identify two **Q**-divisors if their fractional parts coincide and their integer parts are linearly equivalent. By the same token, we will deal with "hybrid" objects of the form $L + D$ where L is a line bundle (defined up to isomorphism) and D is a **Q**-divisor.

In the hope of conveying right away some feeling for how Theorem 6.2 is used, we prove a criterion extending Exercise 6.1. It asserts roughly that the existence of a divisor in $|kL|$ with an "almost isolated" singularity of high multiplicity gives rise to a non-vanishing section of $\mathcal{O}_X(K_X + L)$. This result will play an important role in our proof of Reider's theorem. In the argument below, we will make use of the fact — to be established in Exercise 6.6 — that on a surface, one can ignore the normal crossing hypothesis in (6.2).

Proposition 6.4. *Let L be a nef and big line bundle on a smooth projective surface X. Fix a point $x \in X$, and an integer $s \geq 0$. Suppose that for some $k > 0$ there exists a divisor*

$$D \in |kL| \quad \textit{with} \quad q =_{\text{def}} mult_x(D) > (s+2)k,$$

plus an open neighborhood $U \ni x$ of x in X such that

$$mult_y(D) < q/(s+2) \quad \textit{for all} \quad y \in U - \{x\}.$$

Then

$$H^1(X, \mathcal{O}_X(K_X + L) \otimes \mathcal{I}_x^{s+1}) = 0,$$

i.e. $|K_X + L|$ *generates* s*-jets at* x.

Proof. Write $D = \sum d_i D_i$, where the D_i are distinct prime divisors. The upper bound on $\text{mult}_y(D)$ implies that if D_i passes through x, then $d_i < q/(s+2)$. For simplicity we assume for the time being that every component of D passes through x; the changes necessary in general will be sketched at the end of the proof. Consider as before the blowing-up $f : Y \longrightarrow X$ of X at x. Let $E \subset Y$ be the exceptional divisor, and denote by $D_i' \subset Y$ the proper transform of D_i. Thus $f^*D = qE + \sum d_i D_i'$. The idea of the proof is to study the $\mathbf{Q}$-divisor

$$(*) \qquad M = f^*L - \left(\frac{s+2}{q}\right) f^*D = f^*L - (s+2)E - \sum \left(\frac{s+2}{q}\right) d_i D_i'.$$

The first point to note is that by hypothesis

$$\left(\frac{s+2}{q}\right) d_i < 1 \text{ for all } i,$$

and therefore

$$K_Y + \ulcorner M \urcorner = K_Y + f^*L - (s+2)E.$$

On the other hand, one has the numerical equivalence:

$$M \equiv f^*L - \left(\frac{s+2}{q}\right) f^*D \equiv \left(1 - \left(\frac{s+2}{q}\right) k\right) f^*L,$$

and as $q > k(s+2)$, it follows that M is nef and big. Since we are on a surface we don't need to worry about normal crossings, and therefore Theorem 6.2 gives $H^1(Y, \mathcal{O}_Y(K_Y + f^*L - (s+2)E)) = 0$. We conclude with Lemma 5.1.

It remains to treat the possibility that not all of the components of D pass through x. In this case, we only know that $d_i < q/(s+2)$ for those i such that $D_i \ni x$. Defining M as in (*), it follows that $K_Y + \ulcorner M \urcorner = f^*L - (s+2)E - N'$, where $N' \subset Y$ is an effective (or zero) integral divisor whose support is disjoint from E. Let $N = f_*N'$ be the corresponding divisor on X, so that $x \notin \text{supp } N$. Then as above vanishing gives $H^1(X, \mathcal{O}_X(K_X + L - N) \otimes \mathcal{I}_x^{s+1}) = 0$, i.e. we can find a section of $\mathcal{O}_X(K_X + L - N)$ with arbitrarily prescribed s-jet at x. But one has an inclusion $H^0(\mathcal{O}_X(K_X + L - N)) \subset H^0(\mathcal{O}_X(K_X + L))$, and since N doesn't pass through x, it follows that $|\mathcal{O}_X(K_X + L)|$ also generates s-jets at x, as required. □

Remark 6.5. There is an analogous statement in higher dimensions:

(6.5.1). *Let* V *be a smooth projective variety of dimension* n*, and let* L *be an ample line bundle on* V*. Given a point* $x \in V$*, assume that there exists a divisor* $D \in |kL|$ *with*

$$\text{mult}_x D > (n+s)k$$

but $\text{mult}_y D \leq k$ *for* y *in a punctured neighborhood of* x*. Then* $|K_X + L|$ *generates* s*-jets at* x*.*

We learned of (6.5.1) from Siu. Esnault and Viehweg give a proof in [**EV2**], (7.5), (7.7). The argument is more involved than in the surface case, because in general one has to pass to an embedded resolution of D in order to apply Kawamata-Viehweg vanishing. Note also that in order to verify the upper bound on $\mathrm{mult}_y(D)$ in dimension three or higher, it is not enough to know simply that the components of D appear with low multiplicity.

As we have already remarked, it is a happy fact that one can ignore the normal crossing hypothesis in Theorem 6.2 when working on surfaces:

Exercise 6.6. **(Sakai's Lemma, [Sak1],** cf. **[EL2]**, §1.) Let X be a smooth surface, and let M be any big and nef $\mathbf{Q}$-divisor on X. Then

$$H^i(X, \mathcal{O}_X(K_X + \ulcorner M \urcorner)) = 0 \quad \text{for } i > 0.$$

[A simple if somewhat inefficient argument proceeds as follows. By a sucession of blowings up at points, one constructs a map $\phi : X_1 \longrightarrow X$ such that the fractional part $\{\phi^* M\}$ of $\phi^* M$ is supported on a normal crossing divisor, and hence $H^i(X_1, \mathcal{O}_{X_1}(K_{X_1} + \ulcorner \phi^* M \urcorner)) = 0$ for $i > 0$. Working step by step down from X_1, it is then enough to prove the following:

> Let $f : Y \longrightarrow X$ be the blowing up of a smooth surface at a point $x \in X$, and let M be a $\mathbf{Q}$-divisor on X. If $H^i(Y, \mathcal{O}_Y(K_Y + \ulcorner f^* M \urcorner)) = 0$ for some $i > 0$, then $H^i(X, \mathcal{O}_X(K_X + \ulcorner M \urcorner)) = 0$.

In fact, show that

$$K_Y + \ulcorner f^* M \urcorner = f^*(K_X + \ulcorner M \urcorner) - pE$$

for some $p \geq -1$. Then

$$H^i(Y, \mathcal{O}_Y(K_Y + \ulcorner f^* M \urcorner)) = H^i(X, \mathcal{O}_X(K_X + \ulcorner M \urcorner) \otimes \mathcal{I}_x^p)$$

for all i, where we make the convention that $\mathcal{I}_x^p = \mathcal{O}_X$ if $p = -1$. The assertion follows.] □

Exercise 6.7. **(Algebro-Geometric Multiplier Ideals, [EV2]**, (7.4), (7.5).) Let M be any nef and big $\mathbf{Q}$-divisor on a smooth projective variety V of dimension n. Prove that there exists an ideal sheaf $\mathcal{J}_M \subset \mathcal{O}_V$, with $\mathcal{O}_V/\mathcal{J}_M$ supported in codimension ≥ 2, such that

$$H^i(V, \mathcal{O}_V(K_V + \ulcorner M \urcorner) \otimes \mathcal{J}_M) = 0 \quad \text{for } i > 0. \tag{6.7.1}$$

[Let $f : W \longrightarrow V$ be an embedded resolution of $\mathrm{supp}\{M\}$. One can write

$$K_W + \ulcorner f^* M \urcorner = f^*(K_V + \ulcorner M \urcorner) + P - N,$$

where P and N are relatively prime effective f-exceptional divisors on W. Then set $\mathcal{J}_M = f_* \mathcal{O}_W(-N)$. For (6.7.1), show first that $f_* \mathcal{O}_P(P) = 0$. Then argue e.g. as in the proof of Grauert-Riemenschneider vanishing in **[Kol1]**, Corollary 11, to prove that $R^i f_*(K_W + \ulcorner f^* M \urcorner) = 0$ for $i > 0$.] Esnault and Viehweg show that $\mathcal{J}_M$ is actually independent of the resolution chosen. This is the algebro-geometric analogue — and a special case of — the multiplier ideal sheaf associated to a line bundle with a singular metric (cf. **[De2]** and **[Siu2]**), and (6.7.1) is a special case of Nadel's vanishing theorem **[Nad]**. One thinks of $\mathcal{J}_M$ as measuring the singularities of the fractional part $\{M\}$ of M. Its co-support is contained in the locus of points at which $\operatorname{supp}\{M\}$ fails to be a normal crossing divisor. (Compare [**Kol1**], Theorem 19′.) □

Exercise 6.8. (Singularities of Plane Curves.) In this exercise, $C \subset \mathbf{P}^2$ is a reduced plane curve of degree d.

(i). Let $\Sigma = \operatorname{Sing} C$ be the singular locus of C, considered as a reduced subscheme of $\mathbf{P}^2$. Use Kawamata-Viehweg vanishing to prove the classical theorem that Σ imposes independent conditions on curves of degrees $k \geq d-2$, i.e.

$$(*) \qquad H^1(\mathbf{P}^2, \mathcal{I}_\Sigma(k)) = 0 \text{ for } k \geq d-2.$$

[Write $\Sigma = \{x_1, \ldots, x_m\}$, and consider the blowing-up $f : Y \longrightarrow \mathbf{P}^2$ of $\mathbf{P}^2$ along Σ, with $E = \sum E_i$ the exceptional divisor, and H the pull-back to Y of the hyperplane divisor on $\mathbf{P}^2$. Let $C' \subset Y$ be the proper transform of C, and put $q_i = \operatorname{mult}_{x_i}(C)$, so that $f^*C = C' + \sum q_i E_i$. If $q_i \geq 3$ for all i, then one can argue much as in the proof of Proposition 6.4. In case some $q_i = 2$ one has to make small perturbations of the divisors in question, as follows. Fix first a small rational number $0 < \epsilon \ll 1$ such that $H - \epsilon E$ is ample. Then choose $a < 1$ such that $a > 2/(q_i + \epsilon)$ for all i, and consider the $\mathbf{Q}$-divisor $M = (d+1)H - aC' - \sum a(q_i + \epsilon)E_i$. The numerical equivalence

$$M \equiv (d+1)(1-a)H + a(H - \sum \epsilon E_i) + a(dH - C' - \sum q_i E_i)$$

shows that M is big and nef (in fact ample). Now apply (6.2).] Observe that this argument shows that one can replace the ideal $\mathcal{I}_\Sigma$ in (*) by $\mathcal{J} = \mathcal{I}_{x_1}^{q_1 - 1} \cdots \mathcal{I}_{x_m}^{q_m - 1}$.

(ii). Suppose now that C has a certain number of simple cusps (i.e. $z^2 = w^3$ in local analytic coordinates): let $\Delta = \{x_1, \ldots, x_\kappa\}$ be the set of such. Prove the theorem of Zariski [**Z**], §6, that

$$H^1(\mathbf{P}^2, \mathcal{I}_\Delta(k)) = 0 \text{ for } k > \frac{5}{6}d - 3.$$

It follows for example that a septic curve can have no more than ten simple cusps. [Work on a blowing-up $f : Y \longrightarrow \mathbf{P}^2$ which is an embedded resolution of C over the cusps.] It is interesting to note that Zariski proves this result by studying the irregularity of cyclic coverings of $\mathbf{P}^2$ branched along C. One can see the approach to Kawamata-Viehweg vanishing via covering constructions as a vast generalization of this idea. Building in part on work of Esnault [**Es**], Sakai [**Sak3**] has applied these and other techniques to study the singularities of plane curves. □

Exercise 6.9. (Variant of Proposition 6.4.) Let L be an ample line bundle on a smooth projective surface X, and suppose that $Z \subset X$ is a finite set. Assume given a divisor $D \in |kL|$ for some $k > 0$, say $D = \sum d_i D_i$, plus integers $s \geq 0$ and $q > (s+2)k$, such that $\operatorname{mult}_x(D) \geq q$ for all $x \in Z$, and with $(s+2)d_i \leq q$ whenever D_i meets Z. Prove that then $H^i(X, \mathcal{O}_X(K_X + L) \otimes \mathcal{I}_Z^{s+1}) = 0$ for $i > 0$. [If $(s+2)d_i < q$, proceed as in the proof of Proposition 6.4. In general, as in Exercise 6.8.(i), one can introduce small perturbations, as follows. Let $f : Y \longrightarrow X$ be the blowing-up of Z, with exceptional divisor E, and fix $\epsilon > 0$ such that $f^*L - \epsilon E$ is ample. For rational numbers $0 < b \ll 1$ and $0 < 1 - a \ll 1$, consider

$$\begin{aligned} M &= f^*\left(L - a\left(\frac{s+2}{q}\right)D\right) - b\epsilon E \\ &\equiv \left((1-b) - \frac{a(s+2)}{q}k\right) f^*L + b\,(f^*L - \epsilon E)\,. \end{aligned}$$

If $0 < b < 1 - k(s+2)/q$ and $1 - a < b\epsilon/(s+2)$, then M is ample, and every component of E has coefficient $\leq -(s+2)$ in M.] □

Reider's Theorem Revisited.

Our next goal is to use Kawamata-Viehweg vanishing to give another proof of part of Reider's theorem. Specifically, we focus on the following statement:

Proposition 6.10. *Let X be a smooth projective surface, $x \in X$ a fixed point, and L a nef line bundle on X. Assume that $L^2 \geq 5$, and that*

$$L \cdot C \geq 2 \ \textit{ for all curves } C \subset X \ \textit{ such that } C \ni x.$$

Then $\mathcal{O}_X(K_X + L)$ has a section which doesn't vanish at x.

The plan is this: the hypothesis $L^2 \geq 5$ allows us to construct a divisor $D \in |kL|$ ($k \gg 0$) with high multiplicity at x. If D has small multiplicity at nearby points, then Proposition 6.4 applies. In the contrary case, D contains a component D_0 appearing with high multiplicity. The philosophy of the KRS method is that vanishing still gives a useful surjectivity statement. The problem is then reduced to producing a section on D_0, and this is attacked using the lower bound on $L \cdot D_0$. We remark that Sakai [**Sak2**], using some ideas of Serrano, has given a cohomological proof of Reider's theorem via Miyaoka's vanishing theorem for Zariski decompositions of linear series on surfaces. While the present approach has the advantage of using mainly general techniques from higher dimensional geometry, it would be interesting to understand more clearly than one does at the moment the precise connections between the two proofs. (One can see the construction below as the first step in producing the Zariski decomposition of the relevant linear series.)

Proof of Proposition 6.10. The first step is to show that for $k \gg 0$ there exists a divisor

$$D \in |kL| \ \text{ with } \ q =_{\text{def}} \text{mult}_x D > 2k.$$

This is an elementary parameter count. In fact, consider the exact sequence

$$0 \longrightarrow H^0(\mathcal{O}_X(kL) \otimes \mathcal{I}_x^{2k+1}) \longrightarrow H^0(\mathcal{O}_X(kL)) \longrightarrow H^0(\mathcal{O}_X(kL) \otimes \mathcal{O}_X/\mathcal{I}_x^{2k+1}).$$

We are looking for the divisor of a non-zero section in the group on the left. Since L is nef and $L^2 \geq 5$, a standard argument using Riemann-Roch (cf. [**H2**], V.1.8, or [**SS**], p. 146) shows that:

$$h^0(\mathcal{O}_X(kL)) = \frac{k^2L^2}{2} + o(k^2) \geq \frac{5}{2}k^2 + o(k^2).$$

On the other hand,

$$h^0(\mathcal{O}_X(kL) \otimes \mathcal{O}_X/\mathcal{I}_x^{2k+1}) = \binom{2k+2}{2} = \frac{4k^2}{2} + o(k^2).$$

Hence $H^0(\mathcal{O}_X(kL) \otimes \mathcal{I}_x^{2k+1}) \neq 0$ for $k \gg 0$, as required.

Fix such a divisor D, and write

$$D = \sum d_i D_i + F,$$

where $\{D_i\}$ are the components of D passing through x, and F is the effective (or zero) divisor consisting of those components of D disjoint from x. If $q > 2d_i$ for all i, then Proposition 6.4 applies, and we are done.

Assume next that $q < 2d_i$ for some i. Note that

$$q = \sum d_i \ \mathrm{mult}_x(D_i),$$

so in the first place $q \geq d_i$ for all i. It follows also that there is a *unique* component D_i — call it D_0 — of maximal mutiplicity $d_0 > q/2$. Furthermore, D_0 is necessarily smooth at x. Consider now the $\mathbf{Q}$-divisor $M = L - \frac{1}{d_0}D$ on X. Then

$$K_X + \ulcorner M \urcorner = K_X + L - D_0 - N,$$

where $N = [\frac{1}{d_0}F]$ is an effective divisor supported away from x. We have the numerical equivalence

$$M \equiv L - \left(\frac{1}{d_0}\right) kL \equiv \left(1 - \frac{k}{d_0}\right) L,$$

and since $d_0 > q/2 > k$, it follows that M is nef and big. Keeping in mind that we don't need to worry about normal crossings, Theorem 6.2 gives the vanishing $H^1(X, \mathcal{O}_X(K_X + L - D_0 - N)) = 0$. Therefore the restriction

$$(6.11) \qquad H^0(X, \mathcal{O}_X(K_X + L - N)) \longrightarrow H^0(D_0, \mathcal{O}_X(K_X + L - N) \mid D_0)$$

is surjective.

Observe next that it is enough to show that $\mathcal{O}_X(K_X + L - N) \mid D_0$ has a section $\bar{t}$ which does not vanish at x. For then thanks to the surjectivity of (6.11), $\bar{t}$ lifts to a section $t \in H^0(X, \mathcal{O}_X(K_X + L - N))$ with $t(x) \neq 0$. Since $x \notin$ supp N, t gives rise to a section $s \in H^0(X, \mathcal{O}_X(K_X + L))$ which is non-zero at x, as desired.

The existence of the required section $\bar{t}$ will in turn follow if we verify:

$$(6.12) \qquad (L - D_0 - N) \cdot D_0 > 1.$$

For then, since $(L - D_0 - N) \cdot D_0$ is in any event an integer, it will follow that the restriction $\mathcal{O}_X(K_X + L - N) \mid D_0$ is of the form $\mathcal{O}_{D_0}(K_{D_0} + B)$ for some line bundle B of degree ≥ 2, and hence is free at x thanks to Theorem 1.1. (Note that the possible singularities of D_0 don't cause any problems here since D_0 is Gorenstein and x is a smooth point.) As for (6.12), note that $\ulcorner M \urcorner - M = \sum_{i\geq 1} \frac{d_i}{d_0} D_i + \Delta$, where $\Delta = \ulcorner \frac{1}{d_0} F \urcorner - \frac{1}{d_0} F$ is an effective divisor which meets D_0 properly. Thus

$$L - D_0 - N = \ulcorner M \urcorner \equiv M + \sum_{i \geq 1} \frac{d_i}{d_0} D_i + \Delta.$$

But $L \cdot D_0 \geq 2$ by assumption, $D_i \cdot D_0 \geq mult_x(D_i)$ for $i \geq 1$, and $M \equiv (1 - \frac{k}{d_0})L$. Thus:

$$\begin{aligned}(L - D_0 - N) \cdot D_0 &= \left(1 - \frac{k}{d_0}\right) L \cdot D_0 + \sum_{i \geq 1} \left(\frac{d_i}{d_0}\right) D_i \cdot D_0 + \Delta \cdot D_0 \\ &\geq 2\left(1 - \frac{k}{d_0}\right) + \sum_{i \geq 1} \frac{d_i}{d_0} \operatorname{mult}_x D_i \\ &= 2\left(1 - \frac{k}{d_0}\right) + \left(\frac{q - d_0}{d_0}\right) = 1 + \left(\frac{q - 2k}{d_0}\right).\end{aligned}$$

Recalling that $q > 2k$, (6.12) follows.

It remains to treat the possibility that $q = 2d_i$ for some i. When L is ample one can invoke Exercise 6.9 (i.e. introduce small perturbations of the divisors in question and argue as in Proposition 6.4). In general, apply Exercise 6.13 below. We leave details to the reader. □

Remark. This is the model of the argument used in **[EL2]**, **[ELM]** and **[Fuj2]** to study global generation of adjoint linear series on threefolds. Given an ample line bundle L on a smooth projective threefold V, plus a point $x \in V$, one starts by taking a divisor $D \in |kL|$ ($k \gg 0$) with high multiplicity at x. If D has an (almost) isolated singularity at x, then one concludes at once as in Remark 6.5. Otherwise — very roughly speaking — the KRS approach reduces one to producing a section on the "most singular locus" Z of D, which in the case at hand is either a curve or a component appearing with particularly high multiplicity in D. Then one applies Reider-type statements for $\mathbf{Q}$-divisors on Z. However there are considerable technical difficulties stemming in part from the fact that one has to start by passing to an embedded resolution of D, since already in dimension three one can't ignore the normal crossing hypothesis in (6.2). Unfortunately, these problems have so far blocked the possibility of extending the argument to dimensions four or more.

Exercise 6.13. Let R be a big and nef $\mathbf{Q}$-divisor on a smooth projective surface X, and let $E_1, \ldots, E_k$ be distinct irreducible curves on X which do not appear in the fractional part of R. Assume that $R \cdot E_i > 0$ for all $1 \leq i \leq k$. Prove that then

$$H^i(X, \mathcal{O}_X(K_X + \ulcorner R \urcorner + E_1 + \cdots + E_k)) = 0$$

for $i > 0$. [Use induction on k.] □

Exercise 6.14. **(Reider-type Theorems for Q-divisors, [EL2], §2.)** The proof of Proposition 6.9 can be extended to give analogous statements for (round-ups of) **Q**-divisors. Let X be a smooth projective surface.

(i). Suppose that M is an ample **Q**-divisor on X such that

$$M^2 > 4 \text{ and } M \cdot C \geq 2 \text{ for all irreducible curves } C \subset X.$$

Show that then $\mathcal{O}_X(K_X + \ulcorner M \urcorner)$ is globally generated. [Fix $x \in X$ and then a divisor $D \in |kM|$ for sufficiently divisible $k \gg 0$ such that $q = \text{mult}_x(D) > 2k$. Write $D = \sum d_i D_i + F$ and $M = \ulcorner M \urcorner - \sum a_i D_i - G$ $(0 \leq a_i < 1)$ where the D_i all pass through x, and F, G are effective divisors disjoint from x. Let $p = \text{mult}_x(\ulcorner M \urcorner - M) = \sum a_i \, \text{mult}_x D_i$. If $(2-p)/q < (1-a_i)/d_i$ for all i, blow up x and argue as in the proof of Proposition 6.4. If the reverse inequality holds for some i, adapt the proof given for Proposition 6.10.]

(ii). For applications, it is useful to have a relative statement. Thus consider a surjective map $h : X \longrightarrow X_0$ where X_0 is a complete irreducible surface. Let M be a big and nef **Q**-divisor on X, and fix a point $x_0 \in X_0$. Suppose that $\beta_1, \beta_2 > 0$ are positive rational numbers such that

$$\begin{aligned} &M^2 > (\beta_2)^2 \\ &M \cdot C \geq \beta_1 \ \forall \text{ curves } C \subset X \text{ s.t. } h(C) \text{ is a curve through } x_0. \end{aligned}$$

Assume that $\mathcal{O}_X(K_X + \ulcorner M \urcorner)|\Gamma \cong \mathcal{O}_\Gamma$ for every effective divisor $\Gamma \subset X$ such that $h(\Gamma) = x_0$. Suppose also that

$$\beta_2 \geq 2, \quad \beta_1 \left(1 - \frac{2}{\beta_2}\right) \geq 1.$$

Show that then $\mathcal{O}_X(K_X + \ulcorner M \urcorner)$ has a section which is non-vanishing at some point $x \in h^{-1}(x_0)$. [See [**EL2**], Theorem 2.3.] □

Exercise 6.15. **(Reider-type Theorem for Normal Surfaces, [ELM], §1, compare [Sak2].)** Let S be a complete normal surface. Recall that Mumford has defined a **Q**-valued intersection product for Weil divisors on S. [In brief, let $f : T \longrightarrow S$ be a resolution of S. Given a Weil divisor D on S, let D' denote its proper transform on T. Then there is a unique f-exceptional **Q**-divisor Δ on T such that

$$(D' + \Delta) \cdot E = 0$$

for all f-exceptional divisors E on T. Mumford first defines $f^*D = D' + \Delta$. Given two Weil-divisors D_1, D_2 on S, one then sets $D_1 \cdot D_2 = f^*D_1 \cdot f^*D_2 \in \mathbf{Q}$.] In particular the usual definition for nefness makes sense for Weil divisors on S. Similar considerations of course hold for **Q**-divisors on S. Recall also that the canonical divisor K_S exists as a Weil divisor (class) on S.

Now suppose that M is a nef **Q**-divisor on S, and let $\beta_1, \beta_2 > 0$ be rational numbers such that

$$M^2 > (\beta_2)^2 \ , \quad M \cdot C \geq \beta_1 \ \forall \text{ curves } C \subset S.$$

Assume that $K_S + \ulcorner M \urcorner$ is Cartier, and that

$$(*) \qquad \beta_2 \geq 2 \ , \ \beta_1 \left(1 - \frac{2}{\beta_2}\right) \geq 1.$$

Prove that then $\mathcal{O}_S(K_S + \ulcorner M \urcorner)$ is globally generated. The inequalities (*) are satisfied for example if $M^2 > 16$ and $M \cdot C \geq 2$ for all curves C. We don't know whether the conclusion holds assuming only that $M^2 > 4$ and $M \cdot C \geq 2$. [Let $f : T \longrightarrow S$ be the minimal resolution of S, and define a divisor W on T by the equation $K_T + \ulcorner f^*M \urcorner = f^*(K_S + \ulcorner M \urcorner) - W$. Prove that W is integral and exceptional. Show that every component of $f^*\ulcorner M \urcorner - \ulcorner f^*M \urcorner$ has coefficient > -1, and deduce that W is effective (or zero). Now apply Exercise 6.14.(ii).] □

Exercise 6.16. (Very ampleness for Q-divisors, [Mas].) Let X be a smooth projective surface, and let M be a $\mathbf{Q}$-divisor on X. Assume that

$$M^2 > 18 \quad \text{and} \quad M \cdot C \geq 3 \text{ for all effective curves } C \subset X.$$

Prove that then $\mathcal{O}_X(K_X + \ulcorner M \urcorner)$ is very ample. [The linear series in question is globally generated thanks to Exercise 6.14.(i), and the first step is to argue that given distinct points $x, y \in X$, one can find a divisor $\Lambda \in |K_X + \ulcorner M \urcorner|$ passing through one but not the other of the points. To this end, choose $D \in |kM|$, for k very large and divisible, such that $q_x = \mathrm{mult}_x(D) > 3k$ and $q_y = \mathrm{mult}_y(D) > 3k$. Write $\Delta =_{\mathrm{def}} \ulcorner M \urcorner - M = \sum a_j D_j$ and $D = \sum r_j D_j$, set $\mu_x = \mathrm{mult}_x(\Delta)$, $\mu_y = \mathrm{mult}_y(\Delta)$, and put

$$c_x = \min \left\{ \frac{2 - \mu_x}{q_x}, \frac{1 - a_j}{r_j} \;\middle|\; D_j \ni x \right\},$$

with c_y defined similarly. Define $c = \max\{c_x, c_y\}$, and consider the $\mathbf{Q}$-divisor $M - cD = \ulcorner M \urcorner - \Delta - cD$. Assuming for concreteness that $c = c_x$, one argues in cases according to whether the value of c_x is determined by the multiplicities at x, or whether it is accounted for by a component D_j – say D_0 – of D passing through x. In the latter instance, one further distinguishes between whether or not D_0 goes also through y. The argument for separating tangent directions at x is similar, but requires a little more thought. See [**Mas**] for details, as well as for a statement involving more general numerical hypotheses in the spirit of Exercise 6.14.(ii).] □

Exercise 6.17. (Shokurov's Non-Vanishing Theorem, [Sho], cf. [**CKM**], Lecture 13.) In this exercise, we outline the proof of a "toy" version of Shokurov's Non-Vanishing Theorem. In the case of surfaces treated here, the statement is a consequence of Riemann-Roch. However the argument we indicate, suitably modified, also works in higher dimensions. (As the reader will note, it was the inspiration for the proof of Proposition 6.10 as well as [**EL2**].) The result we aim for is the following:

Theorem. *Let X be a smooth projective surface, and let L be a nef line bundle on X. Assume that*

$$A =_{\mathrm{def}} L - K_X \quad \textit{is ample.}$$

Then $H^0(X, \mathcal{O}_X(mL)) \neq 0$ for all $m \gg 0$.

(i). If L is numerically trivial, show that the statement follows directly from a computation of $\chi(X, \mathcal{O}_X(mL))$. Hence we may assume that L is not numerically trivial. Deduce that then the intersection number $(pL - K_X)^2$ is an increasing function of p.

(ii). Fix a general point $x \in X$. Show that one can take p sufficiently large so that for $k \gg 0$ there exists a divisor $D \in |k(pL - K_X)|$ such that $q =_{\text{def}} \text{mult}_x(D) > 2k$. Write $D = \sum d_i D_i$, and set $d = \max\{d_i\}$. Given a rational number $c > 0$, consider the $\mathbf{Q}$-divisor $N(m,c) = mL - K_X - cD$. The numerical equivalence

$$mL - K_X - cD \equiv (1 - ck)\left((L - K_X) + (p-1)L\right) + (m-p)L$$

shows that $N(m,c)$ is ample if $m \geq p$ and $ck < 1$.

(iii). If $q > 2d$, let $f : Y \longrightarrow X$ be the blowing-up of X at x. Argue as in Proposition 6.4 with the $\mathbf{Q}$-divisor $f^*N(m, \frac{2}{q})$ on Y to deduce the desired non-vanishing.

(iv). Assume $q \leq 2d$. Let D_0 be a component of D of maximal multiplicity $d_0 = d$, and consider the $\mathbf{Q}$-divisor $N(m, \frac{1}{d_0})$. Deduce from Vanishing and Exercise 6.13 that if $m \geq p$, then the restriction map

$$(*) \qquad H^0(X, \mathcal{O}_X(mL)) \longrightarrow H^0(D_0, \mathcal{O}_{D_0}(mL))$$

is surjective.

(v). In the situation of (iv), show that if $m \geq p$, then

$$(**) \qquad \mathcal{O}_{D_0}(mL) = \mathcal{O}_{D_0}(K_{D_0} + A_0)$$

for some ample line bundle A_0 on D_0. Conclude that $H^0(D_0, \mathcal{O}_{D_0}(mL)) \neq 0$ for $m \geq p$, and show that the theorem follows. [If $p_a(D_0) \geq 1$ the required non-vanishing follows from Riemann-Roch. When D_0 is rational, note that the bundle on the right in (**) can't have degree -1 since its degree is divisible by $m \gg 0$.]□

Exercise 6.18. (Kawamata's Basepoint-free Theorem, [K2], [K3], cf. [CKM] Lecture 10.) This exercise is devoted to a stripped-down version on surfaces of a fundamental theorem of Kawamata:

Theorem. *Let X be a smooth surface, and let L be a nef line bundle on X such that $A =_{\text{def}} L - K_X$ is ample. Then the linear series $|mL|$ is free for all $m \gg 0$.*

(i). Observe to begin with it follows from the Non-Vanishing Theorem (Exercise 6.17) that $H^0(X, \mathcal{O}_X(mL)) \neq 0$ for all $m \gg 0$. Let $B(m)$ denote the reduced base locus of $|mL|$. Noting that $B(c^a) \subseteq B(c^b)$ whenever $a > b$ and $c \gg 0$, show that the sequence of subsets $B(c^n)$ stabilizes for $n \gg 0$. Denoting the limit by $B(c)$, show that it is enough to prove that $B(c) = \emptyset$ for every $c \gg 0$.

(ii). Suppose first that for some $m \geq 1$ the linear series $|mL|$ has only isolated base-points. Show that then $|(3m+a)L|$ is free for all $a \geq 1$. [If the base-locus of $|mL|$ consists of a finite set $Z \subset X$, then there exists a reduced divisor $D \in |3mL|$ with $\text{mult}_x(D) \geq 3$ for every $x \in Z$. Let $f : Y \longrightarrow X$ be the blowing-up of Z, and consider the pull-back to Y of the $\mathbf{Q}$-divisor

$$(3m+a)\,L - \frac{2}{3}D \equiv K_X + A + (m+a-1)L.$$

Use vanishing to deduce that $H^1(X, \mathcal{O}_X((3m+a)L \otimes \mathcal{I}_Z)) = 0$.]

Next, fix some large integer p and write

$$|pL| = |M| + \sum r_i F_i,$$

where $F = \sum r_i F_i$ is the fixed divisor of $|pL|$, and $|M|$ is the moving part of $|pL|$, so that $|M|$ has at most isolated fixed points. Assuming $F \neq 0$, we will argue that if F_0 is a component appearing with maximal multiplicity in F, then F_0 is no longer a fixed component of $|mL|$ for $m \gg 0$. In view of (i) and (ii), the theorem will follow. Turning to details, let $D \in |pL|$ be a general divisor. Then $D = D_1 + \sum r_i F_i$ where D_1 is reduced. After re-indexing if necessary we can suppose that $r_0 = \max r_i$. Consider the **Q**-divisor $N(m) = mL - K_X - \frac{1}{r_0}D$. Since L is nef, the numerical equivalence

$$N(m) \equiv \left(m - 1 - \frac{p}{r_0}\right) L + A$$

shows that $N(m)$ is ample if $m > p + 1$.

(iii). Keeping the notation just introduced, use Vanishing and Exercise 6.13 to show that the restriction map

$$(*) \qquad H^0(X, \mathcal{O}_X(mL)) \longrightarrow H^0(F_0, \mathcal{O}_{F_0}(mL))$$

is surjective when $m > p + 1$. Arguing as in Exercise 6.17.(v), show that the group on the right in (*) is non-vanishing. Conclude that F_0 is not in the base locus of $|mL|$ for $m > p + 1$. □

Exercise 6.19. (**Pluricanonical Series.**) With some extra work, one can show that the conclusion of Kawamata's Basepoint-free theorem (Exercise 6.18) remains valid assuming only that $L - K_X$ is nef and big (cf. [**CKM**], Lecture 10). We grant this fact here.

(i). Show that if X is a minimal surface of general type, then $\mathcal{O}_X(mK_X)$ is globally generated for all $m \gg 0$. Deduce that there exists a normal surface X_0, a surjective birational morphism $h : X \longrightarrow X_0$, plus an ample line bundle L_0 on X_0 such that $\mathcal{O}_X(K_X) = h^* L_0$. ($X_0$ is called the *canonical model* of X. Compare [**EV2**], §7.1, and [**W**].)

(ii). With X as in (i), use Exercise 6.14.(ii) to prove that $\mathcal{O}_X(mK_X)$ is globally generated for $m \geq 5$.

This is the model of the arguments used to study pluricanonical series on threefolds in [**EL2**] and [**ELM**]. Note however that the bound in (ii) is slightly less than optimal. □

Bogomolov's Theorem.

Fernández del Busto [**FdB1**] has shown that the sort of argument used to prove Propositions 6.4 and 6.10 leads to a new approach to Bogomolov's Instability Theorem 4.2. This completes in a very nice way the circle of ideas linking linear series on surfaces, vector bundles and vanishing theorems, and we present here an outline of his proof. We shall content ourselves with explaining the principal steps. Therefore we'll make some simplifying assumptions, and relegate some of the numerical calculations to exercises.

To begin with, recall the statement:

Bogomolov's Instability Theorem. *Let E be a rank two vector bundle on a smooth projective surface X. Assume that*

$$c_1(E)^2 - 4c_2(E) > 0.$$

Then there is a saturated invertible subsheaf $A \hookrightarrow E$ such that if $L = \det E$, then

$$(2A - L)^2 > 0 \quad \textit{and} \quad (2A - L) \cdot H > 0 \textit{ for some ample divisor } H.$$

By a saturated subsheaf we mean here that the vector bundle map $A \longrightarrow E$ vanishes only on a finite set. (See Definition 4.1.)

We start with some preliminary reductions and remarks. Note first that since Bogomolov's theorem is invariant under twisting, we are free to tensor E by a high multiple of an ample line bundle. Therefore we may assume that E is globally generated, that its determinant is ample, and that $c_2(E) > 0$. Let $s \in \Gamma(X, E)$ be a general section. Then the zero-scheme $Z = Z(s)$ is reduced, and the Koszul complex (3.6) determined by s realizes E as an extension

$$0 \longrightarrow \mathcal{O}_X \xrightarrow{\cdot s} E \longrightarrow L \otimes \mathcal{I}_Z \longrightarrow 0. \tag{6.20}$$

Here $L = \det(E)$ is an ample line bundle on X, and $c_2(E) = \#Z > 0$.

The strategy of the proof is now very simple. The numerical hypothesis $c_1^2 > 4c_2$ guarantees the existence of a divisor $D \in |kL|$ $(k \gg 0)$ with high multiplicity at the points of Z. If D has small multiplicity away from Z, then the argument of Proposition 6.4 would yield the vanishing of $H^1(X, \mathcal{O}_X(K_X + L) \otimes \mathcal{I}_Z)$. But since E is locally free, Theorem 3.13 shows that the group in question is non-zero. Therefore, as in the proof of (6.10), D must contain some distinguished components appearing with high multiplicity. One uses these components to construct a divisor $\Gamma \subset X$ passing through Z. Vanishing will imply that the inclusion $\mathcal{O}_X(L - \Gamma) \hookrightarrow L \otimes \mathcal{I}_Z$ lifts to an embedding $\mathcal{O}_X(L - \Gamma) \subset E$, and then one argues that the saturation of this subsheaf destabilizes E. One new feature of this argument is that while in Proposition 6.10 we worked with an arbitrary divisor D, here we must be careful to choose s and D rather generally.

Turning to the details, the first point is:

Lemma 6.21. *For any section $s \in \Gamma(X, E)$ as above, with reduced zero-scheme $Z = Z(s)$, there exists a divisor $D \in |kL|$ $(k \gg 0)$ such that $\mathrm{mult}_x(D) > 2k$ for every $x \in Z(s)$. Moreover, by choosing s and D sufficiently generally, we can assume that D satisfies the following*

UNIFORM MULTIPLICITY PROPERTY (UMP). *For any rational number $\delta > 0$, the multiplicity of the integer part $[\delta D]$ is the same at every point of $Z(s)$.*

Proof. For the first assertion, one counts dimensions much as in the proof of Proposition 6.10. Specifically, by Riemann-Roch:

$$h^0(X, \mathcal{O}_X(kL)) = \frac{k^2}{2} L^2 + o(k^2) = \frac{k^2}{2} c_1(E)^2 + o(k^2).$$

On the other hand, the number of conditions required to impose multiplicity $\geq 2k+1$ at each point of Z is:

$$\binom{2k+2}{2}(\#Z) = \binom{2k+2}{2}c_2(E) = \frac{4k^2}{2}c_2(E) + o(k^2).$$

Since $c_1(E)^2 > 4c_2(E)$ by hypothesis, for $k \gg 0$ the required divisor will exist.

Turning to the second assertion, let $S \subset \mathbf{P}H^0(E)$ denote the open subset parametrizing sections of E with finite zero-schemes, and consider the incidence correspondence

$$X \times S \supset \mathcal{Z} = \left\{(x, [s]) \mid x \in Z(s)\right\}.$$

The assumption that E is globally generated implies that $\mathcal{Z}$ is an open subset of a projective bundle over X, and hence is irreducible. Note also that it is finite over S, and in particular $\dim(\mathcal{Z}) = \dim(S)$. By an evident globalization of the parameter count just made, one can construct a divisor $X \times S \supset \mathcal{D} \supset \mathcal{Z}$ such that the fibre $D_s \subset X$ of $\mathcal{D}$ over $s \in S$ is a divisor in the linear series $|kL|$ having multiplicity $> 2k$ at every point of $Z_s = Z(s)$.

To establish the UMP, it is enough to show that given an integer $p > 0$, if $[\delta D_s]$ has multiplicity $\geq p$ at *some* point $x \in Z(s)$ for general $s \in S$, then it has multiplicity $\geq p$ at *every* point $x \in Z(s)$. Consider to this end the Zariski-closed set

$$\mathcal{Z} \supset \mathcal{Z}_p = \left\{(x, [s]) \in \mathcal{Z} \mid \operatorname{mult}_x([\delta\mathcal{D}]_s) \geq p\right\}.$$

By assumption $\mathcal{Z}_p$ dominates S, and hence $\dim(\mathcal{Z}_p) \geq \dim(S) = \dim(\mathcal{Z})$. $\mathcal{Z}$ being irreducible, it follows that $\mathcal{Z}_p = \mathcal{Z}$. Noting that $\operatorname{mult}_x([\delta\mathcal{D}]_s) = \operatorname{mult}_x([\delta D_s])$ for general s, this means exactly that $[\delta D_s]$ has multiplicity $\geq p$ for every $x \in Z(s)$. The Lemma follows. □

Remark. Observe that the second assertion of the Lemma is essentially a monodromy argument in the spirit of [**Har**]. Specifically, fix a general "reference section" $s_0 \in \Gamma(X, E)$ with finite reduced zero-scheme Z_0. Letting s vary over the open subset $U \subset S$ parametrizing sections with reduced zero-loci, one obtains a monodromy action on the points of Z_0. The irreducibility of $\mathcal{Z}$ implies that this action is transitive. Since the divisors D_s also vary with s, it follows that for general $s \in U$, D_s cannot be used to distinguish among the points of $Z(s)$. Therefore D_s must have the same multiplicity at every point of $Z(s)$, and similarly for $[\delta D_s]$.

Returning to the proof of Bogomolov's theorem, fix s and $D \in |kL|$ $(k \gg 0)$, with $D \supset Z = Z(s)$, as in Lemma 6.21. In particular, we suppose that D satisfies the Uniform Multiplicity Property. Therefore D has the same multiplicity at every point of Z, say

$$q = \operatorname{mult}_x(D) \text{ for every } x \in Z.$$

Write $D = \sum d_i D_i$. We henceforth *make the simplifying assumption* that every component D_i of D meets Z. (See Exercise 6.24 for the general case.) Then set

$$d = \max\{d_i\}.$$

We assert next that $2d > q$. In fact, suppose to the contrary that $2d \leq q$. Then an evident generalization of Proposition 6.4 (i.e. Exercise 6.9) implies that

$$(*) \qquad H^1(X, \mathcal{O}_X(K_X + L) \otimes \mathcal{I}_Z) = 0.$$

On the other hand, since E is locally free and $Z \neq \emptyset$, the extension class of the Koszul complex (6.20) must be non-zero. Therefore

$$\mathrm{Ext}^1(L \otimes \mathcal{I}_Z, \mathcal{O}_X) = H^1(X, \mathcal{O}_X(K_X + L) \otimes \mathcal{I}_Z)^* \neq 0,$$

a contradiction. (Compare (3.12), (3.13).) Hence $2d > q$, as claimed.

Consider now the divisor

$$D_0 = \left[\sum \frac{d_i}{d} D_i \right].$$

Then D_0 is reduced, and D_0 meets Z thanks to our assumption that every component D_i passes through at least one point of Z. It follows from the (UMP) that in fact $Z \subset D_0$. Furthermore, D_0 is smooth at every point of Z. D_0 plays the role of the curve Γ appearing in the overview of the proof given above.

Since

$$\frac{1}{d} < \frac{2}{q} < \frac{1}{k},$$

the $\mathbf{Q}$-divisor $L - \frac{1}{d}D$ is ample. Kawamata-Viehweg vanishing (6.2), along with Sakai's Lemma (Ex. 6.6), therefore applies to show that

$$(**) \qquad H^1(X, \mathcal{O}_X(K_X + L - D_0)) = \mathrm{Ext}^1(\mathcal{O}_X(L - D_0), \mathcal{O}_X)^* = 0.$$

Now we can pull back the given extension (6.20) under the natural inclusion $\mathcal{O}_X(L - D_0) \subset \mathcal{O}_X(L) \otimes \mathcal{I}_Z$ to obtain an extension of $\mathcal{O}_X(L - D_0)$ by $\mathcal{O}_X$. But the vanishing (**) implies that the latter extension splits. Therefore the injection $\mathcal{O}_X(L - D_0) \subset \mathcal{O}_X(L) \otimes \mathcal{I}_Z$ lifts to a sheaf monomorphism $\mathcal{O}_X(L - D_0) \hookrightarrow E$:

$$\begin{array}{ccccccccc} & & & & \mathcal{O}_X(L - D_0) & = & \mathcal{O}_X(L - D_0) & & \\ & & & & \downarrow & & \downarrow {\scriptstyle \cdot D_0} & & \\ 0 & \longrightarrow & \mathcal{O}_X & \xrightarrow{\cdot s} & E & \longrightarrow & \mathcal{O}_X(L) \otimes \mathcal{I}_Z & \longrightarrow & 0. \end{array}$$

In other words, we have produced an "unexpected" invertible subsheaf of E. To complete the proof, Fernández del Busto shows that $\mathcal{O}_X(L - D_0)$ is a saturated subsheaf of E, and that it destabilizes E. The destabilization is equivalent to proving the inequalities:

$$(6.22) \qquad \begin{aligned} (L - 2D_0)^2 &> 0 \\ (L - 2D_0) \cdot L &> 0, \end{aligned}$$

and we outline the required calculations in the following Exercise.

Exercise 6.23. Keeping the notations and assumptions made above, we sketch the verification of (6.22).

(i). Suppose given a decomposition $Z = Z_1 \coprod Z_2$ of Z into two disjoint subsets, with $Z_1 \neq \emptyset$. Prove that Z_1 cannot impose independent conditions on the sections of $\mathcal{O}_X(K_X + L)$ vanishing on Z_2, i.e. that the evaluation homomorphism

$$H^0(X, \mathcal{O}_X(K_X + L) \otimes \mathcal{I}_{Z_2}) \longrightarrow H^0(Z_1, \mathcal{O}_{Z_1}(K_X + L))$$

cannot be surjective. [Use Theorem 3.13.]

(ii). Prove that

$$(L - D_0) \cdot D_0 \leq \#Z.$$

[Each point of Z lies on exactly one component of D_0, and every component of D_0 contains at least one point of Z. So it suffices to show that if D' is a component of D_0, and $Z' = D' \cap Z$, then $(L - D_0) \cdot D' \leq \#Z'$. If on the contrary $(L - D_0) \cdot D' >$ deg Z', then the points of Z' impose independent conditions on the linear series $|\mathcal{O}_{D'}(K_{D'} + L - D_0)|$. Use (**) and (i) to arrive at a contradiction.]

(iii). Show that

$$(L - D_0) \cdot D_0 \geq \left(1 - \frac{k}{d}\right) L \cdot D_0 + (\#Z)\left(\frac{q}{d} - 1\right).$$

[$L - D_0 \equiv (1 - \frac{k}{d})L + D^*$, where $D^* = \sum \frac{d_i}{d} D_i - D_0$ is an effective $\mathbf{Q}$-divisor which does not contain any components in common with D_0. Estimate $D^* \cdot D_0$ as in the proof of (6.12).]

(iv). Show that

$$\#Z > \frac{1}{2} L \cdot D_0.$$

[Combine (ii) and (iii).]

(v). Prove that $\mathcal{O}_X(L - D_0)$ is a saturated subsheaf of E, and establish the inequalities (6.22). [For the saturation, use the fact that every component of D_0 contains a point of Z lying only on that component . The first inequality in (6.22) then follows formally from the hypothesis $c_1(E)^2 > 4c_2(E)$ by a suitable computation of $c_2(E)$, much as in (4.8). As for the second, one has $L^2 > 4(\#Z) > 2L \cdot D_0$.]□

Exercise 6.24. Eliminate from the argument above the simplifying assumption that every component of D meets Z. [Take D satisfying Lemma 6.21, and write $D = \sum d_i D_i + F$, where every D_i meets Z and F is disjoint from Z. Defining q and d as above, one still has $2d > q$. Choose a minimal subdivisor $\Delta \subseteq [\frac{1}{d} F]$ for which one has the vanishing

$$H^1(X, \mathcal{O}_X(K_X + L - D_0 - \Delta)) = 0,$$

where $D_0 = \sum \frac{d_i}{d} D_i$. Set $\Gamma = D_0 + \Delta$. Then replace D_0 in the argument above with Γ. See [**FdB1**] for details.] □

§7. Algebro-Geometric Analogue of Demailly's Approach

The method of Kawamata-Reid-Shokurov discussed in the previous section is essentially inductive in nature. To produce a section of $\mathcal{O}_X(K_X + L)$, for example, one starts by constructing a divisor $D \in |kL|$ for $k \gg 0$ with large multiplicity at a given point $x \in X$. If D has small multiplicity at neighboring points $y \in X$ — which is the "good" case — then one gets a vanishing which directly yields the required section (Proposition 6.4). If on the contrary D has a "bad" component D_0 appearing with high multiplicity, then the KRS machine reduces the problem to constructing a section on D_0. Unfortuately, already in the three dimensional situation of [**EL2**] and [**ELM**], this inductive step involves considerable technical difficulties.

Demailly's strategy in [**De1**] is quite different. In effect he puts his efforts into showing that under suitable hypotheses one can produce a divisor $D \in |kL|$ with an "almost isolated" singular point, thereby avoiding any inductions. (We're being somewhat metaphorical here: strictly speaking, Demailly does not work directly with divisors.) The argument of [**De1**] becomes essentially analytic at one point, but it has recently emerged in work with Ein and Nakamaye [**ELN**] that there are quite simple algebro-geometric analogues of Demailly's underlying geometric ideas. In the present section we discuss this approach in the case of surfaces, where (as usual) the picture is particularly clear.

Let X be a smooth projective surface, and consider a fixed point $x \in X$. We start with a definition:

Definition 7.1. Given a line bundle B on X, and a divisor $D \in |kB|$ for some $k > 0$, we say that D has an *almost isolated singularity of index* $> r$ at x if

$$\text{mult}_x(D) > kr,$$

and if there exists a neighborhood $U \ni x$ of x in X such that

$$\text{mult}_y D < k \text{ for } y \in U - \{x\}.$$

In other words, we require that the $\mathbf{Q}$-divisor $\frac{1}{k}D \equiv B$ have multiplicity $> r$ at x, and < 1 in a punctured neighborhood of x. Similarly, one can discuss an almost isolated singularity of index $\geq r$.

We may then rephrase Proposition 6.4 as:

Proposition 7.2. *Let X be a smooth surface, L a big and nef line bundle on X, and $s \geq 0$ an integer. If for some $k > 0$ there exists a divisor $D \in |kL|$ with an almost isolated singularity of index $> s + 2$ at a point $x \in X$, then $|\mathcal{O}_X(K_X + L)|$ generates s-jets at x.* □

So we need to find a way to produce divisors with almost isolated singularities.

The potential obstruction to constructing such divisors is easily described, especially on surfaces. Specifically, in the situation of the Proposition consider the linear series

$$|\mathcal{O}_X(kL) \otimes \mathcal{I}_x^{k(s+2)+1}|$$

of all divisors $D \in |kL|$ having multiplicity $> k(s+2)$ at x. Let F_k denote the fixed divisor of this linear series, and let Σ_k denote its moving part, so that Σ_k has at most isolated base points. Thus

$$(*) \qquad |\mathcal{O}_X(kL) \otimes \mathcal{I}_x^{k(s+2)+1}| = \Sigma_k + F_k,$$

i.e. every divisor $D \in |\mathcal{O}_X(kL) \otimes \mathcal{I}_x^{k(s+2)+1}|$ is of the form $D = D' + F_k$ for some $D' \in \Sigma_k$. Now by Bertini's theorem, a general divisor $D' \in \Sigma_k$ is reduced. Therefore a general element $D \in |\mathcal{O}_X(kL) \otimes \mathcal{I}_x^{k(s+2)+1}|$ will fail to have an almost isolated singular point at x if and only if the fixed divisor F_k has components of multiplicty $\geq k$ that pass through x. So the problem is prove an upper bound on the coefficients of the components of F_k for $k \gg 0$. The strategy roughly speaking will be to use a lower bound on the dimension of the linear series on the left in (*) to deduce an upper bound on the degree of F_k.

It is useful to generalize slightly. Let Y be a smooth projective surface, and let B be a big line bundle on Y. (In the application, Y will be a blow-up of X.) Denote by F_k the fixed divisor of the complete linear series $|kB|$ and put $M_k = kB - F_k$, so that

$$|kB| = |M_k| + F_k$$

for all k. The essential point is:

Proposition 7.3. *Assume that there exists a rational number $\rho > 0$ such that*

$$h^0(Y, \mathcal{O}_Y(kB)) \geq \rho\frac{k^2}{2} + o(k^2) \quad \text{for } k \gg 0.$$

Then for $k \gg 0$:

$$M_k^2 \geq \rho k^2 + o(k^2).$$

To understand the intuition here, suppose it were to happen that $M_k = kM$ for some fixed nef and big divisor M. Then

$$\rho\frac{k^2}{2} + o(k^2) \leq h^0(\mathcal{O}_Y(kB)) = h^0(\mathcal{O}_Y(kM)) = \frac{k^2}{2}M^2 + o(k^2)$$

and the desired inequality follows. The content of the Proposition is that the same inequality holds in general.

The following argument is due to Fernández del Busto [**FdB2**]; the original proof of (7.3) was more cumbersome.

Proof of Proposition 7.3. Fix a very ample divisor H on Y with the property that $K_Y + H$ is very ample, and fix a general divisor $D \in |K_Y + H|$. Then multiplication by D defines for every k an inclusion of sheaves $\mathcal{O}_Y(M_k) \subset \mathcal{O}_Y(K_Y + H + M_k)$. In particular,

$$(*) \qquad h^0(Y, \mathcal{O}_Y(K_Y + H + M_k)) \geq h^0(Y, \mathcal{O}_Y(M_k)) = h^0(Y, \mathcal{O}_Y(kB))$$

for all k. Now M_k is nef, hence $H + M_k$ is ample, so by Vanishing and Riemann-Roch:

$$\begin{aligned} h^0(Y, O_Y(K_Y + H + M_k)) &= \chi(Y, \mathcal{O}_Y(K_Y + H + M_k)) \\ &= \frac{(K_Y + H + M_k)\cdot(H + M_k)}{2} + \chi(Y, \mathcal{O}_Y) \\ &= \frac{M_k \cdot M_k}{2} + \frac{M_k \cdot (K_Y + 2H)}{2} + c(Y, H), \end{aligned}$$

where $c(Y, H)$ is a constant not depending on k. On the other hand, since $K_Y + H$ and H are very ample, there exists a divisor $D' \in |K_Y + 2H|$ which meets F_k properly. Therefore

$$0 \le M_k \cdot (K_Y + 2H) \le kB \cdot (K_Y + 2H)\,.$$

Hence

$$h^0(Y, O_Y(K_Y + H + M_k)) = \frac{M_k \cdot M_k}{2} + o(k^2),$$

and the Proposition follows from (*) and the hypothesis on $h^0(\mathcal{O}_Y(kB))$. □

Remark. One may view Proposition 7.3 as a numerical counterpart to the existence of a Zariski decomposition for B. It would be very interesting to know to what extent an analogue of (7.3) remains true on higher dimensional varieties. The difficulty of course is that the decomposition $kB = F_k + M_k$ of $|kB|$ into fixed and moving parts in general only exists on a blowing-up Y_k of Y, and Y_k will depend on k. While it may be too much to hope for a result for an arbitrary big bundle B, Demailly's work suggests that (7.3) should extend to all dimensions at least for the sort of bundle occurring in the proof of Theorem 7.4. See [**ELN**] for a slightly different approach.

We will apply Proposition 7.3 in the spirit of [**De1**], §8, to prove:

Theorem 7.4. *Let X be a smooth projective surface, $x \in X$ a fixed point, and $s \ge 0$ a non-negative integer. Suppose that L is a nef line bundle on X such that*

$$L^2 \ge (s+2)^2 + 1 \quad \textit{and} \quad L \cdot C \ge s^2 + 3s + 3$$

for all curves $C \subset X$ passing through x. Then for $k \gg 0$ there exists a divisor $D \in |kL|$ having an almost isolated singularity of index $> (s+2)$ at x.

Propositon 7.2 then implies:

Corollary 7.5. *If L is a nef line bundle such that $L^2 \ge 5$ and $L \cdot C \ge 3$ for all curves $C \ni x$, then $\mathcal{O}_X(K_X + L)$ has a section which doesn't vanish at x. More generally, if L satisfies the inequalities in Theorem 7.4, then $|K_X + L|$ generates s-jets at x.* □

See Exercise 7.6 for an analogous criterion for $\mathcal{O}_X(K_X + L)$ to be very ample. Note that the numbers are a little weaker than the optimal bounds (Corollary 2.6) coming from Reider's theorem.

Proof of Theorem 7.4. To simplify the exposition, we first show how to produce a divisor $D \in |kL|$ with an almost isolated singularity of index $\geq (s+2)$ at x. At the end of the argument we will indicate the (routine) changes necessary to get index $> (s+2)$.

Let $f : Y \longrightarrow X$ be the blowing up of X at x, with $E \subset Y$ the exceptional divisor, and put $B = f^*L - (s+2)E$. Then

$$H^0(Y, \mathcal{O}_Y(kB)) = H^0(X, \mathcal{O}_X(kL) \otimes \mathcal{I}_x^{k(s+2)}),$$

so by the usual dimension count (cf. proof of Proposition 6.10):

$$\begin{aligned} h^0(Y, \mathcal{O}_Y(kB)) &\geq \frac{k^2}{2}L^2 - \binom{k(s+2)+1}{2} + o(k^2) \\ &= \left(L^2 - (s+2)^2\right)\frac{k^2}{2} + o(k^2). \end{aligned}$$

Consider as above the decomposition $|kB| = |M_k| + F_k$ of $|kB|$ into its moving and fixed parts. Proposition 7.3 yields the inequality

$$M_k^2 \geq \left(L^2 - (s+2)^2\right)k^2 + o(k^2).$$

Hence by Hodge Index:

$$\begin{aligned} M_k \cdot f^*L &\geq \sqrt{(L^2)\,(M_k^2)} \\ &\geq k\left(\sqrt{(L^2)\,(L^2-(s+2)^2)}\right) + o(k). \end{aligned}$$

Recalling that $F_k = f^*(kL) - k(s+2)E - M_k$, it follows that

(*) $$F_k \cdot f^*L \leq k\left(L^2 - \sqrt{(L^2)\,(L^2-(s+2)^2)}\right) + o(k).$$

Now if $f_s(x)$ denotes the function

$$f_s(x) = x - \sqrt{x\,(x-(s+2)^2)},$$

one finds that $f_s(x) < s^2+3s+3$ when $x \geq (s+2)^2+1$. Hence it follows from (*) that for $k \gg 0$,

(**) $$F_k \cdot f^*L < (s^2+3s+3)k$$

provided that $L^2 \geq (s+2)^2+1$.

On the other hand, consider the image

$$\overline{F}_k = f_*F_k \subset X$$

of F_k in X. Then $\overline{F}_k$ is the fixed divisor of the linear series $|\mathcal{O}_X(kL) \otimes \mathcal{I}_x^{k(s+2)}|$. It follows from (**) that for $k \gg 0$:

$$\overline{F}_k \cdot L < (s^2+3s+3)k.$$

But by hypothesis, every effective curve $F \subset X$ passing through x has L-degree $\geq s^2+3s+3$. In particular, if F is a component of $\overline{F}_k$ through x, then $ord_F(\overline{F}_k) < k$. Therefore Bertini's theorem implies that a general divisor $D \in |\mathcal{O}_X(kL) \otimes \mathcal{I}_x^{k(s+2)}|$

has multiplicity $< k$ in a punctured neighborhood of x, i.e. D has an almost isolated singularity of index $\geq (s+2)$ at x.

It remains to show that we can arrange to produce a divisor $D \in |kL|$ with an almost isolated singularity of index strictly greater than $s+2$. Had we proven a variant of Proposition 7.3 for $\mathbf{Q}$-divisors, one would simply go through the argument just given with the divisor $f^*L - (s+2)E$ replaced by $f^*L - (s+2+\epsilon)E$ for small rational $\epsilon > 0$. To avoid $\mathbf{Q}$-divisors in (7.3) one can equivalently choose a large integer $m \gg 0$ and work with

$$B = f^*(mL) - (m(s+2)+1)\,E.$$

We leave details to the reader. □

Exercise 7.6. **(Criterion for Very Ample Adjoint Bundles.)** Let L be an ample line bundle on a smooth projective surface X such that $L^2 \geq 10$ and $L \cdot C \geq 7$ for all curves $C \subset X$. Show that then $\mathcal{O}_X(K_X + L)$ is very ample. [Given distinct points $x, y \in X$, construct a divisor $D \in |kL|$ ($k \gg 0$) with almost isolated singularities of index > 2 at x and y.] □

Exercise 7.7. **(Effective Matsusaka Theorem on Surfaces, [FdB2]).** Using the results of [**De1**], Siu [**Siu1**] has recently obtained an effective version of Matsusaka's "Big Theorem" (cf. [**LM**], [**KolM**]) on varieties of all dimensions. However the constants that appear in his statement are exponential. Inspired by Siu's general result, Fernández del Busto shows in [**FdB2**] that on a surface one can use Proposition 7.3 to obtain the following considerably sharper bound (having a different shape than Siu's):

Theorem. *Let L be an ample line bundle on the smooth projective surface X, and set*

$$a = L^2 \quad , \quad b = (K_X + 4L) \cdot L.$$

If

$$m > \frac{(b+1)^2}{2a} - 1 \quad \left(\textit{resp. } m > \frac{(b+1)^2}{2a} + 1 \right),$$

then $\mathcal{O}_X(mL)$ is globally generated (resp. $\mathcal{O}_X(mL)$ is very ample).

(i). Let F and G be ample line bundles on X. Show that for $k \gg 0$:

$$h^0\left(X, \mathcal{O}_X(k(F-G))\right) \geq \left(F^2 - 2F\cdot G\right)\frac{k^2}{2} + o(k^2).$$

[In fact, if $F^2 > 2F\cdot G$, then $h^0(\mathcal{O}_X(k(F-G))) \geq (F-G)^2\frac{k^2}{2} + o(k^2)$.] In particular, taking $F = (m+3)L$ and $G = K_X + 4L$, it follows that

$$h^0(\mathcal{O}_X(kB_m)) \geq \rho(m)\frac{k^2}{2} + o(k^2),$$

where:

$$B_m = (m-1)L - K_X,$$
$$\rho(m) = (m+3)^2L^2 - 2(m+3)(K_X+4L)\cdot L.$$

[Siu proves more generally that if F and G are ample line bundles on a smooth projective variety V of dimension n, and if $\rho = F^n - nF^{n-1} \cdot G > 0$, then $h^0(V, \mathcal{O}_V(k(F-G)) \geq \rho \frac{k^n}{n!} + o(k^n)$.]

(ii). Fix $x \in X$, and suppose that m is large enough so that one has the inequalities:

$$(*) \qquad \begin{gathered} \rho(m) > 4 \\ L \cdot B_m - \sqrt{(\rho(m) - 4)(L^2)} < 1. \end{gathered}$$

Show that then for $k \gg 0$ there exists a divisor $D \in |kB_m|$ with an almost isolated singularity of index ≥ 2 at x. Deduce that

$$H^1(X, \mathcal{O}_X(mL) \otimes \mathcal{I}_x) = 0,$$

and conclude that $|mL|$ free at x. [For the existence of D, invoke Proposition 7.3 and argue as in the proof of Theorem 7.4. Then consider the $\mathbf{Q}$-divisor:

$$M = mL - \frac{1}{k}D - K_X \equiv L.$$

Apply vanishing to f^*M on the blowing-up $f : Y \longrightarrow X$ of X at x.]

(iii). Show that the inequality $m > \frac{(b+1)^2}{2a} - 1$ implies (*), which proves the first assertion of the Theorem. Proceed similarly for very ampleness. □

Exercise 7.8. (Bounding singularities via Seshadri Constants.) This exercise presents some further unpublished work of Geng Xu showing how to deduce a variant of Theorem 7.4 from his bound (Exercise 5.16) on the Seshadri constant of an ample line bundle along a finite set of general points.

Let L be an ample line bundle on a smooth projective surface X with $L^2 \geq 5$, and let $x \in X$ be a fixed point. The aim is to show that if $L \cdot C \geq 5$ for all irreducible curves $C \subset X$, then for $k \gg 0$ there exists a divisor $D \in |kL|$ with an almost isolated singularity of index > 2 at x.

(i). Consider as above the decomposition

$$|\mathcal{O}_X(kL) \otimes \mathcal{I}_x^{2k+1}| = \Sigma_k + F_k$$

of $|\mathcal{O}_X(kL) \otimes \mathcal{I}_x^{2k+1}|$ into its moving and fixed parts. Given any $r = L^2 - 5$ points $x_1, \dots, x_r \in X$, show that for $k \gg 0$ there exists a divisor $D' \in \Sigma_k$ with $\mathrm{mult}_{x_i}(D') > k$ for all $1 \leq i \leq r$.

(ii). In the situation of (i), use Exercise 5.16 to prove that if one chooses the x_i sufficiently generally, then

$$L \cdot D' > kr = k(L^2 - 5).$$

Deduce that $L \cdot F_k < 5k$ for $k \gg 0$, and conclude that a general divisor $D \in |\mathcal{O}_X(kL) \otimes \mathcal{I}_x^{2k+1}|$ has an almost isolated singularity of index > 2 at x. □

References

[ABS] M. Andreatta, E. Ballico and A. Sommese, *On the projective normality of adjunction bundles*, Comment. Math. Helv. **66** (1991), 362–367.

[ACGH] E. Arbarello, M. Cornalba, P. Griffiths and J. Harris, *Geometry of Algebraic Curves*, Springer Verlag, 1985.

[BPV] W. Barth, C. Peters and A. Van de Ven, *Compact Complex Surfaces*, Springer Verlag, 1984.

[Bas] B. Basili, *Indice de spécialité des groupes de points de l'espace. Gonalité et indice de Clifford des intersections complètes de l'espace*, Bull. Soc. Math. France (to appear).

[BFS] M. Beltrametti, P. Francia and A. Sommese, *On Reider's method and higher order embeddings*, Duke Math J. **58** (1989), 425–439.

[BS] M. Beltrametti and A. Sommese, *Zero cycles and k-th order embeddings of smooth projective surfaces*, 1988 Cortona proceedings on projective surfaces and their classification, (ed. F. Cantanese), Sympos. Math. 32, Academic Press, 1992.

[BEL] A. Bertram, L. Ein and R. Lazarsfeld, *Vanishing theorems, a theorem of Severi, and the equations defining projective varieties*, Jour. of Amer. Math. Soc. **4** (1991), 587–602.

[BLR] Ch. Birkenhake, H. Lange and S. Ramanan, *Primitive line bundles on abelian threefolds*, Manuscr. Math. (1993).

[Bomb] E. Bombieri, *Canonical models of surfaces of general type*, Publ. Math. IHES **42** (1973), 171–219.

[Bog] F. Bogomolov, *Holomorphic tensors and vector bundles on projective varieties*, Math. USSR Isvestija **13** (1979), 499–555.

[But] D. Butler, *Normal generation of vector bundles over a curve*, J. Diff. Geom **39** (1994), 1–34.

[C] F. Catanese, *Footnotes to a theorem of I. Reider*, Algebraic Geometry: Proceedings, L'Aquila 1988, Lect. Notes. in Math. vol. 1417, 1990, pp. 67–74.

[CL] C. Ciliberto and R. Lazarsfeld, *On the uniqueness of certain linear series on some classes of curves*, Complete Intersections, Acireale 1983, Lect. Notes in Math. vol. 1092, 1984, pp. 198–213.

[CKM] H. Clemens, J. Kollár and S. Mori, *Higher Dimensional Complex Geometry*, Astérisque **166**, 1988.

[DHS] O. Debarre, K. Hulek and J. Spandaw, *Very ample linear systems on abelian varieties*, Math. Ann. **300** (1994), 181-202.

[De1] J.-P. Demailly, *A numerical criterion for very ample line bundles*, J. Diff. Geom **37** (1993), 323–374.

[De2] ———, *Singular Hermitian metrics on positive line bundles*, Complex Algebraic Varieties, Proceedings, 1990, Lect. Notes in Math. vol. 1507, 1992.

[E] L. Ein, *Adjoint linear systems*, Current Topics in Complex Algebraic Geometry (H. Clemens and J. Kollár, eds.), Cambridge University Press, 1995, pp. 87 - 95.

[EKL] L. Ein, O. Küchle and R. Lazarsfeld, *Local positivity of ample line bundles*, J. Diff. Geom. **42** (1995), 193 - 219.

[EL1] L. Ein and R. Lazarsfeld, *Syzygies and Koszul cohomology of smooth projective varieties of arbitrary dimension*, Invent. Math. **111** (1993), 51–67.

[EL2] ———, *Global generation of pluricanonical and adjoint linear series on smooth projective threefolds*, Jour. of Am. Math. Soc. **6** (1993), 875–903.

[EL3] ———, *Seshadri constants on smooth surfaces*, Journées de Géométrie Algébrique d'Orsay, Astér isque **282**, 1993, pp. 177–186.

[ELM] L. Ein, R. Lazarsfeld, V. Maşek, *Global generation of linear series on terminal threefolds*, Internat. J. Math. **6** (1995), 1 - 18.

[ELN] L. Ein, R. Lazarsfeld and M. Nakamaye, *Zero estimates, intersection theory, and a theorem of Demailly*, Higher Dimensional Complex Varieties, de Gruyter, 1996, pp. 183 - 207.

[EGH] D. Eisenbud, M. Green and J. Harris, *Higher Castelnuovo theory*, Journées de Géométrie Algébrique d'Orsay, Astérisque, vol. 282, 1993, pp. 187–202.

[Es] H. Esnault, *Fibre de Milnor d'un cône sur une courbe plane singulière*, Invent. Math. **68** (1982), 477–496.

[EV1] H. Esnault and E. Viehweg, *Logarithmic De Rham complexes and vanishing theorems*, Invent. Math. **86** (1986), 161–194.

[EV2] ———, *Lectures on Vanishing Theorems*, DMV Seminar, Band 20, Birkhäuser Verlag, 1992.

[FdB1] G. Fernández del Busto, *Bogomolov instability and Kawamata-Viehweg vanishing*, J. Alg. Geom. **4** (1995), 693 - 700.

[FdB2] ———, UCLA Thesis, 1994.

[Fuj1] T. Fujita, *On polarized manifolds whose adjoint bundles are not semipositive*, Algebraic Geometry, Sendai, Adv. Stud. Pure Math., vol. 10, T. Oda (ed.), 1987, pp. 167–178.

[Fuj2] ———, *Remarks on Ein-Lazarsfeld criterion of spannedness of adjoint bundles of polarized threefolds* (to appear).

[Gies] D. Gieseker, *On a theorem of Bogomolov on Chern classes of stable bundles*, Amer. J. Math **101** (1979), 77–85.

[Grn] M. Green, *Koszul cohomology and the geometry of projective varieties*, J. Diff. Geom. **19** (1984), 125–171.

[GL] M. Green and R. Lazarsfeld, *On the projective normality of complete linear series on an algebraic curve*, Invent. Math. **83** (1986), 73–90.

[GH1] P. Griffiths and J. Harris, *Principles of Algebraic Geometry*, Wiley and Sons, New York, 1978.

[GH2] ———, *Residues and zero-cycles on algebraic varieties*, Ann. Math. **108** (1978), 461–505.

[Har] J. Harris, *The genus of space curves*, Math. Ann. **249** (1980), 191–204.

[H1] R. Hartshorne, *Ample Subvarieties of Algebraic Varieties*, Lecture Notes in Math. vol. 20, Springer Verlag, 1970.

[H2] ——, *Algebraic Geometry*, Graduate Texts in Math. vol. 52, Springer Verlag, New York, 1977.

[K1] Y. Kawamata, *A generalization of Kodaira-Ramanujam's vanishing theorem*, Math. Ann. **261** (1982), 43–46.

[K2] ——, *On the finiteness of generators of a pluricanonical ring for a 3-fold of general type*, Amer. J. Math. **106** (1984), 1503–1512.

[K3] ——, *The cone of curves of algebraic varieties*, Ann. Math. **119** (1984), 603–633.

[KMM] Y. Kawamata, K. Matsuda and K. Matsuki, *Introduction to the minimal model program*, Algebraic Geometry, Sendai, Adv. Stud. Pure Math., vol. 10, T. Oda (ed.), 1987, pp. 283–360.

[Kod] K. Kodaira, *Pluricanonical systems on algebraic surfaces of general type*, J. Math. Soc. Japan **20** (1968), 170–192.

[Kol1] J. Kollár, *Vanishing theorems for cohomology groups*, Algebraic Geometry, Bowdoin 1985, Proc. Symp. Pure Math. vol 46, 1987, pp. 233–243.

[Kol2] ——, *Effective basepoint freeness*, Math. Ann. **296** (1993), 595–605.

[Kol3] ——, *Shafarevich Maps and Automorphic Forms*, Princeton U. Press, 1995.

[KolM] J. Kollár and T. Matsusaka, *Riemann-Roch type inequalities*, Am. Jour. Math. **105** (1983), 229–252.

[LB] H. Lange and Ch. Birkenhake, *Complex Abelian Varieties*, Springer Verlag, Berlin, 1992.

[L1] R. Lazarsfeld, *Brill-Noether-Petri without degenerations*, J. Diff. Geom **23** (1986), 299–307.

[L2] ——, *A sampling of vector bundle techniques in the study of linear series*, Lectures on Riemann Surfaces, M. Cornalba et. al. (eds), World Scientific Press, 1989, pp. 500–599.

[LM] D. Lieberman and D. Mumford, *Matsusaka's big theorem*, Algebraic Geometry – Arcata 1974, Proc. Symp. Pure Math. vol. 29, 1975, pp. 513–530.

[Mas] V. Maşek, UCLA Thesis, 1994.

[Mka] Y. Miyaoka, *The Chern classes and Kodaira dimension of a minimal variety*, Algebraic Geometry, Sendai, Adv. Stud. Pure Math., vol. 10, T. Oda (ed.), 1987, pp. 449–476.

[Mori] S. Mori, *Classification of higher-dimensional varieties*, Algebraic Geometry, Bowdoin 1985, Proc. Symp. Pure Math. vol. 46, 1987, pp. 269–332.

[Mwk] A. Moriwaki, *Arithmetic Bogomolov-Gieseker's Inequality*, Amer. J. Math. **117** (1995), 1325 - 1347.

[MS] S. Mukai and F. Sakai, *Maximal subbundles of vector bundles on a curve*, Manuscripta math. **52** (1985), 251–256.

[Mfd] D. Mumford, *Curves and their Jacobians*, Univ. of Michigan Press, 1975.

[Nad] A. Nadel, *Multiplier ideal sheaves and the existence of Kähler-Einstein metrics of positive scalar curvature*, Proc. Nat. Acad. Sci. U.S.A. **86** (1990), 7299–7300.

[OSS] C. Okonek, M. Schneider and H. Spindler, *Vector Bundles on Complex Projective Spaces*, Progress in Mathematics, vol. 3, Birkhäuser Boston, 1980.

[Paol] R. Paoletti, *Seshadri constants, gonality of space curves, and restrictions of stable bundles*, J. Diff. Geom. **40** (1994), 329 - 343.

[Ram] S. Ramanan, *Ample divisors on abelian surfaces*, Proc. London Math. Soc. **51** (1985), 231–245.

[Reid] M. Reid, *Bogomolov's theorem* $c_1^2 \leq 4c_2$, Proc. Internat. Symposium on Alg. Geom., Kyoto, 1977, pp. 633–643.

[Rdr] I. Reider, *Vector bundles of rank 2 and linear systems on algebraic surfaces*, Ann. Math. **127** (1988), 309–316.

[Sak1] F. Sakai, *Weil divisors on normal surfaces*, Duke Math. J. **51** (1984), 877–887.

[Sak2] ——, *Reider-Serrano's method on normal surfaces*, Algebraic Geometry, Proceedings: L'Aquila 1988, Lect. Notes in Math. vol. 1417, 1990, pp. 301–319.

[Sak3] ——, *Singularities of plane curves*, Geometry of Complex Projective Vartieties, (Cetraro, 1990), Mediterranean, 1993.

[S-B] N. Shepherd - Barron, *Unstable vector bundles and linear systems on surfaces in characteristic p*, Invent. Math. **106** (1991), 243–262.

[SS] B. Shiffman and A. Sommese, *Vanishing Theorems on Complex Manifolds*, Progress in Math. vol. 56, Birkhäuser, 1985.

[Sho] V. Shokurov, *The non-vanishing theorem*, Math. USSR-Izv. **19** (1985), 591–607.

[Siu1] Y.-T. Siu, *An effective Matsusaka big theorem*, Ann. Inst. Fourier (Grenoble) **43** (1993), 1387 - 1405.

[Siu2] ——, *Effective very ampleness*, Invent. Math. **124** (1996), 563 - 571.

[SBL] A. Sommese, A. Biancofiore and E. Livorni (eds.), *Algebraic Geometry, Proceedings: L'Aquila 1988*, Lect. Notes in Math. vol. 1417, 1990.

[V] E. Viehweg, *Vanishing theorems*, Journ. Reine Ang. Math. **335** (1982), 1–8.

[W] P. Wilson, *Towards birational classification of algebraic varieties*, Bull. London Math. Soc. **19** (1987), 1–48.

[Xu] G. Xu, *Subvarieties of general hypersurfaces in projective space*, J. Diff. Geom **39** (1994), 139-172.

[Z] O. Zariski, *On the irregularity of cyclic multiple planes*, Ann of Math. **32** (1931), 485–511.

Note added in proof (September, 1996): In the two years since these notes were completed there have been a number of developments related to the material discussed here, and we summarize a few of these. Most notably, Angehrn and Siu (Effective freeness and point separation for adjoint bundles, *Invent. Math.* **122** (1995), pp. 291 - 308) have obtained a quadratic polynomial bound for the freeness of adjoint series in all dimensions. As in [**EL2**] and all subsequent work, the basic technique is the method of KRS discussed in §6, but with two innovations. First, instead of explicitly doing an induction (as in 6.10), Angehrn and Siu lift the divisors that one *would* use for an induction to construct a single **Q**-divisor on the ambient manifold X. They show that this divisor has a non-trivial multiplier ideal (cf. Ex. 6.7) provided that one works at a general point. Then they use a limiting argument to pass to an arbitrary point. This paper and others emphasizes the value of the multiplier ideal point of view, of which an introductory account may be found in the paper by Ein and Siu (to appear in the Proceedings of the Santa Cruz Institute). The theorem of Angehrn and Siu has been algebrized and extended by Kollár (Singularities of Pairs, to appear in the Proceedings of the

Santa Cruz Institute). The approach of Anghern and Siu has made it possible to simplify considerably and improve the results of [**EL2**] for threefolds, and to extend them to fourfolds (Ein and Siu, cited above; S. Lee, Remarks on the adjoint and the pluricanonical linear series; Kawamata, On Fujita's conjecture for 3-folds and 4-folds). K. Smith (Fujita's conjecture and local cohomology) has suggested an algebraic approach to Fujita's conjecture. A general result of the sort suggested in the Remark following (7.3) is due to Fujita (Approximating Zariski decomposition of big line bundles, *Kodai Math. J.* **17** (1994), pp. 1 - 3). Some new perspectives on the Seshadri constants discussed in §5 can be found in R. Lazarsfeld, Lengths of periods and Seshadri constants of abelian varieties (to appear in Math. Research Letters). Finally, we would like to call the reader's attention to the nice notes of Demailly (L^2 vanishing theorems for positive line bundles and adjuncton theory, to appear), which discuss some of the same questions as the present lectures, but from an analytic perspective.

Cohomology and Intersection Homology of Algebraic Varieties

Eduard Looijenga

IAS/Park City Mathematics Series
Volume 3, 1997

Cohomology and Intersection Homology of Algebraic Varieties

Eduard Looijenga

Introduction

This is the write up of a series of lectures that intended to give an overview of the algebraic topology of complex algebraic varieties. The most important tool here is the theory of Lefschetz pencils, which serves in algebraic geometry similar purposes as Morse theory in differential topology.

The first four lectures are devoted to the classical aspects of this theory: Lecture 1 deals with dual varieties and it shown there that there are plenty Lefschetz pencils on a smooth projective variety with good properties. Lecture 2 is devoted to the weak Lefschetz theorem, lecture 3 discusses various properties that are equivalent to the hard Lefschetz theorem and lecture 4 is about monodromy and the Picard–Lefschetz formula's. Here I followed to a large extent Lamotke's survey [**12**].

The standard proof of the hard Lefschetz theorem is due to Hodge and is based his theory of harmonic forms. Until the mid-seventies this was the only one. When Deligne proved in [**7**] the corresponding result for varieties defined over finite fields, he obtained at the same time another proof in the complex case (because there is a comparison theorem and a base change property). That proof made ingeneous use of Lefschetz pencils. Ideas of that proof have been used to obtain far reaching generalizations.

One of these generalizations concerns singular projective varieties. It was clear from the outset that the Lefschetz theorems could not hold for such varieties, simply because of the failure of Poincaré duality. However, with the discovery of intersection homology by Goresky-MacPherson, a homology theory for singular spaces had arrived that had Poincaré duality built in. Shortly afterwards, Bernstein, Beilinson, Deligne and Gabber showed that singular projective varieties defined over finite fields satisfy the hard Lefschetz theorem if we replace étale cohomology by the intersection homology analog. They also proved relative versions of this result,

[1]Faculteit Wiskunde en Informatica, Universiteit Utrecht, PO Box 80.010, 3508 TA Utrecht, The Netherlands
E-mail address: `looijenga@math.ruu.nl`

among which the decomposition theorem. This led M. Saito to develop the complex (or rather Kähler) counterpart of these results.

We have attempted to describe some of the ideas that go into this generalization in the last two sections. Lecture 5 is devoted to a detailed study of the Leray spectral sequence of a Lefschetz pencil following Grothendieck and Deligne. We use this to outline a proof of the hard Lefschetz theorem for smooth varieties of which the analytic input is given by Zucker's fundamental theorem on the cohomology of curves with values in polarized Hodge structure with degeneracies. This proof is admittedly more involved than the standard proof, but has the virtue that it generalizes to intersection homology of singular projective varieties. That generalization is the topic of lecture 6.

These lectures were based on a (part of a) course I gave at the University of Utah in the Spring of 1991. During that course, Herb Clemens carefully took notes. These notes were available to me (and the audience) while I gave the lectures and proved very useful in writing this up. I thank him heartily.

A word of warning regarding the bibliography is perhaps in order: this is merely a list of works that we referred to and there is no pretense of completeness of any kind. But we do hope that it may help the reader to find further references.

LECTURE 1
The Dual Variety and Lefschetz Pencils

Throughout this lecture we fix a complex vector space V of dimension $N+1$. We denote by $\mathbb{P} := \mathbb{P}(V)$ be its associated projective space (its points parametrize the one-dimensional subspaces of V) and by $\mathbb{P}^* := \mathbb{P}(V^*)$ the dual (whose points parametrize the linear hyperplanes of V, or equivalently, the projective hyperplanes of $\mathbb{P}$). We write $H_\xi \subset \mathbb{P}$ for the projective hyperplane labeled by $\xi \in \mathbb{P}^*$. Likewise, $x \in \mathbb{P}$ determines a hyperplane H_x of $\mathbb{P}^*$.

The pairs $(x, \xi) \in \mathbb{P} \times \mathbb{P}^*$ with the property that $x \in H_\xi$ (or equivalently, $\xi \in H_x$) make up a closed subvariety W of $\mathbb{P} \times \mathbb{P}^*$, called the *incidence variety*. A choice of a basis for V determines projective coordinates $[x^0 : \dots x^N]$ for $\mathbb{P}$ and $[\xi_0 : \dots \xi_N]$ for $\mathbb{P}^*$; in terms of these coordinates W is given by the single equation $\sum_{i=0}^N x^i \xi_i = 0$. It is a smooth hypersurface in $\mathbb{P} \times \mathbb{P}^*$. Let

$$p : W \to \mathbb{P} \quad , \quad q : W \to \mathbb{P}^*$$

be the two projections.

(1.1) Lemma. *The projections p and q are locally trivial with fibre a projective space of dimension $N-1$.*

The proof is left to you.

Now let $X \subset \mathbb{P}$ be closed irreducible subvariety of dimension $n < N$. If x is a smooth point of X, then let $\mathbb{T}_x X$ denote the projective n-dimensional subspace of $\mathbb{P}$ spanned by the tangent space of X at x. It is clear that H_ξ is tangent to X at x iff $H_\xi \supset \mathbb{T}_x X$. We define the *projectivized conormal bundle* of the smooth part X_{reg} of X by

$$N_{X_{\text{reg}}} := \{(x, \xi) \in X_{\text{reg}} \times \mathbb{P}^* \ : \ H_\xi \supset \mathbb{T}_x X\}.$$

(1.2) Lemma. *$N_{X_{\text{reg}}}$ is a closed irreducible subvariety of $X_{\text{reg}} \times \mathbb{P}^*$ of dimension $N-1$ and the projection $N_{X_{\text{reg}}} \to X_{\text{reg}}$ is locally trivial with fibre a projective space of dimension $N-n-1$.*

Proof. Given $x \in X_{\text{reg}}$, choose a complementary projective subspace K to $\mathbb{T}_x X$ in $\mathbb{P}$. Then the set of $y \in X_{\text{reg}}$ with $\mathbb{T}_y X \cap K = \emptyset$ is a Zariski open neighborhood U of x and we have a natural isomorphism $p^{-1}(U) \cap N_{X_{\text{reg}}} \cong U \times K^*$. This is a local trivialization. $\square$

We write N_X for the Zariski closure of $N_{X_{\mathrm{reg}}}$ in W. This is a subvariety of dimension $N-1$. Clearly, $p(N_X) = X$. The image $q(N_X)$ is called the *dual variety of* X; it is usually denoted by $X^\vee$. It is a closed irreducible subvariety of $\mathbb{P}^*$.

(1.3) Proposition. *The exchange* $\tau : \mathbb{P} \times \mathbb{P}^* \cong \mathbb{P}^* \times \mathbb{P}$ *maps* N_X *onto* $N_{X^\vee}$ *and so* $X^{\vee\vee} = X$.

We put $W_X := p^{-1}(X) \cap W$ and $q_X := q|W_X$. Since $p : W_X \to X$ is locally trivial with smooth fibre, $W_{X,\mathrm{reg}} = W_{X_{\mathrm{reg}}}$. We first prove:

(1.4) Lemma. *For* $(a, \alpha) \in W_{X,\mathrm{reg}}$, *the following are equivalent:*

1. q_X *has not the maximal rank* N *at* (a, α),
2. $(a, \alpha) \in N_X$, *i.e.,* $H_\alpha \supset \mathbb{T}_a X$,
3. $dq(T_{(a,\alpha)} W_X) = T_\alpha H_a$.

Proof. Choose a basis for V such that $a = [1 : 0 : \cdots : 0]$ and H_α is given by $x^N = 0$. Then in terms of the obvious affine coordinates q_X is given by

$$(x; u_1, \ldots, u_{N-1}) \mapsto (-\sum_{i=1}^{N-1} x^i u_i - x^N, u_1, \ldots, u_{N-1}),$$

where $x = [1 : x^1 : \cdots : x^N]$. It is clear that q_X fails to have rank N at (a, α) iff $x_N|X$ has a as a critical point. This last property just means that $H_\alpha \supset \mathbb{T}_a X$. This proves (1) $\Leftrightarrow$ (2).

Since W_X contains $\{a\} \times H_a$, we have

$$T_\alpha H_a \subset dq(T_{(a,\alpha)} W_X).$$

The left-hand side has dimension $N-1$. So if the right-hand side has dimension at most $N-1$, this inclusion must be an equality. This proves (1) $\Leftrightarrow$ (3). □

Proof of (1.3). Let U be an open-dense subset of $N_{X_{\mathrm{reg}}}$ with the property for all $(a, \alpha) \in U$, $\alpha \in X^\vee_{\mathrm{reg}}$ and $dq(T_{(a,\alpha)} N_X) = T_\alpha X^\vee$. By lemma (1.3) we have

$$T_\alpha H_a = dq(T_{(a,\alpha)} W_X)$$

for all $(a, \alpha) \in N_{X_{\mathrm{reg}}}$. So if $(a, \alpha) \in U$, then $T_\alpha H_a \supset T_\alpha X^\vee$. This means that $(\alpha, a) \in N_{X^\vee}$. We have proved that $\tau(U) \subset N_{X^\vee}$. Since both N_X and $N_{X^\vee}$ are closed and irreducible of dimension $N-1$, it follows that $\tau(N_X) = N_{X^\vee}$. □

(1.5) Corollary. *If* X *is smooth, then the set of critical points of* q_X *is equal to* N_X *and hence its set of critical values is* $X^\vee$. *In particular,* q_X *is locally trivial in the* C^∞ *sense over* $\mathbb{P}^* - X^\vee$.

Proof. The first statement is immediate from lemma (1.4). The last clause follows from the Ehresmann fibration theorem. □

(1.6) Corollary. *All smooth hypersurfaces in* $\mathbb{P}^n$ *of a fixed degree* $d > 0$ *are mutually diffeomorphic.*

Proof. Apply the proposition to the image X of the d-fold Veronese embedding $f : \mathbb{P}^n \to \mathbb{P}^N$ (where $N+1$ is the number of monomials of degree d in $n+1$ variables, and f has these monomials as coordinates). □

In the remainder of this lecture we assume that X is smooth.

Let L be line in $\mathbb{P}^*$. It determines a codimension two linear subspace A of $\mathbb{P}$ with the property that

$$\xi \in L \Leftrightarrow A \subset H_\xi.$$

The family of hyperplanes of $\mathbb{P}$ parametrized by L is called a *pencil*. We put

$$\tilde{X} := q_X^{-1}(L) = \{(x, \xi) \in X \times L \ : \ x \in H_\xi\}$$

and we denote the two by projections by $\pi : \tilde{X} \to X$ and $f : \tilde{X} \to X$. We further put $Y := X \cap A$. Observe that then

- $\pi^{-1}Y = Y \times L$; we shall often write $\tilde{Y}$ for this space,
- $\pi : \tilde{X} - \tilde{Y} \to X - Y$ is an isomorphism, the inverse mapping being given by $x \mapsto (x, \mathrm{span}\langle x, A\rangle)$,
- $f^{-1}(\xi) = H_\xi \cap X$; this intersection will also be denoted by X_ξ.

We say that L is *transversal* to $X^\vee$ if L meets $X^\vee$ in its smooth part only and is transversal to $X^\vee_{\mathrm{reg}}$. We first show that such lines are plenty :

(1.7) Lemma. *Given $\alpha \in \mathbb{P}^* - X^\vee$, then the lines in $\mathbb{P}^*$ through α form a projective space $\mathbb{P}_\alpha$ of dimension $N-1$ of which an open dense subset parametrizes $X^\vee$-transversal lines.*

Proof. Let $\phi : X^\vee \to \mathbb{P}_\alpha$ be the map which assigns to $\xi \in X^\vee$ the line spanned by ξ and α. Then $\xi \in X^\vee_{\mathrm{reg}}$ is a critical point of ϕ iff span(ξ, α) is contained in $\mathbb{T}_\xi X^\vee$. So the set of lines passing through α which are not transversal to $X^\vee$ is the union of $\phi(X_{\mathrm{sing}})$ and the set of critical values of $\phi|X^\vee_{\mathrm{reg}}$, and hence is contained in a proper subvariety. □

The number of points of intersection of an $X^\vee$-transversal line with $X^\vee$ is independent of L; this number is called the *class* of X. If $X^\vee$ is a hypersurface, then the class of X is simply the degree of $X^\vee$; it is zero otherwise.

Let us fix an $X^\vee$-transversal line L.

(1.8) Proposition. *The following properties hold:*

1. *A is transversal to X, so that Y is smooth in X of codimension two,*
2. *$\tilde{X}$ is smooth,*
3. *no critical point of f lies in $\tilde{Y}$ and f maps its critical points bijectively onto $L \cap X^\vee$,*
4. *each critical point $\tilde{x}$ of f is nondegenerate, i.e., the hessian of f at such a point (which is a bilinear map $T_{\tilde{x}}\tilde{X} \times T_{\tilde{x}}\tilde{X} \to T_{f(\tilde{x})}L$) is nondegenerate.*

Proof. (1) Assume A is not transversal to X at x. Then there is a hyperplane H containing $\mathbb{T}_x X \cup A$. So $H = H_\xi$ for some $\xi \in \mathrm{L}$ and $(x, \xi) \in N_X$. Hence $\xi \in V^\vee$. But then $\xi \in X^\vee_{\mathrm{reg}}$ since L is transversal to $X^\vee$. Since $(\xi, x) \in N_{X^\vee}$, it follows that $\mathbb{T}_\xi X^\vee \subset H_x$. Since $x \in A$, we also have $L \subset H_x$. So $\mathbb{T}_\xi X^\vee + L \subset H_x$, which contradicts the fact that L is transversal to $X^\vee$.

(2) Since $\tilde{X} = q_X^{-1}(L)$, it is enough to show that $q : W_X \to \mathbb{P}^*$ is tranversal to L. Let $(a, \alpha) \in \tilde{X}$. If q_X has rank N at (a, α), there is nothing to prove. Else

$(a, \alpha) \in N_X$ and $dq(T_{(a,\alpha)}W_X) = T_\alpha H_a \supset T_\alpha X^\vee$. Since L is transversal to $X^\vee$ at α, it follows that L is also transversal to H_a at α.

(3) The restriction of f to $\tilde{Y} = Y \times L$ is just the second projection, so f has no critical point in $\tilde{Y}$. Let $(a, \alpha) \in \tilde{X}$ be a critical point of f. The fact that $q : W_X \to \mathbb{P}^*$ is tranversal to L implies that (a, α) is then also a singular point of f. So $(a, \alpha) \in N_X$. This implies that $\xi \in X^\vee$. As $\xi \in L$ also, $X^\vee$ must be smooth of dimension $N - 1$ at ξ. But $N_X \cong N_{X^\vee} \to X^\vee$ is an isomorphism over the smooth part of $X^\vee$ (why?), and so over α there is no other critical point than (a, α).

(4) Let $(a, \alpha) \in \tilde{X}$ be a critical point of f. We just saw that then $(a, \alpha) \in N_X$ and that $q|N_X$ has rank $N - 1$ at (a, α). Choose projecive coordinates such that: $a = [1 : 0 : \cdots : 0]$, $H_\alpha = \{x^N = 0\}$ and $A = \{x^0 = x^N = 0\}$. Now $T_aX \subset T_aH_\alpha$. Let $(t^1, \ldots, t^n)$ be local analytic coordinates for X near a, so that $X \subset \mathbb{P}$ is given at a by

$$(t^1, \ldots, t^n) \to [1 : x^1(t) : \ldots x^N(t)],$$

and x^N has order ≥ 2 at $t = 0$. Parametrize W_X near (a, α) by

$$(t^1, \ldots, t^n; u_1, \ldots, u_N) \to ([1 : x^1(t) : \cdots : x^N(t)], [g(t, u) : u_1 : \ldots ; u_{N-1} : 1]),$$

where $g(t, u) = -\sum_{\nu=1}^{N-1} x^\nu(t)u_\nu - x^N(t)$. In local coordinates, $f(t) = -x^N(t) = g(t, 0)$. So we must show that

$$(\frac{\partial^2 g}{\partial t^i \partial t^j}(0,0))$$

is nonsingular.

To this end, we first observe that

$$(\frac{\partial^2 g}{\partial t^i \partial u_\nu}(0,0))$$

has rank n since, by the explicit formula for g, it equals $(-\partial z^\nu/\partial t^i)$. This implies that the derivatives $\partial g/\partial t^1, \ldots, \partial g/\partial t^n$ are part of a system of local coordinates for W at (a, α). But notice that their common zero set is precisely the locus where q_X is not of maximal rank. By lemma (1.5) this is just N_X.

Since $q|N_X$ has rank $N - 1$ it follows that the map

$$(\partial g/\partial t^1, \ldots, \partial g/\partial t^n, g, u_1, \ldots, u_{N-1})$$

must have rank $n + N - 1$ in $(0, 0)$. As its jacobian at this point is

$$\begin{pmatrix} (\partial^2 g/\partial t^i \partial t^j)(0,0) & * \\ (\partial g/\partial t^j)(0,0) = -(\partial x^N/\partial t^j)(0) = 0 & -x^\nu(0) = 0 \\ 0 & I_{N-1} \end{pmatrix},$$

it follows that $(\partial^2 g/\partial t^i \partial t^j)(0, 0)$ is nonsingular. □

The collection of hyperplane sections $X_\xi := \{X \cap H_\xi\}_{\xi \in L}$ is called a *general Lefschetz pencil*; these sections are just the fibres of the map f, which itself is called a *Lefschetz fibration*. The subspace $A \subset \mathbb{P}$ is called its *axis* and Y its *fixed point locus*. We will see that a general Lefschetz pencil is an algebro-geometric analog of a Morse function on a manifold; it is an important tool to analyze the topology of X.

LECTURE 2
The Weak Lefschetz Theorem

We will need some results from homotopy theory. We just state them here and refer for proofs to a standard text such as G.W. Whitehead's Elements of Homotopy theory (Springer Graduate Text in Mathematics). For the purposes of homotopy theory, the most suitable class of spaces to work with are those which are *compactly generated*, that is, spaces that satisfy the Hausdoff axiom and have the property that a subset is closed iff its intersection with every compact subset is compact. We shall therefore assume that all spaces in this lecture belong to this class.

We say that a topological pair(X, A) is a *k-cellular extension* if X is obtained from A by simultaneously attaching a number of k-cells. Such a pair is $(k-1)$-connected.

A *relative cell complex of dimension* $\leq d$ is a filtered space

$$X_\bullet = \{X_{-1} \subset X_0 \subset X_1 \subset \cdots \subset X_d\}$$

such that (X_k, X_{k-1}) is a k-cellular extension for $k = 0, \ldots, d$; (X_d, X_{-1}) is then said to admit the structure of a relative cell complex of dimension $\leq d$. If $X_{-1} = \emptyset$, we omit the adjective *relative*. A large class of examples consist of the finite dimensional polyhedra with subcomplexes.

Relative cell complexes have several nice properties of which we list a few.

Proposition A. *Let (X, A) admit the structure of a relative cell complex and assume that $i : A \subset X$ is a homotopy equivalence. Then $A \subset X$ is a deformation retract, i.e., there exists a continuous family of maps $\{r_t : X \to X\}_{0 \leq t \leq 1}$ with $r_t i = i$ for all t, $r_0 = \mathbf{1}_X$ and $r_1(X) = A$.*

If (X, A) admits the structure of a relative cell complex with $A = X_{k-1}$, then by iterated use of the homotopy exact sequence for triples we find that (X, A) is $(k-1)$-connected. There is a homotopy converse which says that if (X, A) admits the structure of a relative cell complex and is $(k-1)$-connected, then (X, A) is homotopy equivalent relative A to a pair (X', A) which admits the structure of a relative cell complex with only cells in dimension $\geq k$. This practically implies the following two propositions.

Proposition B. *Let $f : (X, A) \to (Y, B)$ be a relative homeomorphism of pairs, i.e., B is closed in Y and f maps $X - A$ homeomorphically onto $Y - B$. If (X, A)*

is $(k-1)$-connected and admits the structure of a relative cell complex, then the same is true for (Y, B).

Proposition C. *Let (X, A) resp. (Y, B) be $(k-1)$-connected resp. $(l-1)$-connected and admit the structure of a relative cell complex. Then $(X, A) \times (Y, B) := (X \times Y, X \times B \cup A \times Y)$ is $(k+l-1)$-connected.*

This finishes our short homotopical review. The main result of this lecture is this:

(2.1) Theorem (Weak Lefschetz theorem). *Let $X \subset \mathbb{P}$ be a closed smooth projective variety of dimension n, and let $X_\xi = X \cap H_\xi$ be a tranversal hyperplane section. Then*

1. *(X, X_ξ) is $(n-1)$-connected and*
2. *$X - X_\xi$ has the homotopy type of a finite cell complex of dimension $\leq n$.*

The second statement is actually a special case of the more general fact that any affine variety of complex dimension $\leq n$ has the homotopy type of a finite cell complex of dimension $\leq n$.

(2.2) Corollary. *If X is as in the previous theorem and $Z \subset \mathbb{P}$ is a linear subspace transversal to X, then $(X, X \cap Z)$ is $\dim(X \cap Z)$-connected.*

Proof. Iterated application of part (1) of the weak Lefschetz theorem. □

So X is connected and of dimension ≥ 2, then any tranversal linear section $X' \subset X$ of dimension ≥ 1 is also connected and induces a surjection on fundamental groups (even a bijection if $\dim X' \geq 2$).

The proof of the weak Lefschetz theorem requires some preparation. We begin with recalling the Morse lemma:

(2.3) Lemma. *If $f : (\mathbb{C}^n, 0) \to (\mathbb{C}, 0)$ has a non-degenerate singularity, then after a local-analytic coordinate change, f takes the form $\sum_{i=1}^n (z^i)^2$.*

(2.4) This motivates us to have a closer look at the function $f : \mathbb{C}^n \to \mathbb{C}$ defined by $f(z) = \sum_{i=1}^n (z^i)^2$.

Choose $\epsilon > 0$ and let B_ϵ be the set $z \in \mathbb{C}^n$ with $|z| \leq \epsilon$. It is easily verified that $f|\partial B_\epsilon$ has no singular point z with $|f(z)| < \epsilon^2$. Choose $0 < \eta < \epsilon^2$, and let

$$D := \{t \in \mathbb{C} : |t| \leq \eta\}, \quad B := f^{-1}D \cap B_\epsilon, \quad \dot{B} := f^{-1}D \cap \partial B_\epsilon.$$

From now on we reserve the symbols f resp. $\dot{f}$ for the restrictions $f : B \to D$ resp. $f : \dot{B} \to D$. Notice that $\dot{f}$ is a proper smooth mapping without singularities and hence locally C^∞ trivial. Since D is a disk, $\dot{f}$ is then (globally) C^∞ trivial.

We next turn our attention to the fibre $B_\eta := f^{-1}(\eta)$. If $z = x + \sqrt{-1}y \in \mathbb{C}^n$, then $z \in B_\eta$ iff $|x|^2 - |y|^2 = \eta$, $x \cdot y = 0$ and $|x|^2 + |y|^2 \leq \epsilon^2$. This implies that $|x|^2 = |y|^2 + \eta$ and $|y|^2 \leq \frac{1}{2}(\epsilon^2 + \eta)$. So if we make the substitutions

$$u := \frac{x}{|x|} \quad , \quad v := \frac{y}{\sqrt{\frac{1}{2}(\epsilon^2 + \eta)}},$$

then $|u| = 1$, $|v| \leq 1$ and $u \cdot v = 0$. We thus obtain a diffeomorphism from B_η onto the unit disk bundle of the tangent bundle of S^{n-1}.

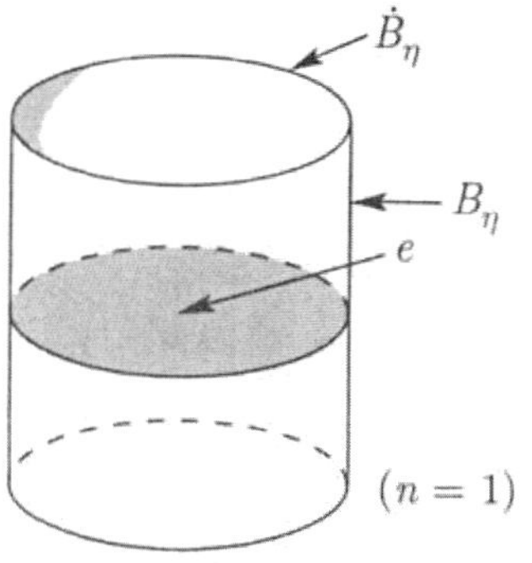

Figure 1. Thimble of an ordinary singularity

The real part of B is the set of $x \in \mathbb{R}^n$ with $|x|^2 \leq \eta$; we denote it by e and call it the *thimble* in B. Notice that $e \cap B_\eta = \partial e$ and that this intersection corresponds to zero section of B_η under the above diffeomorphism.

(2.5) Lemma. *The inclusions*

$$e \subset (B_\eta \cup e) \subset (\dot{B} \cup B_\eta \cup e) \subset B$$

are all deformation retracts.

Proof. Since $e \cap B_\eta \subset B_\eta$ is a deformation retract, so is $e \subset B_\eta \cup e$. Since $\dot{f}$ is trivial, $\dot{B}_\eta \subset \dot{B}$ is a deformation retract. As $\dot{B} \cap (B_\eta \cup e) = \dot{B}_\eta$, it follows that the second inclusion is also a deformation retract. That the last inclusion has this property can probably be shown directly, but tedious work can be avoided by showing that B is starlike and hence contractible. Since e is contractible, so is $\dot{B} \cup B_\eta \cup e$, and hence the last inclusion is a homotopy equivalence. Now apply proposition (A). □

We apply this lemma to the case when f is part of a more global situation:

(2.6) Lemma. *Suppose that we are given a compact $2n$-manifold Z with boundary ∂Z which contains B and a smooth mapping $F : Z \to D$ which extends f. We assume that $F^{-1}(\partial D) = \partial Z$, that $\dot{B}$ meets ∂Z transversally, and that F has no singular point $\neq 0$. Then $Z_\eta \cup e \subset Z$ is a deformation retract, so that up to homotopy, Z is obtained from Z_η by attaching an n-cell.*

Proof. Put $\dot{Z} := Z - (B - \dot{B})$. This is a manifold with corners. The restriction $\dot{F} : \dot{Z} \to D$ of F has no critical points, even when we restrict to the boundary piece ∂B. The Ehresmann fibration theorem applies and we find that $\dot{F}$ is C^∞ trivial. So $\dot{Z}_\eta \subset \dot{Z}$ is a deformation retract. Also, $\dot{Z}_\eta \subset \dot{Z}_\eta \cup \dot{B}$ is a deformation retract, because $\dot{B}_\eta \subset \dot{B}$ is. This implies that the inclusion $\dot{Z}_\eta \cup \dot{B} \subset \dot{Z}$ is a homotopy equivalence. Proposition (A) applies here, so this inclusion is actually a deformation retract. This in turn, implies that $B \cup Z_\eta \subset Z$ is a deformation retract. The lemma now follows if we combine this with the fact that $B_\eta \cup e \subset B$ is a deformation retract. □

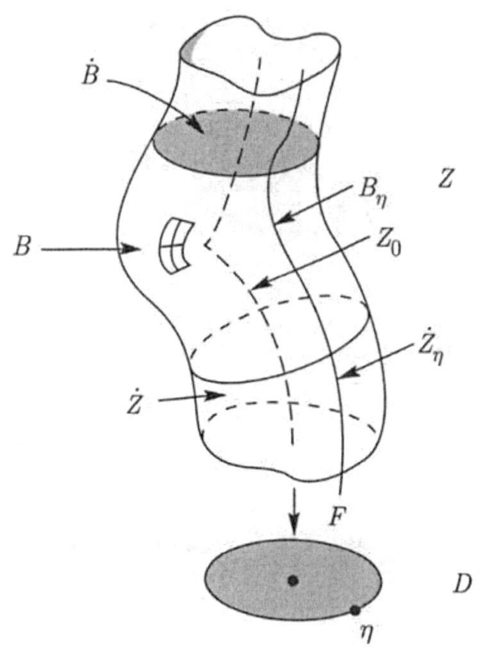

Figure 2. Milnor's fibration

(2.7) Let us now put ourselves in the situation of the weak Lefschetz theorem. According to (1.7) there exists a line $L \subset \mathbb{P}$ passing through ξ which is transversal to $X^\vee$. This defines a general Lefschetz pencil on X. We employ the notation of the previous lecture. So we have a Leschetz fibration $f : \tilde{X} \to X$ satisfying the properties listed in (1.8). If $\xi(1), \dots, \xi(r)$ are the distinct points of $L \cap X^\vee$, then choose an affine coordinate w on L such that $w(\xi) = 1$ and $|w(\xi(\rho))| < 1$, $\rho = 1, \dots, r$. Let $L(0) \subset L$ and $L(\infty) \subset L$ be the hemispheres defined by $|w| \le 1$ and $|w| \ge 1$ respectively. Let $(x(\rho), \xi(\rho))$ be the unique critical point of f over $\xi(\rho)$. Choose (in terms of local analytic coordinates at this point for which is there a constant plus a sum of squares) a neighborhood $B(\rho)$ of $\xi(\rho)$ as in (2.4) whose image under f is a closed disk $D(\rho)$ in the interior of $L(0)$ centered at $\xi(\rho)$. Let $\eta(\rho)$ the point of $D(\rho)$ with maximal real part so that the thimble $e(\rho)$ in $B(\rho)$ has its boundary $\partial e(\rho)$ in $B(\rho)_{\eta(\rho)}$. Draw arcs $\Gamma(\rho)$ in $L(0)$ connecting ξ with $\eta(\rho)$, $\rho = 1, \dots, r$, as in the picture below. Since $\tilde{X}$ is locally trivial over $\Gamma(\rho)$, the embedding $\partial e(\rho) \subset X_{\eta(\rho)}$ extends to an embedding of $\partial e(\rho) \times \Gamma(\rho)$ in $\tilde{X}_{\Gamma(\rho)}$ which commutes with the projection to $\Gamma(\rho)$. Let $\tilde{e}(\rho)$ denote the union of $e(\rho)$ and the image of this embedding. This is a n-cell whose boundary lies in X_ξ. Put

$$X^\# := X_\xi \cup \bigcup_{\rho=1}^{r} \tilde{e}(\rho).$$

This is an n-cellular extension and hence $(n-1)$-connected.

(2.8) Lemma. *$X^\# \subset \tilde{X}_{L(0)}$ is a deformation retract.*

Proof. Let E be the union of the discs $D(\rho)$ and the arcs $\Gamma(\rho)$. Then $E \subset L_0$ is a deformation retract. Since f is locally trivial over $L(0) - \operatorname{int}(E)$, the homotopy

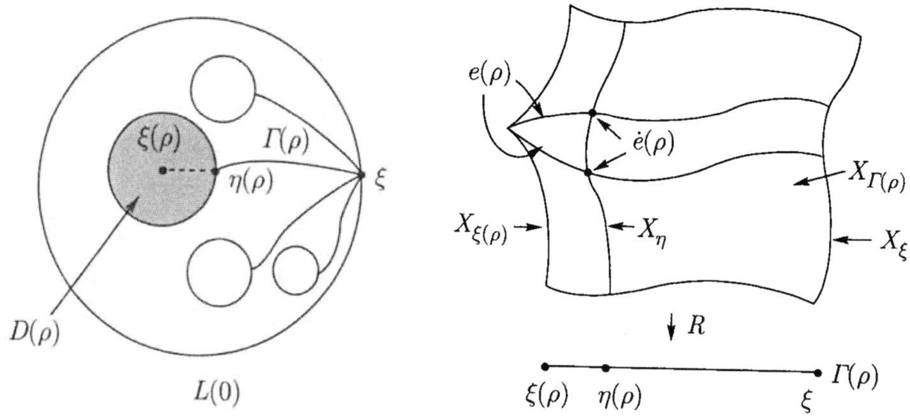

Figure 3. Deformation retract of $X_{L(0)}$

lifting property assures that $\tilde{X}_E \subset X_{L(0)}$ is a deformation retract. According to (2.6), $X_{\eta(\rho)} \cup e(\rho) \subset \tilde{X}_{D(\rho)}$ is a deformation retract. A similar argument shows that

$$X_\xi \cup \tilde{e}(\rho) \subset \tilde{X}_{D(\rho) \cup \Gamma(\rho)}$$

is a deformation retract. The lemma then follows. □

Proof of the weak Lefschetz theorem. We first show (1). For $n = 0, 1$ the statement is trivial. So we assume $n \geq 2$ and proceed by induction. We use the triple

$$(A, B, C) := (\tilde{X}, \tilde{X}_{L(0)} \cup \tilde{Y}, X_\xi \cup \tilde{Y}).$$

Note that in

$$(B, C) = (\tilde{X}_{L(0)} \cup \tilde{Y}, X_\xi \cup \tilde{Y}) \supset (\tilde{X}_{L(0)}, X_\xi \cup \tilde{Y}_{L(0)}) \supset (\tilde{X}_{L(0)}, X_\xi) \supset (X^\#, X_\xi)$$

the first inclusion is a relative homeomorphism. The second is a relative homotopy equivalence since $\tilde{Y}$ is locally trivial over $L(0)$, and the third is a deformation retract by (2.8). Since $(X^\#, X_\xi)$ is $(n-1)$-connected, so is (B, C).

Next observe that

$$(A, C) = (\tilde{X}, X_\xi \cup \tilde{Y}) \to (X, X_\xi)$$

is also a relative homeomorphism.

Finally, consider

$$(A, B) = (\tilde{X}, \tilde{X}_{L(0)} \cup \tilde{Y}) \supset (\tilde{X}_{L(\infty)}, \tilde{X}_{\partial L(\infty)} \cup \tilde{Y}_{L(\infty)}) \cong (X_\xi, Y) \times (L(\infty), \partial L(\infty)),$$

where the isomorphism comes from the local triviality of f over $L(\infty)$. The induction hypothesis tells us that (X_ξ, Y) is $(n-2)$-connected. Since $(L(\infty), \partial L(\infty))$ is 1-connected, it follows from proposition (C) that the product is n-connected.

The inclusion in the above display is a relative homeomorphism. Proposition (B) applies, and so (A, B) is also n-connected.

Feeding these properties in the exact homotopy sequence of the triple (A, B, C) gives that (X, X_ξ) is $(n-1)$-connected.

The proof of (2) also proceeds with induction, but this time starting with n=0. So assume that $n > 0$. Note that all the maps in

$$X - X_\xi \overset{\cong}{\leftarrow} \tilde{X} - \tilde{Y} - X_\xi \cong \tilde{X} - \tilde{Y} - X_{w=\infty} \supset \tilde{X}_{L(0)} - \tilde{Y}_{L(0)} \subset X^\# - Y$$

are homotopy equivalences. Now $(X^\# - Y, X_\xi - Y)$ is an n-cellular extension and by our induction hypothesis, $X_\xi - Y$ has the homotopy type of a finite cell complex of dimension $\leq n-1$. It follows that $X^\# - Y$, and hence $X - X_\xi$, has the homotopy type of a finite cell complex of dimension $\leq n$. □

We now look at the homological aspects of the triple (A, B, C) that appeared in the proof of the weak Lefschetz theorem. Our homotopy discussion and excision yield

$$H_k(B, C) = H_k(X^\#) = \oplus_{\rho=1}^r H_k(\tilde{e}(\rho), \partial\tilde{e}(\rho)) = \begin{cases} 0 & \text{if } k \neq n; \\ \cong \mathbb{Z}^r & \text{else,} \end{cases}$$

$$H_k(A, C) = H_k(X, X_\xi)$$

$$H_k(A, B) = H_k((X_\xi, Y) \times (L(\infty), \partial L(\infty)) \cong H_{k-2}(X_\xi, Y),$$

where the last isomorphism follows from the Künneth formula. So the exact sequence of this triple becomes

$$\to \oplus_\rho H_k(\tilde{e}(\rho), \partial\tilde{e}(\rho)) \to H_k(X, X_\xi) \xrightarrow{L'} H_{k-2}(X_\xi, Y) \to \oplus_\rho H_{k-1}(\tilde{e}(\rho), \partial\tilde{e}(\rho)) \to .$$

The homomorphism L' can be geometrically understood as follows. Choose $\xi' \in L - X^\vee$ in a convex neighborhood ξ (but $\neq \xi$). Then the intersection of $H_{\xi'}$ with (X, X_ξ) is transversal and gives the pair $(X_{\xi'}, Y)$. Represent $z \in H_k(X, X_\xi)$ by a k-chain Z which is transversal to H_ξ. Then its intersection with $H_{\xi'}$, $Z \cdot H_{\xi'}$, is a $(k-2)$-chain on $X_{\xi'}$ with boundary in Y. Move ξ' to ξ along a straight line and drag along $Z \cdot H_{\xi'}$. This produces a cycle on $(X, X_{\xi'})$ representing $L'(z)$.

(2.9) Corollary.

1. $H_k(X, X_\xi) = 0$ *if* $k < n$,
2. $L' : H_k(X, X_\xi) \to H_{k-2}(X_\xi, Y)$ *is an isomorphism for* $k \neq n, n+1$,
3. *we have an exact sequence*

$$0 \to H_{n+1}(X, X_\xi) \xrightarrow{L'} H_{n-1}(X_\xi, Y) \to \oplus_\rho H_n(\tilde{e}(\rho), \partial\tilde{e}(\rho)) \to H_n(X, X_\xi) \to 0.$$

Proof. (1) is immediate from the weak Lefschetz theorem, whereas (2) and (3) follow from the exact sequence above. □

Another corollary is a formula for euler characteristics:

(2.10) Corollary. *The euler characteristics of* X, X_ξ *and* Y *are related by*

$$e(X) = -e(Y) + 2e(X_\xi) + (-1)^n r,$$

where r *is the class of* X.

Proof. Use the equalities

$$e(B,C) = (-1)^n r,$$
$$e(A,C) = e(X,X_\xi) = e(X) - e(X_\xi),$$
$$e(A,B) = e(X_\xi,Y) = e(X_\xi) - e(Y),$$

and the fact that the alternating sum of these numbers must be zero. □

Example. Let X be a smooth plane curve of degree d. Then X_ξ is a finite set of d points and $Y = \emptyset$. The class of X is the degree of the dual curve, which is $d(d-1)$. We thus find $e(X) = -d^2 + 3d$. Since $e(X) = 2 - 2g(X)$, this gives us the well-known formula for the genus: $g(X) = \frac{1}{2}(d-1)(d-2)$.

There are corresponding statements for cohomology:

(2.11).
1. $H^k(X, X_\xi) = 0$ *if* $k < n$,
2. $L' : H^k(X_\xi, Y) \to H^{k+2}(X, X_\xi)$ *is an isomorphism for* $k \neq n, n+1$,
3. *we have an exact sequence*

$$0 \to H^n(X,X_\xi) \to \oplus_\rho H^n(\tilde{e}(\rho), \partial\tilde{e}(\rho) \to H^{n-1}(X_\xi, Y) \xrightarrow{L'} H^{n+1}(X, X_\xi) \to 0,$$

in particular, $H^n(X, X_\xi)$ *is torsion free.*

We omit the proofs, since they are similar to their homological counterparts.

Application. Let $W \subset \mathbb{P}^n$ be a smooth hypersurface of degree d. Regard W as a hyperplane section of $\mathbb{P}^n$ with respect to the d-fold Veronese embedding. Then

$$\begin{aligned} i \leq n-2: &\quad H_i(W) \cong H_i(\mathbb{P}^n) \\ i = n-1: &\quad H_{n-1}(W) \to H_{n-1}(\mathbb{P}^n) \quad \text{is onto} \\ i \geq n: &\quad H_i(W) \cong H^{2n-2-i}(W) \cong H^{2n-2-i}(\mathbb{P}^n). \end{aligned}$$

It is well-known that $H_i(\mathbb{P}^n)$ and $H^i(\mathbb{P}^n)$ are equal to $\mathbb{Z}$ for $i = 0, 2, 4, \ldots, 2n$, and are zero otherwise. So $H_{n-1}(W)$ is really the only interesting homology group. The interesting part of that homology group is $V := \ker(H_{n-1}(W) \to H_{n-1}(\mathbb{P}^n)$. Clearly,

$$e(W) = n - 1 + (-1)^{n-1}\text{rk}(V),$$

If we combine this formula with (2.10), then we see that $\text{rk}(V)$, and hence the $(n-1)$th Betti number of X can be computed inductively.

LECTURE 3
The Hard Lefschetz Theorem

(3.1) We shall need a few facts from algebraic topology. As some of them are not so standard, we list them here.

First, let X be any space. The the cup product gives its total cohomology $H^\bullet(X)$ the structure of a graded-commutative ring with unit. The cap product gives its homology $H_\bullet(X)$ the structure of a graded unital module over $H^\bullet(X)$. So $1 \cap a = a$ and $\alpha \cap (\beta \cap a) = (\alpha \cup \beta) \cap a$. (We adopt the convention to denote homology classes by Latin letters and cohomology classes by Greek letters.) If X is path connected, then $H_0(X) = \mathbb{Z}$, and the usual pairing of homology with cohomology can be written as

$$\langle\ ,\ \rangle : H^k(X) \times H_k(X) \xrightarrow{\cap} H_0(X) = \mathbb{Z}.$$

The universal coefficient theorem for cohomology implies that this pairing is perfect if we mod out the torsion.

The preceding notions are functorial: if $f : X \to Y$ is a continuous map of spaces, then

$$f^* : H^\bullet(Y) \to H^\bullet(X)$$

is a homomorphism of graded-commutative unital rings, and

$$f_* : H^\bullet(X) \to H^\bullet(Y)$$

is homomorphism of $H^*(Y)$-modules. So

$$f_*(f^*\omega \cap a) = \omega \cap f_*a.$$

Now let M be a compact m-manifold which is oriented by means of an orientation (=fundamental) class $[M] \in H_m(M)$. Then

$$\cap[M] : H^k(M) \to H_{m-k}(M)$$

is an isomorphism (Poincaré duality). This gives rise to the intersection pairing

$$H_{m-k}(M) \times H_k(M) \cong H^k(M) \times H_k(M) \xrightarrow{\langle\ ,\ \rangle} \mathbb{Z}, \quad (a, b) \mapsto a \cdot b.$$

Suppose now that P is another compact oriented manifold, of dimension p and let $f : M \to P$ be continuous. Put $c := p - m$. Define the *Gysin homomorphism* as the composite

$$f_! : H^k(M) \overset{PD}{\cong} H_{m-k}(M) \xrightarrow{f_*} H_{m-k}(P) \overset{PD}{\cong} H^{k+c}(P).$$

This map is adjoint to f^* in the sense that

$$f_!\alpha \cdot \zeta = \alpha \cdot f^*\zeta.$$

(We could have defined $f_!$ this way if we were working with rational coefficients.) The map $f_!$ is also a $H^\bullet(P)$-homomorphism:

$$f_!(f^*\zeta \cup \alpha) = \zeta \cup f_!\alpha.$$

We will only need the special case $\alpha = 1$:

$$f_!f^*\zeta = \zeta \cup f_!(1).$$

The class $f_!(1) \in H^c(P)$ is easy to understand geometrically in case f is an embedding: it is then represented by the cocycle which assigns to an M-tranversal c-chain its intersection product with M. The Gysin maps behave well under transversal restriction, and are functorial with respect to compositions. This implies for instance, that if $i_1 : M_1 \subset P$ and $i_2 : M_2 \subset P$ are mutually transversal submanifolds with intersection $i : M \subset P$, then

$$i_!(1) = i_{1!}(1) \cup i_{2!}(1).$$

The corresponding dual notion is

$$f^! : H_k(P) \overset{PD}{\cong} H^{p-k}(P) \xrightarrow{f^*} H^{p-k}(M) \overset{PD}{\cong} H_{k-c}(M),$$

which is adjoint to f_*. We have

$$f_*(\alpha \cap f^!z) = (-1)^c f_!(\alpha) \cap z,$$

so that in particular

$$f_*f^!z = (-1)^c f_!(1) \cap z.$$

This finishes our review of this bit of algebraic topology.

We first illustrate these notions by means of a simple example. If $i : H \subset \mathbb{P}$ is a hyperplane, then we have defined $i_{H!}(1) \in H^2(\mathbb{P})$. Since all such hyperplane sections are isotopic, this class does not depend of the choice H; it is called the *hyperplane class*, and we shall denote it by η.

(3.2) Lemma. *Let $i_Z : Z \subset \mathbb{P}$ be a linear subspace of codimension k. Then $i_{Z!}(1) = \eta^k$.*

Proof. For $k = 1$ there is nothing to prove. We proceed with induction on k, and assume $k \geq 2$. Then Z is of the form $Z' \cap H$, where Z' is of codimension $k-1$, it follows that

$$i_{Z!}(1) = i_{Z'}(1) \cup i_H(1) = \eta^{k-1} \cup \eta = \eta^k.$$

□

Remarks. 1. It is well-known that $H^\bullet(\mathbb{P})$ is additively generatered by $1, \eta, \dots \eta^N$.

2. If $X \subset \mathbb{P}$ is a closed smooth subvariety and $Z \subset X$ is an X-transversal codimension k linear subspace, and $i : X \cap Z \subset X$ the corresponding linear section, then the preceding lemma implies that $i_!(1) = \eta_X^k$, where η_X is the restriction of η to X.

(3.3) We return to the situation studied in the previous lecture, where we are given $\mathbb{P} \supset X \supset X_\xi \supset Y$. Recall that for every extended thimble $\tilde{e}(\rho)$, its boundary is an embedded sphere in X_ξ. If we pick an orientation of the thimble, then its boundary gets also orientated, and may thus be regarded as an $(n-1)$-cycle on X_ξ. This is called a *vanishing cycle*. The submodule of $H_{n-1}(X_\xi)$ generated by these cycles is called the *vanishing homology* (of the pair (X, X_ξ)); we shall denote it by V. Put

$$(u, \partial u) : \cup_\rho(\tilde{e}(\rho), \partial\tilde{e}(\rho)) \to (X, X_\xi),$$

and look at the commutative diagram below:

$$\begin{array}{ccc} \oplus_\rho H_n(\tilde{e}(\rho), \partial e(\rho)) & \xrightarrow{\partial_*} & \oplus_\rho \tilde{H}_{n-1}(\partial\tilde{e}(\rho) \\ \downarrow u_* & & \partial u_* \downarrow \\ H_n(X, X_\xi) & \xrightarrow{\partial_*} & H_{n-1}(X-\xi). \end{array}$$

The vanishing homology is just the image of the right vertical arrow ∂u_*. It follows from (2.10) that the left vertical arrow u_* is surjective. Since the upper boundary map is an isomorphism, it follows that V is equal to the image of the lower boundary map. In particular, V only depends on the pair (X, X_ξ).

Now consider the diagram

$$\begin{array}{ccccccc} H_n(X, X_\xi) & \xrightarrow{\partial_*} & H_{n-1}(X_\xi) & \xrightarrow{i_*} & H_{n-1}(X) & \longrightarrow & 0 \\ & & \| & & & & \\ & & H_{n-1}(X_\xi) & \xleftarrow{i^!} & H_{n+1}(X) & & \\ & & PD \uparrow\cong & & PD \uparrow\cong & & \\ H^n(X, X_\xi) & \xleftarrow{\partial^*} & H^{n-1}(X_\xi) & \xleftarrow{i^*} & H^{n-1}(X) & \longleftarrow & 0. \end{array}$$

The upper and lower rows are exact (the zeroes in these sequences come from the vanishing of $H_{n-1}(X, X_\xi)$ and $H^{n-1}(X, X_\xi)$). We see that $V = \ker(i_*)$. We define the *invariant homology* (of the pair (X, X_ξ)) as the image of $i^! : H_{n+1}(X) \to H_{n-1}(X_\xi)$. These are the classes which are representable as intersections of X_ξ with X_ξ-transversal $(n+1)$-cycles on X.

(3.4) Lemma. *A class in $H_{n-1}(X_\xi)$ is invariant iff its intersection product with every vanishing cycle is zero.*

Proof. Let $V^\perp$ denote the set of classes in $H_{n-1}(X_\xi)$ whose intersection product with every vanishing cycle is zero. Since $i^!$ is the adjoint of i_* the image of $i^!$ and the perp of $\ker i_*$ coincide after tensoring with $\mathbb{Q}$. This means that $I \subset V^\perp$ and that the quotient is torsion. But this quotient can be identified with a subgroup of $H^n(X, X_\xi)$, which, by (2.11), is torsion free. So $I = V^\perp$. □

(3.5) Proposition. *The following six properties are equivalent:*

1. $V_{\mathbb{Q}} \cap I_{\mathbb{Q}} = 0$,
2. $H_{n-1}(X_\xi;\mathbb{Q}) = V_{\mathbb{Q}} \oplus I_{\mathbb{Q}}$,
3. $V_{\mathbb{Q}}$ *is a nondegenerate subspace of* $H_{n-1}(X_\xi;\mathbb{Q})$,
4. $I_{\mathbb{Q}}$ *is a nondegenerate subspace of* $H_{n-1}(X_\xi;\mathbb{Q})$,
5. i_* *maps* $I_{\mathbb{Q}}$ *isomorphically onto* $H_{n-1}(X;\mathbb{Q})$,
6. $i_*i^! : H_{n+1}(X;\mathbb{Q}) \to H_{n-1}(X;\mathbb{Q})$ *is an isomorphism.*

Proof. The mutual equivalence of the first four properties is elementary.

To see that (1) $\Leftrightarrow$ (5), observe that $\dim H_{n-1}(X;\mathbb{Q}) = \dim I_{\mathbb{Q}}$ (see the large diagram in (3.3)). Then note that $V_{\mathbb{Q}} \cap I_{\mathbb{Q}}$ is the kernel of $i_*|I_{\mathbb{Q}}$

Finally, $i_*|I_{\mathbb{Q}}$ is an isomorphism iff $i_*i^!$ is. □

(3.6) Theorem (Hard Lefschetz theorem). *The equivalent assertions of the preceding proposition are all true.*

Lefschetz asserted this theorem in his famous monograph *l'Analysis Situs et la Géometrie Algebrique* (1924). It is called this way, because no one has been able to understand his (geometric) proof. The first accepted proof was not geometric, but analytic in character (based on the harmonic representation of the complex cohomology), and is largely due to Hodge. Another proof was given in the seventies by Deligne [**7**]. He first proves it for varieties defined over a finite field, and then invokes a comparison theorem to pass to the characteristic zero case. It is geometric, but in a rather roundabout way. In this course we shall follow a third approach (mainly due to P. Deligne, S. Zucker and M. Saito) which uses a mixture of geometric and analytic tools. It has the advantage that it is prototypical for a generalization to the case of singular varieties (with intersection homology replacing homology). In the remainder of this lecture we derive some consequences of the hard Lefschetz theorem (we indicate this dependence by an asterisk).

(3.7) Theorem*. *For $k \geq 0$, the maps*

$$\eta_X^k \cap : H_{n+k}(X;\mathbb{Q}) \to H_{n-k}(X;\mathbb{Q}), \quad \eta_X^k \cup : H^{n-k}(X;\mathbb{Q}) \to H^{n+k}(X;\mathbb{Q})$$

are isomorphisms.

In more geometric terms: intersecting $(n+k)$-cycles on X transversally with a fixed codimension k linear subspace of $\mathbb{P}$ defines an isomorphism of $H_{n+k}(X;\mathbb{Q})$ onto $H_{n-k}(X;\mathbb{Q})$.

Proof. The second assertion follows from the first by applying Poincaré duality, so we only need to prove the first assertion. For this we choose transversal linear sections $X_0 \subset X_1 \subset X$ of dimension $n-k$ resp. $n-k+1$. Name the inclusions as follows: $i : X_0 \subset X_1$, $i_1 : X_1 \subset X$, $i_0 = i_1 i : X_0 \subset X$. Then $\eta_X^k = i_{0!}(1)$, and so $\eta_X^k \cap = i_{0!}(1)\cap = i_{0*}i_0^! = i_{1*}i_*i^!i_1^!$. If we compose $i_1^! : H_{n+k}(X) \to H_{n-k+2}(X_1)$ on both sides with the Poincaré duality isomorphisms we get by definition the natural map $i_1^* : H^{n-k}(X) \to H^{n-k}(X_1)$, and (2.11) implies that this is an isomorphism. Similarly, (2.9) implies that $i_{1*} : H_{n-k}(X_1) \to H_{n-k}(X)$ is an isomorphism. It remains to see that $i_*i^!$ is an isomorphism, at least if we tensor with $\mathbb{Q}$. But this is just equivalent to the hard Lefschetz theorem for X_1 and its hyperplane section X_0). □

Let us simply write H^r for $H^r(X,\mathbb{Q})$ and η for η_X. Define for $k \geq 0$ a bilinear $(-)^{n-k}$-symmetric form on H^{n-k} by

$$\Psi : H^{n-k} \times H^{n-k} \to \mathbb{Q}, \quad \Psi(\alpha,\beta) := \langle \alpha \cup \beta \cup \eta^k, [X] \rangle.$$

The preceding theorem implies that Ψ is nondegenerate. Note also that $\eta\cup : H^{n-k-2} \to H^{n-k}$ respects this form, so that its image will be a nondegenerate subspace.

We now define the *primitive cohomology* (resp. *primitive homology*) by

$$P^{n-k}(X) := \ker(\eta^{k+1}\cup : H^{n-k} \to H^{n+k+2}),$$
$$P_{n+k}(X) := \ker(\eta^{k+1}\cap : H_{n+k} \to H_{n-k-2}).$$

(3.8) Lemma*. *$P^{n-k}(X)$ is the perp of $\eta \cup H^{n-k-2}$ with respect to Ψ; in particular, it is a nondegenerate subspace of H^{n-k}.*

Proof. Let $\alpha \in H^{n-k}$. Then $\alpha \in P^{n-k}(X) \Leftrightarrow \alpha \cup \eta^{k+1} = 0 \Leftrightarrow \alpha \cup \eta^{k+1} \cup \beta = 0$ for all $\beta \in H^{n-k-2} \Leftrightarrow \Psi(\alpha, \eta \cup \beta) = 0$ for all $\beta \in H^{n-k-2}$. □

(3.9) Corollary* (Lefschetz decomposition). *We have a natural isomorphism of $\mathbb{Q}[\eta]$-modules*

$$H^\bullet(X;\mathbb{Q}) \cong \oplus_{k=0}^n P^{n-k}(X)[\eta]/(\eta^{k+1}).$$

This decomposition is orthogonal with respect to Ψ.

Proof. In view of (3.7) it suffices to show that for $0 \leq q \leq n$ we have an orthogonal decomposition

$$H^q X;\mathbb{Q}) = P^q(X) \perp (\eta \cup P^{q-2}(X)) \perp (\eta^2 \cup P^{q-4}(X)) \perp \cdots$$

But this follows easily from the preceding lemma. □

Remark. The endomorphism of $H^\bullet(X;\mathbb{Q})$ defined by $\eta_X\cup$ is often denoted by L. If B is the endomorphism of $H^\bullet(X;\mathbb{Q})$ which multiplies classes in degree q by $n-q$, then it is easily checked that $[L,B] = -2L$. The Lefschetz decomposition implies the existence and uniqueness of an operator Λ of degree -2 with the property that $[\Lambda, L] = B$. This means that the $\mathbb{Q}$-span of L, B and Λ is closed under the bracket, and that as a Lie algebra it is isomorphic to $sl_2(\mathbb{Q})$ via

$$\Lambda \leftrightarrow \begin{pmatrix} 0 & 1 \\ 0 & 0 \end{pmatrix} \quad B \leftrightarrow \begin{pmatrix} 1 & 0 \\ 0 & -1 \end{pmatrix} \quad L \leftrightarrow \begin{pmatrix} 0 & 0 \\ -1 & 0 \end{pmatrix}.$$

Thus $H^\bullet(X;\mathbb{Q})$ becomes a representation space of $sl_2(\mathbb{Q})$. The summands of the Lefschetz decomposition are just its isotypical components: the subrepresentation generated by any nonzero $a \in P^{n-k}(X)$ is irreducible of dimension $k+1$ (and equal to the span of $a, \eta \cup a, \ldots \eta_X^k \cup_X a$). It is not known whether there exists an algebro-geometric definition for Λ as there is for L (more precisely, the question is whether Λ can defined by an algebraic cycle on $X \times X$). One of Grothendieck's 'standard' conjectures asserts that this should be the case.

LECTURE 4
Monodromy

(4.1) Suppose we are given a locally trivial map of topological spaces $f : E \to B$. If $\gamma : [0,1] \to B$ is a path from $p \in B$ to $q \in B$, then the pull-back $\gamma^*E \to [0,1]$ is also locally trivial. A finite number of local trivializations of this map cover $[0,1]$ and they can be composed to give a global trivialisation. This defines a continuous family of homeomorphisms $\{h_t : E_p \to E_{\gamma(t)}\}_{0\le t\le 1}$ such that h_0 is the identity. The family is not unique, but the isotopy class of each h_t is. In particular, we get a well-defined isotopy class of homeomorphisms $E_p \to E_q$. Changing γ within its homotopy class (modulo its end points), does not change this isotopy class. Furthermore, composition of homotopy classes of paths corresponds to composition of isotopy classes. In particular, we get a homomorphism from the fundamental group $\pi_1(B,p)$ to the group of isotopy classes of self-homeomorphisms of E_p (we adopt the convention that the product uv in $\pi_1(B,p)$ stands for: traverse first v, then u.) This homomorphism is called the *geometric monodromy* of f relative p. Isotopy classes of of self-homeomorphisms of E_p act on the homology and the cohomology of E_p. The ensueing representation of $\pi_1(B,p)$ on $H_k(E_p)$ or $H^k(E_p)$ is called the *monodromy representation* (of f relative p).

Let us now consider the special case that B is the unit circle S^1. So if F denotes the fiber E_1, then we can find a continuous family of homeomorphisms $\{h_\theta : F \to E_{\exp(\sqrt{-1}\theta)}\}_{0\le\theta\le 2\pi}$, such that h_0 is the identity and $h := h_{2\pi}$ represents the geometric monodromy corresponding to the natural generator of $\pi_1(S^1,p)$; in this case we will simply refer to h as the geometric monodromy of f.

Suppose now that we are given a subspace $\dot{E} \subset E$ plus a trivialisation $u : \dot{E} \to \dot{F}$ of $f|\dot{E}$. Assume that the local trivialisations of f can be chosen to be compatible with u. Such local trivialisations will yield an h which is the identity on $\dot{F}$. Its relative isotopy class relative $\dot{F}$ is again independent of choices. This relative isotopy class remains unaltered if we move u inside the same homotopy class of trivialisations. There is now a well-defined action of h on the long exact (co)homology sequence of the pair $(F,\dot{F})$. But we can extract finer homological information out of h: if z is a k-chain on F whose boundary is supported by $\dot{F}$, then $h_*(z)$ has the same boundary as z, so that $h_*(z) - z$ is a k-cycle. This induces a homomorphism

$$\mathrm{var}(h)_* : H_k(F,\dot{F}) \to H_k(F),$$

called the *variation homomorphism*. It determines the action of h on both $H_k(F)$ and $H_k(F, \dot{F})$, for if $j_* : H_k(F) \to H_k(F, \dot{F})$ is the natural map, then these actions are given by $\mathrm{var}(h)_* j_*$ and $j_* \mathrm{var}(h)_*$ respectively.

(4.2) Now let us go back to the model $f : B \to D$ of the quadratic function introduced in (2.4). In the discussion that follows we are more concerned with the fibers of f then with its total space, which is why we find it convenient to let n denote the fiber dimension (so that B has complex dimension $n+1$). We will also assume that $n \geq 1$. Recall that $f|\dot{B}$ is trivial. A trivialisation of $f|\dot{B}$ is unique up to homotopy (because D is contractible), and so we have a well-defined isotopy class of trivialisations of $\dot{B}_{\partial D} \to \partial D$. The map $(B_{\partial D}, \dot{B}_{\partial D}) \to \partial D$ is a locally trivial fibration of manifolds with boundary, and local trivialisations can be chosen to be compatible with a given trivialisation of $\dot{B}_{\partial D} \to \partial D$. So the preceding discussion applies to this situation and we find a geometric monodromy $h : B_\eta \to B_\eta$ which is the identity on $\dot{B}_\eta$. In (2.4) we constructed a diffeomorphism of B_η onto the unit disk bundle of the n-sphere. Via this diffeomorphism one can give an explicit representative of h (cf. Lamotke [**12**]), but we shall not do this here. We content ourselves with a formula for the associated variation homomorphism. If we orient the thimble e, we get an orientation of its boundary; we denote the corresponding (vanishing) cycle by Δ and its class in $H_n(B_\eta)$ by $[\Delta]$. With respect to the natural orientation of the tangent bundle of S^n, the self-intersection of its zero-section is just the Euler characteristic of S^n, i.e., $1+(-1)^n$. A short computation shows that the diffeomorphism of B_η onto the unit disk bundle in TS^n has degree $(-1)^{\frac{1}{2}n(n-1)}$, so that

$$[\Delta] \cdot [\Delta] = (-1)^{\frac{1}{2}n(n-1)}(1 + (-1)^n).$$

It is clear that $H_k(B_\eta)$ is trivial for $k \neq 0, n$ and is freely generated by $[\Delta]$ for $k = n$. Let Δ^* be an n-chain on B_η which under the diffeomorphism corresponds to a fibre of the unit disk bundle over S^n, and is oriented in such a way that $\Delta \cdot \Delta^* = 1$ (the dot product is here taken with respect to the complex orientation of B_η). We let $[\Delta^*]$ denote its class in $H_n(B_\eta, \dot{B}_\eta)$. We have $H_k(B_\eta, \dot{B}_\eta) \cong H^{2n-k}((B_\eta)$ and so this group is zero for $k \neq n, 2n$, and is freely generated by $[\Delta^*]$ for $k = n$. So the variation homomorphism $\mathrm{var}(h)_* : H_k(B_\eta, \dot{B}_\eta) \to H_k(B_\eta)$ is necessarily trivial for $k \neq n$. For $k = n$ it must map $[\Delta]^*$ to a scalar multiple of $[\Delta]$. The coefficient was computed by Lefschetz:

(4.3) Proposition. *$var(h)_*[\Delta^*] = -(-1)^{\frac{1}{2}n(n-1)}[\Delta]$.*

Now let us put ourselves in a situation comparible to that of lemma (2.6): we assume we are given a space Z containing B and a continuous mapping $F : Z \to D$ extending f. It is assumed that the pair $(\dot{Z} := Z - (B - \dot{B}), \dot{B})$ is locally trivial over D. We further assume that the situation is reasonable: we want the inclusion $(B_\eta, \dot{B}_\eta) \subset (Z_\eta, \dot{Z}_\eta)$ to be excisive. Since F is locally trivial over ∂D, there is defined a geometric monodromy $\sigma : Z_\eta \to Z_\eta$. We can (and will) take it to be h on B_η and the identity on $\dot{Z}_\eta$.

(4.4) Corollary (The Picard–Lefschetz formula's). *The monodromy σ acts on the cohomology of Z_η as the identity in degrees $\neq n$; in degree n it is given by*

$$\sigma_*(a) = a - (-1)^{\frac{1}{2}n(n-1)}(\delta \cdot a)\delta \quad \textit{resp.} \quad \sigma^*(\alpha) = \alpha - (-1)^{\frac{1}{2}n(n-1)}\langle \delta, \alpha \rangle \delta^*,$$

where $\delta \in H_n(Z_\eta)$ denotes the class of Δ in $H_n(Z_\eta)$, and $\delta^* \in H^n(Z_\eta)$ is the class that assigns to a Δ-transversal n-cycle z the value $\Delta \cdot z$.

Proof. We can factorize $\sigma_* - \mathbf{1}$ as follows:

$$H_k(Z_\eta) \to H_k(Z_\eta, \dot{Z}_\eta) \cong H_k(B_\eta, \dot{B}_\eta) \xrightarrow{\mathrm{var}(h)_*} H_k(B_\eta) \to H_k(Z_\eta),$$

where the extremal maps are the natural ones. Since $\mathrm{var}(h)_*$ is trivial in degree $k \neq n$, it follows that σ_* is the identity in these degrees. Now let $a \in H_n(Z_\eta)$. Its image in $H_k(B_\eta, \dot{B}_\eta)$ will be multiple λ of $[\Delta^*]$. Using $[\Delta] \cdot [\Delta^*] = 1$, we see that $\lambda = [\Delta] \cdot \lambda[\Delta^*] = \delta \cdot a$. According to (4.2) the image of $[\Delta^*]$ in $H_n(Z_\eta)$ is equal to $-(-1)^{\frac{1}{2}n(n-1)}\delta$. The formula for σ_* now follows. The proof of the formula for σ^* is similar. □

(4.5) In the case that Z is part of a complex $(n+1)$-manifold, then the dot product on $H_(Z_\eta)$ is preserved by the monodromy (for σ respects the complex orientation). This dot product is symmetric for even n and skew for odd n. When n is even, $\delta, \delta = \pm 2$ (in particular, δ is not a torsion class) and σ_* is the reflection with respect to δ relative the dot product. When n is odd, the dot product is skew and σ_* is a symplectic transvection; in this case it could happen that δ is torsion.

(4.6) It is not difficult to deduce from our assumptions that $Z_0 \subset Z$ is a deformation retract. We also know that $Z_\eta \cup e \subset Z$ is a deformation retract, so that $H_k(Z, Z_\eta) \cong H_k(e, \partial e) \cong \tilde{H}_{k-1}(\partial e)$ is trivial for $k \neq n+1$ and free of rank one for for $k = n$ (in that case, $[\Delta]$ determines a generator). If we substitute this in the exact sequence of the pair (Z, Z_η) we get:

$$H_k(Z_0) \cong H_k(Z) \cong H_k(Z_\eta), \quad \text{for } k \neq n, n+1$$

and an exact sequence

$$0 \to H_{n+1}(Z_\eta) \to H_{n+1}(Z_0) \to \mathbb{Z} \to H_n(Z_\eta) \to H_n(Z_0) \to 0.$$

The image of the generator of $\mathbb{Z}$ in $H_n(Z_\eta)$ is δ. It follows that $H_n(Z_0; \mathbb{Q})$ can be identified with the cokernel of $\sigma_* : H_n(Z_\eta; \mathbb{Q}) \to H_n(Z_\eta; \mathbb{Q})$. This is the largest quotient of $H_n(Z_\eta; \mathbb{Q})$ on which σ_* acts trivially, and is for that reason called the *space of co-invariants* of σ in $H_n(Z_\eta; \mathbb{Q})$.

We now distinguish two cases according to whether δ is torsion:

The usual case: $\delta_{\mathbb{Q}} \neq 0$. Then we have a short exact sequence

$$0 \to \mathbb{Q}\delta \to H_n(Z_\eta; \mathbb{Q}) \to H_n(Z_0; \mathbb{Q}) \to 0$$

and $H_{n+1}(Z_0; \mathbb{Q}) \cong H_{n+1}(Z_\eta; \mathbb{Q})$. So $H_\bullet(Z_0; \mathbb{Q})$ can be identified with the space of σ-co-invariants in $H_\bullet(Z_\eta; \mathbb{Q})$

The special case: $\delta_{\mathbb{Q}} = 0$. This can only happen when n is odd. Then $H_n(Z_\eta; \mathbb{Q}) \cong H_n(Z_0; \mathbb{Q})$ and we have a short exact sequence

$$0 \to H_{n+1}(Z_\eta; \mathbb{Q}) \to H_{n+1}(Z_0; \mathbb{Q}) \to H_{n+1}(e, \partial e; \mathbb{Q}) \to 0$$

There is parallel cohomological discussion which we do not bother to explicate. For instance,

$$H^k(Z_0) \cong H^k(Z) \cong H_k(Z_\eta), \quad \text{for } k \neq n, n+1,$$

and $H^n(Z_0; \mathbb{Q})$ can be identified with the space of σ-invariants in $H^n(Z_\eta; \mathbb{Q})$.

(4.7) We can now explain why in (3.3) the image of $i^!$ was called the invariant homology. Let in the situation of (2.7) $\tilde{\delta}(\rho) \in H_{n-1}(X_\xi)$ be the class defined by an orientation of $\partial\tilde{e}(\rho)$. Put $U := L - X^\vee$, and let $\gamma(\rho)$ be the loop in U based at ξ that traverses first $\Gamma(\rho)$ from ξ to $\eta(\rho)$, next encircles $\xi(\rho)$ along $\partial D(\rho)$ in the counter clockwise direction and then goes back again to ξ via $\Gamma(\rho)$. The monodromy along $\gamma(\rho)$ can only be nontrivial in degree n, and in that degree it is given by $a \mapsto a \pm (\tilde{\delta}(\rho) \cdot a)\tilde{\delta}(\rho)$. So a is invariant under all these monodromy transformations iff $\tilde{\delta}(\rho) \cdot a = 0$ for all ρ. This in turn, is according to (3.4) equivalent to: $a \in I$. Notice that the images of these loops in $\pi_1(U, \xi)$ generate the latter, so we may restate the preceding by saying that I is the set of classes in $H_{n-1}(X_\xi)$ that are left invariant by the monodromy action of $\pi_1(U, \xi)$.

We mention two other properties equivalent to those of (3.5). The proofs are not difficult, but we won't discuss them here (they can be found for instance in the U. of U.-notes).

(4.8) Proposition. *Each of the properties of (3.5) is also equivalent to*

1. *$V_\mathbb{C}$ is a simple representation of $\pi_1(U, \xi)$ (i.e., $V_\mathbb{C}$ has no nonzero proper $\pi_1(U, \xi)$-invariant subspace) and is not the one-dimensional trivial representation, or*
2. *$H_{n-1}(X_\xi; \mathbb{C})$ is a semi-simple representation of $\pi_1(U, \xi)$ (this means that every $\pi_1(U, \xi)$-invariant subspace of $H_{n-1}(X_\xi; \mathbb{C})$ has a $\pi_1(U, \xi)$-invariant complement).*

Remark. One can show that for $d > 0$ sufficiently large, in any general Lefschetz pencil of degree d-hypersurface sections of X the special case of (4.6) does not occur (see P. Deligne, Lemme 3.3.7 in Exposé XV of [**8**]). (Perhaps we should point out that a degree d-hypersurface section of X becomes a hyperplane section if we embed X in a higher dimensional projective space by means of the d-fold Veronese map.)

LECTURE 5
The Leray Spectral Sequence of a Lefschetz Fibration

(5.1) We begin with a definition. Let M be a locally contractible space. A *local system* over M is a sheaf on M of finite dimensional $\mathbb{Q}$-vector spaces which is locally constant. Local systems are closed under taking pull-back and the binary operations $\otimes$ and $\mathcal{H}om$. In particular, if $\mathcal{F}$ is a local system, then so is its dual $\mathcal{F}^\vee := \mathcal{H}om(\mathcal{F}, \mathbb{Q}_M)$.

If $\gamma : [0,1] \to M$ is a path from p to q, then $\gamma^*\mathcal{F}$ is locally constant and hence constant. This determines an isomorphism

$$h_\gamma : \mathcal{F}_p \cong (\gamma^*\mathcal{F})_0 \cong H^0([0,1], \gamma^*\mathcal{F}) \cong (\gamma^*\mathcal{F})_1 \cong \mathcal{F}_q.$$

This isomorphism only depends on the homotopy class of γ, and behaves in the expected manner with respect to composition and inverses of paths. So if M is path connected and p is a base point, then we have a representation of $\pi_1(M,p)$ on the stalk $\mathcal{F}_p$. Conversely, a representation of $\pi_1(M,p)$ on a finite dimensional $\mathbb{Q}$-vector space F determines a local system as follows: let $\tilde{M} \to M$ be the universal covering relative p with its action of $\pi_1(M,p)$ as deck transformations, and let $\pi_1(M,p)$ act on $F \times \tilde{M}$ diagonally (we give F the discrete topology). The orbit space is then the geometric realization of a local system on M. These two constructions are inverses of each other (up to a natural isomorphism), so that after a choice of base point p, giving a local system on M is essentially the same thing as a giving representation of $\pi_1(M,p)$ on a finite dimensional $\mathbb{Q}$-vector space.

Examples of local systems arise as follows. Let $f : Z \to M$ be a map of spaces The *q-th direct image* (of the constant sheaf $\mathbb{Q}_Z$ on Z) is the sheaf associated to the presheaf

$$U \overset{\text{open}}{\subset} M \mapsto H^q(f^{-1}U; \mathbb{Q})$$

One denotes this sheaf by $R^q f_* \mathbb{Q}_Z$. Since M is locally contractible, we have that the stalk at $p \in M$ is just

$$(R^q f_* \mathbb{Q}_Z)_p = \lim_{U \ni p} H^q(f^{-1}U; \mathbb{Q}) \cong H^q(Z_p; \mathbb{Q}).$$

If f happens to be locally trivial, then this sheaf is locally constant, and if moreover the fibres have finite qth Betti number, then we actually have a local system.

(5.2) In the remainder of this lecture, we let C be a compact connected smooth curve and $S \subset C$ a finite subset. We denote $C - S$ by U and name the inclusions

$i : S \subset C$ and $j : U \subset C$. We also choose a base point $\xi \in C$ and we put $\pi_1 := \pi_1(U, \xi)$.

Suppose $\mathcal{F}$ is a local system on U. The direct image $j_*\mathcal{F}$ restricted to U is $\mathcal{F}$ again. If $s \in S$ and D is small disk in C about s, and γ is the loop that traverses ∂D once in the positive direction (starting and ending in some $\eta \in \partial D$), then we have isomorphisms

$$j_*\mathcal{F}_s \xleftarrow{\cong} H^0(D, j_*\mathcal{F}) \xrightarrow{\cong} H^0(\partial D, j_*\mathcal{F}) \xrightarrow{\cong} (\mathcal{F}_\eta)^\gamma.$$

The following lemma gives a relatively concrete interpretation of the cohomology groups of $j_*\mathcal{F}$. (Since C has real dimension two, its cohomology groups in degree > 2 vanish.)

(5.3) Lemma. *We have natural isomorphisms*

$$H^0(j_*\mathcal{F}) \cong (\mathcal{F}_\xi)^{\pi_1},$$
$$H^1(j_*\mathcal{F}) \cong im(H^1_c(U, \mathcal{F} \to H^1(U, \mathcal{F}),$$
$$H^2(j_*\mathcal{F}) \cong (\mathcal{F}_\xi)_{\pi_1},$$

where $H^k_c(U, \mathcal{F})$ denotes cohomology with compact support, and the natural pairings

$$H^k(j_*\mathcal{F}) \times H^{2-k}(j_*\mathcal{F}^\vee) \to H^2(C; \mathbb{Q}) \cong \mathbb{Q}$$

are perfect ($k = 0, 1, 2$).

Proof. We have $H^0(j_*\mathcal{F}) = H^0(U, \mathcal{F})$ and it is not difficult to see that the restriction map $H^0(U, \mathcal{F}) \to \mathcal{F}_\xi$ is an isomorphism onto $(\mathcal{F}_\xi)^{\pi_1}$. We next invoke a standard exact sequence:

$$\cdots \to H^{k-1}(S, i^*j_*\mathcal{F}) \to H^k_c(U, \mathcal{F}) \to H^k(j_*\mathcal{F}) \to H^k(S, i^*j_*\mathcal{F}) \to \cdots.$$

Since $i^*j_*\mathcal{F}$ is a sheaf on the discrete set S it has no higher cohomology, and so $H^2_c(U, \mathcal{F}) \cong H^2(j_*\mathcal{F})$. Poincaré duality for $\mathcal{F}$ implies that

$$H^k(U, \mathcal{F}^\vee) \times H^{2-k}_c(U, \mathcal{F}) \to H^2_c(U, \mathbb{Q}) \cong \mathbb{Q} \tag{1}$$

is a perfect pairing. Since $H^0(U, \mathcal{F}^\vee) \cong H^0(j_*\mathcal{F}^\vee)$, it follows that the pairing in the lemma is perfect for $k = 0$. Applying these assertions to $\mathcal{F}^\vee$ yields the description of $H^2(j_*\mathcal{F})$ plus the assertion that the pairing is perfect for $k = 2$.

It remains to do the case $k = 1$. The exact sequence above gives a surjection

$$H^1_c(U, \mathcal{F}) \to H^1(j_*\mathcal{F}). \tag{2}$$

The Leray spectral sequence of $(j, \mathcal{F})$,

$$E_2^{p,q} := H^p(C, \mathcal{R}^q j_*\mathcal{F}) \Rightarrow H^{p+q}(U, \mathcal{F})$$

gives an injection

$$H^1(j_*\mathcal{F}) \to H^1(U, \mathcal{F}). \tag{3}$$

The composite of (2) and (3) is the natural map, and yields the description of $H^1(j_*\mathcal{F})$. Finally, the homomorphism $H^1_c(j_*\mathcal{F}) \to H^1(j_*\mathcal{F})$ equals minus its transpose with respect to the Poincaré pairing (1). Elementary linear algebra then implies that the Poincaré pairing induces a nondegenerate form on the image of this homomorphism. □

(5.4) Let now be given a closed irreducible subvariety Z of $C \times \mathbb{P}$ of dimension $n+1 \geq 2$ with the property that the projection $f : Z \to C$ has all its critical points nondegenerate and with distinct values. We let S be its set of critical values. We abbreviate $R^k f_* \mathbb{Q}_Z$ by $\mathcal{R}^k$, so that for every $t \in C$ the stalk $\mathcal{R}^k_t$ can be identified with $H^k(Z_t; \mathbb{Q})$. Since f is locally trivial over U, its restriction $j^* \mathcal{R}^k$, is a local system.

As for any sheaf on C, there is a natural homomorphism

$$\mathcal{R}^k \to j_* j^* \mathcal{R}^k.$$

Over U this is clearly an isomorphism, so let us explicate this homomorphism on the stalk at some $s \in S$. Choose a small disk D about s, $\eta \in \partial D$, and γ as before. Then the homomorphism $\mathcal{R}^k_s \to j_* j^* \mathcal{R}^k_s$ corresponds to the map

$$H^k(Z_s; \mathbb{Q}) \cong H^k(Z_D; \mathbb{Q}) \to H^k(Z_\eta; \mathbb{Q})^\gamma.$$

We encountered this map in (4.6); we showed there that it is an isomorphism for $k \neq n+1$, and that for $k = n+1$ it is either an isomorphism (the usual case) or a surjection with one-demensional kernel (the special case). We conclude:

(5.5) Lemma. *The natural homomorphism $\mathcal{R}^k \to j_* j^* \mathcal{R}^k$ is an isomorphism for $k \neq n+1$, and for $k = n+1$ it is surjective with kernel a (skyscraper) sheaf $\mathcal{K}$ whose support is a subset of S.*

Notice that $\mathcal{R}^k \to j_* j^* \mathcal{R}^k$ induces an isomorphism on cohomology in positve degree. The following theorem is essentially due to Deligne and Grothendieck (Thms. 5.6 and 5.6.8 of Exposé XVIII in [**8**]).

(5.6) Theorem. *If the hard Lefschetz theorem holds for the smooth fiber Z_ξ of f, (relative its embedding in $\mathbb{P}$), then the rational Leray spectral sequence*

$$E_2^{p,q} := H^p(C, \mathcal{R}^q) \Rightarrow H^{p+q}(Z; \mathbb{Q})$$

of f (with rational cofficients) degenerates and the natural map $\mathcal{R}^{n+1} \to j_* j^* \mathcal{R}^{n+1}$ has a canonical section.

In somewhat more concrete terms, this means the following: the rational cohomology of Z comes with a natural filtration, the *Leray filtration*:

$$0 \subset F_f^2 H^\bullet(Z; \mathbb{Q}) \subset F_f^1 H^\bullet(Z; \mathbb{Q}) \subset F_f^0 H^\bullet(Z; \mathbb{Q}) = H^\bullet(Z; \mathbb{Q})$$

and the theorem says that we have natural isomorphisms

$$\mathrm{Gr}^k_{F_f} H^\bullet(Z; \mathbb{Q}) \cong H^k(C, \mathcal{R}^{\bullet - k}), \quad k = 0, 1, 2.$$

The subspace $F_f^2 H^k(Z; \mathbb{Q})$ of $H^k(Z; \mathbb{Q})$ can be made explicit: the composite

$$\begin{aligned} &H^k(Z; \mathbb{Q}) \supset F_f^2 H^k(Z; \mathbb{Q}) \\ &\qquad \cong H^2(C, \mathcal{R}^{k-2}) \cong H^2(C, j_* j^* \mathcal{R}^{k-2}) \cong H^{k-2}(Z_\xi; \mathbb{Q})_{\pi_1} \leftarrow H^2(Z_\xi; \mathbb{Q}) \end{aligned}$$

is just the Gysin homomorphism of $Z_\xi \subset Z$, so that $F_f^2 H^k(Z; \mathbb{Q})$ is the image of this Gysin homomorphism. Similarly, the composite

$$\begin{aligned} &H^k(Z; \mathbb{Q}) \to \mathrm{Gr}^0_{F_f} H^k(Z; \mathbb{Q}) \\ &\qquad \cong H^0(C, \mathcal{R}^k) \to H^0(C, j_* j^* \mathcal{R}^k) \cong H^0(Z_\xi; \mathbb{Q})^{\pi_1} \subset H^k(Z_\xi; \mathbb{Q}) \end{aligned}$$

is given by restriction.

Proof of the theorem. Since $E_2^{p,q}=0$ unless $p=0,1,2$, we only need to show that

$$d_2 : H^0(\mathcal{R}^q) = E_2^{0,q} \to E_2^{2,q-1} = H^2(\mathcal{R}^{q-1})$$

is the trivial map ($q=1,2,\dots$).

Let $\eta_Z \in H^2(Z)$ be the pull-back of the hyperplane class of $\mathbb{P}$. Taking the cup product with this class defines a sheaf-homomorphism

$$L : \mathcal{R}^k \to \mathcal{R}^{k+2}.$$

Since L is already defined on the cochain level (for the cup product is), it will commute with the differentials of the spectral sequence. By assumption, $L^k : \mathcal{R}^{n-k} \to \mathcal{R}^{n+k}$ is an isomorphism at ξ ($k\geq 0$). Since this homomorphism is locally constant on U, this is in fact true over U. So if we introduce the *primitive sheaves*

$$\mathcal{P}^{n-k} := \ker(L^{k+1} : j^*\mathcal{R}^{n-k} \to j^*\mathcal{R}^{n+k+2}),$$

we get a Lefschetz decomposition

$$j^*\mathcal{R}^\bullet \cong \bigoplus_{k\geq 0} \mathbb{Q}[L]/(L^{k+1}) \otimes \mathcal{P}^{n-k}. \qquad (*)$$

This gives us the natural section of $\mathcal{R}^{n+1} \to j_*j^*\mathcal{R}^{n+1}$, namely

$$j_*j^*\mathcal{R}^{n+1} \overset{L^{-1}}{\cong} j_*j^*\mathcal{R}^{n-1} \cong \mathcal{R}^{n-1} \xrightarrow{L} \mathcal{R}^{n+1}$$

It also follows that $d_2^{0,n-k} : H^0(\mathcal{R}^{n-k}) \to H^2(\mathcal{R}^{n-k-1})$ kills the subspace $H^0(\mathcal{P}^{n-k}) \subset H^0(j_*j^*\mathcal{R}^{n-k}) = H^0(\mathcal{R}^{n-k}))$, for clearly $L^{k+1}d_2^{0,n-k} = d_2^{0,n-k}L^{k+1}$ does so and

$$L^{k+1} : H^2(\mathcal{R}^{n-k-1}) \cong H^2(j_*j^*\mathcal{R}^{n-k-1}) \overset{L^{k+1}}{\cong} H^2(j_*j^*\mathcal{R}^{n+k+1}) \cong H^2(\mathcal{R}^{n+k+1}).$$

If view of the Lefschetz decomposition $(*)$ and the fact that d_2 and L commute, it follows that d_2 is zero except possibly

$$d_2^{0,n+1} : H^0(\mathcal{R}^{n+1}) = E_2^{0,n+1} \to E_2^{1,n-1} = H^2(\mathcal{R}^n) \cong H^2(j_*j^*\mathcal{R}^n).$$

We deal with this case by invoking Poincaré duality on both Z and the smooth fibers of f. The kth Betti number of Z is given by $\dim E_\infty^{0,k} + \dim E_\infty^{1,k-1} + \dim E_\infty^{2,k-2}$. If $(p,q) \neq (0,n+1),(2,n-1)$, then $\dim E_\infty^{p,q} = \dim E_2^{p,q} = \dim H^p(j_*j^*\mathcal{R}^q)$, whereas $\dim E_\infty^{2,n-1} = \operatorname{coker} d_2^{0,n+1}$. By Poincaré duality the nth and $(n+2)$nd Betti number of Z are equal. Writing this out gives

$$\begin{aligned}\dim H^0(j_*j^*\mathcal{R}^n) + &\dim H^1(j_*j^*\mathcal{R}^{n-1}) + \dim H^2(j_*j^*\mathcal{R}^{n-2})\\ &= \dim H^0(j_*j^*\mathcal{R}^{n+2}) + \dim H^1((j_*j^*\mathcal{R}^{n+1}) + \dim \operatorname{coker} d_2^{0,n+1}.\end{aligned}$$

Poincaré duality on the smooth fibers of f implies that $j^*\mathcal{R}^{n+k}$ can be identified with the dual of $j^*\mathcal{R}^{n-k}$, so that by (5.5) the spaces $H^0(j_*j^*\mathcal{R}^{n+k})$ and $H^2(j_*j^*\mathcal{R}^{n-k})$ can be regarded as each others dual. In particular, they have the same dimension. Since $L : j^*\mathcal{R}^{n-1} \to j^*\mathcal{R}^{n+1}$ is an isomorphism, we also have $\dim H^1(j_*j^*\mathcal{R}^{n-1}) = \dim H^1(j_*j^*\mathcal{R}^{n+1})$. If we feed this in the displayed equality, we find that $\dim \operatorname{coker} d_2^{0,n+1} = \dim H^2(j_*j^*\mathcal{R}^n)$. Hence $d_2^{0,n+1}$ is the zero map. □

It follows from previous theorem that $\mathrm{Gr}^0_{F_f} H^{n+1}(Z;\mathbb{Q}) \cong H^0(\mathcal{R}^{n+1})$ is canonically isomorphic to the direct sum of $H^0(\mathcal{K})$ and $H^0(j_*j^*\mathcal{R}^{n+1})$ so that there is a natural surjection $\mathrm{Gr}^0_{F_f} H^{n+1}(Z;\mathbb{Q}) \to H^0(\mathcal{K})$. Compose this surjection with the evident projection

$$H^\bullet(Z;\mathbb{Q}) \to H^{n+1}(Z;\mathbb{Q}) \to \mathrm{Gr}^0_{F_f} H^{n+1}(Z;\mathbb{Q}),$$

and denote the kernel by $'H^\bullet(Z;\mathbb{Q})$ (so $'H^k(Z;\mathbb{Q}) = H^k(Z;\mathbb{Q})$ if $k \neq n+1$). This subspace is invariant under the operator L. Let $F^k_f{}'H^\bullet(Z;\mathbb{Q}) := F^k_f H^\bullet(Z;\mathbb{Q}) \cap {}'H^\bullet(Z;\mathbb{Q})$. Then for all k and m we have

$$Gr^k_{F_f}{}'H^m(Z;\mathbb{Q}) \cong H^k(j_*j^*\mathcal{R}^{m-k}).$$

(5.7) Proposition. *The subspace $'H^\bullet(Z;\mathbb{Q})$ of $H^\bullet(Z;\mathbb{Q})$ is nondegenerate with respect to Poincaré duality, so that its perp (which is a subspace of $H^{n+1}(Z;\mathbb{Q})$) projects isomorphically onto $H^0(\mathcal{K})$. Moreover, the filtration $F^\bullet_f{}'H^\bullet(Z;\mathbb{Q})$ is self-dual, in the sense that $F^1_f{}'H^\bullet(Z;\mathbb{Q})$ and $F^2_f{}'H^\bullet(Z;\mathbb{Q})$ are each others annihilator.*

Proof. The Leray filtration on $H^\bullet(Z;\mathbb{Q})$ respects the cup product in the sense that $F^k_f \cup F^l_f \subset F^{k+l}_f$. So F^1_f and F^2_f are perpendicular to each other with respect to Poincaré duality. This implies that Poincaré duality induces pairings

$$Gr^k_{F_f}{}'H^m(Z;\mathbb{Q}) \times Gr^{2-k}_{F_f}{}'H^{2n+2-m}(Z;\mathbb{Q}) \to \mathbb{Q}.$$

We must show that these are perfect. But this follows from the fact that this is just the Poincaré duality map

$$H^k(j_*j^*\mathcal{R}^m) \times H^{2-k}(j_*j^*\mathcal{R}^{2n-m}) \to \mathbb{Q}$$

of lemma (5.5). □

(5.8) At this point we need the basic results of Hodge theory. We shall give a very brief review; good surveys (and much more) can be found in [**11**].

Giving a *Hodge structure of weight* $w \in \mathbb{Z}$ on a finite dimensional $\mathbb{Q}$-vector space H is simply giving a decomposition of its complexificion: $H_{\mathbb{C}} = \oplus_{p+q=w} H^{p,q}$ with the property that $\overline{H^{p,q}} = H^{q,p}$. (Here p and q run over the integers.) This is equivalent to giving a descending filtration $F^\bullet = \cdots F^p \supset F^{p+1} \cdots$ of $H_{\mathbb{C}}$ which begins with H and ends with $\{0\}$ such that $F^p \oplus \overline{F}^{w+1-p} \to H_{\mathbb{C}}$ is an isomorphism for all p; we call this the *Hodge filtration* of the Hodge structure. (One passes from one set of data to the other by means of the relations $F^p = \oplus_{p'\geq p} H^{p',w-p'}$ and $H^{p,w-p} = F^p \cap \overline{F}^{w-p}$.) There is an evident notion of a morphism of Hodge structures $\phi : H \to {}'H$: we want that ϕ maps $H^{p,q}$ to $'H^{p,q}$; this is equivalent to the apparently weaker condition that $\phi(F^p) \subset {}'F^p$ for all p, provided that the two Hodge structures have the same weight. There is also the more general notion of a morphism of bidegree (r, r), whose definition you will guess. Practically all the familiar operations with vector spaces have their analogue for Hodge structures, for the category of Hodge structures is abelian and comes with the binary operations Hom and $\otimes$.

A very simple Hodge structure of dimension one is the one-dimensional rational vector space $2\pi\sqrt{-1}\mathbb{Q}$ whose complexification $\mathbb{C}$ has been given bidegree $(-1,-1)$. This is called the Tate Hodge structure and is denoted $\mathbb{Q}(1)$. Note that complex conjugation acts on it as minus the identity. More generally, $\mathbb{Q}(k)$, $k \in \mathbb{Z}$, is the vector space $(2\pi\sqrt{-1})^k\mathbb{Q}$ with bidegree $(-k,-k)$, so that $\mathbb{Q}(k)\otimes\mathbb{Q}(l)\cong\mathbb{Q}(k+l)$. If we twist a weight m-Hodge structure H with $\mathbb{Q}(k)$: $H(k) := H\otimes\mathbb{Q}(k)$, then the result is a weight $(m-k)$-Hodge structure; the underlying complex space looks the same, and is the same if we are prepared to fix a square root of -1. But this one often wants to avoid, and for this reason Tate Hodge structures help us to do the bookkeeping of weights and the action of $\mathrm{Gal}(\mathbb{C}|\mathbb{R})$.

Given a Hodge structure H of weight w, then the *Weil operator* $J : H_\mathbb{C} \to H_\mathbb{C}$ is defined as multiplication by $\sqrt{-1}^{p-q}$ on $H^{p,q}$. This operator commutes with complex conjugation, so is real. A *polarization* of H is a morphism of Hodge structures $\psi : H\otimes H \to \mathbb{Q}(-w)$ such that

1. ψ is $(-)^w$-symmetric and
2. $(\alpha,\beta)\mapsto\psi(\alpha,J\bar\beta)$ is a positive Hermitean form.

The basic results due to Hodge assert among other things the following: if X is a projective manifold (of dimension n, say), then $H^w(X;\mathbb{Q})$ carries a functorial Hodge structure of weight w. The operator $L = \eta_X\cup$ of lecture (3) is actually a morphism of Hodge structures $H^w(X) \to H^{w+2}(X)(1)$. In particular, the cohomology group of top degree $H^{2n}(X;\mathbb{Q})$ is canonically isomorphic to $\mathbb{Q}(-n)$. If one defines $P_{n-k}(X)$ and Ψ as in that lecture—one does not need the hard Lefschetz theorem to make these definitions—then $\psi := (2\pi\sqrt{-1})^w\Psi$ is a polarization up to sign. This, of course, implies the nondegeneracy of Ψ and hence the hard Lefschetz theorem.

Griffiths found that if X varies in a holomorphic, C^∞ locally trivial family of projective manifolds X_t, then the Hodge structures on $H^w(X_t;\mathbb{Q})$ have some remarkable properties concerning their dependence on t. This motivated the following definition:

Let M be a complex manifold. A variation of Hodge structure (VHS) of weight w on M is going to be a local system $\mathcal{H}$ on M plus a Hodge structure of weight w on every stalk $\mathcal{H}_x$ that varies in a nice way. Here 'nice' means:

1. the Hodge filtration varies holomorphically, i.e., we are given a filtration of $\mathcal{O}_M\otimes\mathcal{H}$ by locally free $\mathcal{O}_M$-submodules $\mathcal{F}^\bullet$ which induce in each stalk a Hodge structure of weight m,
2. the Hodge filtration does not vary too much: if $\mathcal{H}$ is trivial over an open connected $U\subset M$, so that $\mathcal{F}^\bullet|U$ can be thought of as a holomorphic family $\{F_t^\bullet\}_{t\in U}$ of filtrations of $H := H^0(U;\mathcal{H})$, then for any a holomorphic section s of $\mathcal{F}^p|U$, given as $t\mapsto s(t)\in F_t^p$, the derivative of s at $t\in U$ maps inside F_t^{p-1}.

Now if $Z\subset M\times\mathbb{P}$ is a closed complex submanifold such that the projection $f : Z\to M$ is C^∞ locally trivial, then it is a fact that the higher direct images $\mathcal{R}^w f_*\mathbb{Q}_Z$ come naturally with a variation of Hodge structure of weight m.

A *polarization* of a VHS $\mathcal{H}$ of weight w is a homomorphism of local systems $\mathcal{H}\otimes\mathcal{H}\to\mathbb{Q}_M(-w)$ which induces on every stalk a polarization as above.

A key result in the abstract theory of variations of Hodge structure is the following theorem, due to Deligne (cases $k=0,2$) and Zucker (case $k=1$):

(5.9) Theorem. *Let $j : U\subset C$ be as before, and let $\mathcal{H}$ be a polarized variation of Hodge structure of weight w over U. Then the cohomology group $H^k(j_*\mathcal{H})$ carries a canonical polarized Hodge structure of weight $m+k$. For $k=0$, the polarization* *arises from the isomorphism* $H^0(j_*\mathcal{H}) \cong (\mathcal{H}_\xi)^{\pi_1}$ *and the polarization on* $\mathcal{H}_\xi$, *for* $k = 2$ *the polarization is similarly defined (it comes from the isomorphism* $H^2(j_*\mathcal{H}) \cong (\mathcal{H}_\xi)_{\pi_1}(-1)$*) and for* $k = 1$ *it is given by* $H^1(j_*\mathcal{H}) \otimes H^1(j_*\mathcal{H}) \to H^2(C, \mathbb{Q}_C(-w)) \cong \mathbb{Q}(-w-1)$.

This theorem implies that for every $t \in U$, $\mathcal{H}_t^{\pi_1}$ is a Hodge substructure of $\mathcal{H}_t$, is constant in t (as a Hodge structure), and is nondegenerate relative the polarization. In view of property of (4) of (3.5) and the interpretation (4.7), we could regard this as an abstract version of the hard Lefschetz theorem.

We will also need the more general notion of mixed Hodge structure. A mixed Hodge structure on a finite dimensional $\mathbb{Q}$-vector space H consists of two filtrations: one increasing filtration $W_\bullet$ on H which begins with $\{0\}$ and ends with H (the *weight filtration*) and a decreasing filtration $F^\bullet$ of $H_\mathbb{C}$ (the *Hodge filtration*) which induces on every quotient W_m/W_{m-1} a Hodge structure $F^\bullet(W_m/W_{m-1})$ of weight m (here $F^p(W_m/W_{m-1})$ is the image of $F^p \cap W_{m,\mathbb{C}}$). Given mixed Hodge structures $(H, W_\bullet, F^\bullet)$ and $('H, 'W_\bullet, 'F^\bullet)$, then a morphism between them is simply a homomorphism $\phi : H \to 'H$ which sends W_m to $'W_m$ and F^p to $'F^p$. It is not-so-trivial fact from linear algebra that this defines an abelian category.

The raison d'être of this definition is the theorem of Deligne which asserts that for every complex-algebraic variety X of finite type, $H^k(X;\mathbb{Q})$ comes with a functorial mixed Hodge structure having $W_{-1} = 0$ and $W_{2k} = H^k(X;\mathbb{Q})$; if X is smooth resp. compact, then we even have $W_{k-1}H^k(X;\mathbb{Q}) = 0$ resp. $W_k = H^k(X;\mathbb{Q})$. He also proved such results for local cohomology groups, as long as they can be defined algebraically.

(5.10) We return to our general Lefschetz pencil $f : \tilde{X} \to L$, where X has dimension $n \geq 2$. We wish to determine the implications of theorem (5.6) and (5.7) for $H^\bullet(X;\mathbb{Q})$. We first show that $H^\bullet(X;\mathbb{Q})$ is a direct summand of $H^\bullet(\tilde{X};\mathbb{Q})$.

(5.11) Lemma. *We a have a natural isomorphism*

$$H^\bullet(\tilde{X};\mathbb{Q}) \cong H^\bullet(X;\mathbb{Q}) \perp H^{\bullet-2}(Y;\mathbb{Q})(-1)$$

of duality spaces. (The Tate twist (-1) *changes the sign of Poincaré pairing of* $H^\bullet(Y;\mathbb{Q})$.*)*

Proof. The projection $\pi : (\tilde{X}, \tilde{Y}) \to (X, Y)$ induces a morphism of long exact cohomology sequences. This morphism induces an isomorphism on the relative cohomology groups (for π is a relative homeomorphism), whereas the induced map $\pi^* : H^k(Y) \to H^k(\tilde{Y}) = H^k(Y \times L)$ is injective with cokernel isomorphic to $H^{k-2}(Y)$ (by the Künneth formula). This gives us the a short exact sequence

$$0 \to H^\bullet(X;\mathbb{Q}) \xrightarrow{\pi^*} H^\bullet(\tilde{X};\mathbb{Q}) \to H^{k-2}(Y;\mathbb{Q}) \to O.$$

The first map preserves the intersection form and admits $\pi_!$ as a perpendicular section. Using the fact that the normal bundle of $\tilde{Y} = Y \times L$ is the pull-back of $\mathcal{O}_L(-1)$, one may complete the proof. □

Let $V(Y)$ be the kernel of the Gysin map $H^{n-2}(Y;\mathbb{Q}) \to H^n(X_\xi;\mathbb{Q})$. Using Poincaré duality on Y and X_ξ, we see that this space can be identified with the kernel of $H_{n-2}(Y;\mathbb{Q}) \to H_{n-2}(X_\xi;\mathbb{Q})$, that is with the vanishing homology of the pair (X_ξ, Y). (This explains the notation.)

(5.12) Proposition. *Suppose that X and its transversal linear sections satisfy the hard Lefschetz theorem. Then we have a natural isomorphism of duality spaces*

$$H^1(\mathcal{R}^{n-1}f_*\mathbb{Q}_{\tilde{X}}) \perp H^0(\mathcal{K}) \cong P^n(X) \perp V(Y)(-1).$$

Proof. It follows from the discussion following the statement of (5.6) and (5.7), that the left-hand side is the perp of the image of $i_{\xi!} : H^{n-2}(X_\xi;\mathbb{Q}) \to H^n(\tilde{X})$ in the kernel of $i_\xi^* : H^n(\tilde{X}) \to H^n(X_\xi;\mathbb{Q})$ where $i_\xi : X_\xi \subset \tilde{X}$. We decompose $H^n(\tilde{X};\mathbb{Q})$ as

$$\begin{aligned} H^n(\tilde{X};\mathbb{Q}) =& H^n(X;\mathbb{Q}) \perp H^{n-2}(Y;\mathbb{Q})(-1) \\ =& P^n(X) \perp H^{n-2}(X_\xi;\mathbb{Q})(-1) \perp H^{n-2}(X_\xi;\mathbb{Q})(-1) \perp V(Y)(-1). \end{aligned}$$

One verifies that $i_{\xi!}$ corresponds to the diagonal embedding on the middle two summands, whereas i_ξ^* corresponds to taking the difference of the components of the middle summands followed by the isomorphism $H^{n-2}(X_\xi;\mathbb{Q})(-1) \cong H^n(X_\xi;\mathbb{Q})$. The proposition follows from this. □

Remember that with the notation of lecture (3),

$$H^0(\mathcal{K}) = \oplus_\rho \ker(H^n(X_{\xi(\rho)}) \to H^n(X_{\eta(\rho)})).$$

So in a sense the left hand side of (5.12) pertains to the cohomology of hyperplane sections of X only. What makes this proposition interesting is that on the other side of this equality appears as a direct summand that part of the cohomology of X that does not appear in its general hyperplane sections, i.e., $P^n(X)$.

This observation may serve as a point of departure for an inductive, though rather indirect, construction of Hodge structures on projective manifolds together with a proof of the hard Lefschetz theorem. The idea is this: suppose we have been able to develop (mixed) Hodge theory in dimension $< n$ (including the relative case of families of projective manifolds of dim $< n$). The induction step then requires us to prove the hard Lefschetz theorem for X and to construct (among other things) an intrinsic Hodge structure on $P^n(X)$ which is (up to a factor) polarized by the intersection product. The hard Lefschetz theorem follows from (5.9), as it implies that $H^{n-1}(X_\xi;\mathbb{Q})$ is a semi-simple representation of π_1. The last proposition puts the Hodge structure on $P^n(X)$. However, one does not know a priori it is intrinsic. This is only an apparent issue because one proceeds in fact in more specific way: from the outset one has attached to any variety a filtered resolution of the constant complex sheaf of which one wishes to show that the resulting filtration on its cohomology yields the Hodge filtration. The presence of the summand $H^0(\mathcal{K})$ forces us to develop mixed Hodge theory at the same time, although that could have been avoided in view of remark at the end of lecture (4).

The reason for discussing this rather roundabout way of developing Hodge theory is that it generalizes well: it is this approach that has been followed by M. Saito to prove that the intersection cohomology groups of complete irreducible varieties (to be discussed in the next lecture) enjoy similar properties as the cohomology of projective manifolds (i.e., that they have a Hodge structure, that the Hard Lefschetz theorem is valid, etc.). It is similar to the proofs of the corresponding results for varieties defined over finite fields that were obtained earlier by Beilinson–Berstein–Deligne. The germ of this idea is already present in Grothendieck's discussion of the Hodge conjecture.

LECTURE 6
Intersection Cohomology

(6.1) Let X be a locally compact Hausdorff space. A *stratification* $\mathcal{X}$ of X is a partition of X into connected topological manifolds (called *strata*) such that the closure of each stratum is a union of strata; the pair $(X, \mathcal{X})$ is then called a *stratified space*. We have $\dim X = \sup_{S \in \mathcal{X}} \dim S$. If X is a complex algebraic (or analytic) variety, then one usually wants the stratification to consist of smooth subvarieties (locally closed for the Zariski topology); if that is the case, we say that the stratification is *algebraic* (resp. *analytic*).

We will be only interested in stratified spaces that have a simple local structure. In order to state the relevant property, recall that the *open cone* over a space L is what one gets if in $L \times [0, \infty)$, the subspace $L \times \{0\}$ is identified to a point; we denote it by $c(L)$. If L is compact Hausdorff, then a stratification of L determines one of the open cone over it.

We say that a stratification $\mathcal{X}$ of X is *locally trivial* if for every stratum $S \in \mathcal{X}$ there exists a neighborhood U_S of S in X, a retraction $r_S : U_S \to S$, and a compact stratified space $(L_S, \mathcal{L}_S)$ with the the property that r_S is locally trivial (with local trivializations respecting the strata) such that any fiber (with its induced stratification) is homeomorphic to the open cone over $(L_S, \mathcal{L}_S)$. We actually want that the stratification $\mathcal{L}_S$ is also locally trivial; since $\dim L < \dim X$, this makes the definition inductive rather than circular.

It is clear that then every $p \in S$ will have a basis of neighborhoods homeomorphic to $\mathbb{R}^{\dim S} \times c(L_S)$; we shall call any such neighborhood a *standard neighborhood* of p.

The following theorem is due to Whitney.

(6.2) Theorem. *Any complex algebraic (resp. analytic) variety X admits an algebraic (resp. analytic) stratification which is locally trivial.*

Actually the stratifications produced by Whitney possess other interesting properties as well; one of them is a generalization of the Ehresmann fibration theorem due to Thom: if f is a proper morphism from X to a smooth variety M such that the restriction of f to any stratum is without critical points, then f is topologically locally trivial. We will refer to these stratifications as *Whitney stratifications*; for a precise definition, see for example [**9**].

The next theorem makes the definition of intersection cohomology possible; the one following it provides the main reason for its interest. Both are due to Goresky–MacPherson, with modifications in statement and proof due to Deligne.

(6.3) Theorem. *Let $(X, \mathcal{X})$ be a locally trivial stratified space. Assume that all the open strata all have the same dimension m, and let $\mathcal{F}$ be local system defined on their union U. Suppose furthermore that all strata have even codimension. Then there exists a complex of injective sheaves of $\mathbb{Q}_X$-modules on X,*

$$\mathcal{I}^\bullet : 0 \to \mathcal{I}^0 \to \mathcal{I}^1 \to \mathcal{I}^2 \to \cdots ,$$

plus a sheaf homomorphism $\alpha : \mathcal{F} \to \mathcal{I}^0|U$ such that the following three properties are satisfied:

1. *$0 \to \mathcal{F} \to \mathcal{I}^0|U \to \mathcal{I}^1|U \to \mathcal{I}^2|U \cdots$ is exact, in other words, $\mathcal{I}^\bullet|U$ is a resolution of $\mathcal{F}$,*
2. *(a finiteness property): if $V' \subset V$ are standard neighborhoods of $p \in S$, $S \in \mathcal{X}$, then the morphism of complexes*
$$\Gamma(V, \mathcal{I}^\bullet) \to \Gamma(V', \mathcal{I}^\bullet)$$
induces an isomorphism of cohomology, and
3. *in degrees $\geq \frac{1}{2}\operatorname{codim} S$, the homology of $\Gamma(V, \mathcal{I}^\bullet)$ is zero, whereas in degree $< \frac{1}{2}\operatorname{codim} S$ it maps isomorphically onto the cohomology of $\Gamma(V \setminus S, \mathcal{I}^\bullet)$.*

Moreover, the homotopy class of $\mathcal{I}^\bullet$ only depends on X and $\mathcal{F}$, not on $\mathcal{X}$.

It follows that in this situation the following spaces are all topological invariants of the pair $(X, \mathcal{F})$: the *intersection cohomology* of $(X, \mathcal{F})$:

$$IH^\bullet(X; \mathcal{F}) := H^\bullet(\Gamma(X, \mathcal{I}^\bullet)),$$

the *intersection cohomology with compact supports* of $(X, \mathcal{F})$:

$$IH_c^\bullet(X; \mathcal{F}) := H^\bullet(\Gamma_c(X, \mathcal{I}^\bullet)),$$

and for a closed subset $Y \subset X$, the relative *intersection cohomology* of $(X, \mathcal{F})$:

$$IH^\bullet(X, Y; \mathcal{F}) := H^\bullet(\Gamma_{X-Y}(X, \mathcal{I}^\bullet)).$$

(Here we are only taking sections of $\mathcal{I}^\bullet$ with support in $X - Y$.)

Examples. 1. If $X = U$, the we can take for $\mathcal{I}^\bullet$ an injective resolution of $\mathcal{F}$, and we find that intersection cohomology is the ordinary cohomology of $\mathcal{F}$: $IH^\bullet(X; \mathcal{F}) = H^\bullet(X; \mathcal{F})$.

2. More generally, if X is manifold and $j : U \subset X$ is the complement of a closed codimension two submanifold, then an injective resolution of $j_*\mathcal{F}$ has the required properties and we find that $IH^\bullet(X; \mathcal{F}) = H^\bullet(X; j_*\mathcal{F})$.

(6.4) Theorem. *Suppose that in the the situation of (6.3) U is oriented. Then we have a natural (Poincaré) pairing*

$$IH^k(X; \mathcal{F}) \times IH_c^{m-k}(X; \mathcal{F}^\vee) \to \mathbb{Q}$$

which is perfect.

So the Poincaré pairing puts the space $IH^\bullet(X, Y; \mathcal{F})$ in duality with the space $IH^\bullet(X - Y; \mathcal{F}^\vee)$.

In view of the second example, the last assertion of lemma (5.3) is a special case of the above theorem.

We list some other useful properties.

1. If $j : V \subset X$ is open, then we have restriction map $j^* : IH^\bullet(X;\mathcal{F}) \to IH^\bullet(V;\mathcal{F})$ and a pushforward $j_! : IH^\bullet_c(V;\mathcal{F}) \to IH^\bullet_c(X;\mathcal{F})$,
2. If $\pi : E \to X$ is locally trivial $\mathbb{R}^d$-bundle, then π and the zero section induces an isomorphism $IH^\bullet(E;\pi^*\mathcal{F}) \cong IH^\bullet(X;\mathcal{F})$.
3. Let us say that a closed subspace $i : Y \subset X$ is *transversally embedded of codimension d*, if it admits a neighborhood in X which retracts onto Y as an $\mathbb{R}^d$-bundle. Then properties (1) and (2) can be combined to yield a natural restriction map $i^* : IH^\bullet(X;\mathcal{F}) \to IH^\bullet(Y;i^*\mathcal{F})$. It fits in long exact sequence

$$\cdots \to IH^k(X,Y;\mathcal{F}) \to IH^k(X;\mathcal{F}) \to IH^k(Y;i^*\mathcal{F}) \to IH^{k+1}(X,Y;\mathcal{F}) \to \cdots.$$

   The map and the sequence dualize under Poincaré duality to give the *Gysin homomorphism* $i_! : IH^{\bullet-d}(Y;i^*\mathcal{F}) \to IH^\bullet(X;\mathcal{F})$ and the long exact sequence

$$\cdots \to IH^{k-1}(X-Y;\mathcal{F}) \to IH^{k-d}(Y;i^*\mathcal{F}) \to IH^k(X;\mathcal{F}) \to IH^k(X-Y;\mathcal{F}) \to \cdots.$$

   If Z is an arbitrary closed subspace of Y, there are also long exact sequences for the triple (X,Y,Z)
4. We have a Mayer-Vietoris sequence for open subsets of X. (More generally, one can set up a Cech spectral sequence for an open covering of X which converges to the intersection cohomology of $(X,\mathcal{F})$.)
5. If f is a continuous map to a space Y, then we have defined the higher direct image sheaf $\mathcal{R}^k f_*\mathcal{I}^\bullet$, $k = 0,1,\ldots$, associated to the presheaf

$$V \overset{open}{\subset} Y \mapsto IH^k(f^{-1}V;\mathcal{F})$$

   and there is a Leray spectral sequence

$$E_2^{p,q} := H^p(Y,\mathcal{R}^k f_*\mathcal{I}^\bullet) \Rightarrow IH^{p+q}(X;\mathcal{F}).$$

6. If $S \in \mathcal{S}$, and $p \in S$, then it follows from property (6.3.2) that $IH^k(V;\mathcal{F})$ is independent of the choice of standard neighborhood of V of p; we shall call this the local kth intersection cohomology group of $(X,\mathcal{F})$ at p.
7. Property (6.3.3) in combination with property (2) allows us to express the local intersection cohomology groups at $p \in S$ in terms of those of the link L_S: if V is a standard neighborhood of $p \in S$, then

$$IH^k(V;\mathcal{F}) = \begin{cases} IH^k(V \setminus S;\mathcal{F}) = IH^k(L_S;\mathcal{F}) & \text{if} \quad k \leq d-1; \\ 0 \quad \text{if } k \geq d. \end{cases}$$

   Applying this to $\mathcal{F}^\vee$ and dualizing gives

$$IH^k_c(V;\mathcal{F}) = \begin{cases} 0 \quad \text{if } k \leq m-d; \\ IH^{(k-m-1+d)+d}(L_S;\mathcal{F}) & \text{if} \quad k \geq m+1-d. \end{cases}$$

   So the first half of the intersection cohomology of L_S appears as intersection cohomology of V and the second half appears as intersection cohomology of V with compact supports.

We say that the stratified space $(X, \mathcal{X})$ *is of finite type* if there exists a compact $K \subset X$ such that $X - K$ is a s a stratified space homeomorphic to the product of a compact stratified space and an open interval. An algebraic Whitney stratification of a quasi-projective variety is always of finite type, but an analytic set need not admit such a stratification (take a countable discrete subset of $\mathbb{C}$).

We will sometimes use the notation $IH^\bullet(X; \mathcal{F})$ in a case where $\mathcal{X}$ some strata have odd codimension; we will only do this in case the strata of odd codimension make up a 'boundary' of X: their union $\dot{X}$ should be a closed subspace of X of codimension one and have a neighborhood in X homeomorphic to $\dot{X} \times [0, \infty)$ as stratified spaces (with the given stratification on $\dot{X}$ and $[0, \infty)$ stratified by $\{0\}$ and $(0, \infty)$). We then let $IH^\bullet(X) := IH^\bullet(X - \dot{X})$. Notice that if X is compact, then $X - \dot{X}$ is of finite type as a stratified space.

We can now state the weak Lefschetz theorem for intersection cohomology.

(6.5) Weak Lefschetz theorem WL(n). *Let X a closed analytic subset of an open convex subset Ω of a complex affine space which is of pure dimension n. Let $\mathcal{X}$ be Whitney stratification of X of finite type and let $\mathcal{F}$ be a local system defined on the union of its open strata. Then $IH^k(X; \mathcal{F}) = 0$ for $k > n$.*

(6.6) Corollary. *Let $X \subset \mathbb{P}$ be a projective variety of pure dimension n, $\mathcal{F}$ a local system on a Zariski open-dense subset of X, and $H \subset \mathbb{P}$ a hyperplane. Then $IH^k(X, X \cap H; \mathcal{F}) = 0$ for $k > n$. If moreover H is transversal to a Whitney stratification of X, then $i : X \cap H \subset X$ induces a restriction map $i^* : IH^k(X; \mathcal{F}) \to IH^k(X \cap H; \mathcal{F})$ which is an isomorphism for $k \leq n-2$ and is injective for $k = n-1$.*

Proof. The first statement follows from $WL(n)$ and the fact that

$$IH^k(X, X \cap H; \mathcal{F}) = IH^k_c(X - X \cap H; \mathcal{F}) \cong IH^k_c(X - X \cap H; \mathcal{F}^\vee)^*,$$

and the second from this and the exact sequence of the pair $(X, X \cap H)$. □

The proof of $WL(n)$ will be by induction on n. The induction process yields an interesting local companion result. To state it we need a bit of discussion first.

(6.7) Let X be an analytic set at $0 \in \mathbb{C}^N$ of pure dimension n, $\mathcal{X}$ a Whitney stratification of X, $\mathcal{F}$ a local system on the union of its open strata and $f : X \to \mathbb{C}$ an analytic function with $f(0) = 0$. We assume that f has in 0 an *isolated singularity* at 0 in the sense that there is a neigborhood Ω of $0 \in X$ such that for any stratum $S \in \mathcal{X}$, $f|S \cap (\Omega - \{0\})$ has no singular point. We construct what we shall call a *good model* for the germ of f at 0 as follows.

The assumptions (and the properties of a Whitney stratification) imply that $X_0 \cap (\Omega - \{0\})$ is transversally embedded in $X \cap (\Omega - \{0\})$ and that $\mathcal{X}$ induces a Whitney stratification on it. It is a well-known fact that for every smooth analytic subset T of Ω (locally closed for the Zariski topology), there exists an $\epsilon > 0$ such that $(0, \epsilon]$ does not contain a critical value of the restriction of the norm function $|z|$ to T. Choose $\epsilon > 0$ such that this is true for all strata of both $\mathcal{X}$ and $\mathcal{X}|X_0 \cap (\Omega - \{0\})$ and assume in addition that the closed ϵ-ball $B(\epsilon)$ is contained in Ω.

Now for every $S \in \mathcal{X}$, $\partial B(\epsilon) \cap S$ and $X_0 \cap S$ are submanifolds of S which meet transversally. So 0 is not a critical value of $f|\partial B(\epsilon) \cap S$. Choose $\eta > 0$ such that the closed η-disk $D(\eta)$ in $\mathbb{C}$ consists of regular values of $f|\partial B(\epsilon)\cap S$ for all S. Put

$$\begin{aligned} D &:= D(\eta), \\ B &:= B(\epsilon)\cap f^{-1}(D(\eta)), \\ \dot{B} &:= \partial B(\epsilon)\cap f^{-1}(D(\eta)), \\ \partial B &:= (\partial B(\epsilon)\cap f^{-1}(D(\eta)))\cup(B(\epsilon)\cap f^{-1}(\partial D(\eta))). \end{aligned}$$

On all these spaces we induce the stratification $\mathcal{X}$. The fiber $(B_\eta, \dot{B}_\eta)$ is a Whitney stratified space with boundary. We will call $f : B \to D$ a *good model* of the germ of f at 0.

It can be shown that B is homeomorphic to the closed cone over its boundary. This implies that

$$IH^k(B;\mathcal{F}) = \begin{cases} IH^k(\partial B;\mathcal{F}) & \text{if} \quad k\le n-1; \\ 0 \quad \text{if } k\ge n \end{cases}$$

and

$$IH^k_c(B;\mathcal{F}) = \begin{cases} 0 \quad \text{if } k\le n; \\ IH^{k-1}(L_S;\mathcal{F}) & \text{if} \quad k\ge n+1. \end{cases}$$

Thom's generalization of the Ehresmann fibration theorem implies that $\dot{B}$ and B are topologically locally trivial (in a stratified sense) over D and $D-\{0\}$ respectively. We therefore have a geometric monodromy $h : (B_\eta, \dot{B}_\eta) \to (B_\eta, \dot{B}_\eta)$ which is the identity on $\dot{B}_\eta$. We can even take it to be the identity on a neighborhood of $\dot{B}_\eta$ and then it is not hard to define a variation homomorphism for intersection cohomology

$$\mathrm{var}(h)^k : IH^k(B_\eta;\mathcal{F}) \to IH^k(B_\eta, \dot{B}_\eta;\mathcal{F})$$

using the sheaf complex $\mathcal{I}^\bullet$. We will give another definition below which is somewhat more useful for our purpose.

For this let ∂_+B, ∂_-B and ∂_0B be the set of points of ∂B where $\mathrm{Re}(f)$ is ≥ 0, ≤ 0 and $=0$ respectively. Define similarly subspaces ∂_+D, ∂_-D and ∂_0D of ∂D. Notice that the pair $(\partial_\pm B, \partial_0 B)$ is homeomorphic to the product $(B_{\pm\eta}, \dot{B}_{\pm\eta}) \times (\partial_\pm D, \partial_0 D)$, under a homeomorphism whose isotopy class is natural. So we have natural identifications

$$\begin{aligned} IH^k(\partial_\pm B;\mathcal{F}) &\cong IH^k(B_{\pm\eta};\mathcal{F}) = IH^k(B_{\pm\eta}-\dot{B}_{\pm\eta};\mathcal{F}), \\ IH^k(\partial_\pm B, \partial_0 B;\mathcal{F}) &\cong IH^{k-1}(B_{\pm\eta}, \dot{B}_{\pm\eta};\mathcal{F}) = IH_c^{k-1}(B_{\pm\eta}-\dot{B}_{\pm\eta};\mathcal{F}) \end{aligned}$$

Now consider the exact sequence of the pair $(\partial B, \partial_+B)$. Since we have

$$IH^{k+1}(\partial B, \partial_+B;\mathcal{F}) \cong IH^{k+1}(\partial_-B, \partial_0B;\mathcal{F}) \cong IH^k(B_{-\eta}, \dot{B}_{-\eta};\mathcal{F}),$$

the boundary map of this sequence can be identified with a homomorphism

$$IH^k(B_\eta;\mathcal{F}) \to IH^k(B_{-\eta}, \dot{B}_{-\eta};\mathcal{F}).$$

This is essentially the variation homomorphism: each semi-circle in ∂D connecting $-\eta$ with η gives a homeomorphism of $B_{-\eta}$ onto B_η, and the two coincide on $\dot{B}_{-\eta}$. The above map is induced by the difference of their actions on the intersection cohomology complex. So if we use one of them to identify $IH^k(B_\eta;\mathcal{F})$ with $IH^k(B_{-\eta};\mathcal{F})$, then we get the variation homorphism. The preceding discussion shows:

(6.8) Lemma. *We have a natural long exact sequence*

$$\cdots \to IH^k(\partial B;\mathcal{F}) \to IH^k(B_\eta;\mathcal{F}) \overset{\mathrm{var}(h)^k}{\longrightarrow} IH^k(B_\eta,\dot{B}_\eta;\mathcal{F}) \to IH^{k+1}(\partial B;\mathcal{F}) \to \dots.$$

We are now ready to state the local companion alluded to above

(6.9) Proposition VAR(n). *We have a natural isomorphism*

$$IH^k(B,\partial_+B;\mathcal{F}) \cong \mathit{Image}\big(\mathit{var}(h)^{k-1} : IH^{k-1}(B_\eta;\mathcal{F}) \to IH^{k-1}(B_\eta,\dot{B}_\eta;\mathcal{F})\big).$$

and both members are trivial for $k \neq n$. In particular, the monodromy action on $IH^k(B_\eta;\mathcal{F})$ is the identity if $k \neq n-1$. Finally,

$$IH^k(B_\eta;\mathcal{F})\begin{cases} =0 & \text{for } k \geq n;\\ \cong IH^k(\partial B;\mathcal{F}) & \text{for } k \leq n-2.\end{cases}$$

Remark. So the above proposition implies that if X is smooth, then $H^k(B_\eta;\mathcal{F}) = IH^k(B_\eta;\mathcal{F})$ is trivial unless $k = 0, n-1$. This is an old result of Milnor.

Proof that $WL(n-1) \Rightarrow VAR(n)$. The hypotheses of $WL(n-1)$ are fulfilled by the Whitney stratified space $B_\eta - \dot{B}_\eta$, and so $IH^k(B_\eta;\mathcal{F}) = 0$ for $k \geq n$ and $IH^k(B_\eta,\dot{B}_\eta;\mathcal{F}) = 0$ for $k \leq n-2$. We identify the variation homomorphism with the boundary map above, and factor this boundary map as

$$IH^k(\partial_+B;\mathcal{F}) \overset{\delta}{\longrightarrow} IH^{k+1}(B,\partial_+B;\mathcal{F}) \overset{\gamma}{\longrightarrow} IH^{k+1}(\partial B,\partial_+B;\mathcal{F}).$$

Either map fits in an exact sequence:

$$IH^k(\partial_+B;\mathcal{F}) \overset{\delta}{\longrightarrow} IH^{k+1}(B,\partial_+B;\mathcal{F}) \to IH^{k+1}(B),$$

$$IH^{k+1}(B,\partial B;\mathcal{F}) \to IH^{k+1}(B,\partial_+B;\mathcal{F}) \overset{\gamma}{\longrightarrow} IH^{k+1}(\partial B,\partial_+B;\mathcal{F}).$$

So δ is surjective if $k \geq n-1$ and γ is injective for $k \leq n$. The assertions of $VAR(n)$ all follow from this. □

Proof that $WL(n-1)+VAR(n) \Rightarrow WL(n)$. The proof will be very similar to the discussion (2.7). But here we must pay attention to the topological type because intersection homology is not a homotopy invariant. We need a technical result which we merely state:

(6.10) Technical lemma. *In the situation of $WL(n)$, there exists a compact convex set $C \subset \Omega$ with smooth boundary ∂C which is transversal to the stratification $\mathcal{X}$, an affine-linear form f, and a convex disk $D \subset \mathbb{C}$ with smooth boundary ∂D such that*

1. *with respect to the stratification induced on $\partial C \cap X$, $f|\partial C \cap X$ has no critical value in D,*
2. *with respect to the stratification induced on $\mathit{int}(C) \cap X$, $f|\mathit{int}(C) \cap X$ has only finitely many critical points over D, and these critical points have distinct values, which are all contained in $\mathit{int}(D)$,*
3. *if we put $B := C \cap f^{-1}D \cap X$, then $X - B$ is homeomorphic to $\partial B \times \mathbb{R}$ as stratified spaces.*

Let $S \subset \text{int}(D)$ be the finite set of critical values of $f : B \to D$. If x_s is the critical point over $s \in S$, then choose neighborhoods $B(s) \ni x_s$ in $X \cap \text{int}(C) \cap f^{-1}D$ and $D(s) \ni s$ in $\text{int}(D)$ such that $f : B(s) \to D(s)$ is a good model of the germ of $f|X$ at x_s. We also suppose that the disks $D(s)$ are disjoint. Choose a smoothly embedded disk $E \subset D$ with the property that for every $s \in S$, $E \cap D(s)$ is the right semi-disk $D(s)_+$ of $D(s)$. Put

$$B' := B_E \cup \bigcup_{s \in S} B(s), \quad \text{and} \quad B'' := \text{clos}(B_E \setminus \bigcup_{s \in S} B(s))$$

One proves that X is obtained from B' by putting an open collar over its boundary. So $IH^k(X; \mathcal{F}) \cong IH^k(B'; \mathcal{F})$. Similarly, if $\eta \in E - S$, then B'' is homeomorphic to B_η cross a disk. So $IH^k(B''; \mathcal{F}) \cong IH^k(B_\eta; \mathcal{F})$. We may apply $WL(n-1)$ to the relative interior of B_η and find that the latter group is zero for $k \geq n$. Furthermore,

$$IH^k(B', B''; \mathcal{F}) \cong \bigoplus_{s \in S} IH^k(B(s), \partial B(s)_+; \mathcal{F})$$

and according to $VAR(n)$, this group is trivial unless $k = n$. If we feed this in the exact sequence of the pair (B', B''), the assertion $WL(n)$ follows. □

We have only a partial analogue of the results mentioned in (4.6). Let C be a smooth curve, $s \in C$, $Z \subset \mathbb{P} \times C$ a closed subvariety of pure dimension n equipped with a Whitney stratification $\mathcal{Z}$ and $\mathcal{F}$ a local system over the union of open strata of $\mathcal{Z}$. Denote the projection $f : Z \to C$, and assume that f has exactly one critical point z over s. Let $B \ni z_0$ be a closed neighborhood of z_0 in Z such that $f : B \to f(B) := D$ is a good representative of the germ of f at z_0, and assume that D is so small that it does not contain any critical value besides t.

(6.11) Corollary. *The restriction map $IH^k(X_D; \mathcal{F}) \to IH^n(X_\eta; \mathcal{F})$ is an isomorphism for $k \neq n-1, n$ and we have an exact sequence*

$$\begin{aligned} 0 \to IH^{n-1}(Z_D; \mathcal{F}) \to & IH^{n-1}(Z_\eta; \mathcal{F}) \\ & \to IH^n(B, B_\eta; \mathcal{F}) \to IH^n(Z_D; \mathcal{F}) \to IH^n(Z_\eta; \mathcal{F}) \to 0. \end{aligned}$$

Proof. This follows from (6.10) and the fact that we have an isomorphism

$$IH^k(Z_D, Z_\eta, \mathcal{F}) \to IH^k(B, B_\eta; \mathcal{F}).$$

□

The first statement of this corollary can be restated as follows: the sheaf on C associated to the presheaf

$$(V \overset{\text{open}}{\subset} C) \mapsto IH^k(f^{-1}V; \mathcal{F})$$

is constant on D if $k \neq n-1$. In the missing dimension one would like to have an *invariant cycle theorem* stating that the image of the map $IH^{n-1}(Z_D; \mathcal{F}) \to IH^{n-1}(Z_\eta; \mathcal{F})$ is just the part that is pointwise fixed under the monodromy transformation. (We proved this in case Z is smooth at z_0 and f has there a nondegenerate singularity.) This is true in case $\mathcal{F}$ underlies a variation of Hodge structure. However, the proof is difficult, and appears as a byresult of the proof of the hard Lefschetz theorem for intersection cohomology.

Consider the situation of (6.7), where $X \subset \mathbb{P}$ is a projective variety of pure dimension n, $\mathcal{F}$ a local system defined on a smooth Zariski open-dense subset U of X and $H \subset \mathbb{P}$ be a hyperplane transversal with respect to U and a Whitney stratification of $X - U$. The inclusion $i_H : X \cap H \subset X$ induces

$$i^* : IH^{n-1}(X;\mathcal{F}) \to IH^{n-1}(X \cap H;\mathcal{F}), \quad i_! : IH^{n-1}(X \cap H;\mathcal{F}) \to IH^{n+1}(X;\mathcal{F}).$$

Generalizing terminology for the smooth case, we call the image of the first map the *invariant intersection cohomology* and the kernel of the second map the *vanishing intersection cohomology* of $X \cap H$. Then proposition (3.5) and its extension (4.8) hold also for this case. The hard Lefschetz theorem in the present context asserts that these equivalent properties all hold if the local system underlies a polarized VHS; it is due to Morihiko Saito:

(6.12) Theorem (Kähler package for intersection cohomology). *Let $X \subset \mathbb{P}$ be a projective variety of pure dimension n, $\mathcal{H}$ a polarized VHS of weight w on a smooth Zariski open-dense subset U of X and $H \subset \mathbb{P}$ be a hyperplane transversal with respect to U and a Whitney stratification of $X - U$. Then*

1. *$IH^k(X;\mathcal{H})$ carries a Hodge structure of weight $w + k$, which has all the expected functoriality properties: it only depends on $(X, \mathcal{H})$ (so not on the embedding $X \subset \mathbb{P}$), is functorial in $\mathcal{H}$, is compatible with the Poincaré pairing in the sense that*

$$IH^k(X;\mathcal{H}) \times IH^{2n-k}(X;\mathcal{H}^\vee) \to \mathbb{Q}(-n)$$

is a morphism of Hodge structures,and is functorial with respect to restriction to a transversally embedded closed subvariety of pure dimension. In particular, the map $L := i_! i^* : IH^k(X;\mathcal{H}) \to IH^{k+2}(X;\mathcal{H})(1)$ is a morphism of Hodge structures of weight $w + k$.

2. *For $k \geq 0$, the map $L^k : IH^{n-k}(X;\mathcal{H}) \to IH^{n+k}(X;\mathcal{H})(k)$ is an isomorphism of Hodge structures, and the resulting map*

$$\psi : IH^{n-k}(X;\mathcal{H}) \times IH^{n-k}(X;\mathcal{H}) \to \mathbb{Q}(k-n), \quad \psi(\alpha, \beta) := L^k(\alpha) \cdot \beta,$$

polarizes the Hodge structure on $IH^{n-k}(X;\mathcal{H})$.

The proof follows essentially the pattern sketched at the end of lecture 5, but to make the induction run requires us to know that if X varies in an algebraic family, then the Hodge structures on the intersection cohomology groups of the fibres determine a variation of Hodge structure over a Zariski open-dense of the parameter space. Once this has been established, then Hodge theory is developed in relative dimension n far enough so that a weight argument can be used to prove the invariant cycle theorem for relative dimension n.

The preceding results admit a powerful generalization to the relative case, the *decomposition theorem* plus the *relative Lefschetz theorem*. For the statements, we refer to Saito's overview [**14**].

References

1. A. Borel (ed.), *Intersection Cohomology*, Progress in Math., Birkhäuser, 1984.
2. A.A. Beilinson, J. Bernstein, P. Deligne, *Faisceaux Pervers*, Astérisque **100** (1982).
3. J. Cheeger, M. Goresky, R. MacPherson, *L^2-Cohomology and intersection homology of singular algebraic varieties*, Seminar on Differential Geometry (S.-T. Yau, ed.), Ann. of Math. Studies, vol. 102, Princeton U.P., 1982.
4. P. Deligne, *Théorème de Lefschetz et critères de dégénérescence de suites spectrales*, Inst. Hautes Études Sci. Publ. Math. **35** (1968), 259–126.
5. P. Deligne, *Théorie de Hodge I*, Congrès Intern. Math. Nice, 1970, pp. 425–430; II, Inst. Hautes Études Sci. Publ. Math. **40** (1971), 5–58; III, Inst. Hautes Études Sci. Publ. Math. **44** (1975), 6–77.
6. P. Deligne, *Poids dans la Cohomologie des Variétés Algebriques*, Intern. Congress Math. Vancouver, 1974, pp. 79–85.
7. P. Deligne, *La conjecture de Weil II*, Inst. Hautes Études Sci. Publ. Math. **52** (1980), 137–252.
8. P. Deligne and N. Katz (eds.), *SGA VII*, Lecture Notes in Math., vol. 340, Springer Verlag, 1973.
9. C.G. Gibson et al., *Notes on Topological Stability*, Lecture Notes in Math., vol. 552, Springer Verlag, 1976.
10. M. Goresky and R. MacPherson, *Stratified Morse theory*, Ergebnisse der Mathematik, vol. 14, Springer Verlag, 1987.
11. Ph. Griffiths (ed.), *Topics in transcendental algbraic geometry*, Ann. of Math. Studies, vol. 106, 1984.
12. K. Lamotke, *The topology of complex varieties after S. Lefschetz*, Topology **20** (1986), 15–51.
13. S. Lefschetz, *l'Analysis Situs et la Géometrie Algebrique*, Collection de Monographies sur le théorie des fonctions, Gauthier-Villars et Cie., 1924.
14. M. Saito, *Introduction to mixed Hodge modules*, Astérisque **179-180** (1987), 145–162.
15. S. Zucker, *Hodge theory with degenerating coefficients: L_2 cohomology in the Poincaré metric*, Ann. Math. **109** (1979), 415–476.

Mathematical Aspects of Mirror Symmetry

David R. Morrison

IAS/Park City Mathematics Series
Volume 3, 1997

Mathematical Aspects of Mirror Symmetry

David R. Morrison

Introduction

Mirror symmetry is the remarkable discovery in string theory that certain "mirror pairs" of Calabi–Yau manifolds apparently produce isomorphic physical theories—related by an isomorphism which reverses the sign of a certain quantum number—when used as backgrounds for string propagation. The sign reversal in the isomorphism has profound effects on the geometric interpretation of the pair of physical theories. It leads to startling predictions that certain geometric invariants of one Calabi–Yau manifold (essentially the numbers of rational curves of various degrees) should be related to a completely different set of geometric invariants of the mirror partner (period integrals of holomorphic forms). The period integrals are much easier to calculate than the numbers of rational curves, so this idea has been used to make very specific predictions about numbers of curves on certain Calabi–Yau manifolds; hundreds of these predictions have now been explicitly verified. Why either the pair of manifolds, or these different invariants, should have anything to do with each other is a great mathematical mystery.

The focus in these lectures will be on giving a precise mathematical description of two string-theoretic quantities which play a primary rôle in mirror symmetry: the so-called A-model and B-model correlation functions on a Calabi–Yau manifold. The first of these is related to the problem of counting rational curves while the second is related to period integrals and variations of Hodge structure. A natural mathematical consequence of mirror symmetry is the assertion that Calabi–Yau manifolds often come in pairs with the property that the A-model correlation function of the first manifold coincides with the B-model correlation function of the second, and *vice versa*. Our goal will be to formulate this statement as a precise mathematical conjecture. There are other recent mathematical expositions of mirror symmetry, by Voisin [**15**] and by Cox and Katz [**3**], which concentrate on other aspects of the subject; the reader may wish to consult those as well in order to obtain a complete picture.

I have only briefly touched on the physics which inspired mirror symmetry (in lectures one and eight), since there are a number of good places to read about some of the physics background: I recommend Witten's address at the International Congress in Berkeley [**100**], a book on "Differential Topology and Quantum Field

[1]Department of Mathematics, Box 90320, Duke University, Durham, NC 27708-0320.
E-mail address: drm@math.duke.edu.

Theory" by Nash [**14**], and the first chapter of Hübsch's "Calabi–Yau Manifolds: A Bestiary for Physicists" [**10**]. There are, in addition, three collections of papers related to string theory and mirror symmetry which contain some very accessible expository material: "Mathematical Aspects of String Theory" (from a 1986 conference at U.C. San Diego) [**16**], "Essays on Mirror Manifolds" (from a 1991 conference at MSRI) [**17**], and its successor volume "Mirror Symmetry II" [**7**]. I particularly recommend the paper by Greene and Plesser "An introduction to mirror manifolds" [**136**] and the paper by Witten "Mirror manifolds and topological field theory" [**152**], both in the MSRI volume.

This is a revised version of the lecture notes which I prepared in conjunction with my July, 1993 Park City lectures, and which I supplemented when delivering a similar lecture series in Trento during June, 1994. The field of mirror symmetry is a rapidly developing one, and in finalizing these notes for publication I have elected to let them remain as a "snapshot" of the field as it was in 1993 or 1994, making only minor modifications to the main text to accommodate subsequent developments. I have, however, added a postscript that sketches the progress which has been made in a number of different directions since then.

Acknowledgments

The ideas presented here concerning the mathematical aspects of mirror symmetry were largely shaped through conversations and collaborations I have had with Paul Aspinwall, Brian Greene, Sheldon Katz, Ronen Plesser, and Edward Witten. It is a pleasure to thank them all for their contributions.

I am grateful to Antonella Grassi and Yiannis Vlassopoulos for providing me with copies of the notes they took during the lectures. I am also grateful to Grassi, Katz, Plesser, and Vlassopoulos, as well as to Michael Johnson, Lisa Traynor, and the referee of [**74**], for pointing out errors in the first drafts of these notes.

This research was supported in part by the National Science Foundation under grant number DMS-9103827.

LECTURE 1
Some Ideas From String Theory

1.1. String theory and quantum field theory

The origins of the startling calculations which have led to tremendous interest among mathematicians in the phenomenon of "mirror symmetry" lie in string theory. String theory is a proposed model of the physical world which idealizes its fundamental constituent particles as one-dimensional mathematical objects ("strings") rather than zero-dimensional objects ("points"). In theories such as general relativity, one has traditionally imagined a point as tracing out what is known as a "worldline" in spacetime; the corresponding notion in string theory is of a "worldsheet" which will describe the trace of a *string* in spacetime. We will consider here only "closed string theory" in which the string is a closed loop; the worldsheets are then (locally) closed surfaces: if we look at a portion of a worldsheet which represents the history of several interacting particles over a finite time interval, we will see a closed surface which has a boundary consisting of a finite number of closed loops.

An early version of string theory was proposed as a model for nuclear processes in the 1960's. Those early investigations revealed a somewhat disturbing property: in order to get a sensible physical theory, the spacetime M in which the string is propagating must have dimension twenty-six. Obviously, when we look around us, we do not see twenty-six dimensions. A later variant which incorporates supersymmetry[1] is sensible exactly when the spacetime has dimension ten—again a bit larger than the four-dimensional spacetime which we observe. Partly for this reason, but primarily because a better model for nuclear processes was found, the original research activity in string theory largely died out in the early 1970's.

String theory was subsequently revived in the 1980's when it was shown [**131**] that if the ten-dimensional string theory were used to model things at much smaller distance scales, an apparently consistent quantum theory of gravity could be produced. (In fact, gravity is predicted as an essential ingredient of this theory.) This "anomaly cancellation" result explained how certain potential inconsistencies in the quantum theory are avoided through an interaction between gravity and the other

[1] I shall not attempt to explain supersymmetry in these lectures.

forces present. Tremendous optimism and excitement pervaded this period, particularly since the new model contains a rich spectrum of elementary particles at low energies and exhibits many features one would expect of a "grand unified field theory" which could describe in a single theory all of the forces observed in nature. The "problem" of ten dimensions in this context can be resolved by assuming that the ten-dimensional spacetime is locally a product $M = M^{1,3} \times M^6$ of a macroscopic four-dimensional spacetime and a compact six-dimensional space whose size is on the order of the Planck length (10^{-33} cm). Because this is so small compared to macroscopic lengths, one wouldn't expect to observe the compact space directly, but its effect on four-dimensional physics could be detectable in various indirect ways.

The next step was even more remarkable for mathematicians—a group of string theorists [**122**] calculated that the compact six-dimensional space must have a Ricci-flat metric on it. (The physically relevant metric is actually a perturbation of this Ricci-flat one.) This is a very restrictive property—it implies, for example, that the six-manifold is a complex Kähler manifold of complex dimension three which has trivial canonical bundle; conversely, such Kähler manifolds always admit Ricci-flat metrics. (This had been conjectured by Calabi [**33**] in the late 1950's and proved by Yau [**103**] in the mid 1970's.) These manifolds have since been named "Calabi–Yau manifolds;" finding and studying them become problems in algebraic geometry, thanks to Yau's theorem.

The model being described here of a string propagating in a spacetime (with a specified metric) is generally regarded as a woefully inadequate description of the "true" string theory, a good formulation of which is as yet unknown. Indeed, if string theory is truly a theory of gravity as we observe it, then the theory should approximate general relativity when the distance scale approaches macroscopic levels. Since the metric on spacetime is part of general relativity, it should be a part of that "approximation" which is somehow to be deduced from a solution to the ultimate "string equations," rather than being something which is put in by hand in advance. Even the *topology* of spacetime should be dictated by the string theory. However, neither these "string equations" nor their exact solutions are known at present.

The Calabi–Yau manifolds and their connections with string theory have been studied intensively for more than a decade. In the earliest period, these manifolds were analyzed using standard mathematical techniques, and the results were applied in a string-theoretic context. However, at the same time, other advances were being made in string theory which suggested other ways of looking at certain aspects of the theory of Calabi–Yau manifolds. This eventually led to the discovery of a surprising new phenomenon known as "mirror symmetry," in which it was observed that different Calabi–Yau manifolds could lead to identical physical theories in a way that implied surprising connections between certain geometric features of the manifolds.

To explain this mirror symmetry observation in more detail, we must first describe a few aspects of quantum field theory and its relationship to string theory. In classical mechanics, the worldline representing a particle is required to minimize the "action" (which is the energy integrated with respect to time), or more precisely, to be a stationary path for the action functional. Due to this "stationary action principle," the location of the path in spacetime is completely determined by a knowledge of boundary conditions. Other physically measurable quantities associated to the particle (which are often represented as some kind of "internal variables") will also evolve from their boundary states in a completely predictable manner, again minimizing the action.

In quantum field theory, however, this changes. Only the probability of various possible outcomes can be predicted with certainty, and *all* trajectories—not just the action-minimizing ones—contribute to the measurement of this probability. The probability is calculated from an integral over the space of all possible paths with these initial and final states,[2] and the classical trajectory is recovered as the leading term in a stationary phase approximation to the path integral.

Relativistic quantum field theories are frequently studied by treating the theory as a small perturbation of a simple type of theory—called a free field theory—whose functional integrals are well-understood. For example, the path integral describing the interaction of two charged particles can be expanded in a perturbative series whose terms are described by "Feynman diagrams." The zeroth order term is the diagram

which represents two particles which do not interact at all, the leading perturbative correction is described by the diagram

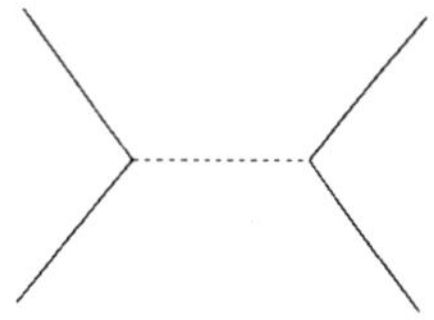

which represents a transfer of momentum from one particle to the other via the emission and absorption of a third particle carrying the force, and higher order corrections involve diagrams with more complicated topologies—loops are allowed, for example. Such diagrams can be cut into simpler pieces, at the expense of performing an integral over all possible intermediate states. For example, the interacting Feynman diagram illustrated above can be decomposed into two more primitive pieces,

each of which represents a fundamental "interaction" vertex.

In string theory, the paths are replaced by surfaces: the interacting diagram might be represented as a sphere with four disks removed (or perhaps as something

[2]There are enormous mathematical difficulties in dealing with these "path integrals" or "functional integrals," and they do not in general have a rigorous mathematical formulation. Nevertheless, in the hands of skilled practitioners they can be used to make predictions which agree with laboratory experiments to a remarkable degree of precision.

with more complicated topology),

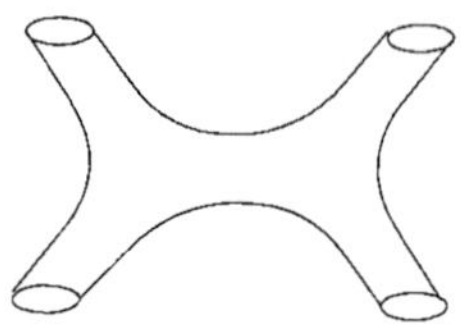

and this could also be decomposed into more primitive pieces

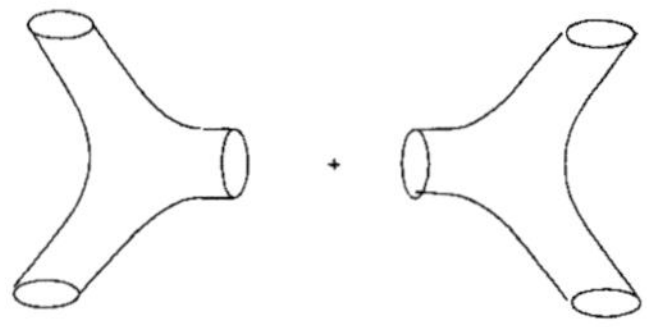

(called "pair of pants" surfaces). One of the advantages of string theory is that this fundamental piece, the "pair of pants" surface, is a smooth surface, in contrast to the interaction vertex which introduced a singularity into the worldline.

The methods of quantum field theory are applied to string theory in a rather interesting way. If we fix the spacetime and consider a string propagating through it, the location of the worldsheet can be viewed as a map from the worldsheet to the spacetime, and we can regard the coordinate functions on the spacetime as functions on the worldsheet. These spacetime coordinate functions are then treated as the "internal variables" of a two-dimensional quantum field theory—formulated on the worldsheet itself—which captures many of the important physical features of the string theory. (The functional integral in this theory involves an integration over all possible metrics on the worldsheet as well as all possible maps from the worldsheet to the spacetime.) The two-dimensional quantum field theories arising from string theory are of a particular type known as a *conformal field theory*; this means that a conformal change of metric on the worldsheet will act as an automorphism of the theory (typically acting linearly on various spaces of "internal fields" of the theory). The formulation in terms of conformal field theory has turned out to be a very fruitful viewpoint for the study of string theory.

1.2. Correlation functions and pseudo-holomorphic curves

The basic quantities which one needs to evaluate in any quantum field theory are the *correlation functions* which determine the probabilities for a specified final state, given an initial state. Specifying the initial and final states means not only specifying positions, but also the values of any "internal variables" which form a part of the theory. The possible initial or final states in a conformal field theory can be represented as operators $\mathcal{O}_P$ on some fixed Hilbert space $\mathcal{H}$, often referred to as "vertex operators."[3] (The label P indicates the position; we should in principle be specifying initial conditions on an entire boundary circle, but in fact it suffices to consider a limit—within the conformal class of the given metric—in which the circles have been shrunk to zero size and the vertex operators are located at points.) The conjugate transpose of an initial state is a final state, so we often don't distinguish

[3]Generally, in quantum field theories states are represented as elements of a Hilbert space $\mathcal{H}$ but in conformal field theories there is also an operator interpretation.

between those in our notation; with these conventions, the correlation function of a number of vertex operators is denoted by

$$\langle \mathcal{O}_{P_1} \mathcal{O}_{P_2} \dots \mathcal{O}_{P_k} \rangle.$$

(Note that the correlation functions are complex-valued and do not directly calculate probabilities, but also include the phase of the quantum-mechanical wavefunction.)

If we fix the topology of the worldsheet we must in general integrate over the choice of metric on that worldsheet. A conformal change of metric leaves the correlation functions invariant, so we only need to integrate over the set of conformal classes of metrics, i.e., over the (finite-dimensional) moduli space $\mathcal{M}_{g,k}$ of k-punctured Riemann surfaces of genus g.

The two-dimensional quantum field theories which are related to mirror symmetry have a subset of their correlation functions whose values do not depend on the position of the points P_j on the worldsheet; these are called *topological* correlation functions. (We don't need to consider an integral over $\mathcal{M}_{g,k}$ in this case.) They will be the primary objects of interest for us. In fact, due to the possibility of decomposing the worldsheet into more primitive pieces, the main case to consider is the case of three vertex operators on surfaces of genus zero, i.e., we take Σ to be the "pair of pants" surface $\mathbb{CP}^1 - \{P_1, P_2, P_3\}$. To evaluate a correlation function

$$\langle \mathcal{O}_{P_1} \mathcal{O}_{P_2} \mathcal{O}_{P_3} \rangle,$$

however, we must still integrate over the infinite-dimensional space $\mathrm{Maps}(\Sigma, M)$ of maps from Σ to the spacetime M.

To proceed further, we need to introduce the "action" functional on the space of maps. We fix Riemannian metrics[4] on both the worldsheet Σ and the spacetime M and define for any sufficiently smooth $\varphi \in \mathrm{Maps}(\Sigma, M)$

$$\mathcal{S}[\varphi] = \int_{\Sigma} \|d\varphi\|^2 \, d\mu$$

using the metrics to define the norm. (In practice, we take M to be the compact six-dimensional manifold rather than the entire space.) The properties of the action functional are easier to analyze if we assume that M has some additional structure—the minimal structure needed is a symplectic form ω and an almost-complex structure J which is ω-tamed. (We will review the definitions of these in lecture three.) When these have been chosen, there is a "d-bar" operator $\overline{\partial}_J$ on maps, and an alternate formula for the action

$$\mathcal{S}[\varphi] = \int_{\Sigma} \|\overline{\partial}_J \varphi\|^2 \, d\mu + \int_{\Sigma} \varphi^*(\omega).$$

A lower bound for $\mathcal{S}[\varphi]$ in any homotopy class of maps is thus given by $\int_{\Sigma} \varphi^*(\omega)$; this bound will be achieved by the so-called pseudo-holomorphic maps—the ones for which $\overline{\partial}_J \varphi \equiv 0$. These have been extensively studied by Gromov [**51**] and others as a natural generalization of complex curves on Kähler manifolds.

This action functional now appears in an integrand which is supposed to be integrated over the infinite-dimensional space of all maps. We will outline the standard manipulations which are made with these functional integrals in physics

[4]This is a "Euclidean" version of the theory, whose correlation functions are related by analytic continuation to those of the "Lorentzian" version in which the worldsheet metric has signature $(1, 1)$.

in order to express the correlation function as an infinite sum of finite-dimensional integrals. We will subsequently use the outcome of those manipulations to make mathematical definitions of the corresponding quantities in the form of a formal sum of these finite-dimensional integrals.

The topological correlation functions we are studying are to be evaluated by a functional integral of the form

$$\begin{aligned}\langle \mathcal{O}_{P_1} \dots \mathcal{O}_{P_k} \rangle &= \int \mathcal{D}\varphi\, \mathcal{O}_{P_1} \dots \mathcal{O}_{P_k}\, e^{-2\pi\, \mathcal{S}[\varphi]} \\ &= e^{-2\pi \int_\Sigma \varphi^*(\omega)} \int \mathcal{D}\varphi\, \mathcal{O}_{P_1} \dots \mathcal{O}_{P_k}\, e^{-2\pi \int_\Sigma \|\overline{\partial}_J \varphi\|^2\, d\mu}.\end{aligned} \tag{1.1}$$

(We are suppressing the "fermionic" part of this functional integral, which is actually very important, but explaining it would take us too far afield.) The "topological" property of these correlation functions turns out to imply [**150, 152**] that if we introduce a parameter t into the exponent of the last functional integral in eq. (1.1) to produce

$$\int \mathcal{D}\varphi\, \mathcal{O}_{P_1} \dots \mathcal{O}_{P_k}\, e^{-2\pi t \int_\Sigma \|\overline{\partial}_J \varphi\|^2\, d\mu},$$

then the resulting expression is independent of t and can be evaluated in a limit in which $t \to \infty$. In such a limit, the only contributions to the functional integral are the maps φ for which $\overline{\partial}_J \varphi \equiv 0$, i.e., the pseudo-holomorphic maps. (This trick for reducing to a finite-dimensional integral is known as the "method of stationary phase.") The space of pseudo-holomorphic maps in a given homotopy class is finite-dimensional, so we have reduced the evaluation of our correlation function to an infinite sum of finite-dimensional integrals, of the form

$$\langle \mathcal{O}_{P_1} \dots \mathcal{O}_{P_k} \rangle = \sum_{\text{homotopy classes}} e^{-2\pi \int_\Sigma \varphi^*(\omega)} \int_{\mathcal{M}} \mathcal{D}\varphi\, \mathcal{O}_{P_1} \dots \mathcal{O}_{P_k}, \tag{1.2}$$

where $\mathcal{M}$ denotes the moduli space of pseudo-holomorphic maps in a fixed homotopy class. It is these finite-dimensional integrals on which we shall eventually base our definitions. The convergence of the infinite sum will remain an issue in our approach, and will lead us to (in some cases) assign a provisional interpretation to this formula as being a formal power series only. From the physics one expects convergence whenever the volume of the corresponding metric is sufficiently large.

1.3. A glimpse of mirror symmetry

If the target space M for our maps is a Calabi–Yau manifold (equipped with a Ricci-flat metric), all of the vertex operators which participate in a given topological correlation functions must be of one of two distinct types. Correlation functions involving vertex operators of the first type are called *A-model correlation functions* while those involving vertex operators of the second type are known as *B-model correlation functions* [**152**]. (These are actually the correlation functions in two "topological field theories" [**150**] which are closely related to the original quantum field theories.) For each type, the vertex operators $\mathcal{O}_P$ in the quantum field theory or topological field theory have a geometric interpretation; we will treat the correlation functions as functions of these geometric objects.

The A-model correlation functions can be defined in a much broader context than Calabi–Yau manifolds: they can be defined for any semipositive symplectic manifold M (where semipositive roughly means that $-c_1(M)$ is nonnegative—we will give the precise definition in lecture three). The vertex operators $\mathcal{O}_{P_j}$ in the topological field theory correspond to harmonic differential forms α_j on M, and the correlation functions $\langle \alpha_1\, \alpha_2\, \alpha_3 \rangle$ take the form of an infinite series whose constant term—corresponding to homotopically trivial maps from Σ to M—is the familiar trilinear function $\int_M \alpha_1 \wedge \alpha_2 \wedge \alpha_3$.

To evaluate the non-constant terms we need an integral over the moduli space of pseudo-holomorphic two-spheres. In the Calabi–Yau case, those two-spheres are expected to be discrete (based on a formal dimension count), so there should be invariants which count the number of rational curves in a given homology class. (There are certain technical difficulties with this, as we shall see in lecture two.) More generally, the non-constant terms in the A-model correlation functions will be related to certain kinds of counting problems for pseudo-holomorphic curves on a semipositive symplectic manifold.

The B-model correlation functions, on the other hand, require a choice of non-vanishing holomorphic n-form Ω on M for their definition, so they are restricted to the Calabi–Yau case. The vertex operators in the topological correlation functions correspond to elements in the space $H^q(\Lambda^p T_M^{(1,0)})$, where we use $T^{(1,0)}$ to denote the holomorphic tangent bundle of an almost-complex manifold. (More precisely, we should use Dolbeault cohomology to describe $H^q(\Lambda^p T_M^{(1,0)})$, and take harmonic representatives to get the vertex operators in the topological field theory.) The "first term" in the correlation function is then defined as a composition of the standard map on cohomology groups

$$H^{q_1}(\Lambda^{p_1} T_M^{(1,0)}) \times H^{q_2}(\Lambda^{p_2} T_M^{(1,0)}) \times H^{q_3}(\Lambda^{p_3} T_M^{(1,0)}) \to H^n(\Lambda^n T_M^{(1,0)})$$

(for $p_1 + p_2 + p_3 = q_1 + q_2 + q_3 = n$) with some isomorphisms depending on the choice of $\Omega^{\otimes 2}$

$$H^n(\Lambda^n(T_M^{(1,0)})) \xrightarrow{\lrcorner\Omega} H^n(\mathcal{O}_M) \cong \left(H^0(K_M)\right)^* \xrightarrow{\otimes\Omega} \mathbb{C},$$

where the middle isomorphism is Serre duality. (This can be written as an integral over M, and so can be thought of as coming from integrating over the moduli space of homotopically trivial maps from Σ to M—this is why we identify it with the first term in an expansion like eq. (1.2).) Remarkably, all of the other terms in the expansion (1.2) of a B-model correlation function are known to vanish on physical grounds [**127, 152**], so we can calculate these correlation functions exactly using geometry, and even use the geometric version of the correlation function as a mathematical definition.

In brief, the idea of mirror symmetry is this. There could be pairs of complex manifolds M, W (each with trivial canonical bundle) which produce identical physics when used for string compactification, except that the rôles of the A-model and B-model correlation functions are reversed. In particular, this would imply the existence of isomorphisms

$$H^q(\Lambda^p(T_M^{(1,0)})^*) \cong H^q(\Lambda^p(T_W^{(1,0)}))$$

(and vice versa), as well as formulas relating the A-model correlation functions on M (which count the number of rational curves) to the B-model correlation functions on W (which are related to period integrals of Ω).

LECTURE 2
Counting Rational Curves

In this lecture we begin the discussion of the problem of counting rational curves on a complex threefold with trivial canonical bundle (a "Calabi–Yau threefold"). These curve-counting invariants will eventually be used to formulate a mathematical version of the A-model correlation functions. In the present lecture, we focus on the problems one encounters in formulating these invariants purely algebraically; we give a number of examples.

Consider the deformation theory of holomorphic maps from $\mathbb{CP}^1 \to M$, where M is a complex projective variety. If we are given such a map $\varphi : \mathbb{CP}^1 \to M$, then a first order variation of that map can be described by specifying in which direction (and at what rate) each point of the image moves. That is, we need to specify a holomorphic tangent vector of M for every point on $\mathbb{CP}^1$, or in other words, a section of $H^0(\mathbb{CP}^1, \varphi^*(T_M^{(1,0)}))$. As might be expected from other deformation problems, the obstruction group for these deformations is $H^1(\mathbb{CP}^1, \varphi^*(T_M^{(1,0)}))$. The moduli problem for such maps will be best-behaved if the obstruction group vanishes, that is, if $h^1(\mathbb{CP}^1, \varphi^*(T_M^{(1,0)})) = 0$. When that is true, the moduli space will be a smooth complex manifold of complex dimension $h^0(\mathbb{CP}^1, \varphi^*(T_M^{(1,0)}))$. More generally, the Euler–Poincaré characteristic

$$\chi(\varphi^*(T_M^{(1,0)})) = h^0(\mathbb{CP}^1, \varphi^*(T_M^{(1,0)})) - h^1(\mathbb{CP}^1, \varphi^*(T_M^{(1,0)}))$$

can be regarded as the "expected complex dimension" of the moduli space.

Although the Euler–Poincaré characteristic can be easily computed from the Riemann–Roch theorem for vector bundles, we shall make a more elementary calculation, based on a structure theorem for bundles on $\mathbb{CP}^1$.

Theorem 2.1 (Grothendieck). *Every vector bundle $\mathcal{E}$ on $\mathbb{CP}^1$ can be written as a direct sum of line bundles:*

$$\mathcal{E} \cong \mathcal{O}(a_1) \oplus \cdots \oplus \mathcal{O}(a_n).$$

Using a Grothendieck decomposition for $\varphi^*(T_M^{(1,0)})$, we can calculate the cohomology directly. For if

$$\varphi^*(T_M^{(1,0)}) \cong \mathcal{O}(a_1) \oplus \cdots \oplus \mathcal{O}(a_n)$$

then using the fact the $h^0(\mathcal{O}(a)) = 1 + a$ we find

$$h^0(\varphi^*(T_M^{(1,0)})) = \sum_j \begin{cases} 1 + a_j & \text{if } a_j \geq -1 \\ 0 & \text{if } a_j < -1 \end{cases}$$

while since $H^1(\mathcal{O}(a_1) \oplus \cdots \oplus \mathcal{O}(a_m)) \cong H^0(\mathcal{O}(-2 - a_1) \oplus \cdots \oplus \mathcal{O}(-2 - a_m))^*$ we have

$$h^1(\varphi^*(T_M^{(1,0)})) = \sum_j \begin{cases} -(1 + a_j) & \text{if } -2 - a_j \geq 0 \\ 0 & \text{if } -2 - a_j < 0 \end{cases}$$

since $1 + (-2 - a_j) = -(1 + a_j)$.

Taking the difference, we find

$$\chi(\varphi^*(T_M^{(1,0)})) = \sum_j (1 + a_j) = n + \sum_j a_j = \dim_{\mathbb{C}} M + \deg \varphi^*(-K_M).$$

Thus, the "expected dimension" is independent of the decomposition. The same result can be obtained from Riemann–Roch.

But our calculation shows more—to ensure vanishing of the obstruction group, we must have $a_j \geq -1$ for all j. In addition to this condition on the a_j's, they must also satisfy $\max\{a_j - 2\} \geq 0$, which is seen as follows. From the exact sequence

$$0 \to T_{\mathbb{CP}^1}^{(1,0)} \to \varphi^*(T_M^{(1,0)}) \to N_\varphi \to 0$$

(where N_φ denotes the normal bundle) and the fact that $T_{\mathbb{CP}^1}^{(1,0)} \cong \mathcal{O}(2)$, we see that there must be a nontrivial homomorphism

$$\mathcal{O}(2) \to \mathcal{O}(a_1) \oplus \cdots \oplus \mathcal{O}(a_m),$$

which implies that $\max\{a_j - 2\} \geq 0$ as claimed. Without loss of generality, we may therefore assume that $a_1 \geq 2$.

In the case relevant to string theory ($K_M = 0$, $\dim_{\mathbb{C}} M = 3$) we then find that in order to have vanishing obstruction group we need

$$0 = a_1 + a_2 + a_3 \geq 2 - 1 - 1 = 0$$

and so $a_1 = 2$, $a_2 = a_3 = -1$. In this case, the moduli space of holomorphic maps will be smooth of dimension three; if we mod out by the automorphism group $\mathrm{PGL}(2, \mathbb{C})$, the moduli space of unparameterized maps will be smooth of dimension 0. The points in that space are what we would like to "count." We discuss some examples, drawn largely from [**59**], to which we refer the reader for more details.

Example 2.2. *Lines on the Fermat quintic threefold.* All of the lines on the Fermat quintic threefold

$$\{x_0^5 + x_1^5 + x_2^5 + x_3^5 + x_4^5 = 0\} \subset \mathbb{CP}^4$$

can be described as follows.[1]

First type (375 lines): The line described by $x_0 + x_1 = x_2 + x_3 = x_4 = 0$, and others whose equations are obtained from these by permutations and multiplication by fifth roots of unity.

---

[1]We use the Fermat quintic because it is an easily-described nonsingular hypersurface, and because it will be related to a mirror symmetry construction later on, *not* because Wiles announced a proof of Fermat's Last Theorem while the 1993 Park City Institute was underway!

Second type (50 one-parameter families of lines): The lines described parametrically by

$$(u, v) \mapsto (u, -u, av, bv, cv)$$

for fixed constants a, b, c satisfying $a^5 + b^5 + c^5 = 0$, and others whose parameterizations are obtained from these by permutations and multiplication by fifth roots of unity.

So we see that the lines are not always finite in number, even for smooth hypersurfaces. (One might have suspected such a "universal finiteness for smooth hypersurfaces" based on experience with cubic surfaces—every smooth cubic surface in $\mathbb{CP}^3$ has precisely twenty-seven lines.)

Example 2.3. *Lines on the general quintic threefold.* However, if we deform from the Fermat quintic threefold to a general one, it is possible to show that the number of lines is finite. The generic number of lines can then be computed as follows. Start from the Grassmannian $\mathrm{Gr}(\mathbb{CP}^1, \mathbb{CP}^4)$ of lines in $\mathbb{CP}^4$. Consider the universal bundle U whose fiber at a line L is the two-dimensional subspace $U_L \subset \mathbb{C}^5$ such that $\mathbb{P}(U_L) = L$. We define a bundle $\mathcal{B} = \mathrm{Sym}^5(U^*)$ whose fibers describe the quintic forms on the lines L. Then every quintic threefold M determines a section $s_M \in \Gamma(\mathcal{B})$: the equation of M is restricted to L to give a homogeneous quintic there. Clearly, the lines contained in M are precisely those whose corresponding points in the Grassmannian are zeros of the section s_M.

The Grassmannian $\mathrm{Gr}(\mathbb{CP}^1, \mathbb{CP}^4)$ has complex dimension six, and the bundle $\mathcal{B}$ has rank six; when things are generic, the section s_M will have finitely many zeros, which can be counted by calculating

$$\#\{L \mid s_M(L) = 0\} = c_6(\mathcal{B}) = 2875.$$

Example 2.2 (bis). Katz [**59**] has found a way to assign multiplicities to each of the isolated lines, and one-parameter families of lines, on the Fermat quintic threefold. His multiplicity assignment for each of the 375 isolated lines is "5," and that for each of the 50 one-parameter families is "20." Thus, the total count is

$$5 \cdot 375 + 20 \cdot 50 = 2875.$$

Katz's methods of assigning multiplicities are not yet completely general,[2] but they do hold out the hope that a "count" of rational curves might be made even in cases when the actual number of curves is not finite.

Example 2.4. *Conics on the general quintic threefold.* We can make a similar calculation for conics on the general quintic threefold. The key observation is that every conic spans a $\mathbb{CP}^2$, so the starting point for describing them is the Grassmannian $\mathrm{Gr}(\mathbb{CP}^2, \mathbb{CP}^4)$. We need the bundle over the Grassmannian whose fiber is the set of conics in the $\mathbb{CP}^2$ in question: this is described by $\mathbb{P}(\mathrm{Sym}^2(U^*))$, where U is the universal subbundle as before.

The space $\mathbb{P}(\mathrm{Sym}^2(U^*))$ contains degenerate conics (pairs of lines, and double lines) as well as smooth conics. However, if M is sufficiently general, then the actual locus of conics which lie in M will be finite in number, and contain only smooth conics.

The vector bundle which will get a section s_M for every quintic M is the bundle $\mathcal{B} := \mathrm{Sym}^5(U^*)/(\mathrm{Sym}^3(U^*) \oplus \mathcal{O}_{\mathbb{P}}(-1))$. This describes the effect of restricting the

[2]See the "Postscript: Recent Developments" section for the current status.

quintic equation to the conic: one gets a quintic equation on the $\mathbb{CP}^2$, but must mod out by those quintics which can be written as the product of a cubic (the $\mathrm{Sym}^3(U^*)$ factor) and the given conic.

We have $\dim_{\mathbb{C}} \mathbb{P}(\mathrm{Sym}^2(U^*)) = \operatorname{rank} \mathcal{B} = 11$, so the computation is made by calculating:

$$\#\{C \mid s_M(C) = 0\} = c_{11}(\mathcal{B}) = 609250.$$

Example 2.5. *Twisted cubics on the general quintic threefold.* The problem gets more difficult for twisted cubics. Again, we can look at the linear span (a $\mathbb{CP}^3$) and begin by considering a Grassmannian $\mathrm{Gr}(\mathbb{CP}^3, \mathbb{CP}^4)$. But this time we must use a bundle $\mathcal{H} \to \mathrm{Gr}(\mathbb{CP}^3, \mathbb{CP}^4)$ whose fibers are isomorphic to the Hilbert scheme of twisted cubics in $\mathbb{CP}^3$. That scheme contains limits which are quite complicated. (For example, there is a limit which is a nodal plane curve with an embedded point at the node which points out of the plane:

see Hartshorne [**9**], pp. 259–260.) Although the bundle $\mathcal{B}$ and the section s_M can be defined and understood at points representing smooth twisted cubics, their extension to the locus of degenerate cubics is by no means easy.

Ellingsrud and Strømme [**40**] have, however, carried this out, and they find that the number of twisted cubics on the general quintic threefold is 317206375.

Clemens [**37**] has conjectured that the general quintic threefold will have only a finite number of rational curves of each degree, and that all of them will satisfy $\varphi^*(T_M^{(1,0)}) = \mathcal{O}(2) \oplus \mathcal{O}(-1) \oplus \mathcal{O}(-1)$. This has been verified up through degree nine by Katz [**58**], Johnsen–Kleiman [**56**] and Nijsse [**80**], and the prospects are good for degrees as high as twenty-four [**57**]. However, as we have seen, making the calculation of the number becomes very difficult past degree two. In fact, for degree greater than three, effective techniques for calculating this number are not presently known.

We now turn to another example which demonstrates that we cannot always expect finiteness, even for the generic deformation of a given threefold with trivial canonical bundle.

Example 2.6. *Rational curves on double solids.* We let M be the double cover of $\mathbb{CP}^3$, branched along a general surface S of degree eight in $\mathbb{CP}^3$; the double cover map is denoted by $\pi : M \to \mathbb{CP}^3$. We let $\pi^*(H)$ be the pullback of a hyperplane H from $\mathbb{CP}^3$; the *degree* of a rational curve C will mean $\pi^*(H) \cdot C$.

To find "lines" on M, that is, curves L with $\pi^*(H) \cdot L = 1$ we consider their images $\pi(L)$. Since $\pi^*(H)$ meets L in a single point P, H meets $\pi(L)$ in the single point $\pi(P)$. Thus, $\pi(L)$ must itself be a line. But its inverse image on M will necessarily have two components: $\pi^{-1}(\pi(L)) = L + L'$. In order to have this splitting into two components, the line $\pi(L)$ must be tangent to S at every point of intersection with S, i.e., it must be four-times tangent to S. Now the Grassmannian $\mathrm{Gr}(\mathbb{CP}^1, \mathbb{CP}^3)$ has dimension four, and it is one condition to be tangent to a surface, so the dimension of the set of four-tangent lines is nonnegative, and can be expected to be equal to zero. (In fact, it turns out to equal zero as expected, when S is general.) The number of such lines in the Grassmannian can be calculated with the Schubert calculus; it turns out to be 14752. The corresponding count of lines on M is 29504.

Finding "conics" on M is a different story, as has been observed by Katz and by Kollár. Given a curve C with $\pi^*(H) \cdot L = 2$, there are two possibilities for $\pi(C)$: it could be a line, or it could be a conic. In the latter case, the conic $\pi(C)$ must be eight-times tangent to S. But in the former case, in order to have an irreducible double cover with a rational normalization, the line $\pi(C)$ must be three-times tangent to S. By our previous dimension count, there is at least a one-parameter family of such lines for any choice of S.

So we won't always have a finite number of things to "count," even if we perturb to a general member of a particular family. And there is an additional difficulty if we wish to count maps from $\mathbb{CP}^1$ to M when multiple covers are allowed, as the next example shows.

Example 2.7. *Multiple covers.* Suppose that $\varphi : \mathbb{CP}^1 \to M$ is generically one-to-one, but that we consider a map $\varphi' := u \circ \varphi$, where $u : \mathbb{CP}^1 \to \mathbb{CP}^1$ is a covering of degree μ. Even if $\varphi^*(T_M^{(1,0)}) = \mathcal{O}(2) \oplus \mathcal{O}(-1) \oplus \mathcal{O}(-1)$, we will get a bad splitting of the pullback via the new map:

$$\varphi'^*(T_M^{(1,0)}) = \mathcal{O}(2\mu) \oplus \mathcal{O}(-\mu) \oplus \mathcal{O}(-\mu).$$

Furthermore, the dimension of the moduli space can be calculated: the moduli space of maps $u : \mathbb{CP}^1 \to \mathbb{CP}^1$ of degree μ has dimension $2\mu+1$. So we see that the dimension of the space of maps will go up and up.

To handle cases such as multiple covers, "virtual" numbers of curves must be introduced; Katz's approach to this is to use excess intersection theory [**6**]. However, this introduction of "virtual" numbers leads to another complication, as our final example shows.

Example 2.8. *Negative numbers of curves* (see [**120**], section 8). There are cases in which the "virtual" number of curves is negative. In general, when the parameter space B for a family of curves is smooth of dimension b, the virtual number of curves should be the top Chern class of the holomorphic cotangent bundle $c_b((T_B^{(1,0)})^*)$. If M is a complex threefold with $K_M = 0$ which contains $\mathbb{CP}^2$ as a submanifold (which can arise from resolving a $\mathbb{Z}/3\mathbb{Z}$-quotient singularity, for example), then the lines on $\mathbb{CP}^2$ are parameterized by $\mathbb{CP}^2$ and have virtual number $c_2((T_{\mathbb{CP}^2}^{(1,0)})^*) = 3$, but the conics on $\mathbb{CP}^2$, being parameterized by $\mathbb{CP}^5$, have virtual number $c_5((T_{\mathbb{CP}^5}^{(1,0)})^*) = -6$. This negative value actually agrees with the predictions of mirror symmetry as shown in [**120**].

LECTURE 3
Gromov–Witten Invariants

3.1. Counting curves via symplectic geometry

The difficulties we encountered in trying to count rational curves on a Calabi–Yau threefold can be avoided by enlarging the category we are considering, and using Gromov's theory of pseudo-holomorphic spheres in symplectic manifolds [**51**]. This approach has the advantage that the almost-complex structure can be slightly perturbed to make the number of such spheres finite, and the finite number so obtained is independent of the choice of small perturbation.

Let (M, ω) be a compact *symplectic manifold* of dimension $2n$. This means that M is a compact oriented differentiable manifold of (real) dimension $2n$ and ω is a closed real two-form on M which is nondegenerate in the sense that its nth exterior power $\omega^{\wedge n}$ is nonzero at every point.

An *almost complex structure* on a manifold M is a map $J : T_M \to T_M$ whose square is -1. If we complexify the tangent spaces, we get $T_{M,p} \otimes \mathbb{C} = T^{(1,0)}_{M,p} \oplus T^{(0,1)}_{M,p}$, the decomposition into $+i$ and $-i$ eigenspaces for J. If these subspaces are closed under Lie bracket, we say that the almost-complex structure is *integrable*; in this case, these subspaces give M the structure of a complex manifold.

If (M, ω) is a symplectic manifold, an almost-complex structure J on M is said to be *ω-tamed* if $\omega(\xi, J\xi) > 0$ for all nonzero $\xi \in T_pM$. If we have fixed an (ω-tamed) almost-complex structure J on M, and φ is a differentiable map from S^2 to M, we define

$$\overline{\partial}_J \varphi = \frac{1}{2}(d\varphi + J\, d\varphi\, J_0),$$

where J_0 is the standard almost-complex structure on S^2.

The main example we have in mind is this: M is a compact Kähler manifold, ω is the Kähler form, and J is an ω-tamed perturbation of the original complex structure on M.

Definition 3.1 (McDuff [**71**]). (M, ω) is *semipositive* if there is no map $\varphi : S^2 \to M$ satisfying

$$\int_{S^2} \varphi^*(\omega) > 0, \quad \text{and} \quad 3 - n \leq \int_{S^2} \varphi^*(-K_M) < 0,$$

where we are writing $-K_M$ as in algebraic geometry to indicate the first Chern class $c_1(M)$, which may be represented as a two-form.

Examples. Here are three ways of producing semipositive symplectic manifolds.

1. If $K_M = 0$ (the Calabi–Yau case) then (M, ω) is semipositive for any ω.
2. If M is a complex projective manifold with $|-K_M|$ ample (a "Fano variety"), then we can take $\omega = -K_M$ to produce a semipositive (M, ω).
3. If $n \leq 3$ then M is automatically semipositive.

Because it is sometimes difficult to check whether a homology class η is represented as the image of a map $\varphi : S^2 \to M$ we also introduce a variant of this property.

Definition 3.2. (M, ω) is *strongly semipositive* if there is no class $\eta \in H_2(M, \mathbb{Z})$ satisfying

$$\omega \cdot \eta > 0, \quad \text{and} \quad 3 - n \leq (-K_M) \cdot \eta < 0.$$

All three of our examples satisfy this stronger property as well.

Fix a homology class $\eta \in H_2(M, \mathbb{Z})$. As we saw in example 2.7, there are technical problems caused by "multiple-covered" maps—maps whose degree onto the image is greater than one. Let us call a map *simple* if its degree onto its image is one. We let $\mathrm{Maps}_\eta^*(S^2, M) \subset \mathrm{Maps}_\eta(S^2, M)$ denote the subset of simple maps with fundamental class η. We also let $\mathrm{Maps}_\eta^*(S^2, M)_{(p)}$ be the set of simple differentiable maps $S^2 \to M$ with fundamental class η whose derivative lies in L_p. Using an appropriate Sobolev norm, $\mathrm{Maps}_\eta^*(S^2, M)_{(p)}$ can be given the structure of a Banach manifold. We can then regard $\overline{\partial}_J$ as a section of the bundle $\mathcal{W} \to \mathrm{Maps}_\eta^*(S^2, M)_{(p)}$ whose fibers are

$$\mathcal{W}_\varphi := H^0_{(p)}(S^2, \mathcal{A}_{S^2}^{(0,1)} \otimes \varphi^*(T_M^{(1,0)})),$$

where the subscript (p) denotes L_p-cohomology, and $\mathcal{A}_{S^2}^{(0,1)}$ denotes the sheaf of $(0,1)$-forms on S^2 (with respect to the complex structure J_0).

The key technical properties we need are summarized in the following two theorems.

Theorem 3.3 (McDuff [**70**]). *If J is generic, then*

$$\widetilde{\mathcal{M}}^*_{(\eta,J)} := \{\varphi \in \mathrm{Maps}_\eta^*(S^2, M)_{(p)} \mid \overline{\partial}_J \varphi = 0\}$$

is a smooth manifold of dimension

$$\dim_{\mathbb{R}} \widetilde{\mathcal{M}}^*_{(\eta,J)} = 2\,\chi(\varphi^*(T_M^{(1,0)})).$$

(The dimension is calculated using the Atiyah–Singer index theorem, which yields the same result as the Riemann–Roch theorem did in algebraic geometry.)

(This theorem would have failed if we had allowed multiple-covered maps to be included.)

The next theorem is due to Gromov [**51**], based on some techniques of Sacks–Uhlenbeck [**88**] and with further improvements by several authors [**84, 102, 104**]. (We refer the reader to those papers for a more precise statement.)

Theorem 3.4. *$\widetilde{\mathcal{M}}^*_{(\eta,J)}$ can be compactified by using limits of graphs of maps; this compactification has good properties.*

In the case relevant to string theory in which M is a projective manifold with $K_M = 0$ of complex dimension three, we find that for generic J, the (real) dimension of $\widetilde{\mathcal{M}}^*_{(\eta,J)}$ is six, and the dimension of

$$\mathcal{M}^*_{(\eta,J)} := \widetilde{\mathcal{M}}^*_{(\eta,J)} / \mathrm{PGL}(2,\mathbb{C})$$

is zero. The space $\mathcal{M}^*_{(\eta,J)}$ itself is already compact in this case; the number of points in that space is our desired invariant. (These points may need to be counted with multiplicity, or with signs.) This invariant counts the number of rational curves (of fixed topological type) on M with respect to its original complex structure, if that number is finite; it can be used as a substitute for that count in the general case.[1]

To describe the invariants in situations more general than complex three-folds with trivial canonical bundle, we must introduce the oriented bordism group $\Omega_*(M)$. The elements of $\Omega_k(M)$ are equivalence classes of pairs (B^k, F) consisting of a compact oriented differentiable manifold B of dimension k (but not necessarily connected), together with a differentiable map $F : B^k \to M$. We say that $(B^k, F) \sim 0$ if there exists an *oriented bordism* (C^{k+1}, H): i.e., a differentiable manifold C of dimension $k{+}1$ and a differentiable map $H : C^{k+1} \to M$ with $\partial C^{k+1} = B^k$ and $H|_{B^k} = F$. We add elements of $\Omega_k(M)$ by means of disjoint union: $(B_1^k, F_1) + (B_2^k, F_2) = (B_1^k \cup B_2^k, F_1 \cup F_2)$; the additive inverse is given by reversing orientation.

The oriented bordism group $\Omega_*(M)$ is a module over the Thom bordism ring Ω (consisting of oriented manifolds modulo oriented bordisms, with no maps to target spaces) via

$$N^j \cdot (B^k, F) = (N^j \times B^k, G)$$

where $G(x,y) = F(y)$.

Theorem 3.5 (Thom [**92**], Conner–Floyd [**2**]).

1. $\Omega_*(M) \otimes \mathbb{Q} \cong H_*(M, \mathbb{Q}) \otimes \Omega$.
2. *If $H_*(M,\mathbb{Z})$ is torsion-free, then $\Omega_*(M) \cong H_*(M,\mathbb{Z}) \otimes \Omega$.*

To describe our basic invariants, we choose three classes α_1, α_2, α_3 in $\Omega_*(M)$ represented by elements $(B_1^{k_1}, F_1)$, $(B_2^{k_2}, F_2)$, $(B_3^{k_3}, F_3)$, and let $Z_j = \mathrm{Image}(F_j)$. We call the invariants defined below the *Gromov–Witten invariants*, since it was Witten [**150**] who pointed out how Gromov's study of $\widetilde{\mathcal{M}}^*_{(\eta,J)}$ could be used in principle to describe invariants relevant in topological quantum field theory. The detailed construction of these invariants was recently carried out by Ruan [**86**]. There are two cases to consider, with one being more technically challenging than the other.[2]

[1]It has not yet been verified that Katz's method of assigning multiplicities to positive-dimensional components in the algebro-geometric context produces the same results as this method from symplectic geometry. Because of the need to include signs in certain circumstances, this invariant can even accommodate the "negative virtual numbers" which occurred in example 2.8.

[2]To simplify the exposition, we have altered Ruan's description of the second case, ignored the necessity of passing to the inhomogeneous $\overline{\partial}$ equation (introduced already by Gromov [**51**]), and built into our definition the so-called "multiple cover formula" expected from the physics [**118, 110**]. (This latter step is now justified thanks to a theorem of Voisin [**98**]; there is also a related result of Manin [**69**].) We are also abusing notation somewhat by using Φ_η in both cases, since the second case is actually related to Ruan's $\widetilde{\Phi}_\eta$ invariant.

Construction (Ruan). *Let $\eta \in H_2(M,\mathbb{Z})$, let $\alpha_j = (B^{k_j}, F_j)$ be a bordism class, and let $Z_j = \mathrm{Image}(F_j)$, for $j = 1,2,3$. Suppose that $\sum_{j=1}^{3}(2n-k_j) = 2n-2K_M\cdot\eta$, where η is the class of the image of φ, and suppose that the almost-complex structure J is generic.*

(a) *If $-K_M\cdot\eta > 0$, then*

$$\{\varphi \in \widetilde{\mathcal{M}}^*_{(\eta,J)} \mid \varphi(0)\in Z_1, \varphi(1)\in Z_2, \varphi(\infty)\in Z_3\}$$

is a finite set. Let $\Phi_\eta(\alpha_1,\alpha_2,\alpha_3)$ denote the signed number of points in this set, with signs assigned according to orientations at the specified points of intersection.

(b) *If $-K_M\cdot\eta = 0$, then there exists an integer $\Phi_\eta(\alpha_1,\alpha_2,\alpha_3)$ which agrees with*

$$\#\{\text{generically injective } \varphi \in \widetilde{\mathcal{M}}^*_{(\eta,J)} \mid \varphi(0)\in Z_1, \varphi(1)\in Z_2, \varphi(\infty)\in Z_3\}$$

(counted with signs) whenever the latter makes sense. (The signs are all positive if the almost-complex structure is integrable.)

These invariants $\Phi_\eta(\alpha_1,\alpha_2,\alpha_3)$ depend only on the bordism classes α_1, α_2, α_3, and do not change under small variation of J.

3.2. Simple properties of Gromov–Witten invariants

In spite of the fact that we needed to pass to bordism to ensure that the Gromov–Witten invariants are well-defined, their dependence on bordism-related phenomena is minimal. In fact, Ruan checks that the invariants are trivial with respect to the Ω-module structure on $\Omega_*(M)$, and so it follows from the theorem of Thom and Conner–Floyd that we get a well-defined $\mathbb{Q}$-valued invariant on rational homology $H_*(M,\mathbb{Q})$. If M has no torsion in homology, we even get an integer-valued invariant on integral homology.

The Gromov–Witten invariants $\Phi_\eta(\alpha_1,\alpha_2,\alpha_3)$ will vanish if α_1 corresponds to a class of real codimension zero or one. This is easy to see—if there are any elements in the set

$$\{\varphi \in \widetilde{\mathcal{M}}^*_{(\eta,J)} \mid \varphi(0)\in Z_1, \varphi(1)\in Z_2, \varphi(\infty)\in Z_3\}$$

then the intersection of the image of φ with the image Z_1 of F_1 has real dimension two or one. By varying the location of $\varphi(0)$, we will produce a two- or one-parameter family of maps. This contradicts the set being finite; thus, the set must be empty and the invariant vanishes.

Note what happens to the Gromov–Witten invariants in the case of interest to string theory ($\dim_{\mathbb{C}} M = 3$, $K_M = 0$): the only relevant invariants are those with $k_1 = k_2 = k_3 = 4$. (This is because $k_j \le 4$ to get a nonzero invariant, so that $6 = \sum(6-k_j) \ge \sum_{j=1}^{3} 2 = 6$, which implies that each k_j is 4.) The possible location of 0 under a generically injective map is easy to spot: the image curve $\varphi(S^2)$ is some rational curve on M, and meets the four-manifold Z_1 in precisely $\#(Z_1\cap\eta) = \alpha_1\cdot\eta$ points; we can choose any of these for the image of 0. Similar remarks about the images of 1 and ∞ lead to the calculation:

$$\Phi_\eta(Z_1,Z_2,Z_3) = (\alpha_1\cdot\eta)(\alpha_1\cdot\eta)(\alpha_1\cdot\eta)\,\#(\mathcal{M}^*_{(\eta,J)}).$$

3.3. The A-model correlation functions

Although we have defined the Gromov–Witten invariants for oriented bordism classes, we will now use them in cohomology instead. As previously remarked, thanks to the triviality of the invariants under the Ω-module structure, if we tensor with $\mathbb{Q}$ we can move the invariants to homology (and then by Poincaré duality, to cohomology). This is at the expense of possibly allowing them to become $\mathbb{Q}$-valued on integral classes. One hopes that they will remain integer valued on integer classes, but this has not yet been established. Therefore, we will give a presentation using $\mathbb{Q}$-coefficients, but the reader should bear in mind that most of the formulas are expected to be valid with integer coefficients if one uses integer cohomology classes.

In brief, the bordism class of $\alpha = (B^k, F)$ gives rise to a homology class $[Z] \in H_k(M, \mathbb{Z})$ (using the image Z of F to represent the class), and by duality to a cohomology class $\zeta = [Z]^\vee \in H^{2n-k}(M, \mathbb{Q})$. (Our retreat to $\mathbb{Q}$-coefficients will be in part because we do not know that every integer cohomology class can be so represented.) We extend the definition of Gromov–Witten invariants to cohomology by defining

$$\Phi_\eta(\zeta_1, \zeta_2, \zeta_3) := \Phi_\eta(\alpha_1, \alpha_2, \alpha_3)$$

when $\alpha_j = (B_j^{k_j}, F_j)$ and $\zeta_j = [\text{Image}(F_j)]^\vee$; then extend by linearity to all of $H^*(M, \mathbb{Q})$.

Our "A-model correlation functions" are then built from the Gromov–Witten invariants, following a calculation from the physics literature [**145, 126, 118, 110**]. There is a certain danger in using the *outcome* of a physics calculation as a *definition*—later, the physicists may become interested in a slightly different problem, whose outcome is radically different from the original one, and we mathematicians will find that our definitions are inadequate.

Nevertheless, we will go ahead and define the A-model correlation functions. These are trilinear functions on the cohomology $H^*(M, \mathbb{Q})$ defined by:

$$\begin{aligned} \langle \zeta_1 \, \zeta_2 \, \zeta_3 \rangle := (\zeta_1 \cup \zeta_2 \cup \zeta_3)|_{[M]} \; &+ \sum_{\substack{\eta \in H_2(M,\mathbb{Z}), \\ -K_M \cdot \eta > 0}} \Phi_\eta(\zeta_1, \zeta_2, \zeta_3)\, q^\eta \\ &+ \sum_{\substack{0 \neq \eta \in H_2(M,\mathbb{Z}), \\ -K_M \cdot \eta = 0}} \Phi_\eta(\zeta_1, \zeta_2, \zeta_3) \sum_{m=1}^{\infty} q^{m\eta} \end{aligned} \tag{3.6}$$

It is sometimes convenient to formally sum the geometric series in the final term, and write $q^\eta/(1-q^\eta)$ in place of $\sum_{m=1}^{\infty} q^{m\eta}$, in which case eq. (3.6) becomes

$$\begin{aligned} \langle \zeta_1 \, \zeta_2 \, \zeta_3 \rangle := (\zeta_1 \cup \zeta_2 \cup \zeta_3)|_{[M]} \; &+ \sum_{\substack{\eta \in H_2(M,\mathbb{Z}), \\ -K_M \cdot \eta > 0}} \Phi_\eta(\zeta_1, \zeta_2, \zeta_3)\, q^\eta \\ &+ \sum_{\substack{0 \neq \eta \in H_2(M,\mathbb{Z}), \\ -K_M \cdot \eta = 0}} \Phi_\eta(\zeta_1, \zeta_2, \zeta_3) \frac{q^\eta}{1-q^\eta} \end{aligned} \tag{3.7}$$

The terms with $K_M \cdot \eta = 0$ have been separated out because they are where the multiple-covered maps cause the greatest difficulty. Heuristically, the coefficients in these functions (as we have defined them) are expected to count the simple maps only.

The symbol q^η which appears in these formulas has not yet been defined. In fact, there are two natural interpretations of eq. (3.6), one algebraic and one geometric, and we consider them in turn in the next two lectures.

LECTURE 4
The Quantum Cohomology Ring

4.1. Coefficient rings

There are several possible ways to interpret the "A-model correlation functions" defined by eq. (3.6). In this lecture, we will focus on the algebraic interpretation, in which the symbol q^η can be regarded as an element of a group ring or semigroup ring.[1] Recall that for any commutative semigroup $\mathcal{S}$ and any commutative ring R, the *semigroup ring of $\mathcal{S}$ with coefficients in R* is the ring

$$R[q;\mathcal{S}] := \left\{ \sum_{\eta\in\mathcal{S}} a_\eta q^\eta \mid a_\eta \in R \text{ and } \{\eta \mid a_\eta \neq 0\} \text{ is finite} \right\}.$$

The symbol q serves as a placeholder, translating the semigroup operation (usually written additively) into a multiplicative structure on a set of monomials. If $\mathcal{S}$ is a group, this coincides with the usual "group ring" construction.

In the case of a Fano variety, the sum in eq. (3.6) defining the A-model correlation function is finite, and we can regard it as taking values in the rational group ring[2] $\mathbb{Q}[q; H_2(M,\mathbb{Z})]$. To be more concrete, if we assume for simplicity that $H_2(M,\mathbb{Z})$ has no torsion, and choose a basis $e_1, \ldots ,e_r$ of $H_2(M,\mathbb{Z})$, then writing $\eta = \sum a^j e_j$ we may associate to η the rational monomial $q^\eta \in \mathbb{Q}(q_1,\ldots,q_r)$ defined by

$$\log q^\eta = \sum a^j \log q_j.$$

(One can also write this multiplicatively:

$$q^\eta = \prod (q_j)^{(a^j)}$$

but then great care is required in distinguishing exponents from superscripts.)

If we choose our basis so that the coefficients a^j are nonnegative for all classes η which have nonvanishing Gromov–Witten invariants $\Phi_\eta(\zeta_1,\zeta_2,\zeta_3)$ for some ζ_1, ζ_2,

[1] I am grateful to A. Givental for pointing out the relevance of group rings.

[2] If we knew that the Gromov–Witten invariants were integers, we could use the integral group ring $\mathbb{Z}[q; H_2(M,\mathbb{Z})]$. But when we passed from bordism to cohomology we lost control of the integer structure.

ζ_3, then each q^η occurring in eq. (3.6) is a *regular* monomial, i.e., q^η belongs to the polynomial ring $\mathbb{Q}[q_1, \dots, q_r]$, and we can calculate eq. (3.6) in that ring.

In the Calabi–Yau case in which $K_M = 0$, the sum in eq. (3.6) is not finite and we must work harder. The simplest interpretation would be to simply allow infinite sums $\sum a_\eta q^\eta$ as formal expressions. However, in order to construct quantum cohomology (which we shall do in the next section) we need the values of the correlation function to lie in a *ring*. In the definition of semigroup rings one restricts to finite sums in order to ensure that the partial sums which occur in the expansion of a product will be finite. That finiteness can still be guaranteed for products of infinite sums if the semigroup satisfies a special property, given below.

We say that a semigroup $\mathcal{S}$ has the *finite partition property* if for every $\eta \in \mathcal{S}$ there are only finitely many pairs $(\eta_1, \eta_2) \in \mathcal{S} \times \mathcal{S}$ such that $\eta = \eta_1 + \eta_2$. For such semigroups, any expression of the form

$$\sum_{\substack{(\eta_1,\eta_2)\ \text{s.t.}\\ \eta_1+\eta_2=\eta}} a_{\eta_1} a_{\eta_2}$$

(for fixed η) will be finite. Thus, infinite sums can be multiplied. So if $\mathcal{S}$ is a semigroup with the finite partition property, we define the *formal semigroup ring of $\mathcal{S}$ with coefficients in R* to be

$$R[[q;\mathcal{S}]] := \{\sum_{\eta\in\mathcal{S}} a_\eta q^\eta\},$$

with the product defined by

$$(\sum_{\eta_1\in\mathcal{S}} a_{\eta_1} q^{\eta_1}) \cdot (\sum_{\eta_2\in\mathcal{S}} a_{\eta_2} q^{\eta_2}) = \sum_{\eta\in\mathcal{S}} (\sum_{\substack{(\eta_1,\eta_2)\ \text{s.t.}\\ \eta_1+\eta_2=\eta}} a_{\eta_1} a_{\eta_2}) q^\eta.$$

The semigroup $H_2(M,\mathbb{Z})$ of interest to us is actually a *group* with a nontrivial free abelian part, and so does not satisfy the finite partition property. However, in eq. (3.6) we are only required to sum over classes which can be realized by pseudo-holomorphic curves—these generate a smaller semigroup. If we are using an integrable almost-complex structure J on M for which M is a Kähler manifold, this smaller semigroup is the *integral Mori semigroup* defined (in the case $h^{2,0} = 0$, for simplicity) as

$$\overline{\mathrm{NE}}(M,\mathbb{Z}) := \{\eta \in H_2(M,\mathbb{Z}) \mid (\omega,\eta) \geq 0\ \forall\, \omega \in \overline{\mathcal{K}}_J\},$$

where $\mathcal{K}_J$ is the Kähler cone and $\overline{\mathcal{K}}_J$ is its closure. The Mori semigroup has the finite partition property (the free part lies in a strongly convex cone, and the torsion part is finite), so we can form the formal semigroup ring $R[[q;\overline{\mathrm{NE}}(M,\mathbb{Z})]]$. Presumably, by using the symplectic version of the Kähler cone, we would find a similar property for the analogous semigroup in the almost-complex case and could form a similar ring in that case.

There is an important variant which we will have occasion to consider. Let $\mathrm{Aut}_J(M)$ be the image in $\mathrm{Aut}\, H_2(M,\mathbb{Z})$ of the group of diffeomorphisms of M compatible with the almost-complex structure J. This group acts on the pseudo-holomorphic curves and so permutes the Gromov–Witten invariants. The values of the A-model correlation function are preserved by the group action, and can be regarded as lying in the ring of invariants

$$R[[q;\overline{\mathrm{NE}}(M,\mathbb{Z})]]^{\mathrm{Aut}_J(M)}.$$

As in the Fano variety case, if we choose an appropriate basis (and assume $H_2(M,\mathbb{Z})$ is torsion-free) then we can regard the correlation function defined in eq. (3.6) as taking values in a formal power series ring $\mathbb{Q}[[q_1,\ldots,q_r]]$. Note that if we set all q_j's to 0, we simply recover the topological trilinear function $(\zeta_1\cup\zeta_2\cup\zeta_3)|_{[M]}$. But although the formal series in eq. (3.6) is expected by the physicists to converge near $q_j = 0$, no convergence properties of the series (as we have defined it) are known at present.

There is an alternative to using the semigroup rings: we could instead use the Novikov rings [**82**] which have played a rôle elsewhere in symplectic geometry [**55**]. For each Kähler class ω, the *Novikov ring* Λ_ω consists of all formal power series

$$\sum_{\eta\in H_2(M,\mathbb{Z})} a_\eta q^\eta$$

such that the set

$$\{\eta \mid a_\eta \neq 0 \text{ and } (\omega,\eta) < c\}$$

is finite for all $c \in \mathbb{R}$. (If it is necessary to specify the ring R in which the coefficients a_η take their values, the notation $\Lambda(\omega, R)$ is used.) The product of two elements of Λ_ω is well-defined, and also belongs to Λ_ω. In the case $H_2(M,\mathbb{Z}) = \mathbb{Z}^r$, Λ_ω is the ring of *generalized Laurent series*

$$\{\sum a_{\vec{k}} q^{\vec{k}} \mid \text{there are only finitely many terms with } \omega\cdot\vec{k} < c \text{ for any } c\in\mathbb{R}\}.$$

4.2. A new algebra structure

The correlation functions defined in the previous lecture can be used to describe a new algebra structure on the cohomology of M, in the following way. Let R be an integral domain (usually we use $R = \mathbb{Z}$ or $R = \mathbb{Q}$), and choose a coefficient ring $\mathcal{R}$ from among

1. the group ring $R[q; H_2(M,\mathbb{Z})]$ (in the case of a Fano variety),
2. the formal semigroup ring with coefficients in R for the Mori semigroup $R[[q; \overline{\mathrm{NE}}(M,\mathbb{Z})]]$ (when this is well-defined, such as in the case of a Kähler manifold),
3. the subring $R[[q; \overline{\mathrm{NE}}(M,\mathbb{Z})]]^{\mathrm{Aut}_J(M)}$ of $\mathrm{Aut}_J(M)$-invariants, or
4. one of the Novikov rings $\Lambda(\omega, R)$.

We introduce a binary operation $\zeta_1 \star \zeta_2$ on $H^*(M,\mathcal{R})$ defined by the requirement

$$((\zeta_1\star\zeta_2)\cup\zeta_3)|_{[M]} = \langle\zeta_1\,\zeta_2\,\zeta_3\rangle.$$

(This is well-defined since the cup product is a perfect pairing.) The class $\mathbb{1} := [M]^\vee \in H^0(M)$ which is dual to the fundamental class $[M] \in H_{2n}(M)$ has the property that the Gromov–Witten invariants $\Phi_\eta(\mathbf{1},\zeta_2,\zeta_3)$ all vanish, hence

$$\langle\mathbb{1}\,\zeta_1\,\zeta_2\rangle = (\zeta_2\cup\zeta_3)|_{[M]};$$

it follows that $\mathbb{1}$ serves as the identity element for the binary operation $\star$.

This interpretation of the correlation function as a binary operation also comes from physics [**141, 151**]. Let us return to the picture we had of the "pair of pants"

surface

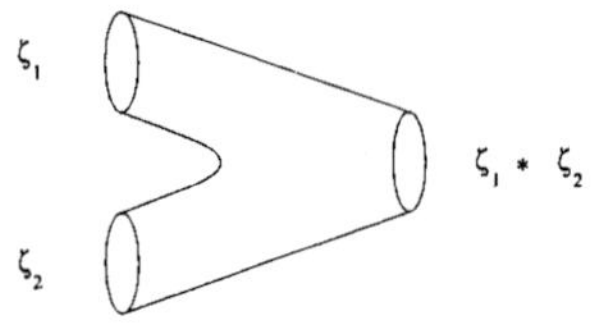

as describing a possible evolution between an initial state with two "incoming" vertex operators ζ_1, ζ_2 on the left and a final state with one "outgoing" vertex operator $\zeta_1 \star \zeta_2$ on the right.

This point of view leads to the remarkable expectation that the binary operation should be associative! A heuristic argument for this runs as follows: the product $(\zeta_1 \star \zeta_2) \star \zeta_3$ is evaluated by means of the surface

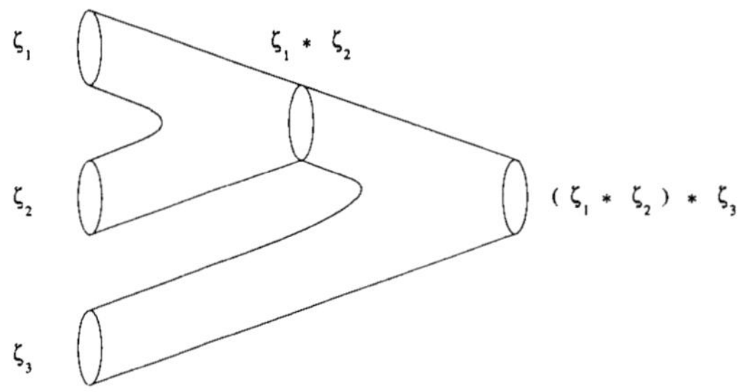

(with an outgoing vertex operator of one piece attached to an incoming vertex operator of the other) while the product $\zeta_1 \star (\zeta_2 \star \zeta_3)$ is evaluated by means of the surface

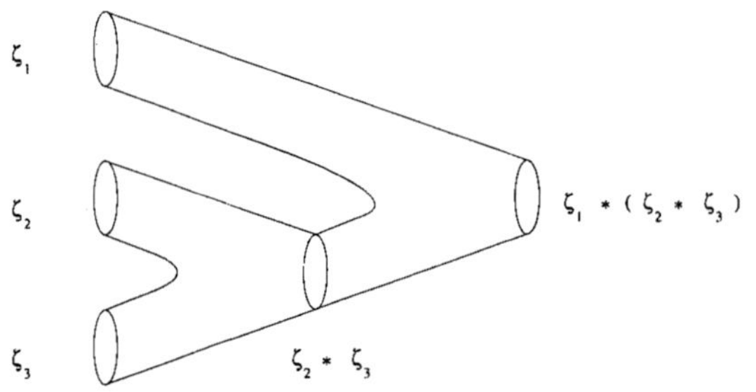

which is a deformation of the first one. So long as the values of the resulting quadrilinear function do not depend on the location of the four points in $\mathbb{CP}^1$ used in defining it, these two products will agree. In fact, as we pointed out in the introduction, the correlation functions we are studying are expected from the physics to be precisely of this "topological" nature which makes them independent of the location of the points [**150**].

This associativity property of the binary operation $\star$ can be rewritten as a set of relations which must be satisfied among the Gromov–Witten invariants themselves. This turns out to be a very deep property, which had not been proved at the time these lectures were delivered (although proofs were given not too long thereafter [**87, 67, 12**]). We have formulated the Gromov–Witten invariants and the binary operation at this level of generality primarily because this associativity property is such an interesting one. However, as we will see in more detail below, for the case of primary interest in mirror symmetry—that of Calabi–Yau threefolds—the associativity is automatic, and there is nothing to prove. (Associativity *does* say something interesting for Calabi–Yau manifolds of higher dimension.)

The $\mathcal{R}$-module $H^*(M, \mathcal{R})$ equipped with the binary operation $\star$ is called the *quantum cohomology ring* of M, or the *quantum cohomology algebra* when we wish to emphasize the $\mathcal{R}$-module structure.

We can give a more geometric description of the new binary operation, by turning each Gromov–Witten invariant itself into a kind of binary operation. Here is a heuristic description of what this construction should look like.

We want a cohomology class $Q_\eta(\zeta_1, \zeta_2)$ with the property that

$$(Q_\eta(\zeta_1, \zeta_2) \cup \zeta_3)|_{[M]} = \Phi_\eta(\zeta_1, \zeta_2, \zeta_3).$$

Consider the set of pseudo-holomorphic curves which satisfy the conditions imposed by ζ_1 and ζ_2 only:

$$\mathcal{M}_\eta(\zeta_1, \zeta_2) := \{\varphi \in \widetilde{\mathcal{M}}^*_{(\eta,J)} \mid \varphi(0) \in Z_1, \varphi(1) \in Z_2\},$$

where $\zeta_j = [Z_j]^\vee$. To count the maps contributing to $\Phi_\eta(\zeta_1, \zeta_2, \zeta_3)$, we must look for all maps in $\mathcal{M}_\eta(\zeta_1, \zeta_2)$ which also send ∞ into Z_3. What subset of M has the property that its intersection with Z_3 is in one-to-one correspondence with such maps? It is the subset consisting of *all possible* points $\varphi(\infty)$ which might be mapping to Z_3. In other words, we can write $Q_\eta(\zeta_1, \zeta_2) = [T_\eta(Z_1, Z_2)]^\vee$, where $T_\eta(Z_1, Z_2)$ is the cycle defined by

$$\begin{aligned} T_\eta(Z_1, Z_2) &:= \{P \in M \mid P = \varphi(\infty) \text{ for some } \varphi \in \mathcal{M}_\eta(\zeta_1, \zeta_2)\} \\ &= \bigcup_{\varphi \in \mathcal{M}_\eta(\zeta_1, \zeta_2)} \operatorname{Image}(\varphi). \end{aligned}$$

Then $T_\eta(Z_1, Z_2) \cap Z_3$ will correspond to the maps counted by $\Phi_\eta(\zeta_1, \zeta_2, \zeta_3)$, where $\zeta_3 = [Z_3]^\vee$. Note that for this heuristic description to work, we need the set $T_\eta(Z_1, Z_2)$ to be of the expected dimension. A better formal definition of $Q_\eta(\zeta_1, \zeta_2)$ would be the pushforward under evaluation at ∞ of the pullback of $\mathcal{M}_\eta(\zeta_1, \zeta_2)$ to the universal family of maps.

Expressed in these terms, then, the binary operation can be written:

$$\begin{aligned} \zeta_1 \star \zeta_2 := \zeta_1 \cup \zeta_2 \;&+ \sum_{\substack{\eta \in H_2(M,\mathbb{Z}),\\ -K_M \cdot \eta > 0}} q^\eta\, Q_\eta(\zeta_1, \zeta_2) \\ &+ \sum_{\substack{0 \neq \eta \in H_2(M,\mathbb{Z}),\\ -K_M \cdot \eta = 0}} \frac{q^\eta}{1 - q^\eta}\, Q_\eta(\zeta_1, \zeta_2) \end{aligned} \tag{4.1}$$

Recall that the Gromov–Witten invariant $\Phi_\eta(\zeta_1, \zeta_2, \zeta_3)$ with $\zeta_j \in H^{\ell_j}(M, \mathbb{Q})$ is zero unless

$$\ell_1 + \ell_2 + \ell_3 = 2n + 2(-K_M \cdot \eta), \quad \text{and} \quad \ell_j \geq 2.$$

It follows that if the cycle $Q_\eta(\zeta_1, \zeta_2)$ is nonzero, we have

$$Q_\eta(\zeta_1, \zeta_2) \in H^{2n - \ell_3}(M, \mathbb{Q}) = H^{\ell_1 + \ell_2 - 2(-K_M \cdot \eta)}(M, \mathbb{Q}).$$

Thus, if $K_M = 0$, then the binary operation $\star$ preserves the grading on cohomology, while if $-K_M \cdot \eta > 0$ the grading is shifted down by $2(-K_M \cdot \eta)$. But note that in any case, the $\mathbb{Z}/2\mathbb{Z}$-grading on cohomology is preserved.

Note also that $\ell_j \leq 2n$ implies $\ell_1 + \ell_2 + \ell_3 \leq 6n$ and hence $-K_M \cdot \eta \leq 2n$.

Exercise. Show that the semipositivity condition $3 - n < -K_M \cdot \eta$ implies that the grading cannot shift up, it can only shift down.

Example 4.2. (cf. [**138, 149**]) We now compute an example of the quantum cohomology ring. Let $M = \mathbb{CP}^n$ (with ω induced from the Fubini–Study metric, which will ensure semipositivity). The formal semigroup ring in this case can be written as $\mathcal{R} = \mathbb{Q}[[q]]$ (or we could use $\mathcal{R} = \mathbb{Q}[q]$ since we know the sums are finite, this being a Fano variety).

If C is any complex curve on M, then $-K_M \cdot C = d(n+1)$, where d is the degree of the curve. Since $-K_M \cdot C \le 2n$, we must have $d = 1$. So only lines (and constant maps) will contribute to our correlation function.

Now the predicted real dimension of the space of maps $\mathbb{CP}^1 \to M$ whose image L has degree one is

$$2n + 2(-K_M \cdot L) = 2n + 2(n+1) = 4n + 2$$

while the actual dimension is

$$\dim_{\mathbb{R}} \mathrm{PGL}(2,\mathbb{C}) + \dim_{\mathbb{R}} \mathrm{Gr}(\mathbb{CP}^1, \mathbb{CP}^n) = 6 + 2 \cdot 2(n-1) = 4n + 2$$

so we should be able to use the given complex structure to compute the invariants.

The Gromov–Witten invariants are evaluated as follows. A basis for $H^*(M,\mathbb{Q})$ is given by the classes $\zeta^k \in H^{2k}(M,\mathbb{Q})$ where ζ is the class of a hyperplane. We choose k_1, k_2, k_3, satisfying

$$2k_1 + 2k_2 + 2k_3 = 4n + 2$$

and find that there is a *unique* line in $\mathbb{CP}^m$ meeting three fixed linear spaces of codimensions k_1, k_2 and k_3. And there is a unique map sending 0, 1, ∞ to the intersection points with the three linear spaces. Thus,

$$\Phi_L(\zeta^{k_1}, \zeta^{k_2}, \zeta^{k_3}) = 1.$$

Expressed in terms of the binary operation, we find that

$$\zeta^{k_1} \star \zeta^{k_2} = \begin{cases} \zeta^{k_1+k_2} & \text{if } k_1 + k_2 \le n \\ \zeta^{k_1+k_2-n-1}\, q & \text{if } k_1 + k_2 \ge n+1 \end{cases}.$$

It follows that the quantum cohomology ring can be described as:

$$\mathcal{R}[\zeta]/(\zeta^{\star(n+1)} - q).$$

Example 4.3. If we consider the case relevant to string theory ($\dim_{\mathbb{C}}(M) = 3$, $K_M = 0$), we find that the only products which differ from the cup product are products $\zeta_1 \star \zeta_2$, with $\zeta_1, \zeta_2 \in H^2(M)$, and these are given by

$$\zeta_1 \star \zeta_2 := \zeta_1 \cup \zeta_2 \;+ \sum_{0 \ne \eta \in H_2(M,\mathbb{Z})} \left(\zeta_1(\eta) \cdot \zeta_2(\eta) \cdot \#(\mathcal{M}^*_{(\eta,J)}) \right) \frac{q^\eta}{1 - q^\eta}\, \eta$$

Here, $\#(\mathcal{M}^*_{(\eta,J)})$ denotes the number of curves in class η (counted with appropriate multiplicity).

Remark. Note that the associativity of the binary operation $\star$ is automatically satisfied by threefolds with trivial canonical bundle, since only one of the products being associated can be different from the cup product.

Example 4.4. Let $\lambda \in H^2(M)$ be represented by L, a submanifold of real codimension two. If we define

$$\mathcal{M}_\eta(\zeta) := \{\varphi \in \widetilde{\mathcal{M}}^*_{(\eta,J)} \mid \varphi(1) \in \zeta\},$$

and

$$\Gamma_\eta(\zeta) := [\{P \in M \mid P = \varphi(\infty) \text{ for some } \varphi \in \mathcal{M}_\eta(\zeta)\}]^\vee,$$

then we can expect that

$$Q_\eta(\lambda, \zeta) = \lambda(\eta) \cdot \Gamma_\eta(\zeta).$$

This is because the image of each φ should meet L in precisely $\lambda(\eta)$ points, any of which may be chosen as $\varphi(0)$.

The binary operation can then be written:

$$\begin{aligned}\lambda \star \zeta := \lambda \cup \zeta \ &+ \sum_{\substack{\eta \in H_2(M,\mathbb{Z}),\\ -K_M \cdot \eta > 0}} \lambda(\eta)\, q^\eta\, \Gamma_\eta(\zeta)\\ &+ \sum_{\substack{0 \neq \eta \in H_2(M,\mathbb{Z}),\\ -K_M \cdot \eta = 0}} \lambda(\eta) \frac{q^\eta}{1-q^\eta} \Gamma_\eta(\zeta)\end{aligned}$$

We regard Γ_η as a map on cohomology, and call it the *Gromov–Witten map.*

4.3. Algebraic properties of the correlation functions

Let K be the field of fractions of our coefficient ring $\mathcal{R}$; tensoring the quantum cohomology ring with K makes it into a K-algebra. This quantum cohomology algebra carries some additional structure which makes it into what is known as a *Frobenius algebra.*[3] By definition this is a K-algebra A with a multiplicative identity element $\mathbb{1}$, such that there exists a linear functional $\varepsilon : A \to K$ for which the induced bilinear pairing $(x, y) \mapsto \varepsilon(x \star y)$ is nondegenerate. There does not seem to be a standard name for such a functional; we call it an *expectation function* (cf. [**143**]). If an expectation function exists at all, then most linear functionals on A can serve as expectation functions. If A is $\mathbb{Z}$-graded, we call ε a *graded expectation function* when $\ker(\varepsilon)$ is a graded subalgebra of A (and we call A a *graded Frobenius algebra* when such a function exists). There is much less freedom to choose graded expectation functions.

The cohomology of a compact manifold M has the structure of a graded Frobenius algebra, with multiplication given by cup product, $\mathbb{1}$ given by the standard generator of $H^0(M)$, and a graded expectation function given by "evaluation on the fundamental class." The quantum cohomology algebra is a deformation of this algebra, with the expectation function given by

$$\varepsilon(\zeta) = \langle \zeta\, \mathbb{1}\, \mathbb{1} \rangle,$$

which again can be interpreted as evaluation on the fundamental class. The induced bilinear pairing

$$(\zeta_1, \zeta_1) \mapsto \varepsilon(\zeta_1 \star \zeta_2) = \langle \zeta_1\, \zeta_2\, \mathbb{1} \rangle$$

[3] We follow standard mathematical usage [**4, 11**] and do not require a Frobenius algebra to be commutative; our definition therefore differs slightly from that in [**129**]. However, we will primarily be interested in the even part $H^{ev}(M)$ of the cohomology of M, on which the quantum product will in fact be commutative.

coincides with the usual cup product pairing. Note that the correlation function is also determined by ε and $\star$, via

$$\langle \zeta_1\, \zeta_2\, \zeta_3 \rangle = \varepsilon(\zeta_1 \star \zeta_2 \star \zeta_3).$$

That is, rather than specifying the correlation function first and using it to determine the quantum product, we can simply work with the quantum product and the expectation function.

For most symplectic manifolds, the Frobenius algebra structure on quantum cohomology is not graded; however, in the Calabi–Yau case we get the structure of a graded Frobenius algebra.

Generally, given any associative K-algebra A with multiplicative identity, and any linear functional φ on A, the kernel of the bilinear form $(x, y) \mapsto \varphi(x * y)$ is an ideal $\mathcal{J}_\varphi$, and the quotient ring $A/\mathcal{J}_\varphi$ is a Frobenius algebra with expectation function induced by φ. If A is itself a Frobenius algebra with an expectation function ε, then by a theorem of Nakayama **[79]** (see **[11]** for a modern discussion), φ takes the form $\varphi(x) = \varepsilon(\alpha * x)$ for some fixed element $\alpha \in A$, and $\mathcal{J}_\varphi$ coincides with the annihilator of α.

Although the correlation functions determine the ring structure, the opposite does not hold in general—there can be many expectation functions on a given algebra. However, if A is a graded Frobenius algebra of finite length as a K-module and all elements of A have nonnegative degree, then the graded expectation functions on A are in one-to-one correspondence with degree 0 elements of A which are not zero-divisors. (This is because they must all be of the form $\varphi(x) = \varepsilon(\alpha * x)$ for some α which is not a zero-divisor, but every element of degree > 0 must be a zero-divisor.) In particular, in the case of the quantum cohomology algebra of a Calabi–Yau manifold M (equipped with a symplectic structure), we have a graded Frobenius algebra of finite length in which the degree 0 elements are just the one-dimensional vector space $H^0(M)$. This means that the graded expectation function is unique up to multiplication by an element of K, and that the ring structure determines the correlation functions up to this overall factor. (It is not hard to see in the Calabi–Yau case that the graded expectation function is nonzero precisely on the top degree piece $H^{2n}(M)$, where $n = \dim_{\mathbb{C}} M$, and that $H^{2n}(M)$ must also be one-dimensional.) We will see this structure again when we study the B-model correlation functions in lecture six.

LECTURE 5
Moduli Spaces of σ-Models

5.1. Calabi–Yau manifolds and nonlinear σ-models

In this lecture, we wish to give a more geometric interpretation to the A-model correlation functions as defined by eq. (3.6). This geometric interpretation is motivated in part by a study of the moduli spaces of the conformal field theories associated to Calabi–Yau manifolds, so we begin with a description of those moduli spaces.

Let M be a Kähler manifold with $K_M = 0$. Underlying M is a differentiable manifold X of real dimension $2n$. We can regard M as consisting of X together with a chosen integrable almost-complex structure J and a Kähler metric g_{ij}, such that $K_M = 0$. (The complex manifold specified by J will then be denoted X_J.) If ω denotes the Kähler form of the metric, then by a theorem of Calabi [**33**] there is at most one Ricci-flat metric whose Kähler form is cohomologous to ω; by a theorem of Yau [**103**] such a Ricci-flat metric always exists. The global holonomy of such a metric is necessarily contained in $\mathrm{SU}(n)$. (The metric being Kähler implies that its holonomy is contained in $\mathrm{U}(n) \subset SO(2n)$; the Ricci-flatness further restricts the holonomy to $\mathrm{SU}(n)$, and also implies that the canonical bundle is trivial. See [**25**] for an account of these holonomy conditions.)

We use the term *Calabi–Yau manifold* to mean a compact connected orientable manifold X of dimension $2n$ which admits Riemannian metrics whose (global) holonomy is contained in $\mathrm{SU}(n)$. You should be aware that there are some places in the literature (including papers of mine [**73**]) where "Calabi–Yau manifold" is used in the more restrictive sense of a Riemannian manifold with holonomy precisely $\mathrm{SU}(n)$. These alternate definitions will often also insist that a complex structure has been chosen on X.

Given a Calabi–Yau manifold X (in our sense) and a metric on it whose holonomy lies in $\mathrm{SU}(n)$, there always exist complex structures on X for which the given metric is Kähler. If $h^{2,0} = 0$, then there are only a finite number of such complex structures. (If the universal cover is a written as a product of indecomposable pieces, one may apply conjugation on the various factors to obtain other complex structures.) When $h^{2,0} > 0$, however, the complex structures depend on parameters. There are some very interesting cases with $h^{2,0} > 0$, including the famous K3 surfaces, but lack of time in these lectures forces us to assume—with regret—that $h^{2,0} = 0$ henceforth.

The physical model discussed in lecture one which considers maps from surfaces to a six-dimensional target space is a special case of a class of physical theories called "nonlinear σ-models." One regards these as quantum field theories on the surfaces themselves, with various vertex operators and correlation functions derived from the space of maps from the surface to the target. The target should be a fixed Riemannian manifold, usually assumed to be compact.

When the Riemannian metric on the target is (a particular perturbation of) one which has holonomy in $\mathrm{SU}(n)$, the resulting "nonlinear σ-model" is believed to be invariant under conformal transformations of the surface. It thus is a type of "conformal field theory"—an even broader class of physical models which have a rich literature devoted to their study (see [**130**] for an introduction and further references). Conformal field theories typically depend on finitely many parameters, and in the case of a nonlinear σ-model those parameters have a direct geometric interpretation. In the Lagrangian formulation of the theory, one must specify the metric g_{ij} on the target X together with an auxiliary harmonic two-form B on X called the "B-field." (To simplify matters, we take our metrics to have holonomy in $\mathrm{SU}(n)$, even though the true metrics of interest in physics will be perturbations of those; we also assume that $H_2(X, \mathbb{Z})$ has no torsion.[1]) The data consisting of the pair (g_{ij}, B) accounts for all local parameters in the conformal field theory moduli space, so we get at least a good local description of moduli if we specify such a pair. More details about these moduli spaces can be found in [**75**].

Two pairs (g_{ij}, B) and (g'_{ij}, B') will determine isomorphic conformal field theories if there is a diffeomorphism $\varphi : X \to X$ such that $\varphi^*(g'_{ij}) = g_{ij}$, and $\varphi^*(B') - B \in H^2_{\mathrm{DR}}(X, \mathbb{Z})$. (We use the notation $H^k_{\mathrm{DR}}(X, \mathbb{Z})$ to denote the image of integral cohomology in de Rham cohomology.) This second condition arises because the appearance of B in the Lagrangian is always in the form $\int_\Sigma B$, and the Lagrangian is exponentiated (with an appropriate factor of $2\pi i$) in every physically observable quantity.

We call the set of all isomorphism classes of such pairs the *semiclassical nonlinear σ-model moduli space*, or simply the *σ-model moduli space* (for short). This may differ from the actual *conformal field theory moduli space* for three reasons.

1. It may happen that the physical theory does not exist for all values of g_{ij} and B. Most of the study of these theories uses perturbative methods, valid near a limit of "large volume" of the metric, but it may be that the theory breaks down when the volume (either of X, or of images of holomorphic maps into X) becomes too small.
2. On the other hand, there may be a sort of analytic continuation of the theory beyond the region where the σ-model description is valid. (This was shown to occur in [**106, 153**].) It was only claimed above that the specification of (g_{ij}, B) gave good *local* parameters for the moduli.
3. Furthermore, there could be subtle isomorphisms between conformal field theories which do not show up in the σ-model interpretation. This is known to happen in the K3 surface case [**113**], for example (which we have no time

[1] The correct description of the moduli space will be slightly different if torsion is included—see section 5.3 below.

to discuss here)—mirror symmetry provides a new identification of conformal field theories.

We will ignore these phenomena for the present, and concentrate on the "σ-model moduli space" which parameterizes pairs (g_{ij}, B) modulo equivalence.

To study this moduli space using the tools of algebraic geometry, we must choose a complex structure on X. In fact, if we consider the set of triples (g_{ij}, B, J) modulo equivalence, with J being an integrable almost-complex structure for which the metric g_{ij} is a Ricci-flat Kähler metric, then the map from the set of equivalence classes of triples to that of pairs is a finite map. (It is a map of degree two if the holonomy is precisely $SU(n)$.)

On the other hand, we can map the set of triples (g_{ij}, B, J) to the moduli space $\mathcal{M}_{\text{cx}}$ of complex structures on X. That moduli space is quite well-behaved, both locally and globally. The local structure is given by the theorem of Bogomolov–Tian–Todorov [**29, 93, 95**], which says that all first-order deformations are unobstructed. (I recommend Bob Friedman's paper [**43**] for a very readable account of this theorem.) Thus, the moduli space $\mathcal{M}_{\text{cx}}$ will be smooth, and the tangent space at $[J]$ can be canonically identified with $H^1(T_{X_J}^{(1,0)})$. Globally, $\mathcal{M}_{\text{cx}}$ is known to be a quasi-projective variety (if one specifies a "polarization") by a theorem of Viehweg [**96**]. We will study the moduli space $\mathcal{M}_{\text{cx}}$ in more detail (using variations of Hodge structure) in the next section.

The fibers of the map

$$\{(g_{ij}, B, J)\}/\!\sim \;\to\; \mathcal{M}_{\text{cx}} \tag{5.1}$$

(from the set of equivalence classes of triples to the moduli space) are spaces of the form $\mathcal{D}/\Gamma$, with

$$\begin{aligned} \mathcal{D} &= H^2(X, \mathbb{R}) + i\,\mathcal{K}_J \\ \Gamma &= H^2_{\text{DR}}(X, \mathbb{Z}) \rtimes \operatorname{Aut}_J(X). \end{aligned}$$

One hopes that the map (5.1) is some kind of fiber bundle (at least generically); this would require that both the family of Kähler cones and the family of automorphism groups are generically locally constant. This has been shown for the Kähler cones in the case of complex dimension three by Wilson [**99**].

The tangent spaces to the fibers of the map (5.1) can be canonically identified with $H^1((T_{X_J}^{(1,0)})^*)$. Mirror symmetry predicts that X should have a mirror partner Y, such that the moduli spaces of conformal field theories on X and Y should be isomorphic, but with a reversal of rôles of $H^1(T_{X_J}^{(1,0)})$ and $H^1((T_{X_J}^{(1,0)})^*)$. That is, under the isomorphism between the conformal field theory moduli spaces, the part of the tangent space corresponding to $H^1((T_{X_J}^{(1,0)})^*)$ on X should map to the part corresponding to $H^1(T_{Y_{J'}}^{(1,0)})$ on Y, and vice versa.

In particular, the rôles of base and fiber in (5.1) should be reversed. This is at first sight a rather peculiar statement, since the base and the fiber do not look much alike: the base $\mathcal{M}_{\text{cx}}$ is a quasi-projective variety, whereas the fiber $\mathcal{D}/\Gamma$ looks much more like a Zariski open subset of a bounded domain—a typical model for the space is $(\Delta^*)^r$, where Δ^* is the punctured disk.

This is in fact one of the indicators that the conformal field theory moduli space must be analytically continued beyond the realm of σ-models, as suggested in point 2 above. We will see further evidence of this at the end of lecture seven.

5.2. Geometric interpretation of the A-model correlation functions

We turn now to a geometric interpretation of the A-model correlation functions, which in the case of Calabi–Yau manifolds will turn out to be closely related to the spaces $\mathcal{D}/\Gamma$ described above.

In the previous lecture, the symbols q^η were treated purely formally, which allowed us to discuss some algebraic aspects of quantum cohomology. Now, however, we would like to make the new product more geometric by giving specific values to the q^η's, thereby making the quantum cohomology ring into a deformation of the usual cohomology ring. Turning algebraic parameters into geometric data is a familiar task for algebraic geometers; however here, we only have formal parameters. We will describe a natural parameter space as a formal completion of a certain geometric space—if some day someone proves that the series (3.6) and (4.1) are convergent power series, then the true parameter space will be a neighborhood (in the classical topology) of the completion point within the geometric space which we will construct.

Let $\mathcal{R} = \mathbb{Q}[[q; \overline{\mathrm{NE}}(X_J, \mathbb{Z})]]$ be the formal semigroup ring of the integral Mori semigroup. If $\overline{\mathrm{NE}}(X_J, \mathbb{Z})$ is finitely generated, then we can take as the geometric space $\operatorname{Spec} \mathbb{C}[q; \overline{\mathrm{NE}}(X_J, \mathbb{Z})]$ (the spectrum of the semigroup ring), and as its completion the formal scheme $\operatorname{Spf} \mathcal{R}_\mathbb{C}$, where $\mathcal{R}_\mathbb{C}$ denotes $\mathcal{R} \otimes_\mathbb{Q} \mathbb{C}$ and Spf is the formal spectrum. More generally, if the ring of $\mathrm{Aut}_J(X)$-invariants $\mathcal{R}^{\mathrm{Aut}_J(X)}$ is the formal completion of a ring of finite type over $\mathbb{Q}$, we take our completed parameter space to be $\operatorname{Spf}(\mathcal{R}_\mathbb{C}^{\mathrm{Aut}_J(X)})$.

In the finitely generated case, this geometric space $\operatorname{Spec} \mathbb{C}[q; \overline{\mathrm{NE}}(X_J, \mathbb{Z})]$ is in a natural way an affine toric variety, and as such admits a rather concrete description: the geometric points are in one-to-one correspondence with the set of semigroup homomorphisms $\mathrm{Hom}_{\mathrm{sg}}(\overline{\mathrm{NE}}(X_J, \mathbb{Z}), \mathbb{C})$, where $\mathbb{C}$ is given the structure of a *multiplicative* semigroup. Any geometric point ξ in the parameter space—regarded as a semigroup homomorphism—specifies compatible values $\xi(q^\eta)$ for the symbols q^η. An important open problem is to decide for which ξ the series expressions (3.6) for the correlation functions converge. If convergent, the correlation functions would become actual $\mathbb{C}$-valued functions on a parameter space (as expected by the physicists), which would be an open subset of $\operatorname{Spec} \mathbb{C}[q; \overline{\mathrm{NE}}(X_J, \mathbb{Z})]$ in the classical topology.

To make this even more concrete, consider the case in which the Mori semigroup is freely generated by elements $e_1, \ldots, e_r$ which also serve as a basis of the lattice $H_2(X, \mathbb{Z})$. In this case, we can define $q_j := q^{e_j}$, and write the ring $\mathcal{R}$ as a formal power series ring $\mathcal{R} = \mathbb{Q}[[q_1, \ldots, q_r]]$. The geometric space $\operatorname{Spec} \mathbb{C}[q_1, \ldots, q_r]$ can then be identified as $\mathbb{C}^r$ with coordinates $q_1, \ldots q_r$. One natural candidate for the open set on which the correlation functions might converge is

$$\{(q_1, \ldots, q_r) \in \mathbb{C}^r \mid 0 \le |q_j| < 1\}.$$

More generally, still assuming that $H_2(X, \mathbb{Z})$ is torsion-free, suppose we choose a basis $e_1, \ldots, e_r$ whose span as a semigroup *contains* $\overline{\mathrm{NE}}(X_J, \mathbb{Z})$. Then the corresponding formal power series ring $\mathbb{Q}[[q_1, \ldots, q_r]]$ contains our coefficient ring $\mathcal{R}$. If we let σ denote the open real cone generated by the dual basis $e^1, \ldots, e^r$, then that formal power series ring can be more canonically described as the formal semigroup ring $\mathcal{R}_\sigma := \mathbb{Q}[[q; \check{\sigma} \cap H_2(X, \mathbb{Z})]]$. The same cone σ can be used to give a canonical description of the open set specified by $0 < |q_j| < 1$ in the form

$$(H^2(X,\mathbb{R}) + i\sigma)/H^2(X,\mathbb{Z}).$$

(To see this, write a general element of $H^2(X,\mathbb{C})$ modulo $H^2(X,\mathbb{Z})$ in the form

$$\frac{1}{2\pi i}\sum(\log q_j)e^j,$$

and note that the condition $0 < |q_j| < 1$ is equivalent to $\mathrm{Im}(\frac{1}{2\pi i}\log q_j) > 0$.) The Mori semigroup $\overline{\mathrm{NE}}(X_J,\mathbb{Z})$ will be contained in the semigroup spanned by $\{e_j\}$ precisely when the cone σ is contained in the Kähler cone of X_J.

We will treat such a choice of cone σ as specifying a coordinate chart on the geometric space we are trying to construct. For any such cone, we define

$$\mathcal{D}_\sigma = H^2(X,\mathbb{R}) + i\,\sigma \subset H^2(X,\mathbb{C})$$

In terms of local coordinates, as pointed out above we have

$$\mathcal{D}_\sigma/H^2(X,\mathbb{Z}) = \{(q_1,\ldots,q_r) \mid 0 < |q_j| < 1\}.$$

The open subset of our desired geometric space will be a partial compactification of this, defined by

$$(\mathcal{D}_\sigma/H^2(X,\mathbb{Z}))^- = \{(q_1,\ldots,q_r) \mid 0 \le |q_j| < 1\}.$$

We call the origin $0 \in (\mathcal{D}_\sigma/H^2(X,\mathbb{Z}))^-$ the *distinguished limit point* in this space.

It is hoped that the expressions for the A-model correlation functions, or for the binary operation $\zeta_1 \star \zeta_2$, will converge in a neighborhood of the distinguished limit point 0 in $(\mathcal{D}_\sigma/H^2(X,\mathbb{Z}))^-$. The different possible choices of σ will correspond to operations—such as blowing up the boundary—which change the compactification without changing the underlying space.

Intrinsically, we can describe $\mathcal{R}_\sigma \otimes \mathbb{C}$ as the formal completion of the local ring of $(\mathcal{D}_\sigma/H^2(X,\mathbb{Z}))^-$ at its distinguished limit point 0.

The geometric space which is emerging from this discussion is very closely related to the space $\mathcal{D}/\Gamma$ which formed part of the nonlinear σ-model moduli space in the case of a Calabi–Yau manifold with $h^{2,0} = 0$. In fact, if $\mathcal{K}_J$ is the Kähler cone of such a Calabi–Yau manifold which can be partitioned into cones σ_α which are spanned by various bases of $H^2(X,\mathbb{Z})$, then $\mathcal{D}/H^2(X,\mathbb{Z})$ is the interior of the closure of the union of the sets $\mathcal{D}_{\sigma_\alpha}/H^2(X,\mathbb{Z})$. Ideally, one could make such a partition in an $\mathrm{Aut}_J(X)$-equivariant way. This would be guaranteed by the following conjecture.

The Cone Conjecture. *Let X be a Calabi–Yau manifold on which a complex structure J has been chosen, and suppose that $h^{2,0}(X) = 0$. Let $\mathcal{K}_J$ be the Kähler cone of X, let $(\mathcal{K}_J)_+$ be the convex hull of $\overline{\mathcal{K}}_J \cap H^2(X,\mathbb{Q})$, and let $\mathrm{Aut}_J(X)$ be the group of holomorphic automorphisms of X. Then there exists a rational polyhedral cone $\Pi \subset (\mathcal{K}_J)_+$ such that $\mathrm{Aut}_J(X).\Pi = (\mathcal{K}_J)_+$.*

A nontrivial case of this conjecture—Calabi–Yau threefolds which are fiber products of generic rational elliptic surfaces with section (as studied by Schoen [**90**])—has been checked by Grassi and the author [**49**]. There are some other pieces of supporting evidence in examples worked out by Borcea [**30**] and Oguiso [**83**].

When this conjecture holds, there is a partial compactification of $\mathcal{D}/\Gamma$ constructed in [**74**] by gluing together the spaces $(\mathcal{D}_{\sigma_\alpha}/H^2(X,\mathbb{Z}))^-$ for an $\mathrm{Aut}_J(X)$-equivariant partitioning of $\mathcal{K}_J$, and modding out by $\mathrm{Aut}_J(X)$. This produces a "semi-toric" partial compactification of the type introduced by Looijenga [**68**]. Because it is covered by explicit coordinate charts, this is a convenient type of compactification for making comparisons of correlation functions.

There is also a "minimal" semi-tori compactification determined from the same data, which partially compactifies $\mathcal{D}/\Gamma$ more directly, adding several new strata but only a single stratum of maximal codimension (the analogue of the "distinguished limit points"). When the cone conjecture holds, the ring of invariants $\mathcal{R}^{\mathrm{Aut}_J(X)}$ is the formal completion of a ring of finite type over $\mathbb{Q}$, and the completion of the local ring of the minimal semi-toric compactification at its distinguished point P coincides with $\mathrm{Spf}(\mathcal{R}_{\mathbb{C}}^{\mathrm{Aut}_J(X)})$. On such a compactification, we will expect

$$\lim_{Q\to P}\langle\zeta_1\,\zeta_2\,\zeta_3\rangle_Q = (\zeta_1\cup\zeta_2\cup\zeta_3)|_{[X]} \tag{5.2}$$

(the "$q_j = 0$ values" in coordinate charts). Such a point is called a "semiclassical limit" in the physics literature [**109**].

5.3. The rôle of torsion in the moduli space

Up to this point, we have not considered the effects of possible torsion in $H_2(X,\mathbb{Z})$ and in fact we have explicitly assumed at several points that there was no torsion. If torsion is present, we can define the formal semigroup ring $\mathcal{R} = \mathbb{Q}[[q;\overline{\mathrm{NE}}(X_J,\mathbb{Z})]]$ as before, and it will have a torsion part $\mathcal{R}_{\mathrm{torsion}}$ whose spectrum is a finite set of geometric points. This can be identified with the set of connected components of our parameter space. It can also be seen in the following description of the σ-model moduli space.

The complete description of the σ-model moduli space (with the torsion included) considers the quantity $e^{2\pi i(B+i\omega)}$ to lie in $\mathrm{Hom}(H_2(X,\mathbb{Z}),\mathbb{C}^*)$. This can be thought of concretely as having a torsion part, together with a free part which lies in the space

$$\mathrm{Hom}(H_2(X,\mathbb{Z})/\mathrm{torsion},\mathbb{C}^*) \cong H^2(X,\mathbb{C}^*) \cong H^2(X,\mathbb{C})/H^2_{\mathrm{DR}}(X,\mathbb{Z})$$

where (as in section 5.1) $H^2_{\mathrm{DR}}(X,\mathbb{Z})$ is the image of $H^2(X,\mathbb{Z})$ in de Rham cohomology, isomorphic to $H^2(X,\mathbb{Z})/\mathrm{torsion}$. A representative of the free part can be written as $B_{\mathrm{free}} + i\omega \in H^2(X,\mathbb{C})$, where ω is the Kähler form and B_{free} is the real two-form which appeared in section 5.1. The torsion part of $e^{2\pi i(B+i\omega)} \in \mathrm{Hom}(H_2(X,\mathbb{Z}),\mathbb{C}^*)$ can be identified with the torsion part of our coefficient ring $\mathcal{R}_{\mathrm{torsion}}$ from the algebraic interpretation. One way to interpret this "B-field with torsion included" is to regard it as an element of $H^2(X,\mathbb{R}/\mathbb{Z})$.

LECTURE 6
Variations of Hodge Structure

6.1. The B-model correlation functions

Our goal in this lecture is to describe the B-model correlation functions and how they are related to variations of Hodge structure. We work with Calabi–Yau manifolds on which complex structures have been chosen. That is, we let W be a complex manifold with $K_W = 0$. The assumption of trivial canonical bundle is needed in order to define the B-model correlation functions.

Let us define

$$H^{-p,q}(W) := H^q(\Lambda^p(T_W^{(1,0)})),$$

and consider all of these groups together:

$$H^{-*}(W) := \bigoplus_{p,q} H^{-p,q}(W).$$

There is a natural ring structure on $H^{-*}(W)$ which can be thought of as a sheaf cohomology version of the cup product pairing:

$$H^q(\Lambda^p(T_W^{(1,0)})) \otimes H^{q'}(\Lambda^{p'}(T_W^{(1,0)})) \to H^{q+q'}(\Lambda^{p+p'}(T_W^{(1,0)})).$$

Note that since these are sheaf cohomology groups, this ring structure is not "topological" in nature; in fact, it depends heavily on the choice of complex structure on W.

Recall that in the case of the A-model correlation functions on a symplectic manifold M, the expectation function which determined the Frobenius algebra structure was a very familiar object, given by evaluating a cohomology class on the fundamental class of M (which determines a canonical map $H^{n,n}(M) \to \mathbb{C}$). By contrast, the ring structure on quantum cohomology was unusual. In this new "B-model" case, however, the ring structure is straightforward but the expectation function is more elusive. To define it, we must choose a nonvanishing global section $\Omega^{\otimes 2}$ of $(K_W)^{\otimes 2}$. This is then used in two steps to specify the expectation function:

$$H^{-n,n}(W) = H^n(\Lambda^n(T_W^{(1,0)})) \xrightarrow{\lrcorner\Omega} H^n(\mathcal{O}_W) \cong \left(H^0(K_W)\right)^* \xrightarrow{\otimes\Omega} \mathbb{C},$$

where the middle isomorphism is Serre duality.

Using this expectation function and the "sheaf cup product" binary operation, we define the B-model correlation functions (in the standard way from the Frobenius algebra structure):

$$\langle \beta_1\, \beta_2\, \beta_3 \rangle = ((\beta_1 \cup \beta_2 \cup \beta_3) \lrcorner\, \Omega) \otimes \Omega.$$

(Once again we have a definition which is inspired by the outcome of a calculation in the physics literature [**147**].) This gives a map

$$H^{-p,q}(W) \times H^{-p',q'}(W) \times H^{-(n-p-p'),n-q-q'}(W) \to \mathbb{C}.$$

Note that as in the A-model case, we actually have a graded Frobenius algebra of finite length, so the expectation function is uniquely defined up to a scalar multiple (which can be absorbed in the choice of $\Omega^{\otimes 2}$.)

In order to relate this correlation function to a more familiar mathematical object, we can proceed as follows: first use the two Ω's to transform two of the arguments, and then use the cup product:

$$\langle \beta_1\, \beta_2\, \beta_3 \rangle = ((\beta_1 \lrcorner\, \Omega) \cup \beta_2 \cup (\beta_3 \lrcorner\, \Omega)).$$

This variant of the correlation function can be regarded as a map

$$H^{n-p,q}(W) \times H^{-p',q'}(W) \times H^{p+p',n-q-q'}(W) \to \mathbb{C},$$

or, if we treat it as a modified "binary operation," as a map

$$H^{n-p,q}(W) \times H^{-p',q'}(W) \to H^{n-p-p',q+q'}(W).$$

This version of the "binary operation" expresses the cohomology $H^*(W)$ as a module over the ring $H^{-*}(W)$. As we shall see, this variant has the pleasant property that it can be directly interpreted in terms of variations of Hodge structure and the differential of the period map. Of course, the original version of the correlation function can be recovered from this, once we have specified $\Omega^{\otimes 2}$.

6.2. Variations of Hodge structure

We now briefly review the theory of variations of Hodge structure, in order to explain the mathematical origin of the B-model correlation functions. Variations of Hodge structure were introduced as a tool for measuring how the complex structure on a differentiable manifold can vary. Good general references for this are Griffiths et al. [**8**], and Schmid [**89**].

There are two primary ways one can view deformations of complex structure. In the first viewpoint, we fix a compact differentiable manifold Y, and consider various integrable almost-complex structures J on Y. Then the set of such, modulo diffeomorphism, is known to be a finite-dimensional space.

In the second viewpoint, we consider proper holomorphic maps $\pi : \mathcal{W} \to S$ with $W_s = \pi^{-1}(s)$ diffeomorphic to Y. Each fiber W_s has an induced structure of a complex manifold. If S is contractable, then π can be trivialized in the C^∞ category, and we can regard π as specifying a family of complex structures. One wants to represent the functor

$$S \mapsto \{\pi : \mathcal{W} \to S\}/(\text{isomorphism}),$$

by maps to a moduli space which has a "universal family." This is generally too much to hope for, but there are often "coarse moduli spaces" whose points are in one-to-one correspondence with the possible complex structures. (The appendices in [**13**] provide good background for moduli problems in general.)

We will study complex structures on Y by studying the Hodge decomposition induced on cohomology by each choice of complex structure. In general, if W_s is a Kähler manifold there is a *Hodge decomposition* of the cohomology:

$$H^k(W_s, \mathbb{C}) \cong \bigoplus_{p+q=k} H^{p,q}(W_s). \tag{6.1}$$

Now in a family over a contractable base, the bundle of $H^k(W_s, \mathbb{C})$'s may be canonically trivialized. Over more general bases S (assumed to be connected), it is convenient to consider $R^k\pi_*\mathbb{C}_{\mathcal{W}}$, which is simply the sheaf whose local sections are topologically constant families of cohomology classes. This sheaf has the structure of a *local system*: it can be characterized by its fiber $H^k(W_s, \mathbb{C})$ at a particular point $s \in S$ together with a representation of the fundamental group

$$\rho : \pi_1(S, s) \to \operatorname{Aut}(H^k(W_s, \mathbb{C}))$$

which specifies what happens when the locally constant sections are followed around loops. There is useful dictionary **[5]** between local systems and pairs $(\mathcal{H}, \nabla)$ consisting of a holomorphic vector bundle $\mathcal{H}$ on S and a flat holomorphic connection

$$\nabla : \mathcal{H} \to (T_S^{(1,0)})^* \otimes \mathcal{H}.$$

The way the dictionary works is this: given a local system $\mathbb{H}$, define $\mathcal{H} = \mathcal{O}_S \otimes \mathbb{H}$, and $\nabla(\sum \varphi_j h^j) = \sum d\varphi_j \otimes h^j$ for $\{h^j\}$ a local basis of sections of $\mathbb{H}$. Conversely, given $(\mathcal{H}, \nabla)$, define $\Gamma(U, \mathbb{H}) = \{h \in \Gamma(U, \mathcal{H}) \mid \nabla(h) = 0\}$ for every open set U.

In the case of the cohomology local system $R^k\pi_*\mathbb{C}_{\mathcal{W}}$, the associated connection ∇ on $\mathcal{H}^k$ is called the *Gauss–Manin connection*. An explicit version of this Gauss–Manin connection goes like this: if we choose a local basis $\alpha^1, \dots, \alpha^r$ for the space of sections $\Gamma(U, R^k\pi_*\mathbb{C}_{\mathcal{W}})$, then any $\beta(s) \in \Gamma(U, \mathcal{H}^k)$ can be written $\beta(s) = \sum f_j(s)\alpha^j$ for some coefficient functions $f_j \in \Gamma(U, \mathcal{O}_S)$. Then

$$\nabla(\beta) = \sum df_j \otimes \alpha^j \in \Gamma(U, (T_S^{(1,0)})^* \otimes \mathcal{H}^k).$$

This can be given an interpretation in terms of classical "period integrals" as follows. The basis $\alpha^1, \dots, \alpha^r$ is dual to some basis $\gamma_1, \dots, \gamma_r \in H_k(W_{s_0}, \mathbb{C})$. Then the coefficient functions are the period integrals $f_j(s) = \int_{\gamma_j} \beta(s)$. (We use integration to denote the pairing between homology and cohomology.)

The great advantage of expressing everything in terms of the Gauss–Manin connection is that the Gauss–Manin connection can be computed algebraically, without knowing the topological cycles in advance.

Although the sheaf $R^k\pi_*\mathbb{C}_{\mathcal{W}}$ of cohomology groups can be locally trivialized over the base S, the Hodge decomposition (6.1) will vary as we vary the complex structure. The properties of this variation are more conveniently expressed using the *Hodge filtration*:

$$F^p(W_s) := \bigoplus_{p' \geq p} H^{p',k-p'}(W_s) \subset H^k(W, \mathbb{C})$$

rather than the Hodge groups $H^{p,q}(W_s)$ directly. The spaces $F^p(W_s)$ in the Hodge filtration vary holomorphically with parameters, fitting together to form a holomorphic subbundle $\mathcal{F}^p \subset \mathcal{H}^k$. One might also try to construct a bundle of $H^{p,q}$'s by the simple procedure

$$\mathcal{H}^{p,q}_{C^\infty} := \bigcup_{s \in S} H^{p,q}(W_s) \subset \mathcal{H}^k.$$

As the notation indicates, this defines a C^∞ bundle, but it is not in general holomorphic. There is a holomorphic bundle $\mathcal{H}^{p,k-p}$ defined by the exact sequence

$$0 \to \mathcal{F}^{p+1} \to \mathcal{F}^p \to \mathcal{H}^{p,k-p} \to 0, \tag{6.2}$$

but this exact sequence *has no canonical splitting*, and $\mathcal{H}^k$ cannot in general be written as a direct sum of these holomorphic $\mathcal{H}^{p,k-p}$ bundles.

The key property satisfied by the Hodge bundles is known as *Griffiths transversality*: when we differentiate with respect to parameters by using the Gauss–Manin connection, the Hodge filtration only shifts by one, i.e.,

$$\nabla(\mathcal{F}^p) \subset (T_W^{(1,0)})^* \otimes \mathcal{F}^{p-1}.$$

To study the totality of complex structures on W, we can map the moduli space, or any parameter space S for a family, to the classifying space for Hodge structures. Each Hodge structure on a fixed vector space H^k determines a point in a flag variety

$$\mathrm{Flags}_{(f_j)} := \{\{0\} \subset F^k \subset \cdots \subset F^0 = H^k \mid \dim F^j = f_j\},$$

with the f_j's specifying the dimensions of the spaces making up the filtration. The group $\mathrm{GL}(f_0, \mathbb{C})$ acts transitively on such flags, and if we fix a reference flag $F_0^\bullet$, then the flag variety can be described as $\mathrm{GL}(f_0,\mathbb{C})/\operatorname{Stab}(F_0^\bullet)$. (The stabilizer $\operatorname{Stab}(F_0^\bullet)$ is the group of block lower triangular matrices.) There are some additional conditions which should be imposed to get a good Hodge structure (cf. [**8, 89**]); these restrict us to an open subset $\mathcal{U}$ of a subvariety[1] of the flag variety on which a discrete group Γ acts, and the desired classifying space for Hodge structures is $\mathcal{U}/\Gamma$. The classifying map $S \to \mathcal{U}/\Gamma$ for a family is often referred to as the *period map*.

The tangent space to the flag variety can be described as

$$\bigoplus_j \mathrm{Hom}(F^j/F^{j+1}, H^k/F^j).$$

So another way of stating Griffiths transversality is to say that the differential of the period map $S \to \mathcal{U}/\Gamma$ sends $T_S^{(1,0)}$ to the subspace

$$\bigoplus_j \mathrm{Hom}(F^j/F^{j+1}, F^{j-1}/F^j) = \bigoplus_j \mathrm{Hom}(H^{j,k-j}(W_s), H^{j-1,k+1}(W_s))$$

of the tangent space.

The differential of the map $S \to \mathrm{Flags}_{(f_j)}$ factors through a map $T_S^{(1,0)} \to H^1(T_W^{(1,0)})$ which describes the first-order deformations represented by S at $[W]$. The map which then induces the differential is the map

$$H^1(T_W^{(1,0)}) \to \bigoplus_j \mathrm{Hom}(H^{j,k-j}(W), H^{j-1,k+1}(W)) \tag{6.3}$$

given by sheaf cup product.

The success of this approach to studying the moduli of complex structures derives from the *local Torelli theorem* for Calabi–Yau manifolds, which states that the map (6.3) is injective. This means that at least locally, the moduli space can be accurately described by using variations of Hodge structure. However, that same

[1] We must pass to a subvariety to restrict to the so-called *polarized* Hodge structures—see [**8**] or [**38**] for an explanation of this.

map can now be given a new interpretation, as a B-model correlation function. That is, *the B-model correlation function*

$$H^1(T_W^{(1,0)}) \times H^{j,k-j}(W) \to H^{j-1,k+1}(W)$$

coincides with the differential of the period map!

We now restrict our attention to the middle-dimensional cohomology $H^n(W,\mathbb{C})$. Stated in terms of the Gauss–Manin connection, we find the following "bundle version" of our correlation function [**73**]: given a vector field θ on the moduli space and sections $\alpha \in \mathcal{F}^j$, $\beta \in \mathcal{F}^{j-1}$, the correlation function is

$$\langle \theta\, \alpha\, \beta \rangle = \int_W \nabla_\theta(\alpha) \wedge \beta$$

(where $\nabla_\theta = \theta \lrcorner \nabla$ denotes the directional derivative in direction θ).

However, as used in physics the correlation function is a specific function rather than a map between bundles. To find this interpretation, we will need to choose specific sections of these bundles on which to evaluate the map. It is this issue to which we now turn.

6.3. Splitting the Hodge filtration

Our method for specifying sections of the Hodge bundles will be given in terms of a choice of splitting for the Hodge filtration on the middle-dimensional cohomology $H^n(W,\mathbb{C})$, i.e., a set of splittings of the exact sequences (6.2) (but defined only locally in the parameter space). We determine such a splitting by means of a filtration on *homology*, which we think of as specifying "which periods to calculate."

Let $\mathbb{S}_\bullet$ be a filtration of the homology local system $\mathrm{Hom}(R^n\pi_*\mathbb{C}_\mathcal{W}, \mathbb{C}_S)$ by sub-local systems, and let

$$\mathbb{S}^\ell := \mathrm{Ann}(\mathbb{S}_{\ell-1}) := \{\alpha \in \mathcal{H}^n \mid \int_\gamma \alpha = 0\ \forall\, \gamma \in \mathbb{S}_{\ell-1}\}.$$

be the associated filtration of annihilators of $\mathbb{S}_\bullet$ in cohomology. We say that $\mathbb{S}_\bullet$ is a *splitting filtration for $\mathcal{F}^\bullet$* if $(\mathcal{H}^n)_s \cong (\mathcal{F}^p)_s \oplus (\mathbb{S}^{n-p+1})_s$ for every $s \in S$ and for every $0 \le p \le n$. (In this case, $\mathbb{S}^\bullet$ and $\mathcal{F}^\bullet$ are called *opposite filtrations of weight n* [**38**].)

One way of producing examples of splitting filtrations is as follows: fix a point $s \in S$, and consider the conjugate of the Hodge filtration at s_0, namely, $\overline{F^q}_{s_0}$. The "opposite" property for these filtrations is easy to check: by definition

$$(\mathcal{F}^p)_s = H^{n,0}(W_s) \oplus \cdots \oplus H^{p,n-p}(W_s)$$

and so

$$\begin{aligned}\overline{(\mathcal{F}^{n-p+1})_s} &= \overline{H^{n,0}(W_s) \oplus \cdots \oplus H^{n-p+1,p-1}(W_s)} \\ &= H^{0,n}(W_s) \oplus \cdots \oplus H^{p-1,n-p+1}(W_s),\end{aligned}$$

where we have used the fact that $\overline{H^{p,q}(W_s)} = H^{q,p}(W_s)$. The Gauss–Manin connection can be used to extend this from a filtration at one point to a filtration of the local system. Although this filtration only coincides with the conjugate of the Hodge filtration at one point in the parameter space, it remains opposite to the Hodge filtration at all points nearby.

Given a splitting filtration $\mathbb{S}_\bullet$, we define

$$\mathcal{H}_{\mathbb{S}}^{p,q} := \mathcal{F}^p \cap \operatorname{Ann}(\mathbb{S}_{q-1}),$$

on any open set on which $\mathbb{S}_\bullet$ is single-valued. Then

$$\mathcal{H} = \bigoplus_{p=0}^{n} \mathcal{H}_{\mathbb{S}}^{p,q} \quad \text{and} \quad \mathcal{F}^p = \bigoplus_{p' \geq p} \mathcal{H}_{\mathbb{S}}^{p',n-p'}.$$

(This is the promised splitting of the Hodge filtration.) More concretely, this space can be described in terms of conditions on the periods as follows. The sections of $\mathcal{H}_{\mathbb{S}}^{p,q}$ over U are

$$\Gamma(U, \mathcal{H}_{\mathbb{S}}^{p,q}) := \{\beta \in \Gamma(U, \mathcal{F}^p) \mid \int_\gamma \beta = 0 \ \forall\, \gamma \in \mathbb{S}_{q-1}\}.$$

We also define a space of *distinguished sections* of $\mathcal{H}_{\mathbb{S}}^{p,q}$ by

$$\Gamma(U, \mathcal{H}_{\mathbb{S}}^{p,q})_{\text{dist}} := \{\beta \in \Gamma(U, \mathcal{H}_{\mathbb{S}}^{p,q}) \mid d\left(\int_\gamma \beta\right) = 0 \ \forall\, \gamma \in \mathbb{S}_q\}.$$

(That is, the period integrals $\int_\gamma \beta$ are constant for all $\gamma \in \mathbb{S}_q$, and vanish for all $\gamma \in \mathbb{S}_{q-1}$.)

For each $\mathbb{S}_\bullet$, then, we can define specific B-model correlation functions, using the Ω coming from the distinguished section of $\mathcal{H}_{\mathbb{S}}^{n,0}$ (which is well defined up to a complex scalar multiple). This has the advantage that the correlation functions have been turned into actual functions on a parameter space (in accord with the physicists' interpretation) rather than sections of a bundle. The disadvantage is that further parameters—in the form of a choice of splitting—have been introduced. However, the necessity of considering further parameters such as these, on which the correlation functions will depend anti-holomorphically rather than holomorphically, was recently realized in the physics literature [**124**].

In addition to the distinguished n-form Ω, our choice of splitting determines a family of distinguished vector fields which when contracted with Ω yield the distinguished sections of $\mathcal{H}_{\mathbb{S}}^{n-1,1}$. These vector fields can be integrated into *canonical coordinates*, well-defined up to a $\mathrm{GL}(r, \mathbb{C})$ transformation. (The flexibility of that final $\mathrm{GL}(r, \mathbb{C})$ choice comes from the constants of integration, which must also be specified in order to completely determine a set of canonical coordinates.)

A bit more explicitly, if γ_0 spans $\mathbb{S}_0$ and $\gamma_0, \gamma_1, \ldots, \gamma_r$ span $\mathbb{S}_1$, then the distinguished Ω satisfies $\int_{\gamma_0} \Omega = \text{constant}$, and the coordinates are given by

$$\int_{\gamma_1} \Omega, \ldots, \int_{\gamma_r} \Omega.$$

If we start with an arbitrary n-form $\widetilde{\Omega}$, we can write the distinguished n-form as

$$\Omega := \frac{\widetilde{\Omega}}{\int_{\gamma_0} \widetilde{\Omega}}$$

and the canonical coordinates as

$$\frac{\int_{\gamma_1} \widetilde{\Omega}}{\int_{\gamma_0} \widetilde{\Omega}}, \ldots, \frac{\int_{\gamma_r} \widetilde{\Omega}}{\int_{\gamma_0} \widetilde{\Omega}}.$$

This is the most general possible form for canonical coordinates (and a distinguished n-form) needed for the physical theory, according to recent work in physics [**115**].

Let us fix a splitting filtration $\mathbb{S}_\bullet$. Consider a basis $\{\beta^i\}$ of $\mathcal{H}^n$ consisting of distinguished sections of the bundles $\mathcal{H}_{\mathbb{S}}^{p,q}$ (ordered so that the basis is also adapted to the Hodge filtration $\mathcal{F}^\bullet$), and a multi-valued basis $\{\gamma_j\}$ of the homology local system $\mathrm{Hom}(R^n\pi_*\mathbb{C}_\mathcal{W}, \mathbb{C}_S)$, adapted to the splitting filtration $\mathbb{S}_\bullet$. Then the period matrix $(\int_{\gamma_j} \beta^i)$ (which has multi-valued entries) will take a block upper triangular form with constant diagonal blocks. And if we calculate the connection matrix in the basis $\{\beta^i\}$, it takes the special form

$$\begin{pmatrix} 0 & A_0^1 & 0 & & \cdots & 0 \\ & 0 & A_1^1 & 0 & \cdots & 0 \\ & & \ddots & \ddots & & \vdots \\ \vdots & \vdots & & & 0 & A_{n-1}^1 \\ 0 & 0 & \cdots & & & 0 \end{pmatrix}$$

in which the only nonzero entries are in the first block superdiagonal of the matrix. The entries A_j^1 precisely contain the data for the B-model correlation functions, calculated in our distinguished basis.

LECTURE 7
The A-Variation of Hodge Structure

7.1. Variations of Hodge structure near the boundary of moduli

In this lecture, we begin by reviewing the asymptotic behavior of a variation of Hodge structure near the boundary of moduli space, and the behavior of the B-model correlation functions there. Comparing to the A-model correlation functions will reveal some similarities—this is one of the hints of mirror symmetry. We make the similarities even more apparent by using the A-model correlation functions to construct a new variation of Hodge structure, which we call the A-variation of Hodge structure.

Let $S = (\Delta^*)^r \subset \overline{S} = \Delta^r$, and suppose we are given a family $\pi : \mathcal{W} \to S$ of complex manifolds. We will assume that there is a way to complete this to a family $\overline{\pi} : \overline{\mathcal{W}} \to \overline{S}$ in which $\overline{\pi}$ is still proper (but no longer smooth). Thus, $0 \in \overline{S}$ is a boundary point in the parameter space. Pick a basepoint $s \in S$; then the fundamental group $\pi_1(S, s)$ is generated by loops $\gamma^{(1)}, \ldots, \gamma^{(r)}$ with $\gamma^{(j)}$ homotopic to the standard generator of $\pi_1(\Delta_j^*)$, where Δ_j^* is the jth factor in $(\Delta^*)^r$.

The Monodromy Theorem (Landman [**65**]). *The action of each generator $\gamma^{(j)}$ gives a quasi-unipotent automorphism $T^{(j)}$ of $H^k(W_s, \mathbb{Q})$, i.e., $(((T^{(j)})^{b_j} - I)^{r_j} = 0$. (This is called* unipotent *if $b_j = 1$.)*

We will restrict attention to the unipotent case. This is partially for technical convenience, but in fact, in the examples which have been calculated for mirror symmetry purposes, only unipotent monodromy transformations have played a rôle.

When $T^{(j)}$ is unipotent, its logarithm can be defined by the following sum (which is finite).

$$N^{(j)} := \log T^{(j)} := (T^{(j)} - I) - \frac{1}{2}(T^{(j)} - I)^2 + \cdots .$$

(Note that the $T^{(j)}$'s and $N^{(j)}$'s all commute.) Let $z_1, \dots, z_r$ be coordinates on $\overline{S}$, with z_j a coordinate on the jth disk. Consider the operator

$$\mathcal{N} := \exp\left(-\frac{1}{2\pi i}\sum \log z_j\, N^{(j)}\right) = \\ I + \left(-\frac{1}{2\pi i}\sum \log z_j\, N^{(j)}\right) + \frac{1}{2!}\left(-\frac{1}{2\pi i}\sum \log z_j\, N^{(j)}\right)^2 + \cdots$$

(also a finite sum). For any section e of the local system $R^k\pi_*(\mathbb{C}_{\mathcal{W}})$, a simple calculation shows that

$$\nabla(\mathcal{N}(e)) = -\frac{1}{2\pi i}\sum \frac{dz_j}{z_j} N^{(j)}(e). \tag{7.1}$$

The key facts about the asymptotic behavior are as follows.

The Nilpotent Orbit Theorem (Schmid [**89**]). *Assume that each monodromy transformation $T^{(j)}$ is unipotent. Let $e_1(s), \dots, e_r(s)$ be a multi-valued basis of $R^k\pi_*(\mathbb{C}_{\mathcal{W}})$, and let $\eta_\ell := \mathcal{N}(e_\ell)$. Then each η_ℓ is a single-valued section of $\mathcal{H}^k$ on S, and together they can be used to generate an extension $\overline{\mathcal{H}}^k$ of $\mathcal{H}^k$ to $\overline{S}$. By eq.* (7.1), *the Gauss–Manin connection extends to a connection on $\overline{\mathcal{H}}^k$ (again denoted by ∇) with* regular singular points, *i.e., the extended connection is a map*

$$\nabla : \overline{\mathcal{H}}^k \to (T^{(1,0)}_{\overline{S}})^*(\log B) \otimes \overline{\mathcal{H}}^k$$

where $(T^{(1,0)}_{\overline{S}})^*(\log B)$ is the free $\mathcal{O}_{\overline{S}}$-module generated by $\frac{dz_j}{z_j}$, $j = 1, \dots, r$. Moreover, the Hodge bundles $\mathcal{F}^p$ have locally free extensions to subbundles $\overline{\mathcal{F}}^p \subset \overline{\mathcal{H}}^k$ such that

$$\nabla(\overline{\mathcal{F}}^p) \subset (T^{(1,0)}_{\overline{S}})^*(\log B) \otimes \overline{\mathcal{F}}^{p-1}.$$

The asymptotic behavior as $z_j \to 0$ of the B-model correlation functions

$$\langle \theta\, \alpha\, \beta\rangle = \int_W \nabla_\theta(\alpha) \wedge \beta$$

can be deduced from this theorem. If we let $\theta_j = 2\pi i\, z_j \frac{d}{dz_j}$ (chosen to remove poles in the asymptotic expression for the correlation function) then the leading term in $\langle \theta_j\, \eta_\ell\, \beta\rangle$ is given by the monodromy:

$$\lim_{z_j \to 0} \langle \theta_j\, \eta_\ell\, \beta\rangle = -\int_W N^{(j)}(e_\ell) \wedge \beta. \tag{7.2}$$

The essential properties of the monodromy are captured by the *monodromy weight filtration* $\mathbb{W}_\bullet$ on the cohomology, which has the properties that $N^{(j)}\mathbb{W}_\ell \subset \mathbb{W}_{\ell-2}$, and that for any positive real numbers $a_1, \dots, a_r$, the operator $N := \sum a_j N^{(j)}$ induces isomorphisms $N^\ell : \mathrm{Gr}^{\mathbb{W}}_{n+\ell} \to \mathrm{Gr}^{\mathbb{W}}_{n-\ell}$. Any splitting filtration which we use to make calculations of B-model correlation functions must be somehow compatible with this monodromy weight filtration, if those calculations are to make sense near the boundary.

If mirror symmetry is going to hold, there must be a correspondence between the limiting behaviors described in eqs. (5.2) and (7.2). In fact, the first thing to notice is that the natural flat coordinates on the A-model moduli space are multiple-valued, with the ambiguity precisely specified by $H^2_{\mathrm{DR}}(M, \mathbb{Z})$. So there must be some part of the monodromy weight filtration which matches that behavior. This motivated the following definition, first given in [**73, 74**] (cf. also [**38**]).

We say that a boundary point is *maximally unipotent* if

$$\mathcal{H}_s = (\mathcal{F}^n)_s \oplus (\mathbb{W}_{2n-2})_s$$

and

$$\mathcal{H}_s = (\mathcal{F}^{n-1})_s \oplus (\mathbb{W}_{2n-4})_s$$

for all s near the point. With this definition, the distinguished holomorphic n-form and the canonical coordinates can be defined as in lecture six.

There is an alternate version of this "maximally unipotent monodromy" condition, which agrees with the original one for Calabi–Yau threefolds, but is more restrictive in higher dimension. We say that a boundary point is *strongly maximally unipotent* if the weight filtration $\mathbb{W}_\bullet$ has nontrivial graded pieces in even degree only, and if the induced filtration on homology defined by

$$\mathbb{S}_\ell := \operatorname{Ann}(\mathbb{W}_{2n-2\ell+2})$$

is a splitting filtration. (Note that the corresponding filtration on cohomology is then

$$\mathbb{S}^\ell := \operatorname{Ann}(\mathbb{S}_{\ell-1}) = \mathbb{W}_{2n-2\ell};$$

this is the filtration which should be opposite to the Hodge filtration.) In this case, we will be able to use distinguished sections to calculate B-model correlation functions, as explained earlier.

At the moment, only the original version of the definition has been justified to the satisfaction of physicists as an appropriate characterization of points which should be useful for mirror symmetry. To completely carry out a mirror symmetry type calculation, though, the second version would seem to be necessary. And as we shall see, that version has been extremely successful in examples.

Actually, even just at the level of the monodromy action, the parallels between the structure of the Lefschetz operators on the cohomology and the action of monodromy are rather striking, as was first observed by Cattani, Kaplan and Schmid [**35**]. The operators $\operatorname{ad}(e^j)$ describe Lefschetz decompositions of the cohomology of M, which have many structural parallels to the monodromy weight filtration at a maximally unipotent point.

7.2. Reinterpreting the A-model correlation functions

Let M be a Calabi–Yau manifold on which a complex structure and Kähler metric have been fixed. Inspired by some of the similarities between the two different types of correlation functions, we wish to improve the analogy by translating the A-model correlation functions into data describing a variation of Hodge structure. Consider the moduli space $\mathcal{D}/\Gamma$ for A-model correlation functions, and a coordinate chart specified by a cone σ:

$$\begin{array}{ccccc} \mathcal{D}/\Gamma & \leftarrow & \mathcal{D}_\sigma/H^2(M,\mathbb{Z}) & \cong & (\Delta^*)^r \\ & & \cap\!\vert & & \cap\!\vert \\ & & (\mathcal{D}_\sigma/H^2(M,\mathbb{Z}))^- & \cong & \Delta^r \end{array}$$

We assume that the cone σ (which we call a *framing*) is generated by a basis e^1, …, e^r of $H^2(M,\mathbb{Z})$. Let t_1, …, t_r be coordinates on $H^2(M,\mathbb{C})$ dual to this basis (so that elements of $H^2(M,\mathbb{C})$ take the form $\sum t_j e^j$). The natural vector fields for making calculations of correlation functions which involve a term from the tangent space $H^2(M,\mathbb{C})$ are the vector fields $\partial/\partial t_j$. These are the analogues of the distinguished vector fields which we had on the B-model side.

On the other hand, natural coordinates on $\mathcal{D}_\sigma/H^2(M,\mathbb{Z})$ are furnished by $q_j = \exp(2\pi i\, t_j)$. Then

$$\frac{\partial}{\partial t_j} = 2\pi i\, q_j \frac{\partial}{\partial q_j},$$

from which we conclude that those correlation functions should naturally be evaluated on the basis $2\pi i\, q_j\, \partial/\partial q_j$ of the sheaf of logarithmic vector fields on the space $(\mathcal{D}_\sigma/H^2(M,\mathbb{Z}))^-$.

We identify $\partial/\partial t_j$ with the operation of taking the quantum product with the basis element $e^j \in H^2(M,\mathbb{Q})$. The resulting map is determined by the correlation functions of the form $\langle e^j\, \alpha\, \beta\rangle$. We had a particularly simple form for these correlation functions, given in example 4.4, in terms of the Gromov–Witten maps Γ_η. We now wish to reinterpret that formula in the following way. We will describe a holomorphic bundle[1] $\mathcal{E} := \left(\bigoplus H^{\ell,\ell}(M)\right) \otimes \mathcal{O}_{(\mathcal{D}_\sigma/H^2(M,\mathbb{Z}))^-}$ with a connection[2] (with regular singular points)

$$\nabla := \frac{1}{2\pi i}\left(\sum d\log q_j \otimes \mathrm{ad}(e^j) + \sum_{0\neq\eta\in H_2(M,\mathbb{Z})} d\log\left(\frac{1}{1-q^\eta}\right)\otimes\Gamma_\eta\right)$$

which was derived from the formulas for $e^j\star$, where $\mathrm{ad}(e^j) : H^k(M) \to H^{k+2}(M)$ is defined by $\mathrm{ad}(e^j)(A) = e^j \cup A$. We also define a "Hodge filtration"

$$\mathcal{E}^p := \left(\bigoplus_{0\leq\ell\leq m-p} H^{\ell,\ell}(M)\right) \otimes \mathcal{O}_{(\mathcal{D}_\sigma/H^2(M,\mathbb{Z}))^-}.$$

This describes a structure we call the *framed A-variation of Hodge structure with framing* σ. To be a bit more precise, we should study "formally degenerating variations of Hodge structure," since the series used to define ∇ is only formal. (We won't formulate that theory in detail here.)

The connection ∇ which we defined from the Gromov–Witten invariants is in fact a *flat* holomorphic connection [**151**]. The flatness follows from the associativity (and commutativity) of the binary operation. In fact, since the directional derivatives with respect to ∇ corresponded to binary products $e^j \star \zeta$ (where e^j describes the direction of the derivative), iterated directional derivatives have the form $e^k \star (e^j \star \zeta)$. We would simply need to know that reversing the order of j and k produces the same result, and this is guaranteed by the commutativity and associativity.

In particular, the flatness is automatic when $\dim_{\mathbb{C}} M = 3$, a case in which there is no issue of associativity. The recent proofs of associativity of quantum cohomology [**87, 67, 12**] guarantee that this connection is flat in arbitrary dimension.

---

[1] There are a few variants to this construction, in which one uses slightly different bundles. Essentially, one can restrict to any subbundle of $\bigoplus H^{\ell,\ell}(M)$ which is preserved by cup products with the part of $H^{1,1}(M)$ which it contains.

[2] I am indebted to P. Deligne for advice which led to this form of the formula (cf. [**38**]).

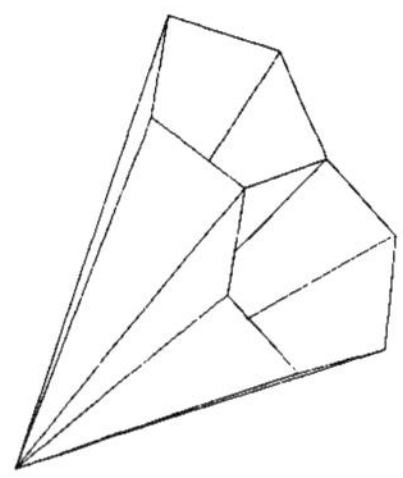

Figure 1. Adjacent Kähler cones

As in the geometric case, there is an additional structure associated to this variation of Hodge structure: a local system. The local system on homology takes the simple form

$$\mathbb{S}_\ell := H_{0,0} \oplus H_{1,1} \oplus \cdots \oplus H_{\ell,\ell},$$

and the corresponding local system on cohomology then becomes

$$\mathbb{S}^\ell = H^{\ell,\ell} \oplus H^{\ell+1,\ell+1} \oplus \cdots \oplus H^{n,n}.$$

The logarithms of the monodromy actions which define these local systems are specified by the topological pairings, and coincide with the cup-product maps

$$H^2(M,\mathbb{Z}) \otimes \mathbb{S}^\ell \to \mathbb{S}^{\ell+1}.$$

In the next lecture, we will formulate a precise conjecture which equates this A-variation of Hodge structure with the geometric variation of Hodge structure on a mirror partner.

7.3. Beyond the Kähler cone

We indicated in lecture five that the conformal field theory moduli space is actually *larger* than the nonlinear σ-model moduli space. We can now explain how this comes about—it is due to an analysis of the effect of flops on the conformal field theory.

Flops are birational transformations among Calabi–Yau threefolds which have been studied extensively as part of the minimal model program (see for example [**1**]). The effect of flops on the Kähler cone of a Calabi–Yau threefold is as follows. Given a Calabi–Yau threefold X with a complex structure J, and a linear system $|L|$ inducing a flopping contraction from X_J to $\widehat{X}_{\widehat{J}}$, the Kähler cones $\mathcal{K}_J$ and $\widehat{\mathcal{K}}_{\widehat{J}}$ share a common wall, which contains the class of $|L|$, as depicted in figure 1. The Kähler cone has already occurred in our discussion of the moduli spaces of σ-models. The natural question arises: suppose we attempt to "attach" the moduli spaces $\mathcal{D}/\Gamma$ and $\widehat{\mathcal{D}}/\widehat{\Gamma}$ along (the images of) their common wall? In fact, it now appears likely that the conformal field theory moduli spaces of X and $\widehat{X}$ are analytic continuations of each other, and that this "attached space" is a part of the full conformal field theory moduli space [**106, 153**]. (This at least seems to happen in examples—the arguments for this rely on mirror symmetry, and involve finding regions in the mirror's moduli space which correspond to the X_J and $\widehat{X}_{\widehat{J}}$ theories, respectively.) One of the consequences of this would be an analytic continuation of correlation functions from $\mathcal{D}/\Gamma$ to $\widehat{\mathcal{D}}/\widehat{\Gamma}$.

Here is a formal calculation from [**153, 108**] which supports this analytic continuation idea (see also [**76**] for a more mathematical treatment). The union of all of the Kähler cones of birational models of X_J is known as the *movable cone* $\operatorname{Mov} X_J$ [**61**]. We compute in the formal semigroup ring $\mathbb{Q}[q; \operatorname{Mov}(X_J)^\vee]$ (which we identify canonically with the same ring for $\widehat{X}_{\widehat{J}}$), and so the computation is purely formal.

Consider the simplest flop: the flop based on a collection of disjoint holomorphic rational curves $\Gamma_i \subset X_J$ (in a common homology class $[\Gamma]$) such that the normal bundle is $N_{\Gamma_i/X_J} = \mathcal{O}(-1)\oplus\mathcal{O}(-1)$. (These curves must be flopped simultaneously in order to ensure that the flopped variety is Kähler.) A reasonable genericity assumption about the *other* rational curves on X_J is this: all (pseudo-)holomorphic curves in classes $\eta \notin \mathbb{R}_{>0}[\Gamma]$ are disjoint from the Γ_i's. Since there is a proper transform map on divisors, the Gromov–Witten invariants (which in this case are determined entirely by intersection properties of η and the number of elements in $\mathcal{M}^*_{(\eta,J)}$) do not change when passing from X_J to $\widehat{X}_{\widehat{J}}$, except for the invariants $\Phi_{[\Gamma]}$ themselves. The cup product can also change.

The A-model correlation functions on X_J can be written in the form

$$\begin{aligned}\langle A B C\rangle = A\cdot B\cdot C + &\frac{q^{[\Gamma]}}{1-q^{[\Gamma]}}\,(A\cdot\Gamma)(B\cdot\Gamma)(C\cdot\Gamma)\,n_\Gamma \\ &+ \sum_{\substack{\eta\in H_2(X,\mathbb{Z})\\ \eta\neq\lambda\Gamma}} \frac{q^\eta}{1-q^\eta}\,\Phi_\eta(A,B,C).\end{aligned}$$

Only the first terms change when passing to $\widehat{X}_{\widehat{J}}$ and in fact we claim that

$$\begin{aligned}A\cdot B\cdot C + &\frac{q^{[\Gamma]}}{1-q^{[\Gamma]}}\,(A\cdot\Gamma)(B\cdot\Gamma)(C\cdot\Gamma)\,n_\Gamma \\ =&\widehat{A}\cdot\widehat{B}\cdot\widehat{C} + \frac{q^{[\widehat{\Gamma}]}}{1-q^{[\widehat{\Gamma}]}}\,(\widehat{A}\cdot\widehat{\Gamma})(\widehat{B}\cdot\widehat{\Gamma})(\widehat{C}\cdot\widehat{\Gamma})\,n_{\widehat{\Gamma}},\end{aligned}$$

where $\widehat{A}$, $\widehat{B}$, and $\widehat{C}$ are the proper transforms of A, B, and C. (In other words, the change in the topological term is precisely compensated for by the change in the $q^{[\Gamma]}$ term.)

We will check this formula in the case in which A and B meet one of the curves Γ transversally at a and b points, respectively, and $\widehat{C}$ meets $\widehat{\Gamma}$ transversally at c points. (The general case can be deduced from this one.) Then C must contain Γ with multiplicity c, and the configuration of divisors is as in figure 2 (which illustrates the case $a = b = c = 1$ for simplicity). A and B have no intersection points along Γ, but both $\widehat{A}$ and $\widehat{B}$ contain $\widehat{\Gamma}$, and they meet $\widehat{C}$. The total number of intersection points of $\widehat{A}$, $\widehat{B}$ and $\widehat{C}$ (counted with multiplicity) which lie in $\widehat{\Gamma}$ is thus abc.

Since a similar thing happens for each curve Γ_i in the numerical equivalence class, we see that

$$\widehat{A}\cdot\widehat{B}\cdot\widehat{C} - A\cdot B\cdot C = abc\,n_\Gamma = -(A\cdot\Gamma)(B\cdot\Gamma)(C\cdot\Gamma)\,n_\Gamma \tag{7.3}$$

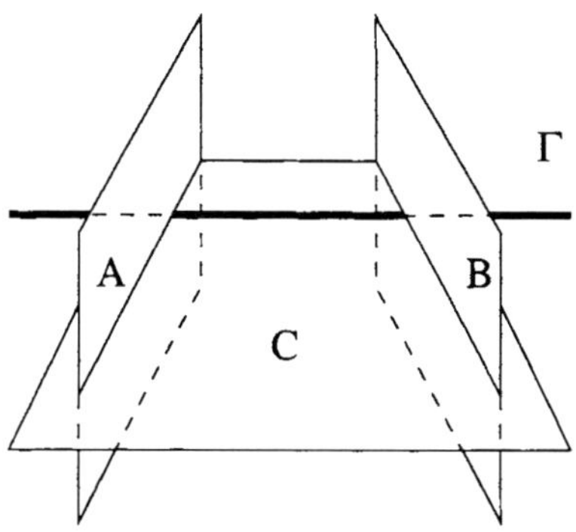

Figure 2a. Before the flop.

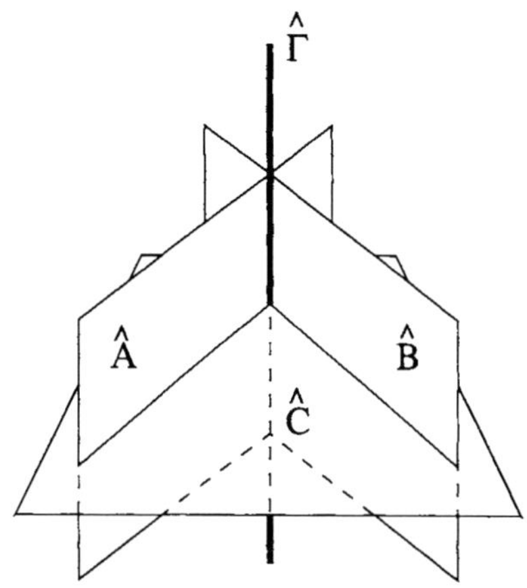

Figure 2b. After the flop.

(using $A \cdot \Gamma = a$, $B \cdot \Gamma = b$, $C \cdot \Gamma = -c$). On the other hand, since $[\widehat{\Gamma}] = -[\Gamma]$ and $n_{\widehat{\Gamma}} = n_\Gamma$, we can compute:

$$\begin{aligned}
&\frac{q^{[\Gamma]}}{1-q^{[\Gamma]}}\,(A\cdot\Gamma)(B\cdot\Gamma)(C\cdot\Gamma)\,n_\Gamma - \frac{q^{[\widehat{\Gamma}]}}{1-q^{[\widehat{\Gamma}]}}\,(\widehat{A}\cdot\widehat{\Gamma})(\widehat{B}\cdot\widehat{\Gamma})(\widehat{C}\cdot\widehat{\Gamma})\,n_{\widehat{\Gamma}} \\
&\quad = \frac{q^{[\Gamma]}}{1-q^{[\Gamma]}}\,(A\cdot\Gamma)(B\cdot\Gamma)(C\cdot\Gamma)\,n_\Gamma + \frac{q^{-[\Gamma]}}{1-q^{-[\Gamma]}}(A\cdot\Gamma)(B\cdot\Gamma)(C\cdot\Gamma)\,n_\Gamma \qquad (7.4)\\
&\quad = \left(\frac{q^{[\Gamma]}}{1-q^{[\Gamma]}} + \frac{1}{q^{[\Gamma]}-1}\right)(A\cdot\Gamma)(B\cdot\Gamma)(C\cdot\Gamma)\,n_\Gamma \\
&\quad = -\,(A\cdot\Gamma)(B\cdot\Gamma)(C\cdot\Gamma)\,n_\Gamma.
\end{aligned}$$

Adding eqs. (7.3) and (7.4) proves the desired formula.

The conclusion from all of this should be that the mirror symmetry phenomenon is really about birational equivalence classes. For if there is any analytic continuation of the correlation function from the region associated to $\mathcal{K}_J$ out into the next cone $\widehat{\mathcal{K}}_{\widehat{J}}$, the calculation above shows that this analytic continuation must in fact reproduce the correlation function of the flopped model $\widehat{X}_{\widehat{J}}$.

It is tempting to think that if we combined the σ-model moduli spaces for all birational models of X we would fill out the entire conformal field theory moduli space. However, some examples that have been worked out by Witten [**153**] and by Aspinwall, Greene and the author [**107**] show that this is not the case. In those examples, there are other regions in the moduli space which correspond to rather different kinds of physical model, including some called *Landau–Ginzburg theories* which will play a rôle again in the next lecture.

LECTURE 8
Mirror Symmetry

8.1. Mirror manifold constructions

The original speculations about mirror symmetry were based on the appearance of arbitrariness of a choice that was made in identifying certain constituents of the conformal field theory associated to a Calabi–Yau manifold with geometric objects on the manifold. The distinction between vertex operators which appear in the A-model and B-model correlation functions is simply a difference in sign of a certain quantum number; if that sign is changed, the geometric interpretation is altered dramatically. This led Dixon [**128**] and Lerche–Vafa–Warner [**140**] to propose that there might be a second Calabi–Yau manifold producing essentially the same physical theory as the first, but implementing this change of sign.

Some time later,[1] an explicit construction was made by Greene and Plesser [**135**] which showed that this phenomenon does indeed occur in physics. The construction rests on a chain of equivalences which are believed to hold among different physical models, as follows.

1. Certain σ-models on Calabi–Yau manifolds are believed to correspond to so-called Landau–Ginzburg theories [**137**]. (It has recently been recognized [**153**] that this correspondence is not direct, but involves analytic continuation on the moduli space.) Roughly speaking, the class of Calabi–Yau manifolds for which this correspondence can be made is the class of ample anti-canonical hypersurfaces in toric varieties. Such a hypersurface will have an equation of the form $\Phi(x_1, \ldots, x_{n+1}) = 0$ (in some appropriate coordinates on the torus), and this same polynomial is used as a "superpotential" in constructing the Landau–Ginzburg theory.
2. Certain Landau–Ginzburg theories—quotients of the ones for which the superpotential is of "Fermat type"

$$\Phi(x_1, \ldots, x_{n+1}) = x_1^{d_1} + \cdots + x_{n+1}^{d_{n+1}}$$

[1] At about the same time, another important piece of evidence for mirror symmetry was given by Candelas, Lynker, and Schimmrigk [**123**], who found an almost perfect symmetry under the exchange $h^{1,1} \leftrightarrow h^{2,1}$ on the set of Hodge numbers coming from Calabi–Yau threefolds which can be realized as weighted projective hypersurfaces.

by certain finite groups Γ—are believed to correspond to yet another type of conformal field theory. This other theory is described in terms of discrete series representations $V^{(k)}$ of the "N=2 superconformal algebra," and it takes the form

$$\left(\bigotimes_j V^{(d_j+2)}\right)/G$$

where G is a slight enlargement of the group Γ. (Note that the case of Γ being trivial is allowed, but then G is not trivial.)

The representation theory of the N=2 superconformal algebra is related to these things by analyzing the conformal field theory on an infinite cylinder. (The superconformal algebra can be described in terms of automorphisms of the cylinder.)

3. By studying the representation theory, Greene and Plesser find a kind of duality among the finite groups G: there is a dual group $\widehat{G}$ and an isomorphism

$$\left(\bigotimes_j V^{(d_j+2)}\right)/G \cong \left(\bigotimes_j V^{(d_j+2)}\right)/\widehat{G}$$

which has the "sign-reversing property" of mirror symmetry.

4. The duality can be extended to the groups Γ, and the mirror Landau–Ginzburg theory of Φ/Γ is $\Phi/\widehat{\Gamma}$. This looks a bit asymmetric, since for example the case Γ trivial leads to a rather large group $\widehat{\Gamma}$. But the group Γ continues to act as a group of "quantum symmetries" on the quotient theory, in a way that restores symmetry to this construction.
5. Finally, the Calabi–Yau which is the quotient of the Fermat hypersurface by Γ should have as its mirror the one which is the quotient by $\widehat{\Gamma}$.

This is called the *Greene–Plesser orbifolding construction.*

There is a conjectural generalization of this construction, which as of yet has no basis in conformal field theory—it is simply a mathematician's guess. This generalization would work for an arbitrary family of Calabi–Yau hypersurfaces in toric varieties. The construction is due to V. Batyrev [**21**].[2]

Take an ample anticanonical hypersurface M in a toric variety V, and let $\{M_t\}$ be the family of such. This family is determined by the Newton polygon of the corresponding equations—that is a polygon $P \subset L_{\mathbb{R}} := L \otimes \mathbb{R}$, where L is the *monomial lattice* of the torus T (of which V is a compactification).

Batyrev shows that the Calabi–Yau condition admits a particularly simple characterization in terms of P: the polyhedron P is *reflexive*, which means that each hyperplane H which supports a face of codimension one of P can be written in the form

$$H = \{y \in L_{\mathbb{R}} \mid (\ell, y) = -1\}$$

for some appropriate vector $v \in \mathrm{Hom}(L, \mathbb{Z})$. (The key property here is the *integrality* of the vector v—there would always be some $v \in \mathrm{Hom}(L, \mathbb{R})$ to define H.)

---

[2] We restrict ourselves to the hypersurface case here; further generalizations—to complete intersections—were subsequently given by Borisov and Batyrev–Borisov [**32, 22**].

Lemma 8.1 (Batyrev). *If P is reflexive, then the* polar polyhedron

$$P^o := \{x \in \mathrm{Hom}(L, \mathbb{R}) \mid (x, y) \geq -1 \text{ for all } y \in P\}$$

is also reflexive.

The conjectured generalization is that the mirror of the family $\{M_t\}$ of hypersurfaces determined by P should be the family $\{W_s\}$ of hypersurfaces (in a compactification of the dual torus of T) determined by the polar polyhedron P^o.

One of the pieces of evidence for this conjecture is

Theorem 8.2 (Batyrev).

$$\dim H^{\pm 1,1}(\widehat{M}) = \dim H^{\mp 1,1}(\widehat{W}),$$

where $\widehat{M}$ and $\widehat{W}$ are $\mathbb{Q}$-factorial terminalizations of M and W respectively.

Batyrev and collaborators have also explored the Hodge structures of these hypersurfaces in considerable detail [**19, 24, 23**].

A refinement of Batyrev's theorem called the *monomial-divisor mirror map* was introduced in [**18**]. This map gives an explicit combinatorial correspondence between (appropriate subspaces of) $H^{\pm 1,1}(\widehat{M})$ and $H^{\mp 1,1}(\widehat{W})$, and is expected to correctly determine the derivative of the mirror map near the large radius limit point. That derivative data is precisely what one needs in order to evaluate the "constants of integration" in finding the canonical coordinates q_j.

There is another mirror manifold construction for a class of threefolds which has been proposed by Voisin [**97**] and Borcea [**31**]. Let S be a K3 surface with an involution ι such that $\iota^*(\Omega) = -\Omega$ for any holomorphic two-form Ω on S, and let E be an elliptic curve. The quotient $\overline{M} = (S \times E)/(\iota \times (-1))$ has singularities along the fixed curves of the involution $\iota \times (-1)$, but they can be resolved by a simple blowing up to produce a Calabi–Yau threefold M.

Involutions of this type on K3 surfaces have been classified by Nikulin [**81**], who found that they fall into a pattern with a remarkable symmetry; when the Hodge numbers of the associated Calabi–Yau threefold are calculated, this symmetry becomes the expected mirror relation among Hodge numbers. The detailed knowledge which is available concerning the variations of Hodge structure on K3 surfaces can be used to study the correlation functions in detail for these models [**97**], which provides further evidence that mirror partners have been correctly identified. In fact, there is also a physics argument explaining why these pairs of conformal field theories are actually mirror to each other [**113**], based on the physics of mirror symmetry for K3 surfaces.

8.2. Hodge-theoretic mirror conjectures

We can now formulate the main conjecture in the mathematical study of mirror symmetry.

The Hodge-Theoretic Mirror Symmetry Conjecture. *Given a boundary point $P \in \overline{\mathcal{M}}_W$ with maximally unipotent monodromy (or perhaps with strongly maximally unipotent monodromy), there should exist a mirror partner M of W, a framing σ of M, a neighborhood U of P in $\mathcal{M}_W$, and a "mirror map"*

$$\mu : U \to (\mathcal{D}_\sigma / L)^-$$

which is determined up to constants of integration by the property that

$$\mu^*(d \log q_j) = d\left(\frac{\int_{\gamma_j} \Omega}{\int_{\gamma_0} \Omega}\right),$$

such that μ induces an isomorphism between appropriate sub-variations of Hodge structure of

1. *the formal completion of the geometric variation of Hodge structure at P, and*
2. *the framed A-variation of Hodge structure with framing σ.*

(The sub-variations of Hodge structure should contain the entire first two terms of the Hodge filtration on both sides.)

There are additional conjectures one wants to make about the relationship between M and W: there should also be isomorphisms

$$H^{p,q}(W) \cong H^{-p,q}(M) \quad p \geq 0,$$

and these should preserve all correlation functions. (In particular, the "reverse" mirror isomorphism should hold, and there should also be isomorphisms between correlation functions which do not come from variations of Hodge structure.) Of course, such isomorphisms only make sense if we have specified the constants of integration. In fact, one wants to conjecture that the entire conformal field theory moduli spaces are isomorphic, but this is a difficult conjecture to make precisely at present since we do not have a complete mathematical understanding of conformal field theory moduli spaces.

If we start with the A-variation of Hodge structure, there is another conjecture we can make.

Converse Conjecture. *Conversely, given (M, σ), the corresponding A-variation of Hodge structure comes from geometry, in the sense that there is a family $\mathcal{Z} \to \overline{S}$ of varieties degenerating at $0 \in \overline{S}$ such that the framed A-variation of Hodge structure is isomorphic to the formal completion at 0 of a (Tate-twisted) sub-variation of Hodge structures of the variation of Hodge structures on some cohomology of Z_s.*

Due to the phenomenon of rigid Calabi–Yau manifolds, we can't assume any stronger properties about Z_s: Calabi–Yau threefolds with $h^{2,1} = 0$ cannot have mirror partners in the usual sense, since such a mirror partner would satisfy $h^{1,1} = 0$, which is absurd. However, there is an example in the physics literature of a rigid Calabi–Yau manifold, known as the "Z-orbifold," which has a mirror physical theory that was worked out recently by Candelas, Derrick and Parkes [**119**] (see also [**105**]). In this example, the variation of Hodge structure associated to the mirror theory can be described by the family of cubic sevenfolds in $\mathbb{P}^8$ (with a suitable Tate twist).

8.3. Some computations

We explain some of the evidence in favor of the mirror symmetry conjectures which has been accumulated through specific computations.[3] We will compute

[3] The computations presented here are taken from the original paper of Candelas, de la Ossa, Green and Parkes [**118**] on the quintic threefold, and a paper of Greene, Plesser and the present author [**133**] on higher dimensional mirror manifolds. A survey of other calculations of this type (and the methods for making them) can be found in [**77**].

with Calabi–Yau hypersurfaces of dimension $n \geq 3$ in ordinary projective space $\mathbb{CP}^{n+1}$; the degree of the hypersurface must be $n+2$. The family of such hypersurface includes a Fermat hypersurface, which is part of the "Dwork pencil" with defining equation:

$$x_0^{n+2} + \cdots + x_{n+1}^{n+2} - (n+2)\psi\, x_0 \cdots x_{n+1} = 0,$$

where ψ is a parameter. The group

$$\Gamma := \{(\alpha_0, \ldots, \alpha_{n+1}) \mid \alpha_j \in \mu_{n+2}, \prod \alpha_j = 1\}/\{(\alpha, \ldots, \alpha)\}$$

acts on the fibers of this family by componentwise multiplication on the coordinates.

Using either the Greene–Plesser orbifolding construction, or Batyrev's polar polyhedron construction, one sees that the family $\{M\}$ of hypersurfaces of degree $n+2$ in $\mathbb{CP}^{n+1}$ has as its predicted mirror the family $\{W\}$ described as the Dwork pencil modulo Γ (living in the quotient space $\mathbb{CP}^{n+1}/\Gamma$). In fact, we can describe the moduli space of this mirrored family in terms of the parameter ψ^{n+2}—the reason for passing to a power is the existence of an additional automorphism, acting on the family as a whole, generated by componentwise multiplication in the x's by $(\alpha, 1, \ldots, 1)$ while simultaneously multiplying ψ by α^{-1}.

It is not difficult to compute where this family becomes singular. The partial derivatives of the defining equation are all of the form

$$(n+2)\left(x_j^{n+1} - x_j^{-1}\psi\, x_0 \cdots x_{n+1}\right)$$

and for these to vanish simultaneously we must have $\psi^{n+2} = 1$. Moreover, the additional automorphism of the family fixed the fiber $\psi = 0$, and so causes additional singularities there. Thus, we can describe the moduli space as $\mathbb{CP}^1 - \{0, 1, \infty\}$, with its natural compactification being $\mathbb{CP}^1$.

What is the monodromy behavior at the boundary points? (We label the monodromy transformations according to the point.) At $\psi^{n+2} = 0$, we find that the monodromy has finite order, at $\psi^{n+2} = 1$ it is unipotent but $(T_1 - I)^2 = 0$ so the order is not maximal (since $n \neq 1$), and at $\psi^{n+2} = \infty$ we find maximal order of unipotency $(T_\infty - I)^n \neq 0$. In fact, this point is maximally unipotent, and even strongly maximally unipotent, in the terminology established earlier.

To compute canonical coordinates and correlation functions near $\psi^{n+2} = \infty$ we need to know the period functions there. These can be found by studying the differential equations which they satisfy. In this case of toric hypersurfaces, we have a special method available—the representation of cohomology by means of residues of differential forms on the ambient space with poles along the hypersurface. A basis for the primitive cohomology can be written (in the affine chart $x_0 = 1$, say) as

$$\beta_j := \operatorname{Res}\left(\frac{\psi^{j+1}\,(x_1 \cdots x_{n+1})^j\, dx_1 \wedge \cdots \wedge dx_{n+1}}{\left(1 + x_1^{n+2} + \cdots + x_{n+1}^{n+2} - (n+2)\psi\, x_1 \cdots x_{n+1}\right)^{j+1}}\right)$$

The connection matrix in this basis can then be found using Griffiths' "reduction of pole order" lemma **[50]** to calculate coefficients θ_{ij} such that

$$\nabla(\beta_i) = \sum \theta_{ij}\beta_j.$$

To find the period matrix from the connection matrix, one must solve some differential equations. For if $\{e_k\}$ is a basis for the local system and we write $e_k = \sum \eta_{ki}\beta_i$

n	n-point function
3	$5 + 2875\,q + 4876875\,q^2 + 8564575000\,q^3 + 15517926796875\,q^4$ $+ 28663236110956000\,q^5 + 53621944306062201000\,q^6$ $+ 101216230345800061125625\,q^7 + 192323666400003538944396875\,q^8$ $+ 367299732093982242625847031250\,q^9$ $+ 704288164978454714776724365580000\,q^{10}$ $+ 1354842473951260627644461070753075500\,q^{11}$ $+ 2613295702542192770504516764304958585000\,q^{12}$ $+ 5051976384195377826370376750184667397150000\,q^{13}$ $+ 9784992122065556293839548184561593434114765625\,q^{14}$ $+ 18983216783256131050355758292004110332155634496875\,q^{15}$ $+ 36880398908911843175757970052077286676680907186572875\,q^{16}$ $+ 71739993072775923425756947313710004388338109828244718125\,q^{17}$ $+ 139702324572802672116486725324237666156179096139345867681250\,q^{18}$ $+ \dots$
4	$6 + 120960\,q + 4136832000\,q^2 + 148146924602880\,q^3 + 5420219848911544320\,q^4$ $+ 200623934537137119778560\,q^5 + 7478994517395643259712737280\,q^6$ $+ 280135301818357004749298146851840\,q^7$ $+ 10528167289356385699173014219946393600\,q^8$ $+ 396658819202496234945300681212382224722560\,q^9$ $+ 14972930462574202465673643937107499992165427200\,q^{10}$ $+ 566037069767251121484562070892662863943365345190400\,q^{11}$ $+ 21424151141341932048068067497996096856724987411324108800\,q^{12} + \dots$
5	$7 + 3727381\,q + 2637885990187\,q^2 + 1927092954108108787\,q^3$ $+ 1425153551321014327663291\,q^4 + 1060347883438857662557634869906\,q^5$ $+ 791661306374088776109692880989252173\,q^6$ $+ 592348256908461616176898022359492565546566\,q^7$ $+ 443865568545713063761643598030194801299861575595\,q^8$ $+ 332947403131697202086626568381790256001850741509664373\,q^9 + \dots$
6	$8 + 106975232\,q + 1672023727001600\,q^2 + 26611692333081695092736\,q^3$ $+ 426129121674687823674948571136\,q^4$ $+ 6842148599241293047857339542861643776\,q^5$ $+ 110018992594692024449889564415904439556898816\,q^6$ $+ 1770551943055574073245974844490813198478975912902656\,q^7$ $+ 28508925683951911989843155602330000507452539542539447947264\,q^8$ $+ \cdots$

Table 1. n-point functions in dimension n

then

$$0 = \nabla(e_k) = \sum d\eta_{ki}\,\beta_i + \sum \eta_{ki}\theta_{ij}\beta_j$$

gives differential equations for the unknown coefficient functions η_{ki}:

$$d\eta_{ki} = -\sum \eta_{k\ell}\theta_{\ell i}.$$

The flatness of ∇ is equivalent to the integrability of these equations, which can therefore be solved.

$$
\begin{aligned}
Y_1^1 = 5 &+ 2875\,\tfrac{1^3 q}{1-q} + 609250\,\tfrac{2^3 q^2}{1-q^2} + 317206375\,\tfrac{3^3 q^3}{1-q^3} + 242467530000\,\tfrac{4^3 q^4}{1-q^4}\\
&+ 229305888887625\,\tfrac{5^3 q^5}{1-q^5} + 248249742118022000\,\tfrac{6^3 q^6}{1-q^6}\\
&+ 295091050570845659250\,\tfrac{7^3 q^7}{1-q^7} + 375632160937476603550000\,\tfrac{8^3 q^8}{1-q^8}\\
&+ 503840510416985243645106250\,\tfrac{9^3 q^9}{1-q^9}\\
&+ 704288164978454686113488249750\,\tfrac{10^3 q^{10}}{1-q^{10}}\\
&+ 1017913203569692432490203659468875\,\tfrac{11^3 q^{11}}{1-q^{11}}\\
&+ 1512323901934139334751675234074638000\,\tfrac{12^3 q^{12}}{1-q^{12}}\\
&+ 2299488568136266648325160104772265542625\,\tfrac{13^3 q^{13}}{1-q^{13}}\\
&+ 3565959228158001564810294084668822024070250\,\tfrac{14^3 q^{14}}{1-q^{14}}\\
&+ 5624656824668483274179483938371579753751395250\,\tfrac{15^3 q^{15}}{1-q^{15}}\\
&+ 9004003639871055462831535610291411200360685606000\,\tfrac{16^3 q^{16}}{1-q^{16}} + \dots
\end{aligned}
$$

Table 2. Three-point function in dimension three

$$
\begin{aligned}
Y_1^1 = 6 &+ 60480\,\tfrac{1^2 q}{1-q} + 440884080\,\tfrac{2^2 q^2}{1-q^2} + 6255156277440\,\tfrac{3^2 q^3}{1-q^3}\\
&+ 117715791990353760\,\tfrac{4^2 q^4}{1-q^4} + 2591176156368821985600 \cdot 5^2\,\tfrac{5^2 q^5}{1-q^5} + \dots
\end{aligned}
$$

Table 3. Three-point function in dimension four

If we work in a local coordinate $z = \psi^{-n-2}$ near $\psi^{n+2} = \infty$, we find that a basis $e_0(z), \dots, e_{n+1}(z)$ of local solutions can be found such that $e_0(z)$ is single-valued near $z = 0$, and

$$e_{j+1}(z) = (\log z)\, e_j(z) + \text{single-valued function.}$$

(This is a consequence of the maximally unipotent monodromy.) The vectors $e_j(z)$ form the columns of the period matrix.

One can then use row operations to put the period matrix in upper triangular form, with constant diagonal elements. (Let us choose the diagonal elements to all be $n+2$.) This implements the change of basis to a basis consisting of distinguished sections of $\mathcal{H}^{p,q}_{\mathbb{S}}$. The nonzero entries A_j^1 in the connection matrix are then calculated by differentiating rows of the period matrix, and writing the result as a multiple of a subsequent row. Each such entry takes the form

$$A_j^1 = Y_j^1\,\frac{dq}{q},$$

and the functions Y_j^1 represent correlation functions $\langle (\partial/\partial t)\,\beta_j\,\beta_{n-j-1}\rangle$.

This can all be done very explicitly, using power series expansions of the unknown single-valued functions, in these examples. (I advise using MAPLE or MATHEMATICA if you would like to try it for yourself.) We show two kinds of calculations in the tables. For the first, only the "maximally unipotent" assumption is required, since the calculation requires only the distinguished n-form and the canonical coordinates. What is computed in table 1 is the "n-point function," which iterates the differential of the period map n times. (This was introduced some years ago in the variation of Hodge structures context by Carlson, Green, Griffiths and Harris **[34]**.)

$$\begin{aligned} Y_1^1 &= 7 + 1009792\,\tfrac{1^2 q}{1-q} + 122239786088\,\tfrac{2^2 q^2}{1-q^2} + 30528671745480104\,\tfrac{3^2 q^3}{1-q^3} \\ &\quad + 1037819950939588615321 6\,\tfrac{4^2 q^4}{1-q^4} + \dots \\ Y_2^1 &= 7 + 1707797\,\tfrac{1\cdot q^1}{1-q^1} + 510787745643\,\tfrac{2\cdot q^2}{1-q^2} + 222548537108926490\,\tfrac{3\cdot q^3}{1-q^3} \\ &\quad + 113635631482486991647224\,\tfrac{4\cdot q^4}{1-q^4} + \dots \end{aligned}$$

Table 4. Three-point functions in dimension five

$$\begin{aligned} Y_1^1 &= 8 + 15984640\,\tfrac{1^2 q}{1-q} + 33397159706624\,\tfrac{2^2 q^2}{1-q^2} + 154090254047541417984\,\tfrac{3^2 q^3}{1-q^3} \\ &\quad + 1000674891265872131899670528\,\tfrac{4^2 q^4}{1-q^4} + \dots \\ Y_2^1 &= 8 + 37502976\,\tfrac{1\cdot q^1}{1-q^1} + 224340704157696\,\tfrac{2\cdot q^2}{1-q^2} + 2000750410187341381632\,\tfrac{3\cdot q^3}{1-q^3} \\ &\quad + 21122119007324663457380794368\,\tfrac{4\cdot q^4}{1-q^4} + \dots \\ Y_2^2 &= 8 + 59021312\,\tfrac{q}{1-q} + 821654025830400\,\tfrac{q^2}{1-q^2} + 12197109744970010814464\,\tfrac{q^3}{1-q^3} \\ &\quad + 186083410628492378226388631552\,\tfrac{q^4}{1-q^4} + \dots \end{aligned}$$

Table 5. Three-point functions in dimension six

The other computations, displayed in tables 2–5, are of three-point functions Y_b^a, read off of the connection matrix in a distinguished basis. (There is a symmetry $Y_b^a = Y_{n-a-b}^a = Y_{n-a-b}^b$ so we only show some of these.) The coefficients in the series expansions are the predicted values of the Gromov–Witten invariants. The three-point function Y_0^1 has the value $n+2$ (a constant, due to the definition of canonical coordinates) and is not shown in the tables. The other functions Y_j^1 come directly from the connection matrix. In dimension six, there is also a "secondary" function, which (by the B-model version of the associativity, which is simply the associativity of the "sheaf cup product" pairing) can be calculated as $Y_2^2 = (Y_2^1)^2/Y_1^1$.

There is a relation between the computations in table 1, and those in tables 2–5, which can be explicitly verified from these tables: it is

$$n\text{-point function} = \frac{Y_0^1 \cdot Y_1^1 \cdot \dots \cdot Y_{n-1}^1}{(n+2)^n}.$$

The functions Y_b^a are predicted to agree with quantum products on the mirror manifolds

$$\zeta^a \star \zeta^b \star \zeta^{n-a-b},$$

where ζ^j is the class of a linear space (in $\mathbb{CP}^{n+1}$) of complex codimension j. In fact, we have displayed things in tables 2–5 with this in mind, writing series in terms of $q^k/(1-q^k)$.

Also in tables 2–5, we have pulled out some factors of the degree of the rational curve. If there are ℓ occurrences of "1" among $\{a, b, n-a-b\}$, then there will be ℓ of the linear spaces of codimension one, and each meets a given rational curve Γ in $\deg(\Gamma)$ points, giving rise to a factor of $(\deg(\Gamma))^\ell$ in the Gromov–Witten invariants. Pulling out those factors makes the comparison with "counting" problems more transparent.

All of the predicted Gromov–Witten invariants in degrees one and two in these tables have been verified by Katz [**60**]; most of the invariants in degree three have been verified by Ellingsrud and Strømme [**40, 41**].

POSTSCRIPT: Recent Developments

As mentioned in the introduction, the subject of mirror symmetry is a rapidly developing one, and much has happened since the lectures on which these notes are based were delivered. We will briefly sketch some of these developments in this postscript.

The Gromov–Witten invariants and their generalizations have been studied particularly intensively. The definition of Ruan [**86**] which we presented in the lectures has been supplanted by other definitions drawn from symplectic geometry (cf. [**12, 87**]) which work directly in cohomology (avoiding the bordism technicalities) and are also more general. In full generality these extended Gromov–Witten invariants are not only associated to curves of genus zero with three vertex operators, but also to curves of arbitrary genus g with k vertex operators (provided that $2g-2+k > 0$) and even to some non-topological correlation functions.[4] There are at least three proofs of the associativity relations for these symplectic Gromov–Witten invariants [**87, 67, 12**], including proofs of a stronger form of associativity known as the Witten–Dijkgraaf–Verlinde–Verlinde (WDVV) equations [**151, 125, 101, 129**] which are relevant in the case of higher genus. As in the genus zero case, these higher genus invariants can be used to encode a kind of quantum cohomology ring (somewhat larger than the one we studied here); it is also possible to interpret the WDVV associativity relation as the flatness of a certain connection [**129**]. A very accessible exposition of this circle of ideas has been written by McDuff and Salamon [**12**].

Parallel to this development, Gromov–Witten invariants have also been defined purely within algebraic geometry. The methods of Katz described in the lectures were developed further (see [**139**] and the appendix to [**114**]), and similar methods based on the construction of a "virtual moduli cycle" were developed independently by Li and Tian [**66**]. The foundations for an algebraic theory of Gromov–Witten invariants were carefully laid by Kontsevich and Manin [**64**] (again, the higher genus

[4]There have also been investigations into the physical interpretation of these higher genus invariants, and how they should transform under mirror symmetry (in the case of Calabi–Yau threefolds) [**114, 115**]. At one time, it had been expected that for Calabi–Yau threefolds the genus zero topological correlation functions would completely determine the conformal field theory, but now it is known that higher genus invariants are needed as well [**111**].

invariants and the WDVV equations play an important rôle), and the program they initiated was ultimately carried out [**28, 27, 26**], producing a definition of Gromov–Witten invariants based on stable maps. (The work of Li–Tian mentioned above [**66**] is also closely related to this program.) Even before this program was complete, Kontsevich had applied it to obtain some spectacular results in enumerative geometry, including a verification of the predicted number 242467530000 of rational quartics on the general quintic threefold [**62**]. The stable map theory is nicely explained, with further references, in [**45**].

Kontsevich has also formulated a "homological" version of the mirror conjecture [**63**] involving what are known as A^∞-categories (cf. [**91**]), which is related to the "extended moduli space" introduced by Witten [**152**]. By a construction of Fukaya [**44**], to every compact symplectic manifold (Y, ω) with vanishing first Chern class, one can associate an A^∞-category whose objects are essentially the Lagrangian submanifolds of Y, and whose morphisms are determined by the intersections of pairs of submanifolds. Kontsevich's conjecture relates the bounded derived category of the Fukaya category of Y (playing the rôle of the B-model) to the bounded derived category of the category of coherent sheaves on a mirror partner X (playing the rôle of the A-model). I must refer the reader to [**63**] for further details concerning this fascinating conjecture.

The art of making predictions about enumerative geometry from calculations with the variation of Hodge structure on a candidate mirror partner has been considerably refined: see [**77**] for a survey and references to the literature. The era of numerical experiments in mirror symmetry seems to be largely over, and has been supplanted by a more analytical period. Witten's analysis of the physics related to Calabi–Yau manifolds which are hypersurfaces in toric varieties [**153**] was further developed in [**143**], where techniques were found—somewhat related to methods introduced by Batyrev [**20**] for the study of quantum cohomology of toric varieties—for precisely calculating a variant of the quantum cohomology ring of the Calabi–Yau manifold. (The variant is derived from enumerative problems on the ambient space rather than directly on the Calabi–Yau manifold.) There is a physics argument, but not a complete mathematics argument, which explains why this variant should coincide with the usual quantum cohomology ring after a change of coordinates in the coefficient ring. This variant *can* be rigorously shown to agree with the correlation functions of the mirror Calabi–Yau manifold, again calculated in the "wrong" coordinates. In this way, the results of [**143**] provided the first analytical proof that some kind of enumerative problem on one side of the mirror could be related to a variation of Hodge structure calculation on the other side. Further development of these ideas in [**144**] led to a preliminary argument to the effect that the physical theories associated to a Batyrev–Borisov pair should actually be mirror to each other.

In a striking recent development, Givental has proved [**46, 47, 48**] that for Calabi–Yau complete intersections in projective spaces, the "predicted" enumerative formulas which one calculates by using a Batyrev–Borisov candidate mirror partner are in fact correct evaluations of the Gromov–Witten invariants. This establishes, for example, the accuracy of *all* of the predictions about the general quintic threefold made by Candelas et al. [**118**] (and which we listed in table 2). Givental's remarkable proof actually has very little to do with mirror symmetry *per se*: in studying an equivariant version of quantum cohomology, he finds enough structure to enable a calculation which is formally similar to (and certainly inspired by) the variation of Hodge structure calculations on the candidate mirror partner.

The last several years have also been a period of dramatic developments in string theory. There are new techniques which go by the names of "duality" and "nonperturbative methods," and a number of the recent results have been closely related to Calabi–Yau manifolds and mirror symmetry. One of the earliest nonperturbative results [**146, 134**] was the discovery[5] that the string theory moduli spaces associated to Calabi–Yau manifolds should be attached along loci corresponding to "conifold transitions"—a process in which a collection of rational curves is contracted to ordinary double points and the resulting space is then smoothed to produce another Calabi–Yau manifold. This new attaching procedure supplements, but is rather different from, the gluing of Kähler cones which we discussed in section 7.3. In the new procedure, a moduli space of a different dimension (corresponding to a Calabi–Yau manifold with different Hodge numbers than the original) is cemented on at the same point where the two like-dimensional pieces (Kähler cones differing by a flop) have been glued together. The "cement" which holds these two spaces together (i.e., the physical process responsible) is a phase transition between charged black holes on one component of the moduli space and elementary particles on the other.

The string theory moduli spaces mentioned above are actually somewhat larger than the conformal field theory moduli spaces which were one of the primary subjects of these lectures. There are two variants of string theory which are relevant, called type IIA and type IIB string theories, and the additional parameters which must be added to the conformal field theory moduli space differs between the two. In the case of type IIA, the extra parameters are a choice of holomorphic 3-form and the choice of an element in the intermediate Jacobian of the Calabi–Yau threefold. (Some of the mathematical structure of these spaces related to the intermediate Jacobians was anticipated in work of Donagi and Markman [**39**].) In the case of type IIB, the new parameters are similar, but related to the even cohomology of the manifold. These two types of parameters should be mapped to each other under mirror symmetry [**112, 142**]. In fact, a large number of other related structures called "D-brane moduli spaces" should also correspond under mirror symmetry—the precise implications of this correspondence (which appears to be connected to Kontsevich's homological mirror symmetry conjecture) are still being worked out.

Finally, in a very exciting recent development, a completely new geometric aspect of mirror symmetry has been discovered by Strominger, Yau and Zaslow [**148**]. A Calabi–Yau manifold X of real dimension $2n$ on which a complex structure J and Kähler form ω have been fixed has a natural class of n-dimensional submanifolds M defined by the property that $\omega|_M \equiv 0$ and $\mathrm{Im}(\Omega)|_M \equiv 0$ for some choice of holomorphic n-form Ω. These *special Lagrangian submanifolds* were introduced by Harvey and Lawson [**53**] as a natural class of volume-minimizing submanifolds; they have many other interesting properties, including an exceptionally well-behaved deformation theory [**72**]. Strominger, Yau and Zaslow argue on physical grounds (using the correspondence of D-brane moduli spaces mentioned above) that whenever X has a mirror partner, then X must admit a map $\rho : X^{2n} \to B^n$ whose generic fiber

[5]This had been anticipated some time earlier in the physics literature [**116, 132, 121, 117**] based on the discovery of and speculations about conifold transitions in the mathematics literature [**36, 42, 54, 94, 85, 43**], but an understanding of the physical mechanism behind the attachment of the moduli spaces was lacking.

is a special Lagrangian n-torus, and which has a section $\sigma : B \to X$ whose image is itself a special Lagrangian submanifold. Given this structure, the mirror partner of X is then predicted to be a compactification of the family of dual tori of the fibers of ρ. (The section specifies a point $p_b := \sigma(b)$ on each torus $T_b := \rho^{-1}(b)$; the dual torus is then $\mathrm{Hom}(\pi_1(T_b, p_b), \mathrm{U}(1))$.) There is also an argument—quite similar in nature to [**144**]—that such a structure should suffice for producing a mirror isomorphism between the corresponding physical theories. A mathematical account of this construction can be found in [**78**], which attempts to make the mathematical implications of this story precise: given a "special Lagrangian m-torus fibration," all of the structure we have seen relating the quantum cohomology and the variation of Hodge structure should (conjecturally) follow as a consequence. For the Voisin–Borcea threefolds, the structure of these special Lagrangian torus fibrations (using a mildly degenerate metric) has been worked out in complete detail by Gross and Wilson [**52**], who find compatibility with the previously observed mirror phenomena in a beautiful geometric form.

BIBLIOGRAPHY

Books

1. H. Clemens, J. Kollár, and S. Mori, Higher-Dimensional Complex Geometry, Astérisque, vol. 166, Société Mathématique de France, 1988.
2. P. E. Conner and E. E. Floyd, Differentiable Periodic Maps, Ergeb. Math. Grenzgeb. N. F., vol. 33, Springer-Verlag, Berlin, Göttingen, Heidelberg, 1964.
3. D. Cox and S. Katz, Mirror Symmetry and Algebraic Geometry, in preparation.
4. C. W. Curtis and I. Reiner, Representation Theory of Finite Groups and Associative Algebras, Interscience Publishers, New York, 1962.
5. P. Deligne, Equations Différentielles à Points Singuliers Réguliers, Lecture Notes in Math., vol. 163, Springer-Verlag, Berlin, Heidelberg, New York, 1970.
6. W. Fulton, Intersection Theory, Ergeb. Math. Grenzgeb. (3), vol. 2, Springer-Verlag, Berlin, New York, 1984.
7. B. Greene and S.-T. Yau (eds.), Mirror Symmetry II, International Press, Cambridge, 1996.
8. P. Griffiths et al., Topics in Transcendental Algebraic Geometry, Annals of Math. Studies, vol. 106, Princeton University Press, Princeton, 1984.
9. R. Hartshorne, Algebraic Geometry, Springer-Verlag, New York, Heidelberg, Berlin, 1977.
10. T. Hübsch, Calabi–Yau Manifolds: A Bestiary for Physicists, World Scientific, Singapore, New Jersey, London, Hong Kong, 1992.
11. G. Karpilovsky, Symmetric and G-Algebras, With Applications to Group Representations, Kluwer Academic Publishers, 1990.
12. D. McDuff and D. Salamon, J-Holomorphic Curves and Quantum Cohomology, University Lecture Series, vol. 6, American Mathematical Society, Providence, 1994.
13. D. Mumford and J. Fogarty, Geometric Invariant Theory, 2nd ed., Ergeb. Math. Grenzgeb. N. F., vol. 34, Springer-Verlag, Berlin, Heidelberg, New York, 1982.
14. C. Nash, Differential Topology and Quantum Field Theory, Academic Press, London, San Diego, 1991.

15. C. Voisin, Symétrie Miroir, Panoramas et Synthèses, vol. 2, Société Mathématique de France, Paris, 1996.
16. S.-T. Yau (ed.), Mathematical Aspects of String Theory, World Scientific, Singapore, 1987.
17. ———, Essays on Mirror Manifolds, International Press, Hong Kong, 1992.

Mathematics Articles

18. P. S. Aspinwall, B. R. Greene, and D. R. Morrison, *The monomial-divisor mirror map*, Internat. Math. Res. Notices (1993), 319–337.
19. V. V. Batyrev, *Variations of the mixed Hodge structure of affine hypersurfaces in algebraic tori*, Duke Math. J. **69** (1993), 349–409.
20. ———, *Quantum cohomology rings of toric manifolds*, Journées de Géométrie Algébrique d'Orsay (Juillet 1992) (A. Beauville et al., eds.), Astérisque, vol. 218, Société Mathématique de France, 1993, pp. 9–34.
21. ———, *Dual polyhedra and mirror symmetry for Calabi-Yau hypersurfaces in toric varieties*, J. Algebraic Geom. **3** (1994), 493–535.
22. V. V. Batyrev and L. A. Borisov, *Dual cones and mirror symmetry for generalized Calabi–Yau manifolds*, Mirror Symmetry II (B. Greene and S.-T. Yau, eds.), International Press, Cambridge, 1996, pp. 71–86.
23. V. V. Batyrev and D. A. Cox, *On the Hodge structure of projective hypersurfaces in toric varieties*, Duke Math. J. **75** (1994), 293–338.
24. V. V. Batyrev and D. van Straten, *Generalized hypergeometric functions and rational curves on Calabi–Yau complete intersections in toric varieties*, Comm. Math. Phys. **168** (1995), 493–533.
25. A. Beauville, *Variétés Kähleriennes dont la première classe de Chern est nulle*, J. Differential Geom. **18** (1983), 755–782.
26. K. Behrend, *Gromov–Witten invariants in algebraic geometry*, preprint, 1996 (alg-geom/9601011).
27. K. Behrend and B. Fantechi, *The intrinsic normal cone*, preprint, 1996 (alg-geom/9601010).
28. K. Behrend and Yu. Manin, *Stacks of stable maps and Gromov–Witten invariants*, Duke Math. J., to appear.
29. F. A. Bogomolov, *Hamiltonian Kähler manifolds*, Dokl. Akad. Nauk SSSR **243** (1978), no. 5, 1101–1104.
30. C. Borcea, *On desingularized Horrocks–Mumford quintics*, J. Reine Angew. Math. **421** (1991), 23–41.
31. ———, *K3 surfaces with involution and mirror pairs of Calabi–Yau manifolds*, Mirror Symmetry II (B. Greene and S.-T. Yau, eds.), International Press, Cambridge, 1996, pp. 717–744.
32. L. Borisov, *Towards the mirror symmetry for Calabi–Yau complete intersections in toric Fano varieties*, preprint, 1993 (alg-geom/9310001).
33. E. Calabi, *On Kähler manifolds with vanishing canonical class*, Algebraic Geometry and Topology, A Symposium in Honor of S. Lefschetz (R. H. Fox et al., eds.), Princeton University Press, Princeton, 1957, pp. 78–89.
34. J. Carlson, M. Green, P. Griffiths, and J. Harris, *Infinitesimal variations of Hodge structure, I*, Compositio Math. **50** (1983), 109–205.
35. E. Cattani, A. Kaplan, and W. Schmid, *L^2 and intersection cohomologies for a polarizable variation of Hodge structure*, Invent. Math. **87** (1987), 217–252.

36. C. H. Clemens, *Double solids*, Adv. Math. **47** (1983), 107–230.
37. ———, *Some results on Abel–Jacobi mappings*, Topics in Transcendental Algebraic Geometry (P. Griffiths, ed.), Princeton University Press, Princeton, 1984, pp. 289–304.
38. P. Deligne, *Local behavior of Hodge structures at infinity*, Mirror Symmetry II (B. Greene and S.-T. Yau, eds.), International Press, Cambridge, 1996, pp. 683–700
39. R. Donagi and E. Markman, *Cubics, integrable systems, and Calabi–Yau threefolds*, Proc. of the Hirzebruch 65 Conference on Algebraic Geometry (M. Teicher, ed.), Israel Math. Conf. Proc., vol. 9, Bar-Ilan University, 1996, pp. 199–221.
40. G. Ellingsrud and S. A. Strømme, *The number of twisted cubic curves on the general quintic threefold*, Math. Scand. **76** (1995), 5–34.
41. ———, *Bott's formula and enumerative geometry*, J. Amer. Math. Soc. **9** (1996), 175–193.
42. R. Friedman, *Simultaneous resolution of threefold double points*, Math. Ann. **274** (1986), 671–689.
43. ———, *On threefolds with trivial canonical bundle*, Complex Geometry and Lie Theory (J. A. Carlson et al., eds.), Proc. Symp. Pure. Math., vol. 53, American Mathematical Society, Providence, 1991, pp. 103–134.
44. K. Fukaya, *Morse homotopy, A^∞-category, and Floer homologies*, Proceedings of GARC Workshop on Geometry and Topology '93, Lecture Notes Ser., 18, Seoul Nat. Univ., Seoul, 1993, pp. 1–102.
45. W. Fulton and R. Pandharipande, *Notes on stable maps and quantum cohomology*, Algebraic Geometry—Santa Cruz 1995, Proc. Symp. Pure Math., American Mathematical Society, Providence, to appear.
46. A. B. Givental, *Homological geometry I: Projective hypersurfaces*, Selecta Math. (N.S.) **1** (1995), 325–345.
47. ———, *Homological geometry and mirror symmetry*, Proc. Internat. Congr. Math. Zürich 1994 (S. D. Chatterji, ed.), vol. 1, Birkäuser Verlag, Basel, Boston, Berlin, 1995, pp. 472–480.
48. ———, *Equivariant Gromov–Witten invariants*, Internat. Math. Res. Notices (1996), 613–663.
49. A. Grassi and D. R. Morrison, *Automorphisms and the Kähler cone of certain Calabi–Yau manifolds*, Duke Math. J. **71** (1993), 831–838.
50. P. A. Griffiths, *On the periods of certain rational integrals, I*, Annals of Math. (2) **90** (1969), 460–495.
51. M. Gromov, *Pseudo-holomorphic curves on almost complex manifolds*, Invent. Math. **82** (1985), 307–347.
52. M. Gross and P. M. H. Wilson, *Mirror symmetry via 3-tori for a class of Calabi–Yau threefolds*, preprint, 1996 (alg-geom/9608004).
53. R. Harvey and H. B. Lawson, Jr., *Calibrated geometries*, Acta Math. **148** (1982), 47–157.
54. F. Hirzebruch (notes by J. Werner), *Some examples of threefolds with trivial canonical bundle*, Gesammelte Abhandlungen, Band II, Springer-Verlag, Berlin, New York, 1987, pp. 757–770.
55. H. Hofer and D. A. Salamon, *Floer homology and Novikov rings*, The Floer Memorial Volume (H. Hofer et al., eds.), Birkhäuser Verlag, Basel, Boston, Berlin, 1995, pp. 483–524.

56. T. Johnsen and S. Kleiman, *Rational curves of degree at most* 9 *on a general quintic threefold*, Comm. Algebra **24** (1996), 2721–2753.
57. ———, *Toward Clemens' conjecture in degrees between* 10 *and* 24, preprint, 1996, (alg-geom/9601024).
58. S. Katz, *On the finiteness of rational curves on quintic threefolds*, Compositio Math. **60** (1986), 151–162.
59. ———, *Rational curves on Calabi–Yau threefolds*, Essays on Mirror Manifolds (S.-T. Yau, ed.), International Press, Hong Kong, 1992, pp. 168–180.
60. ———, *Rational curves on Calabi–Yau manifolds: Verifying predictions of mirror symmetry*, Projective Geometry with Applications, Lecture Notes in Pure and Appl. Math., vol. 166, Marcel Dekker, New York, 1994, pp. 231–239.
61. Y. Kawamata, *Crepent blowing-up of 3-dimensional canonical singularities and its application to degenerations of surfaces*, Annals of Math. (2) **127** (1988), 93–163.
62. M. Kontsevich, *Enumeration of rational curves via torus actions*, The Moduli Space of Curves (R. Dijkgraaf et al., eds.), Progress in Math., vol. 129, Birkhäuser, Boston, 1995, pp. 335–368.
63. ———, *Homological algebra of mirror symmetry*, Proc. Internat. Congr. Math. Zürich 1994 (S. D. Chatterji, ed.), vol. 1, Birkäuser Verlag, Basel, Boston, Berlin, 1995, pp. 120–139.
64. M. Kontsevich and Yu. Manin, *Gromov–Witten classes, quantum cohomology, and enumerative geometry*, Comm. Math. Phys. **164** (1994), 525–562.
65. A. Landman, *On the Picard–Lefschetz transformations*, Trans. Amer. Math. Soc. **181** (1973), 89–126.
66. J. Li and G. Tian, *Virtual moduli cycles and Gromov–Witten invariants of algebraic varieties*, preprint, 1996 (alg-geom/9602007).
67. G. Liu, *Associativity of quantum multiplication*, preprint, 1994.
68. E. Looijenga, *New compactifications of locally symmetric varieties*, Proceedings of the 1984 Vancouver Conference in Algebraic Geometry (J. Carrell et al., eds.), CMS Conference Proceedings, vol. 6, American Mathematical Society, Providence, 1986, pp. 341–364.
69. Yu. I. Manin, *Generating functions in algebraic geometry and sums over trees*, The Moduli Space of Curves (R. Dijkgraaf et al., eds.), Progress in Math., vol. 129, Birkhäuser, Boston, 1995, pp. 401–417.
70. D. McDuff, *Examples of symplectic structures*, Invent. Math. **89** (1987), 13–36.
71. ———, *Symplectic manifolds with contact type boundaries*, Invent. Math. **103** (1991), 651–671.
72. R. C. McLean, *Deformations of calibrated submanifolds*, J. Differential Geom., to appear.
73. D. R. Morrison, *Mirror symmetry and rational curves on quintic threefolds: A guide for mathematicians*, J. Amer. Math. Soc. **6** (1993), 223–247.
74. ———, *Compactifications of moduli spaces inspired by mirror symmetry*, Journées de Géométrie Algébrique d'Orsay (Juillet 1992), Astérisque, vol. 218, Société Mathématique de France, 1993, pp. 243–271.
75. ———, *Mirror symmetry and moduli spaces of superconformal field theories*, Proc. Internat. Congr. Math. Zürich 1994 (S. D. Chatterji, ed.), vol. 2, Birkäuser Verlag, Basel, Boston, Berlin, 1995, pp. 1304–1314.
76. ———, *Beyond the Kähler cone*, Proc. of the Hirzebruch 65 Conference on

Algebraic Geometry (M. Teicher, ed.), Israel Math. Conf. Proc., vol. 9, Bar-Ilan University, 1996, pp. 361–376.

77. ______, *Making enumerative predictions by means of mirror symmetry*, Mirror Symmetry II (B. Greene and S.-T. Yau, eds.), International Press, Cambridge, 1996, pp. 457–482.

78. ______, *The geometry underlying mirror symmetry*, preprint, 1996 (alg-geom/9608006).

79. T. Nakayama, *On Frobeniusean algebras I*, Annals of Math. (2) **40** (1939), 611–633.

80. P. Nijsse, *Clemens' conjecture for octic and nonic curves*, Indag. Math. (N. S.) **6** (1995), 213–221.

81. V. V. Nikulin, *Discrete reflection groups in Lobachevsky spaces and algebraic surfaces*, Proc. Internat. Congr. Math. Berkeley 1986 (A. M. Gleason, ed.), vol. 1, American Mathematical Society, Providence, 1987, pp. 654–671.

82. S. P. Novikov, *Multivalued functions and functionals — an analogue of the Morse theory*, Soviet Math. Dokl. **24** (1981), 222–225.

83. K. Oguiso, *On algebraic fiber space structures on a Calabi–Yau 3-fold*, Internat. J. Math. **4** (1993), 439–465.

84. T. H. Parker and J. G. Wolfson, *Pseudo-holomorphic maps and bubble trees*, J. Geom. Anal. **3** (1993), 63–98.

85. M. Reid, *The moduli space of 3-folds with $K=0$ may nevertheless be irreducible*, Math. Ann. **278** (1987), 329–334.

86. Y. Ruan, *Topological sigma model and Donaldson type invariants in Gromov theory*, Duke Math. J. **83** (1996), 461–500.

87. Y. Ruan and G. Tian, *A mathematical theory of quantum cohomology*, J. Differential Geom. **42** (1995), 259–367.

88. J. Sacks and K. Uhlenbeck, *The existence of minimal immersions of 2-spheres*, Annals of Math. (2) **113** (1981), 1–24.

89. W. Schmid, *Variation of Hodge structure: The singularities of the period mapping*, Invent. Math. **22** (1973), 211–319.

90. C. Schoen, *On fiber products of rational elliptic surfaces with section*, Math. Z. **197** (1988), 177–199.

91. J. D. Stasheff, *Homotopy associativity of H-spaces, I, II*, Trans. Amer. Math. Soc. **108** (1963), 275–312.

92. R. Thom, *Quelques propriétés globales des variétés différentiables*, Comment. Math. Helv. **28** (1954), 17–86.

93. G. Tian, *Smoothness of the universal deformation space of compact Calabi–Yau manifolds and its Peterson–Weil metric*, Mathematical Aspects of String Theory (S.-T. Yau, ed.), World Scientific, Singapore, 1987, pp. 629–646.

94. G. Tian and S.-T. Yau, *Three dimensional algebraic manifolds with $c_1 = 0$ and $\chi = -6$*, Mathematical Aspects of String Theory (S.-T. Yau, ed.), World Scientific, Singapore, 1987, pp. 629–646.

95. A. N. Todorov, *The Weil–Petersson geometry of the moduli space of $SU(n\geq 3)$ (Calabi–Yau) manifolds, I*, Comm. Math. Phys. **126** (1989), 325–246.

96. E. Viehweg, *Weak positivity and the stability of certain Hilbert points, III*, Invent. Math. **101** (1990), 521–543.

97. C. Voisin, *Miroirs et les involutions sur les surfaces K3*, Journées de Géométrie Algébrique d'Orsay (Juillet 1992), Astérisque, vol. 218, Société Mathématique de France, 1993, pp. 273–323.

98. ———, *A mathematical proof of the Aspinwall–Morrison formula*, Compositio Math., to appear.
99. P. M. H. Wilson, *The Kähler cone on Calabi–Yau threefolds*, Invent. Math. **107** (1992), 561–583, *Erratum*, ibid. **114** (1993), 231–233.
100. E. Witten, *Physics and geometry*, Proc. Internat. Congr. Math. Berkeley 1986 (A. M. Gleason, ed.), vol. 1, American Mathematical Society, Providence, 1987, pp. 267–303.
101. ———, *Two dimensional gravity and intersection theory on moduli space*, Surveys in Differential Geom. **1** (1991), 243–310.
102. J. G. Wolfson, *Gromov's compactness of pseudo-holomorphic curves and symplectic geometry*, J. Differential Geom. **28** (1988), 383–405.
103. S.-T. Yau, *On Calabi's conjecture and some new results in algebraic geometry*, Proc. Nat. Acad. Sci. U.S.A. **74** (1977), 1798–1799.
104. R. Ye, *Gromov's compactness theorem for pseudo-holomorphic curves*, Trans. Amer. Math. Soc. **342** (1994), 671–694.

Physics Articles

105. P. S. Aspinwall and B. R. Greene, *On the geometric interpretation of $N = 2$ superconformal theories*, Nuclear Phys. B **437** (1995), 205–230.
106. P. S. Aspinwall, B. R. Greene, and D. R. Morrison, *Multiple mirror manifolds and topology change in string theory*, Phys. Lett. B **303** (1993), 249–259.
107. ———, *Calabi–Yau moduli space, mirror manifolds and spacetime topology change in string theory*, Nuclear Phys. B **416** (1994), 414–480.
108. ———, *Measuring small distances in N=2 sigma models*, Nuclear Phys. B **420** (1994), 184–242.
109. P. S. Aspinwall and C. A. Lütken, *Quantum algebraic geometry of superstring compactifications*, Nuclear Phys. B **355** (1991), 482–510.
110. P. S. Aspinwall and D. R. Morrison, *Topological field theory and rational curves*, Comm. Math. Phys. **151** (1993), 245–262.
111. ———, *Chiral rings do not suffice: N=(2, 2) theories with nonzero fundamental group*, Phys. Lett. B **334** (1994), 79–86.
112. ———, *U-duality and integral structures*, Phys. Lett. B **355** (1995), 141–149.
113. ———, *String theory on K3 surfaces*, Mirror Symmetry II (B. Greene and S.-T. Yau, eds.), International Press, Cambridge, 1996, pp. 703–716.
114. M. Bershadsky, S. Cecotti, H. Ooguri, and C. Vafa, *Holomorphic anomalies in topological field theories* (with an appendix by S. Katz), Nuclear Phys. B **405** (1993), 279–304.
115. ———, *Kodaira–Spencer theory of gravity*, Comm. Math. Phys. **165** (1994), 311–427.
116. P. Candelas, A.M. Dale, C. A. Lütken, and R. Schimmrigk, *Complete intersection Calabi–Yau manifolds*, Nuclear Phys. B **298** (1988), 493–525.
117. P. Candelas and X. C. de la Ossa, *Comments on conifolds*, Nuclear Phys. B **342** (1990), 246–268.
118. P. Candelas, X. C. de la Ossa, P. S. Green, and L. Parkes, *A pair of Calabi–Yau manifolds as an exactly soluble superconformal theory*, Nuclear Phys. B **359** (1991), 21–74.
119. P. Candelas, E. Derrick, and L. Parkes, *Generalized Calabi–Yau manifolds and the mirror of a rigid manifold*, Nuclear Phys. B **407** (1993), 115–154.

120. P. Candelas, A. Font, S. Katz, and D. R. Morrison, *Mirror symmetry for two parameter models – II*, Nuclear Phys. B **429** (1994), 626–674.
121. P. Candelas, P. S. Green, and T. Hübsch, *Rolling among Calabi–Yau vacua*, Nuclear Phys. B **330** (1990), 49–102.
122. P. Candelas, G. Horowitz, A. Strominger, and E. Witten, *Vacuum configuration for superstrings*, Nuclear Phys. B **258** (1985), 46–74.
123. P. Candelas, M. Lynker, and R. Schimmrigk, *Calabi–Yau manifolds in weighted* $\mathbb{P}_4$, Nuclear Phys. B **341** (1990), 383–402.
124. S. Cecotti and C. Vafa, *Topological anti-topological fusion*, Nuclear Phys. B **367** (1991), 359–461.
125. R. Dijkgraaf, E. Verlinde, and H. Verlinde, *Topological strings in* $D < 1$, Nuclear Phys. B **352** (1991), 59–86.
126. M. Dine, N. Seiberg, X.-G. Wen, and E. Witten, *Nonperturbative effects on the string world sheet, I, II*, Nuclear Phys. B **278** (1986), 769–789, ibid. **289** (1987), 319–363.
127. J. Distler and B. Greene, *Some exact results on the superpotential from Calabi–Yau compactifications*, Nuclear Phys. B **309** (1988), 295–316.
128. L. J. Dixon, *Some world-sheet properties of superstring compactifications, on orbifolds and otherwise*, Superstrings, Unified Theories, and Cosmology 1987 (G. Furlan et al., eds.), World Scientific, Singapore, 1988, pp. 67–126.
129. B. Dubrovin, *Integrable systems in topological field theory*, Nuclear Phys. B **379** (1992), 627–689.
130. P. Ginsparg, *Applied conformal field theory*, Fields, Strings, and Critical Phenomena, Les Houches Summer School in Theoretical Physics, 1988 (E. Brézin and J. Zinn-Justin, eds.), North-Holland, Amsterdam, 1990, pp. 1–168.
131. M. Green and J. Schwarz, *Anomaly cancellations in supersymmetric* D=10 *gauge theory and superstring theory*, Phys. Lett. B **149** (1984), 117–122.
132. P. S. Green and T. Hübsch, *Connecting moduli spaces of Calabi–Yau threefolds*, Comm. Math. Phys. **119** (1988), 431–441.
133. B. R. Greene, D. R. Morrison, and M. R. Plesser, *Mirror manifolds in higher dimension*, Comm. Math. Phys. **173** (1995), 559–598.
134. B. R. Greene, D. R. Morrison, and A. Strominger, *Black hole condensation and the unification of string vacua*, Nuclear Phys. B **451** (1995), 109–120.
135. B. R. Greene and M. R. Plesser, *Duality in Calabi–Yau moduli space*, Nuclear Phys. B **338** (1990), 15–37.
136. ———, *An introduction to mirror manifolds*, Essays on Mirror Manifolds (S.-T. Yau, ed.), International Press, Hong Kong, 1992, pp. 1–27.
137. B. R. Greene, C. Vafa, and N. Warner, *Calabi–Yau manifolds and renormalization group flows*, Nuclear Phys. B **324** (1989), 371–390.
138. K. Intriligator, *Fusion residues*, Mod. Phys. Lett. A **6** (1991), 3543–3556.
139. S. Katz, *Gromov–Witten invariants via algebraic geometry*, Trieste Conference on S-Duality and Mirror Symmetry, Nuclear Phys. B Proc. Suppl., vol. 46, 1996, pp. 96–107.
140. W. Lerche, C. Vafa, and N. P. Warner, *Chiral rings in* $N = 2$ *superconformal theories*, Nuclear Phys. B **324** (1989), 427–474.
141. G. Moore and N. Seiberg, *Classical and quantum conformal field theory*, Comm. Math. Phys. **123** (1989), 177–254.
142. D. R. Morrison, *Mirror symmetry and the type II string*, Trieste Conference on S-Duality and Mirror Symmetry, Nuclear Phys. B Proc. Suppl., vol. 46,

1996, pp. 146–155.

143. D. R. Morrison and M. R. Plesser, *Summing the instantons: Quantum cohomology and mirror symmetry in toric varieties*, Nuclear Phys. B **440** (1995), 279–354.
144. ———, *Towards mirror symmetry as duality for two-dimensional abelian gauge theories*, Trieste Conference on S-Duality and Mirror Symmetry, Nuclear Phys. B Proc. Suppl., vol. 46, 1996, pp. 177–186.
145. A. Strominger, *Yukawa couplings in superstring compactification*, Phys. Rev. Lett. **55** (1985), 2547–2550.
146. ———, *Massless black holes and conifolds in string theory*, Nuclear Phys. B **451** (1995), 96–108.
147. A. Strominger and E. Witten, *New manifolds for superstring compactification*, Comm. Math. Phys. **101** (1985), 341–361.
148. A. Strominger, S.-T. Yau, and E. Zaslow, *Mirror symmetry is T-duality*, Nuclear Phys. B **479** (1996), 243–259.
149. C. Vafa, *Topological mirrors and quantum rings*, Essays on Mirror Manifolds (S.-T. Yau, ed.), International Press, Hong Kong, 1992, pp. 96–119.
150. E. Witten, *Topological sigma models*, Comm. Math. Phys. **118** (1988), 411–449.
151. ———, *On the structure of the topological phase of two-dimensional gravity*, Nuclear Phys. B **340** (1990), 281–332.
152. ———, *Mirror manifolds and topological field theory*, Essays on Mirror Manifolds (S.-T. Yau, ed.), International Press, Hong Kong, 1992, pp. 120–159.
153. ———, *Phases of N=2 theories in two dimensions*, Nuclear Phys. B **403** (1993), 159–222.